Plant Stress Physiology

Edited by Akbar Hossain

Published in London, United Kingdom

IntechOpen

Supporting open minds since 2005

Plant Stress Physiology
http://dx.doi.org/10.5772/intechopen.88761
Edited by Akbar Hossain

Contributors
Akbar Hossain, Mst. Tanjina Islam, Tofazzal Islam, Raghad S. Mouhamad, Michael Alabboud, Shandrea Stallworth, Te-Ming Tseng, Mary Gracen Fuller, Brooklyn Schumaker, Sankar Narayan Sinha, Momezul Haque, Karabi Biswas, Shalim Uddin, Abu Sayeed Md. Hasibuzzaman, Farzana Akter, Shamim Ara Bagum, Nilima Hossain, Tahmina Akter, Ayman EL Sabagh, Murat Erman, Fatih Çiğ, Muhammad Mubeen, Celaleddin Barutçular, Ömer Konuşkan, Mohammad Sohidul Islam, Habib-ur-Rehman Athar, Oksana Sytar, Shah Fahad, Ram Swaroop Meena, Muhammad Aamir Iqbal, Allah Wasaya, Hakki Akdeniz, Ferhat Kizilgeci, Marian Brestic, Tasmiya Jabeen, Maham Asif Bukhari, Faraz Azeem, Muhammad Ikram, Sobhy Sorour, Wajid Nasim, Mabrouk Elsabagh, Muhammad Rizwan, Akihiro Ueda, Liyun Liu, Hirofumi Saneoka, Saima Muzammil, Sumreen Hayat, Asma Ashraf, Bilal Aslam, Rizwan Asif, Muhammad Waseem, Imran Riaz Malik, Mohsin Khurshid, Muhammad Afzal, Muhammad Saqalein, Muhammad Hussnain Siddique, Aqsa Muzammil, Sumera Sabir, Muhammad Asif Zahoor, Magda H. M. Youssef, Lobna F. F Wahman, Marwa M. Abd Rabo, Amany Hanafy M. Elgoly, Hussain Sajid, Shazia Iqbal, Muhammad Abdul Qayyaum, Muhammad Ashraf, Saifullah, Sami Ullah Khan, Alvina Gul, Zulfiqar Ali Gurmani, Waseem Ahmed, Shahzad Ahmed, Yoko Kato, Ana Carolina Feitosa De Vasconcelos, Kemal Yuce, Ahmet Ismail Ozkan, Natalia Rudenko, Lyudmila K. Ignatova, Tatiana P. Fedorchuk, Elena M. Nadeeva-Zhurikova, Maria M. Borisova-Mubarakshina, Boris N. N Ivanov, Huseyin Arslan, Ejaz Waraich, Mustafa Ceritoğlu, Doğan Arslan, Disna Ratnasekera, Narendra Kumar, Analía Llanes, Sajjad Hussain, Hany Gharib, Mehtab Muhammad Muhammad Aslam, Kashif Akhtar, Joseph K. Karanja, Noor-ul- Ain, Fasih Ullah Haider, Kallingil Gopi Divya, Muhammad Zukiffal, Javed Ahmed, Aneela Ahsan, Muhammad Musa, Amna Kanwal, Muhammad Saleem, Javed Anwar, Aziz Ur Rehman, Sadia Ajmal, Saima Gulnaz, Muhammad Makky Javaid, Moin Ahmad Khan, M. Badruzzaman Siddiqui, Tushar Ranjan, Sudhir Kumar, Shampa Purkyastha, Chandan Roy, Rakesh Deo Ranjan, Nnaemeka Success Esiobu, Chinedu Gilbert Onubuogu, Blessing Chidinma Nwachukwu, Sylvarlene Munachim Njoku

First published in London, United Kingdom, 2021 by IntechOpen
IntechOpen is the global imprint of INTECHOPEN LIMITED, registered in England and Wales, registration number: 11086078, 5 Princes Gate Court, London, SW7 2QJ, United Kingdom
Printed in Croatia

British Library Cataloguing-in-Publication Data
A catalogue record for this book is available from the British Library

Additional hard and PDF copies can be obtained from orders@intechopen.com

Plant Stress Physiology
Edited by Akbar Hossain
p. cm.
Print ISBN 978-1-83962-526-8
Online ISBN 978-1-83962-527-5
eBook (PDF) ISBN 978-1-83962-528-2

For EU product safety concerns:
IN TECH d.o.o., Prolaz Marije Krucifikse Kozulić 3, 51000 Rijeka, Croatia, info@intechopen.com or visit our website at intechopen.com.

Meet the editor

Dr. Akbar Hossain is presently working as a senior agronomist at the Bangladesh Wheat and Maize Research Institute, Bangladesh. Dr. Hossain finished his Ph.D degree from the Institute of Ecology and Biological Sciences (particularly stress physiology), Astrakhan State University, Russia. The research interests and expertise of Dr. Hossain are plant physiology, conventional and molecular breeding, biotechnology, genetic engineering, crop and weed management, weed biology and ecology, conservation agriculture, climate change impact assessment on field crops through modeling. He has authored more than 190 journal articles. He is currently serving as an MS student supervisor and also a voluntary reviewer/editor for different journals. Dr. Hossain is also linked with several international projects. Details of Dr. Hossain are available in the links: https://www.researchgate.net/profile/Akbar_Hossain2?ev=hdr_xprf

Contents

Preface

The concept of plant stress physiology has been well-established over the past 60 years due to the increasing trends of extreme environmental events. Any significant unfavorable condition that interrupts or alters the metabolic process of plants and ultimately hinders the growth and development of plants is called plant stress. It is established that plant stress is linked to the physiological process of plants. Plant physiology perceptions contribute to analyzing past achievements of plant breeding in increasing potential and stability of crop yield and also resource use efficiency and productivity through recognizing mechanisms that have been indirectly exaggerated by the selection process of desirable crop plants. Crop physiology has been found to have a strategic association among phenotypic traits and crop performance as it pursues to elucidate and forecast the multifaceted connections between appropriate traits and/or the environment for effective crop production in the modern era of climate change. Researchers found that crop stress physiology has an association with two main areas of research, one is concerned with agronomic research, the other is concerned with plant breeding. Therefore, to meet the food demand of the increasing population, the contents of the current book highlight the integration of both breeding and agronomy strategies to ensure agricultural productivity and environmental safety under changing climate.

As an academic editor of the book, I want to thank all the authors for their wonderful contributions, as well as the speed and efficiency in the delivery of their chapters. Special gratitude should be mentioned to the excellent team from the Editorial Board of IntechOpen, particularly to Ms. Sandra Maljavac, Author Service Manager, for their continued support and final condition of this book.

Akbar Hossain, PhD
Senior Agronomist,
Bangladesh Wheat and Maize Research Institute,
Dinajpur, Bangladesh

Section 1

Plant Response to Abiotic Stress

Chapter 1

Heat and Drought Stresses in Wheat (*Triticum aestivum* L.): Substantial Yield Losses, Practical Achievements, Improvement Approaches, and Adaptive Mechanisms

Muhammad Zulkiffal, Aneela Ahsan, Javed Ahmed, Muhammad Musa, Amna Kanwal, Muhammad Saleem, Javed Anwar, Aziz ur Rehman, Sadia Ajmal, Saima Gulnaz and Muhammad Makky Javaid

Abstract

The major wheat-producing countries have heterogeneous and fragile agro climatic surroundings but frequently restraining wheat yield and quality losses are predominant under heat and drought prone agriculture exclusively when both stresses occur in blend, which looms the food security globally. However, many suggested examples are available in these countries for the mitigation of these two stresses by using different conventional and modern improvement and agronomic approaches. In addition to these approaches, morphological, physiological, anatomical, biochemical, phenological, and physiochemical vicissitudes, which trigged during these stresses, have also been elucidated. There complete deliberation in combination for wheat improvement is still a contest, but a win-win option is a holistic attitude in future.

Keywords: heat, drought, yield losses, achievements, improvement, mechanisms, major countries, wheat

1. Introduction

The global inhabitants expansion proportions have been projected to upsurge and the domain people will grasp 8 billion by 2025 and strength be a slight greater than 9 billion by 2050 Hence, to encounter the ever-growing hassles of the population world sustenance fabrication desires to be doubled by the year 2050 [1].

The aftermath of a universal climate alteration has brought about the enlargement of extreme events. Among these events heat and drought are most multidimensional, vibrant, and shoddier stresses whose occurrences are unpredictable at any stage affecting wheat productivity. Today, wheat is an essential staple food for more than 2 billion people and is grown on more terrestrial zone than any other marketable produce. The major wheat-producing countries have sundry and flimsy agro climatic circumstances, typologies, wheat production schemes and thereby having an erratic consequence of heat and drought stresses. To combat these resilient, formulations of short- and long-term strategies are dire need for these countries. Heat stress occur when air and soil temperature become beyond a threshold level while drought stress takes place when ambient air temperature is high, soil and atmospheric humidity is low.

This review intended at revealing some of the foremost features and some probable heat and drought tolerance pointers of wheat related to heat and drought robust, which are relevant for agronomic and genetic improvement in wheat. Nevertheless, general triumph of wheat management and improvement depends on the intensive exertions of molecular biologists, physiologists, and modelers in addition to agronomist, geneticist, and breeder since these improvements are incremental in nature due to compound genomes and polygenic traits in wheat.

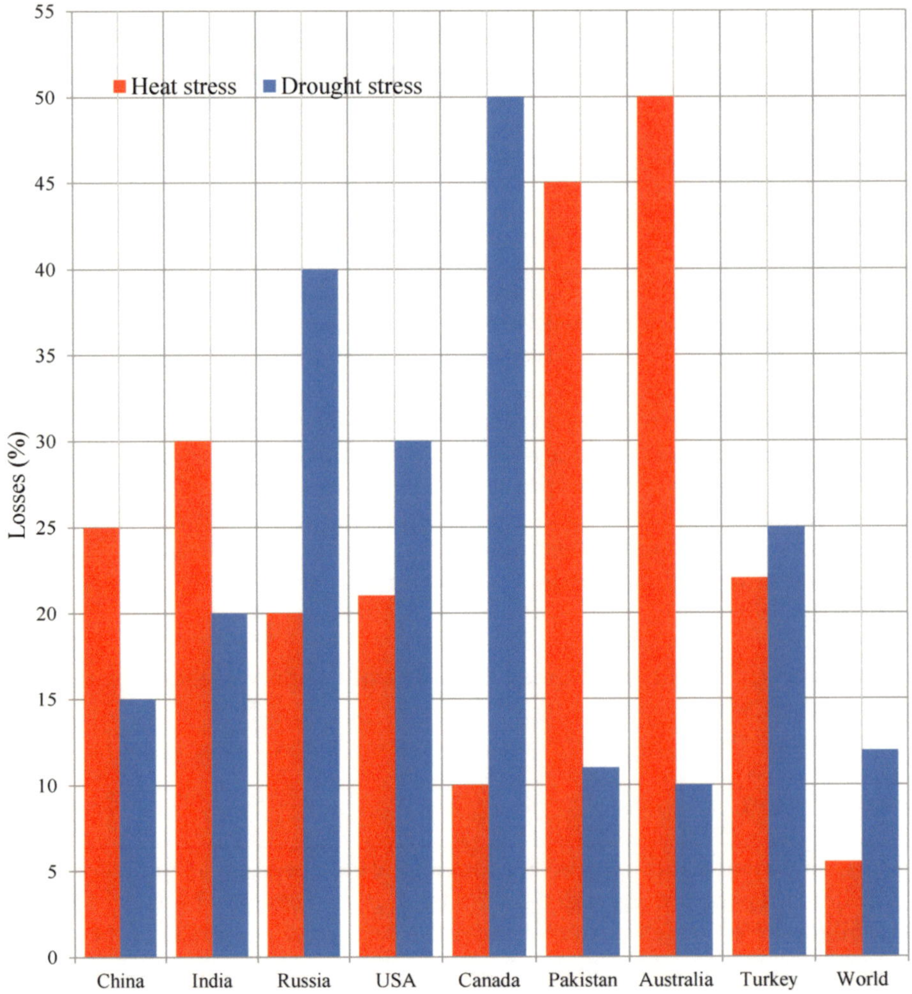

Figure 1.
Estimated yield losses (%) of foremost wheat-producing countries due to heat and drought stresses.

2. Substantial wheat yield losses

Wheat throughput is vanished predominant exclusively or jointly due to abiotic stresses primarily heat and drought with a large portion of potential in major wheat-producing nations and at the same time, globally. Mostly curtailing losses are prevalent due to sensitivity at reproductive phase under heat and drought prone agriculture, which threatens the food security worldwide. Therefore, this urges that these harms should be curtailed in major area of distress for all nations. Yield is an endpoint inclined by stressor therefore it is used as a yard stick for measuring these stresses. A middling appraisal of 50% yield losses in agricultural crops is caused by abiotic dynamics, heat in low latitude zones and drought stress common in most arid and semi-arid zones. Together heat and drought have persuasive effect in Mediterranean climate [2].

The internationally wheat losses due to heat and drought stresses encompasses, 5.5 and 12%, respectively. The actual losses however, varied substantially by region in the foremost wheat-producing countries viz; China [3], India [4], Russia [5], the USA [6], Canada [7], Pakistan [8], Australia [9], Turkey [10], and World [3] (**Figure 1**). The appearance of heat stress accredited to the extraordinary yield loss in Australia and Pakistan followed by India and China while stemming effect for Canada, Russia, USA, and Turkey are visible. Likewise, Canada, Russia, and USA extremely hit by drought followed by Turkey, India, and China while Pakistan and Australia remained at par. The frequency and magnitude of these losses may increase in future because the projections advocate that global temperatures may upsurge by 0.6–2.5°C by 2050 and 1.4–5.8°C by 2100 escort by increased severity of drought condition [11].

3. Practical achievements

Wheat yield increased under productive conditions, but in the regions where heat and drought condition prevail practical, achievements are less prolific due

Countries	Heat resistant	Drought resistant	References
China	Xifeng 9, Qingxuan 15, Changle 5, Bonong 7023, Zhangchun 9, Xinchun 2, Xinchun 3, and Changchun 2		[12–17]
	Gaoyou 9415, Hemai 13, Jimai 22, Kexin 9, Shannong 8355, Taishan 23, Yannong 5286, Zimai 7, and Nongda Heng	Keyi26, Nongda 36, Nongda 183, Shijiazhuang 407, Huabei 187, Taigu 49, Yulin3, Mazhamai, Xuzhou 14, Jinmai 33, Jinmai 2148, Hezuo 2, Hezuo 4, Hezuo 7, Minn 2761, Kefeng 2, Kefeng3, Gaoyuan 602, Xindong 7, Lunkan6, Lunkan 7, Lumai 14, Heimangchunmai, Datouchunmai, Xindong 2, Jinmai 33, Kehan 9, Xinkehan 9, Inmai 47, Shijiazhuang 8, Cang 6001, Cangmai 02, Cangmai 6005, JM-262, Xihan No. 2, Longchun 23, Luhan7, Luhan 2, and Yannong 19.	
India	CPAN 4079 and Nepal 38, Arnej, Ajanta, and Gomti		[4, 18–23]
	HD 4502 (Malvika), AKW 1071, Purna, Parnhani-51, K9644, Atal, K 7903 Halna, DDK 1029, Ventnor, HS-240, K-0-307, and Raj 3765	Shekhar, WH 1142, HD1467, Harshita, N59, and BRW 3723	
Russia	Dustlik, H-104, Sanzar-8, Sanzar4, Hasan-Orif, Bayaut1, Oasis, and Gul DU	Sarrubra, Sarrosa, Saratovskaya 29, Svetlana, Milturum, and Cesium	[24–26]

Countries	Heat resistant	Drought resistant	References
USA	Long Branch		[27, 28]
		Greer, Joe, Plains Gold Avery, SY Monument, Tatanka, WB-Grain field, TAM112, White Sonora LCS Chrome, LCS Mint, and T158	
Canada	Pelissier	Stettler, Lillian, AC Barrie, and Strongfield	[29–31]
Pakistan	Gold-16, Punjab-11, Fakhar-e-Bhakkar	Chakwal50, NARC 2009, Tijaban-10, Dharabi-11, NRL 2017, Pakistan-13, Pakistan-13, Shahkar-13 and NIFA-Lalma, Shahkar-13, NIFA-Lalma, Hashim-8, Ghanemat-2015, BARS-09, Tatara, AZRC-1, Siran-2007, Raj, Chakwal-87, Rawal-97, Pothwar-93, Kohsar-95, Chakwal-97, GA-2002, Ehsan-16, Barani-17, Fateh Jang-16	[6, 32–45]
Australia	Longsword		[46–51]
	Suntop, Spitfire, GBA Hunter, Livingstone and EGA Gregory	1:ZIZ12, 12:ZIZ12, 56:ZIZ12, 134:ZWB12, Allora Spring, Farmer's Friend, King's Jubilee, Steinwedel, Kord CL Plus, drysdale, Wyalkatchem, and Estoc	
Turkey	Bayraktar 2000	Karahan-99, Gerek-79 and Alka quality, Saricanak-98, Altay-2000, Dagdas-94, Katea-1, and Kirac-66	[52–54]

Table 1.
Development of few heat- and drought-resistant wheat varieties by foremost wheat-producing countries.

to tired inconsistency, large genomic decoration with multigenic trait, and multifaceted environmental dealings to these hassles. Even then, in the global milieu, copious efforts to mitigate heat and drought through breeding wheat-resilient varieties are under way and much remarkable advancement has been reported. The most sustainable, encouraging, economically effective, and communally adequate approach is the developing of wheat varieties with in-built heat and drought tolerance. To alleviate these two stresses, the **Table 1** demonstrates most evoked examples of the triumph stories in chief wheat-producing countries.

4. Improvement strategies

4.1 Conventional breeding

Conventional breeding approaches have been tremendously effectual in the development of heat and drought tolerant wheat cultivars on the globe. Conventional plant breeding typically trusts upon fortuitous by hybridization, the succeeding phenotypically selection for loftier desirable traits using Mendelian and quantitative genetics approaches in filial generations and final multi-locational trials valuation. The nitty-gritties of wheat improvement by conventional approaches for these two stresses rely on the varied scale usage of biodiversity, which include wild relatives, landraces, exotic material, advanced lines, isogenic lines, mapping population, and cytogenetic stocks. Among all sources, wild relative and landrace of wheat are potentially most significance for traits of stress adoptive due to the accretion of genes for tolerance to stresses. Therefore, this narrow wheat genetic diversity for higher tolerance to heat and drought stresses can be boost by the use of wild relatives and local land races. Among wild relatives, *Aegilops squarrosa* is more heat, while *Aegilops tauschii* and *Triticum dicoccoides* are more drought tolerant. Conventional plant breeding has had discriminatory conquest in dying both stresses instantaneously, which may be due to the hurdle linked with traits stressed by polygenic inheritance, masking effect, and environmental interaction. As a magnitude of these confines of conventional breeding, additional genetic advances for developing tolerance against given wheat resilient are becoming progressively problematic (**Figure 2**).

4.2 Mutation breeding

Mutation breeding does not stance any moral matters regarding human health and sustainability as it become an customary tool in improvement of genepool have momentous impression. In nature, variation occurs chiefly as a consequence of mutations; that is swhy mutation-based breeding increases desirable variability,

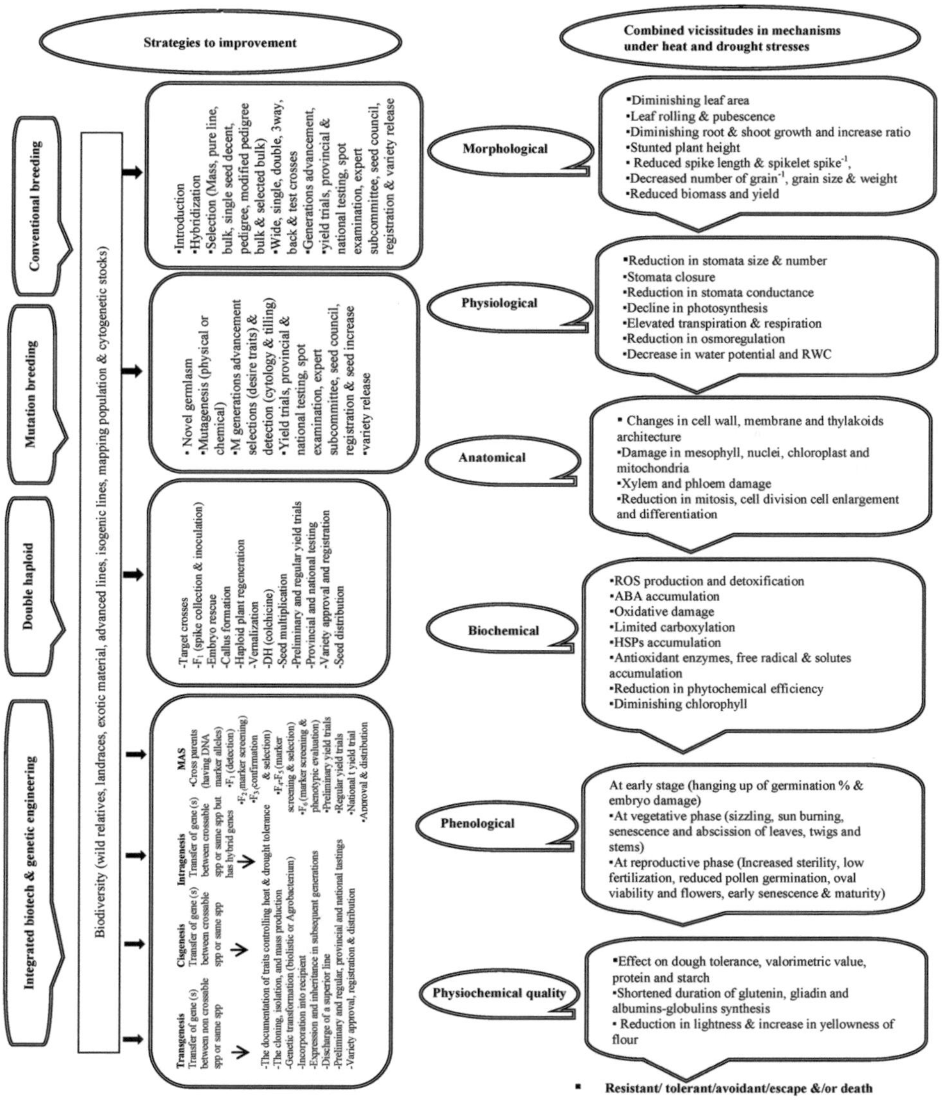

Figure 2.
Combined changes in mechanisms and improvement strategies under heat and drought stresses.

which is not found in nature especially with the help of various physical (X rays, gamma rays, UV light, proton, neutron, alpha and beta particles) and chemical mutagens (alkylating agents, nitrous acid, acridine, base analogue, azide, and antibiotics). By the use of mutation breeding, 254 superior bread wheat varieties including abiotic stress tolerant (26 resistant to drought) have been released globally. Three significant economic impact wheat varieties (Jauhar-78, Soghat-90 and Kiran-95) released through induced mutagenesis in Pakistan [55]. Al-Naggar and Shehab-El-Deen [56] endeavored to induce (gamma rays and EMS) drought-tolerant mutants in six Egyptian bread wheat varieties. These mutants surpassed 20% grain yield over parents under drought condition. Laghari et al. [57] developed and gaged two wheat mutants capable of earlier maturity and higher grain yield than the checks. The mutation wheat varieties (Kievsky and Novosibivskaya 67) were characterized by upright productivity and resistant to lodging in Ukrain [58]. This exhilarated further work on mutation breeding, leading to the release of mutant wheat cultivars expressly the traits linked to heat and drought tolerance. Mutant byzantine screening, difficulty in regulatory the direction and nature of variation, low beneficial mutant occurrence and mutagenic efficacy, mutation breeding approaches have curb and are facing challenges. Owing to this bottleneck, induced mutations have also evidenced valuable in the preparation of genetic maps that will ease molecular marker-assisted plant breeding for developing heat- and drought-tolerant wheat varieties in the upcoming.

4.3 Double haploid

Double haploid counterpart the conventional breeding programs to hasten the release of new varieties tolerant to heat and drought stresses by rapid generation advancement. Therefore, double haploid approach should be unified with convention approaches perceptively for food security. From heterozygous individual haploids are made and converted to diploid, which create instant homozygous lines, which evade fertility obstacles inherent to wide crosses as it is genotypes sovereign. For heat and drought tolerance improvement double haploid have been broadly used to judge allelic variation as it deliver great level of polymorphism by using limited quantity of tested lines through molecular mapping as targets for transformations. Two drought-resistant wheat varieties (Jinghua 1 and Jinghua 764) in China, one in France (Florin), one in Hungary (Gk Delibab), one in Morocco (Malika) were developed and released with the help of doubled haploid technique [30, 59–61]. Moreover, double haploid genotypes (DH1 and 2) under drought and (DH132 and 133) under heat and (DH136, 210, 236, 248, 257 and 263) under both conditions found superior than checks in Egypt, Iran, and USA respectively [62–64]. The high production cost, know-how, restriction on number of crosses and low haploid generation facilities restricted the use of double haploid.

4.4 Integrated genetic engineering and biotechnology approaches

Additional consideration need to be paid now to develop high-yielding wheat varieties by integrated genetic engineering and biotechnology approaches under heat and drought stresses. This will open new opportunities for enhancing existence of narrow genetic base and help for understanding the genetic mechanisms for theses stresses. But this requires the identification of key tolerance genetic determinants underlying these two stresses and introducing into wheat. This introduction includes transgenesis, which holds artificial regulatory order, sexually incompatibility with no barrier but may be safe while cisgenesis and intragenesis are sexually compatibility, contain natural regulatory order and new combination of regulatory order (hybrid genes), respectively with barrier but are safe. Due to large wheat genome size (17,000 Mb) the complete sequencing is challenging. However, there are few examples for introduction of several stress-inducible genes into wheat, which increased tolerance to heat and drought stresses but due risk of other undesirable traits transfer and reproductive barrier, strategy for tolerance is not much successful yet. Khurana et al. [65] identified and characterized large number of high temperature-responsive genes aiming to functionally validate them in wheat transgenic. Zang et al. [66] identified heat stress-responsive gene (TaPEPKR2), transformed into another wheat cultivar, observed that the transgenic lines exhibited enhanced heat and drought stresses tolerance and suggested that it could be utilized as a candidate gene in transgenic breeding. Karolina et al. [67] analyzed wheat gene (P5CS and P5CR) expression in response to drought stress and found that they have a significant function in controlling tolerance to water deficits. Hua et al. [68] identified target genes in response to drought stress in wheat (*Triticum aestivum* L.) and suggested that these could be exploited via genetic engineering to improve drought tolerance in wheat. Overexpressing of TaNAC69, HVA1, CAT TaDREB2, and TaDREB3 genes the transgenic wheat produced more shoot biomass, yield, and improved water use efficiency drought conditions, which suggested that these have potential for wheat engineering for drought tolerance [69].

Concerning quality, [70] recognized 26 genes to gauge their function and transcript levels for starch synthesis in wheat. Rooke et al. [71] produced and then confirm wheat transgenicline (B73-6-1), which holds additional genes (Glu-1D-1) for high-molecular weight gluten sub units (HMW-GS). Similarly, [72] investigate interactive effects between the transgenically wheat line (B102-1-2/1) with HMW-GS and suggested that by using transgenic wheat lines expressing HMW-GScan improve dough properties. Parallel interpretations are described in transgenic wheat lines (B72-8-11b and B102-1-2) for HMW-Gs by [73]. Ashraf et al. [74] transforms and detects HMW-gene (Dy10) in Egyptian wheat and scrutinized that transgenic grain own higher levels of glutenin compared to control. Alvarez et al. [75] transformed the HMW-GS genes (1Ax1 and 1Dx5) into wheat and revealed that overexpression of 1Dx5 gene upsurges overall protein content.

Genomic assistance breeding (GAB) established on the application of marker assistance selection, which discriminate genetically sundry phenotypes elucidated by the markers were scrutinized. The wheat breeding understanding for abiotic stress tolerance is restricted due to their complex inheritance and wide range of environmental interaction with respect to rate, intensity, timing, duration, and increased genetic gain at early phenological phases of wheat. This issue bounds all conventional breeding efforts. In this situation genomic assistance breeding is very prompt, cost effective, and accurate in harsh unpredictable and unapproachable environmental situation in which individual targeted wheat plants can be selected on phenotypic basis. Furthermore, genomic approaches are welcome for condemnation from social sectors hesitant to the use of transgenic breeding expressly on wheat crop. In amalgamation with the conventional breeding, this unlocked up tangible forecasts for new schemes in wheat breeding for injurious stress tolerance as it bounce greater genetic reply for QTL inveterate in multi-environments because it resulted in the development of next generation sequencing methods. A wide range of population structures can be used for QTL mapping, backcrossing, recombinant inbred, double haploid and F_2 selfing or heterozygous inter crossing of major genes is repeatedly used to lessen the association around the target gene and to recuperate the recurrent parent by using less number of filial generations. Similarly recombinant inbred lines and double haploid, which can be sustained and produced permanently have been extensively used to judge allelic variation as it provide high level of polymorphism by using limited number of tested lines through molecular

Source	QTLs	Markers with co-localization	Ref	Source	QTLs	Markers with co-localization	Ref
Heat stress							
F_2 population from PBW743/ WH1081 under terminal heat stress	Membrane thermostability	Xgwm156-3B	[76]	251 recombinant inbred lines from (heat-tolerant/ susceptible) HD2808/ HUW510	Heat susceptibility index	gwm122-2A	[63]
F_1–F_2 population from (heat-tolerant/sensitive) Debra/Yecora Rojo	Grain filling rate	Xgwm132-6A Xgwm577-6B Xgwm617-D	[77]	111 recombinant inbred lines from (heat-tolerant/ susceptible) WH 730/Raj 4014	Grain filling rate	Xgwm314-6B	[78]
F_1–F_3 population from (heat-tolerant/susceptible) ventnor/ Karl 92	Grain filling duration	gwm11-5A gwm 293-1B	[79]	143 recombinant inbred lines from Kauz/MTRWA116	Heat stress susceptibility index	gwm190-1B gwm133-5B gwm63-7B	[80]
BC_1 F_2 population from (heat-tolerant/susceptible) HD2733/ WH730 and HD2733/HI150	Canopy temperature	barc68-5A	[81]	106 recombinant inbred lines from NW1014/HUW468	Grain weight heat environment	Xgwm972-7D1	[82]
Bi-parental F_2 population	Stay green under prolong heat	Xgwm533-3B3	[83]	121 recombinant inbred lines from (heat-tolerant/ susceptible) Halberd/Karl92	Canopy temperature depression	barc84-3B gwm154-5A gwm179-5A1	[84]
205 F2 population from (heat-sensitive/tolerant) YecoraRojo/ Ksu106	Grain filling rate under heat stress	wmc326-3B wmc25-2A wmc327-5A	[85]	148 recombinant inbred lines from NW1014/HUW468	Canopy temperature depression	Xgwm1025-7BL	[86]
144 doubled haploid wheat populations bare to heat hassle	Chlorophyll loss and shoot weight reduction due to heat treatment	Xgwm1034-3B Xbarc75-6B	[87]	Genetic population (heat-tolerant accession) derived from Karl 92	Plasma and thylakoid membrane damage	Xbarc113-6A Xbarc121-7A Xbarc49-1D	[23]
RAC875/Kukri doubled haploid population	Canopy temperature	barc0075-3B1	[88]	25 wheat genotypes exposed to heat stress	Kernel weight and grain filling duration	gwm11-5A gwm293-1B	[89]

Source	QTLs	Markers with co-localization	Ref	Source	QTLs	Markers with co-localization	Ref
Drought stress							
F_3 and F_4 population derived from drought-tolerant (Oste-Gata) and sensitive (Massara-1) populations	Thousand grain weight	Xgwm408-2B	[90]	127 recombinant lines from (drought-tolerant/sensitive) DharwarDry/Sitta	Grain yield	Xwmc420-4AL	[91]
Two $F_{8–9}$ recombinant inbred lines population from (resistant/susceptible) Luohan 2/Weimai 8 and annong19/Weimai 8	Seedling traits under drought	Xmag3356-5D Xbarc158-3B	[17]	167 recombinant inbred lines plus parents derived from (drought-tolerant) Seri/Babax	Canopy temperature under water deficit	gwm388-1B	[92]
Near isogenic lines from (tolerant/susceptible) C306/Dharwar Dry	Grain yield under post anthesis drought	gwm368-4B	[93]	118 recombinant inbred lines from Tabassi/Taifun	Yield under drought stress	Xgwm194-7B	[94]
A panel of 100 lines	Root traits under drought	Xwmc175-2B	[95]	154 accessions development under irrigated and drought-stressed conditions	Plant height	Xgwm495-4B	[96]

Table 2.
Identified QTLs for different traits with lined markers under heat and drought stress environments in wheat.

Water conservation techniques	Residual and nutrients management	Planting time	Biological control	Chemical control	Adaptation mechanism
• Improve water harvesting techniques • Minimum tillage • Laser leveling • Weed management • Raised bed planting • Seed priming	• Residue retention alone or in combination with nitrogen and phosphors fertilizers • Straw mulching • Balanced use of nutrients at proper time and stage	• Early sowing or as soon as adequate rainfall/moisture available or plant early maturing cultivars	• Inoculation of arbuscular mycorrhizal fungi • Ameliorating plant growth and development	• Exogenous application of hormones, antioxidant enzymes, biochemical solutes, and osmoprotectants to seed or growing wheat	• Wheat crop modeling • Meteorological decision support schemes

Table 3.
Agronomy management for heat and drought mitigation.

mapping for heat and drought tolerance improvement. MAS for target traits relied on finding markers linked to quantitative traits loci (QTL). A large number of quantitative trait loci (QTLs) mapping studies have been magnificently applied as a tool for genetic analysis for wheat under heat and drought tolerance. These abolish perplexing effects of the environment throughout selection, permits for unintended selection of traits governing these buoyant and provide footprints of domestication construction on early victories.

Although numerous reports are obtainable for the use of association mapping approaches to categorize the QTLs, linked markers associated with co-localization in wheat for heat (heat stability index, canopy temperature, membrane thermostability, stay green, grain filling duration etc.) and drought (relative water contents, stomatal conductance, grain weight etc.) yet their prosperous placement in the development of superior cultivar has had only limited success (**Table 2**). As the efficiency of MAS is effected by population size with broader genetic base, loci numbers, complexity of traits, and selection approaches, the breeder must evaluate before applying it especially for quantitative traits regarding resilient stressors under single environment because it analyzed one trait and less efficient to determine the effect of each QTL. Therefore, there is an immense need to develop more effective markers associated with the agronomic traits under these stresses for MAS.

4.5 Agronomic approaches

Promotion of agronomic practices can increased wheat productivity and farm income by sustaining and/or defending the production system against heat and drought resilient along with genetic approaches joined with breeding procedures. To overcome the hostile effect of climate change the notion of climate smart agriculture has been anticipated, which embraces many of the agronomic practices based on sustainable crop production and field management. Some of the winning agronomic approaches that alleviate heat and drought in wheat have been presented in **Table 3**.

Among these, water conservation techniques increase water use efficiency and time saving; residual and nutrients management; moderates soil temperature, reduces evaporation losses, reduce pollution by reducing greenhouse gases and CH_4 and N_2O emission; chemical and biological control protect the wheat from damage that resulted due to high temperature and drought stress; pick up of least perilous growing period; adaptation mechanisms are effective for right time by operational management by translating weather information.

5. Adaptive mechanisms

The heat and drought stresses in wheat crop triggers a wide variety of responses, which cause morphological, physiological, anatomical, biochemical, phenological, and physiochemical vicissitudes changes individually or in combination due to direct or indirect injury that leads to significant loss in yield potential. General elucidation of the all discussed mechanisms by wheat reply to heat and drought are deliberated, but there complete consideration in combination for wheat improvement is still a contest.

5.1 Morphological vicissitudes

Under heat and drought stresses most morphological traits (leaf size, plant height, grain size and weight, root length, shoot length, root shoot length ratio,, number of tiller plant^{-1}, spike length, spikelet spike^{-1}and biomass) show decreasing trends. The first and prime effect harshly diminish sprouting and seedling. After seedling emerged, cell division, cell enlargement, and differentiation are badly affected due to these stresses, which afterward affect the leaf size and plant height. The retrieval does not take place at the advanced periods but may take place at initial phase of both stresses. Remarkably, the decline in leaf size can oblige as heat and drought avoidance mechanism because it abridged transpiration.

Under serious water shortage, cell elongation is subdued by interruption of water drift from the xylem to the development cells resulted in reduced growth due to decrease in mitosis course. This reduction leads to thwart the development of flower production, grain development, and filling due to a attenuation in the activities of sucrose and starch synthesis enzymes.

Core cause of grain size and weight is the expansion of maternal cells throughout the grain filling phase, which is major parameter upsetting grain size and weight and it rest on the ear and flag leaf and stem reserves as they deliver the pivotal element (carbon). Heat and drought stresses, causes the reduction of spike length, spikelet spike^{-1} and biomass and positively correlated with each other and also with grain yield. Jaiswal et al. and Hafiz et al. [97, 98] observed a reducing drift in root length ranged from 7.2 to 23.0 cm (normal) and from 5.3 to 17.7 cm (drought) and shoot length from 13.2 to 29.2 cm (normal) and from 11.0 to 25.2 cm (drought) while the root/shoot length ratio ranged from 0.27 to 0.94 (normal) and from 0.29 to 0.92 (drought) among all tested genotypes. Hasan et al. [99] observed root and shoot length under different temperature regimes, the lowest values (2.8 and 1.14 cm) was attained at 15°C, optimum (11.5 and 9.03 cm) at 25°C and thereafter decreased trend (6.41 and 7.53 cm) at 35°C, respectively in all tested wheat genotypes. The increase in temperature, the shoot-to-root ratio was also increased because the adverse effect of higher temperature (35°C) on root length was more than that on shoot length.

5.2 Physiological vicissitudes

Normalized vegetation index (NDVI) and canopy temperature (CT) are good physiological pointer of a genotype's suitability against heat and drought stress environment and these traits may be used as morphological selection tools for developing heat and drought stress-tolerant genotypes. For appraisal of physiological diversity in wheat genotypes under heat and drought environments, [100] revealed the positive correlation of yield with NDVI at booting and anthesis and negative correlation with CT at same stages. A positive association of NDVI advocated the existence of stay-green while negative array of CT at both stages supported cooler canopies genotypes. Likewise, [101] clarified same results for these physiological traits while working on wheat local land races for consecutive 3 years as genetic resources for yield potential and heat tolerance.

Under heat and drought stress conditions, wheat plants improve canopy temperature by closing their stomata swiftly, which resulted in reduced transpiration and water loss. This reduction in stomatal opening causes low amount of CO_2 fixation that lead to reduction in photosynthesis and ultimately chlorophyll content. This reduction resulted due to structural and then adjacent changes in chloroplasts, which ultimately disrupt chlorophyll synthesis and photosynthesis. As compared to 100% control, heat, drought, and combine stress reduces photosynthesis rate by 19, 11, and 79%. Relative water content, membrane stability, and osmotic potential are maintained by osmoregulation physiological mechanism, which losses their viability under both stresses. As an indicator of water status, relative water content is the meaningful determinant of heat and drought tolerance because it signifying the membrane stability and balance between water supply and evapotranspiration. The relative water content was reduced by 55, 26, and 61% under drought, heat, and combined stress, respectively. Membrane stability index was affected most by combination of drought and heat stress (60%) than by heat stress (55%) and finally by drought stress (43%). Transpiration rate under high temperature stress compared to control slightly increased. However, drought stress decreased transpiration rate while under combine effect the reduction rate is 60–63% [102, 103]. Photosynthesis is also extremely sensitive under heat and drought prone conditions as the reduction in the ratio and quantity of chlorophyll (a and b) and carotenoid occurred upon increasing intensity of heat and drought.

5.3 Anatomical vicissitudes

Anatomical changes like reduced leaf anatomy, cell size, damage in mesophyll, cell membrane stability, plasma membrane permeability, chloroplast, nuclei, changes in xylem and phloem are vital reflection under both stresses. Cell membrane stability shields the plant from ROS that causes significantly decrease in membrane stability under both stresses. Tolerant and susceptible genotypes retain more than 70% and less than 50%, cell membrane stability values, correspondingly. Leaf anatomy under heat stress causes development of higher leaf area with thinner leaves while leaves that develop under drought generally have smaller cells with higher stomatal density. Under heat stress chloroplasts become round and stretched from ellipse-shaped with destroyed wrappers and fully developed grana lamella become loosely organized with abundant layer on it. The appearance of more osmiophilic particles occurred, thylakoids also become inflated and resultantly chloroplasts swelled to altered extents and some of their external membranes vanished entirely at advanced periods of pressure. While in the drought stress there was decrease in the number of granal thylakoids of chloroplasts. No starch granules in chloroplast stroma were found under combine stress. Concerning mitochondria, a few multi vesicular body's lipids are formed due to appearance of spoiled double membranes mitochondria, which signposted the process of mitochondria degradation. These discrete membrane variations also befell in nuclei, representing augmented senescence process under heat stress. Under drought, leaf mitochondria were less preserved than normal conditions. But in combine stresses large size mitochondria, devoid of cristae and similar to vacuoles were observed than individual stress.

5.4 Biochemical vicissitudes

Biochemical traits are another important constituent for developing heat-and drought-tolerant genotypes with higher yield and disease resistant. But the mechanisms of these mutually stresses on a biochemical basis is not relatively well-understood, research on this voyage in wheat is desirable in future. Temperature stress causes membrane injury to wheat due to of reactive oxygen species (ROS). To cope with ROS under heat stress, wheat plant own sequence of detoxification systems to limit oxidative damage by breaking toxic with the help of antioxidant enzymes (peroxidase, superoxide dismutase, catalase, and glutathione reductase), metabolites (glutathione, carotenoids, and ascorbic acid), and biochemical solutes (proline, glycine betaine, salicylic acid, starch, potassium, and abscisic acid). The buildup of these shields the damage caused by oxidative stress. Moaed et al. [104] estimated antioxidant enzymes and metabolites at three stages of wheat. The varieties that showed significant increase in the activity of these during vegetative and anthesis phase (in the late and very late planting) showed minimum reduction in membrane injury index. Likewise under heat and drought stresses, superoxide dismutase, peroxidase enzymes protect the cellular systems of plants from cytotoxic effects of the active oxygen species. A significant increase effect of superoxide dismutase (12–52% and 28%) and peroxidase (40–44% and 21%) enzymes was renowned under heat and drought stresses, respectively [105, 106]. Likewise, biochemical solutes are accumulated that gives advantage to wheat plant against under heat and drought stresses. Among all, proline, glycine betaine, and salicylic acid are key biochemicals that are significantly accumulated in plants including wheat when exposed to heat and drought. The higher accumulation of three forages reactive oxygen species conveys strong antioxidant defense system, increased relative water content, reduces the rate of transpiration and membrane injury. That is why, to reduce the effect of heat and drought stress, exogenous application of glycine betaine and salicylic acid has been found [107]. Amarshettiwar and Berad [108] revealed that biochemical and yield traits of wheat were significantly influenced by heat stress with regard to values of increase in proline contents and decrease in starch contents albumins, globulins, and yield contributing traits. ABA is a naturally occurring compound that helps to regulate plant growth and development. The ABA level increased during heat and especially drought stresses and is therefore an essential arbitrator as it refunded the plant to pre stress condition. Quarrie and Jones [109] exogenously applied ABA to investigate its effects on the changing penalties of water under stress and found that ABA application decreased the mean cell size, increased the production of trichomes, and reduced the number of stomata. These changes reduce the transpiration rate and ultimately bound the water losses. Likewise under heat stress, little is known about ABA accumulation in wheat regardless of the fact that its level is increased however, enhanced levels of ABA in leaves increased leaf resistance under high (38°C) air temperature, which play an important role in thermo-tolerance. Zhao et al. [110] six heat-induced MYB genes in wheat and studies their gene regulation by exogenous abscisic acid under heat stress scenario. By heat stress (40°C), the expression of the two MYB out of six was not vividly up delimited by application of exogenous ABA levels.

In addition, internal and external signals were the chief basis of transit surge in the calcium concentrations inside the cytosol in supporting the normal level of Ca^{2+} under heat stress· This sustainability resulted in transduction of heat shocks proteins (calmodulin, calcineurin, and annexin), which induces the thermos tolerance defensive ability in wheat. A total of 39 heat shock proteins and 33 drought stress-responsive proteins are identified in different wheat cultivars, which trigger, maintain, and recover stresses [111]. The heat shock proteins are further classified in to five groups (Hsp100, Hsp90, Hsp70, Hsp60, and small Hsps) on the bases of their molecular masses. Late embryogenesis abundant protein represent a wide range adaptation to water deficit involved in desiccation tolerance and slow down the rate of water losses under drought condition These are accumulated at later stages and are classified in to seven groups on the basis of specific domain. Transgenic approaches showed that over expression of these proteins improve abiotic stresses especially drought in wheat. However, their exact and precise molecular function is not clear yet.

5.5 Phenological vicissitudes

To stirring heat and drought stresses multi-modeling collaborative phenological approaches were experienced. The acquaintance of the duration, timing, and sequence of growing changes in wheat is vital for effective management else it has generous errors. Many models can predict phenology accurately built on the main driver of temperature and/or directly spoke these retorts to drought and appropriate photoperiod. Under heat and drought conditions, phenological vicissitudes are utmost significant attribute intricate in adaptation and final yield because these stresses effects are apparent at all development stages of wheat. Wheat threshold temperature at germination (10–30°C), vegetative, reproductive (15°C), and post anthesis (35°C) phases cause irrevocable hurt to plant growth and development. During the first week of growth, under heat stress (45°C), hang-up of germination leads to cell death and embryo damage. At vegetative phase sizzling, sun-burning, senescence and abscission of leaves, twigs, stems, stunted plant height and less tillers and finally reduced biomass results. During reproductive phase of terminal heat stress intense discount occur in fertilization efficiency due to pollen grains damage, reduced number and weight of grains spike^{-1} due to less anthesis, reduces the grain filling period and early maturity which finally resulted in reduced harvest index [112]. Akbar et al. [5] found cutback in grain yield from 7.7 to 15.7% for every 1°C ascend in mean air temperature during booting to maturity phasic development. Similarly under drought stress condition, water-use efficiency increased at early stage of stress. At vegetative stage causes multiple effects are visible such as stomatal closure, reduced swelling, loss of leaves, reduction in tillering and sheath and prevention of some tillers from producing spikes. At reproductive phase, flowering occurs starting in the apical part of the spike chiefly on the main stem and decline in transpiration due to relative evapotranspiration deficit and the period of maturation eventually resulted in reduced number, weight of grains spike^{-1} and yield. Oviedo et al. [113] estimated the grain production was reduced 23, 42, and 9% at water stress tillering, booting, and grain filling phenological stages.

The correlation of growing degree days with the phenology of wheat plant is a best climate impact indicator. High temperature attached with increase magnitude dry spell causes sweeping changes on wheat phenology reliant upon stage, time, duration, and rate of stresses occurrence. Heat shocks and early monsoon shifted the wheat sowing as compared to past scenario. For instances, under both heat and drought shortening the length of vegetative and reproductive phases allow the crop to escape the stresses. Therefore, early flowering, long grain filling period and late maturity period should be taking into account while selecting under these stresses on phenological bases.

5.6 Physiochemical quality vicissitudes

Heat and drought are determinant factors on wheat end-use quality. Under amplified temperature protein quantity, which persisted high due to intensity of essential amino acids, sedimentation index, and condense effect. Dough strength however is reduced due to early maturity, which resulted in shortened duration of glutenin synthesis [114]. Similarly, under drought condition, valorimetric value, protein, and starch are negatively affected, which ultimately effect dough properties for bread making [115]. Balla et al. [116] found that both drought and heat in combination or drought alone have a much greater influence on a better protein ratio than heat alone. In case of drought alone a noteworthy negative correlation was pragmatic between granule sizes of starch and relative protein content telltale that this parameter contributes significantly for the baking quality of the flour because heat stress can reduce grain set and combined with abscisic acid build up can increase the response compared to just one stress. All this suggest that effects of heat and drought stresses are beneficial for some quality traits like ash and protein but on the outlay of seed yield because quality and quantity have inversely proportional with each other. Therefore, evaluation, selection, and development under these three environments should be done with average good quality traits to meet end user requirement. Among protein components (glutenin, gliadin, and albumins-globulins), albumins-globulins have only a trivial impact on the dough quality but glutenin and gliadin are responsible for the flexibility and extensibility of the dough. They reported reduction in the glutenin and gliadin proportion of the flour while the ratio of albumins and globulins did not increase proportionately in response to heat, drought, and in combination after anthesis.

6. Conclusions

1. The best step forward and future predominant ultimate approaches are imperative to develop new wheat varieties for more tolerant against these two robust episodes. The identification, characterization, and screening of broad based genetic resources through conventional breeding along with the use of modern genetics protocols and agronomic management will pave the way for efficient and accurate screening at each phonological stage of wheat. Controlling patterns for accountabilities of risk management and valuation must be framed regarding transgenic wheat development for these two stress factors.

2. Crop modeling system testing (in natural and artificial buoyant environments) for susceptible zones are still a big room for understanding the genetic and environmental interactions and improvement of all the mechanisms which bloom in these syndromes.

3. Heat and drought are major drivers of climate variability, can last much longer than other weather events and cannot be detect easily especially in combination. For understanding their vigilance, a reliable decision-support system and forecasting should be used.

4. For exploiting reliability and genetic stability for wheat yield both stresses should be contemplated together for traits having the main influence on yield. Therefore, a win-win possibility is a holistic attitude in future. In more prone areas however, if the problem does not resolve, relocating to new areas and growing different crops are the alternative range of options.

Acknowledgement

The authors gratefully acknowledge financial support from textbook and academic association and harvest plus international.

Author details

Muhammad Zulkiffal*, Aneela Ahsan, Javed Ahmed, Muhammad Musa, Amna Kanwal, Muhammad Saleem, Javed Anwar, Aziz ur Rehman, Sadia Ajmal, Saima Gulnaz and Muhammad Makky Javaid
Ayub Agricultural Research Institute, Faisalabad, Pakistan

*Address all correspondence to: zulkiffal@yahoo.com

References

[1] Population Reference Bureau (PRB). World Population Data Sheet. Washington, DC: Population Reference Bureau (PRB); 2008. Available from: https://www.prb.org/2008wpds/

[2] Wang JP, Raman H, Zhou MX, Ryan PR, Delhaize E, Hebb DM, et al. Interaction between drought and chronic high temperature during kernel filling in wheat in a controlled environment. Annals of Botany. 2007;**90**:469-476

[3] Liu B, Liu L, Cao W, Zhu Y, Asseng S. Post heading heat stress and yield impact in winter wheat in China. Global Change Biology. 2014;**20**(20):s372-381

[4] Sareen S, Tyagi BS, Sarial AK, Tiwari V, Sharma I. Trait analysis, diversity, and genotype × environment interaction in some wheat landraces evaluated under drought and heat stress conditions. Chilean Journal of Agricultural Research. 2014;**74**(2):135-142

[5] Akbar H, Jaime A, Teixeira DS, Marina VL, Vacheslav PZ. High temperature combined with drought affect rainfed spring wheat and barley in South-Eastern Russia: Phenology and growth. Saudi Journal of Biological Sciences. 2012;**19**:473-487

[6] Anonymous. Barani Agricultural Research Institute, Chakwal. 2019. Available from: https://aari.punjab.gov.pk/crop_varieties_wheat

[7] Lesk C, Rowhani P, Ramankutty N. Influence of extreme weather disasters on global crop production. Nature. 2016;**529**:84-87

[8] Jamil M, Ali A, Ghafoor A, Gul A, Akbar KF, Bashir H, et al. Yield reduction analysis of bread wheat under heat stress at two different environments in Pakistan. Fresenius Environmental Bulletin. 2017;**26**(7):4602-4605

[9] Asseng S, Foster I, Turner NC. The impact of temperature variability on wheat yields. Global Change Biology. 2011;**17**:997-1012

[10] Castro M, Peterson CJ, Rizza MD, Dellavalle PD, Vazquez D, Ibanez V, et al. Influence of heat stress on wheat grain characteristics and protein molecular weight distribution. In: Buck HT, Nisi JE, Salomon N, editors. Wheat Production in Stressed Environment. Dordrecht: Springer; 2007. pp. 365-371

[11] Intergovernmental Panel on Climate Change (IPCC). Fourth Assessment Report: Climate Change. Geneva; 2007

[12] Ling H, Yan X, Shoujin F, Zongshuai W, Fahong W, Bin Z, et al. Comparative analysis of root transcriptome profiles between drought-tolerant and susceptible wheat genotypes in response to water stress. Plant Science. 2018;**272**:276-293. DOI: 10.1016/j.plantsci.2018.03.036

[13] Yu L, Niu L, Fu J, Wang F, Zhao S, Lu L, et al. Selection and breeding of drought resistant, water-saving and high-yield wheat variety cangmai 028. Asian Agricultural Research. 2017;**9**:33-38. DOI: 10.22004/ag.econ.257326

[14] Li Q, Wang Z, Li D, Wei J, Qiao W, Meng X, et al. Evaluation of a new method for quantification of heat tolerance in different wheat cultivars. Journal of Integrative Agriculture. 2018;**17**(4):786-795

[15] Zhao SS, Wang FZ, Lu L, Zhang HY, Zhang XY. Breeding and selection of drought resistant and salt-tolerant wheat variety Cang 6001. Acta Agriculturae Boreali-Sinica. 2000;**15**(S1):113-117

[16] Yu L, Wang W, Niu L, Wang W, Lu L, Wang FW, et al. A New Cultivation Technique of Cangmai 6005 for High Yield in Cangzhou Dry-Alkali Land. Asian Agricultural Research: USA-China Science and Culture Media Corporation; 2018. Available from: https://ideas.repec.org/a/ags/asagre/273103.html

[17] Zhang HF, Wang L, Junli DA, Zhao C, Bao Y, Yang WH. Conditional and unconditional qtl mapping of drought-tolerance-related traits of wheat seedling using two related RIL populations. Journal of Genetics. 2013;**92**(2):213-231

[18] Anonymous. Status Paper on Wheat 2014-15. Kamla Nehru Nagar, Ghaziabad, UP, India: Directorate of wheat development, Ministry of Agriculture, C.G.O. Complex; 2015

[19] Degu WT. High yielding wheat varieties with heat and drought tolerance. Research brief 2 Science matter ICARDA, Lebanon; 2015

[20] Dhyani K, Ansari MW, Rao YR, Verma RS, Shukla A, Tuteja N. Comparative physiological response of wheat genotypes under terminal heat stress. Plant Signal And Behavior. 2013;**8**(6):e24564

[21] Gupta A, Singh C, Kumar V, Tyagi BS, Tiwari V, Chatrath R, et al. Wheat Varieties Notified in India Since 1965. Karnal-132001, India: ICAR-Indian Inst. Wheat & Barley Research; 2018. pp. 41-54

[22] Gupta A, Kumar V, Singh C, Tiwari V. Development and release of new wheat and barley varieties for different zones and states. Journal of Wheat Research. 2017;**9**(1):68-71

[23] Shyamal KT, Md A, Kolluru V, Jesse P, Pagadala VVP, Robert B, et al. Mapping QTL for the traits associated with heat tolerance in wheat (Triticum aestivum L). BMC Genetics. 2014;**15**(97):1-13

[24] Iljina LG, Galkin AN, Kuzmenko AL. Catalogue of Spring Wheat Varieties Bred. Leningrad VIR: Agricultural Research Institute of Southeast; 1986. p. 126

[25] Morgounov A, McNab A, Campbell KG, Parada R. Increasing wheat production in central Asia through science and international cooperation. In: Proceeding of the First Central Asian Wheat Confer. 2003. pp. 1-181. ISBN: 970-648-130-3

[26] Morgunov AI. Wheat and Wheat Breeding in Former USSR. Wheat Special Report No. 13. CIMMYT: Mexico, DF; 1992

[27] Jourdan MB, Rudd J, Trostle C, Neely C. Wheat variety Characteristics Varieties Planted in the Texas High Plains Uniform Variety Trials. Texas A & M. Agri Life Extension; 2016-17. Available from: https://amarillo.tamu.edu/files/2017/08/2016-2017-Wheat-Variety-Charactersitics.pdf

[28] Reddy SK, Liu S, Rudd JC, Xue Q, Payton P, Finlayson SA, et al. Physiology and transcriptomics of water-deficit stress responses in wheat cultivars TAM 111 and TAM 112. Journal of Plant Physiology. 2014;**171**(14):1289-1298

[29] Ashe P, Shaterian H, Akhov L, Kulkarni M, Selvaraj G. Contrasting root and photosynthesis traits in a large-acreage Canadian durum variety and its distant parent of Algerian origin for assembling drought/heat tolerance attributes. Frontiers in Chemistry. 2017;**5**(121):1-10

[30] De B, Henry J, Lonnet Y, Hertzog P, Hespel R. Florin: A doubled haploid wheat variety developed by the anther culture method. Plant Breeding. 1987;**98**:53-56

[31] Fleury D. Accelerating drought tolerance in wheat. Agronomy and plant breeding. Top crop manager. Plant Biology. 2018;**41**:1261-1269

[32] Ahmad Z, Mujahid MY, Khan MA, Yasmin S, Asif M, Qamar M. NARC-2009: A high yielding wheat variety for rainfed areas of Pakistan. Pakistan Journal of Agricultural Research. 2010;**23**(1):1-4

[33] Akhtar A, Kamran AJ, Akmal M. Yield comparison of potential wheat varieties by delay sowing as rainfed crop for Peshawar climate. Sarhad Journal of Agriculture. 2017;**33**(3):480-488

[34] Farooq A, Khan AJ, Ali A, Muhammad T. NRL 2017: A high a yielding drought tolerant wheat strain for rainfed areas of NWFP. Sarhad Journal of Agriculture. 2007;**23**(4):895-898

[35] Ghulam MS, Hussain M, Javed A, Javed A, Muhammad T, Sher BK. Anew high yielding, stress tolerant wheat variety Punjab-2011. Journal of Agricultural Research. 2014;**52**(3):317-328

[36] Khan J, Khan S, Khetran MA, Amanullah SN, Islam M, Hanan A, et al. Tijaban-10-a drought tolerant and high yielding wheat variety for rainfed/sailaba areas of Balochistan. Pakistan Journal of Botany. 2013;**45**(4):1357-1362

[37] Khan MH, Din NU, Khakwani AA, Baloch MS, Zubair M, Khan S, et al. Hashim-8: A short duration, high yielding and disease resistant wheat variety for rain-fed areas of Pakistan. International Journal of Agriculture and Biology. 2011;**13**:956-960

[38] Krishna DJ, Attiq UR, Ghullam U, Mian FN, Mahreen ZJA, Muhammad K, et al. Acceptance and competitiveness of new improved wheat varieties by smallholder farmers. Journal of Crop Improvement. 2017;**31**(4):608-627

[39] Mahmood A, Mian MA, Ihsan M, Ijaz M, Rabbani G, Iqbal MS. Chakwal-50: A high yielding and disease resistant wheat variety for rainfed region. The Journal of Animal & Plant Sciences. 2013;**23**(3):833-839

[40] Muhammad I, Zubeda P, Abdul G, Niaz H, Muneer A, Muhammad A, et al. Fakhar-E-Bhkkar—a high yielding, temperature stress tolerant and rust resistant spring bread wheat variety. International Journal of Advanced Research in Biological Sciences. 2018;**5**(8):36-45

[41] Saifullah K, Jahangir K. Drought tolerant wheat cultivar (raj) for rainfed areas of KPK, Pakistan. Pakistan Journal of Agricultural Sciences. 2010;**47**(4):355-359

[42] Sajid R, Waheed A, Sheraz A, Manzoor H, Muhammad T, Abid M, et al. S-09: A high yielding and rust resistant wheat (*Triticum aestivum* l.) variety for rainfed areas of Punjab. Journal of Agricultural Research. 2012;**50**(2):189-200

[43] Shamsul I. Scientists Develop 'Gold-16': A Heat-Tolerant, High-Yield Wheat Variety. The Express Tribune; 2017. Available from: https://tribune.com.pk/story/1463405/scientists-develop-gold-16-heat-tolerant-high-yield-wheat-variety/

[44] Tariq M, Mahmood A, Mian MA, Cheema NM, Sabar M, Ihsan M, et al. Dharabi-11: A new high yielding drought and disease tolerant wheat variety. International Journal of Agriculture and Biology. 2013;**15**:701-706

[45] Waheed A, Ali N, Shiraz A, Muhammad Z, Muhammad IK, Amina B, et al. Fatehjang-2016: A high yielding and rust resistant wheat (*Triticum aestivum* l.) variety for rainfed areas of Punjab. Journal of Agricultural Research. 2018;**56**(3):173-179

[46] Condon AG, Richards RA, Rebetzke GJ, Farquhar GD. Breeding for high water-use efficiency.

Journal of Experimental Botany. 2004;**55**(407):2447-2460

[47] De P, Townley RM, Humphreys TF, Knox RE, Clarke FR, Clarke JM. Lillian hard red spring wheat. Canadian Journal of Plant Science. 2005;**85**:397-401

[48] Dirk HRS. Wheat Varieties Growing 19th Century Australia-a Hand List of Varieties-Waggawagga. Sydney: N.S.W. The Farrer Centre, Faculty of Science and Agriculture, Charles Sturt University; 2001

[49] Donald GM, Leary RO. Drought Tolerance of Wheat Varieties. Australian Government: Grain Research and Development Corporation; 2016. Available from: https://grdc.com.au/resources-and-publications/grdc-update-papers/tab-content/grdc-update-papers/2016/02/drought-tolerance-fo-wheat-varieties

[50] Peter M, Don M. NWS Winter Crop Variety Sowing Guide. NSW DPI Management. Grain Research and Development Corporation. 2019. Available from: https://www.dpi.nsw.gov.au

[51] Trethowan R, Thistlethwaite R, Watson IA. The Heat Tolerance of some Northern Bread Wheat Varieties. Australia: Grains Research Centre, The University of Sydney; 2016. Available from: https://grdc.com.au/resources-and-publications/grdc-update-papers/tab-content/grdc-update-papers/2016/03/the-heat-tolerance-of-some-northern-bread-wheat-varieties

[52] Cheng L, Wang Y, He Q, Li H, Zhang X, Zhang F. Comparative proteomics illustrates the complexity of drought resistance mechanisms in two wheat (*Triticum aestivum* L.) cultivars under dehydration and rehydration. BMC Plant Biology. 2016;**16**(188):1-23

[53] Ahmed M, Kofi NA, Mesut K, Alexey M, Kenan P, Ahmet B, et al. Adoption and Impacts of Improved Winter and Spring Wheat Varieties in Turkey. ICARDA: Syria; 2009. pp. 1-45

[54] Yoruk E, Keles EN, Sefer O, Eraslan M. Salinity and drought stress on barley and wheat cultivars planted in Turkey. Journal of Environmental Biology. 2018;**39**:943-950

[55] Ahmed N, Alizai NA, Kakar AH, Shah R, Ali M. Mutation breeding: A tool to improve wheat yield and yield components. Life Science International Journal. 2015;**9**(1):3274-3327

[56] Al-Naggar AMM, Shehab-Eldeen MT. Predicted and actual gain from selection for early maturing and high yielding wheat genotypes under water stress conditions. Egypt Journal of Plant Breeding. 2012;**16**(3):73-92

[57] Laghari KA, Sial MA, Arain MA, Khanzada SD, Channa SA. Evaluation of stable wheat mutant lines for yield and yield associated traits. Pakistan Journal of Agriculture, Agricultural Engineering and Veterinary Science. 2012;**28**(2):124-130

[58] Shkvarnikov PK, Kulik MI. Induction of mutations in wheat. Procedding Indian National Science Academy. 1975;**41**(3):204-217

[59] Hu D, Tang Y, Yuan Z, Wang J. The induction of pollen sporophytes of winter wheat and the development of thenew variety Jinghua No. 1. Scientia Agricultura Sinica. 1983;**1**:29-35

[60] Hu Y, Bao RR, Xue XY. The new strain '764' of spring wheat by pollen haploid technique from anther culture. Genetic Manipulation in Crops Newsletter. 1988;**4**:70-85

[61] Pauk J, Kertesz Z, Beke B, Bona L, Csosz M, Matuz J. New winter wheat variety: 'GK Delibab' developed via combining conventional breeding and in vitro andoogenesis. Cereal Research Communications. 1995;**23**:251-256

[62] Abdelsamad A, El-Sayed OE, Ibrahim HF. Development of drought tolerant double haploid wheat using biochemical genetic markers on in vitro culture. Journal of Applied Scientific Research. 2007;**3**(11):1589-1599

[63] Bakhshi N, Sarial AK, Sharma P, Sareen S. Mapping QTLs for grain yield components in wheat under heat stress. PLoS One. 2017;**12**(12):e0189594

[64] Poudel P. Screening of winter wheat double haploid population 'buster' under heat and drought stress [MSc thesis]. USA: Faculty of the Graduate College of the Oklahoma State University; 2016. pp. 1-118

[65] Khurana P, Chauhan H, Khurana N. Characterization and expression of high temperature stress responsive genes in bread wheat. Czech Journal of Genetics and Plant Breeding. 2011;**47**(Special Issue):S94-S97

[66] Zang X, Geng X, He KWF, Tian X, Xin M, Yao Y, et al. Overexpression of the wheat (*Triticum aestivum* L.) Tapepkr2 gene enhances heat and dehydration tolerance in both wheat and arabidopsis. Frontier. Plant Science. 2018;**9**:1710

[67] Karolina D, Magdalena Z, Andreas B, Hubert S, Krzysztof K, Michał N. Analysis of wheat gene expression related to the oxidative stress response and signal transduction under short-term osmotic stress. Scientific Reports. 2019;**9**:2743

[68] Hua Y, Zhang C, Shi W, Chen H. High-throughput sequencing reveals micro RNAs and their targets in response to drought stress in wheat (*Triticum aestivum* L.). Biotechnology and Biotechnological Equipment. 2019;**1314-3530**:1-7

[69] Pellegrineschi A, Reynolds M, Pacheco M, Brito RM, Almeraya R, Kazuko YS, et al. Stress-induced expression in wheat of the arabidopsis thaliana DREB1A gene delays water stress symptoms under greenhouse conditions. Genome. 2004;**47**:493-500

[70] Guo ZK, Wei X, Guo QL, Xiao QP, Tian CG. Comprehensive analysis of the transcription of starch synthesis genes and the transcription factor RSR1 in wheat (Triticum aestivum) endosperm. Genome. 2013;**56**(2):115-122

[71] Rooke L, Beke F, Fido R, Barro F, Gras P, Tatham A, et al. Over expression of a gluten protein in transgenic wheat results in greatly increased dough strength. Journal of Cereal Science. 1999;**30**(2):115-120

[72] Mao X, Li Y, Zhao S, Zhang J, Lei Q, Meng D, et al. The interactive effects of transgenically overexpressed 1Ax1 with various HMW-GS combinations on dough quality by introgression of exogenous subunits into an elite Chinese wheat variety. PLoS One. 2013;**8**(10):e78451. DOI: 10.1371/journal.pone.0078451

[73] Sanhe LLJ, Wang N, Wang Y, Yang G, Fang J, Guangyuan HE. Inheritance and expression of copies of transgenes *1Dx5* And*1Ax1* In elite wheat (*Triticum aestivum* L.) varieties transferred from transgenic wheat through conventional crossing. Acta Biochimica et Biophysica Sinica. 2007;**39**(5):377-383

[74] Ashraf HF, Khaled SA, Mohamed A. Integration and expression of the high-molecular-weight glutenin subunit DY10 gene into Egyptian wheat. Arab Journal of Biotechnology. 2007;**10**(1):49-56

[75] Alvarez ML, Guelman S, Halford NG, Lustig S, Reggiardo MI, Ryabushkina N, et al. Silencing of HMW glutenins In transgenic wheat expressing extra HMW subunits. Theoretical and Applied Genetics. 2000;**100**:319-327

[76] Gupta M, Chawla V, Garg P, Yadav N, Munjal R, Sharma B. Genetic analysis of yield and heat stress related traits in wheat (*Triticum aestivum L. em. Thell*) using microsatellite markers. Journal of Applied and Natural Science. 2015;**7**(2):739-744

[77] Barakat MN, Al-Doss AA, Elshafei AA, Moustafa KA. Bulked segregant analysis to detect quantitative trait loci (QTLl) related to heat tolerance at grain filling rate in wheat using simple sequence repeat (SSR) markers. African Journal of Biotechnology. 2012;**11**(61):12436-12442

[78] Girish CP, Jagadish R, Sindhu S, Priyanka S, Singh NK, Ratan T. Molecular investigations on grain filling rate under terminal heat stress in bread wheat (*Triticum aestivum L.*). African Journal Of Biotechnology. 2013;**12**(28):4439-4445

[79] Yang J, Sears RG, Gill BS, Paulsen GM. Quantitative and molecular characterization of heat tolerance in hexaploid wheat. Euphytica. 2002;**126**:275-282

[80] Mohammadi V, Zali AA, Bihamta MA. Mapping QTLs for heat tolerance in wheat. Journal of Agricultural Science and Technology. 2008;**10**:261-267

[81] Prabhu KV. Validation of SSR markers linked with drought and heat tolerant QTLs in bread wheat (*Triticum aestivum L. em.Thell.*). International Journal Pure Applied Biosciences. 2017;**5**(5):700-705

[82] Kumari S, Jaiswal V, Mishra VK, Paliwal R, Balyan HS, Gupta PK. QTL mapping for some grain traits in bread wheat (*Triticum aestivum* L.). Physiology and Molecular Biology. 2018;**Pl24**(5):909-920

[83] Sharma DK, Torp AM, Rosenqvist E, Ottosen C, Andersen SB. QTLs and potential candidate genes for heat stress tolerance identified from the mapping populations specifically segregating forFv/Fmin wheat. Frontiers in Plant Science. 2017;**8**:1668

[84] Mason RE, Mondal S, Francis W, Beecher FW, Hays DB. Genetic loci linking improved heat tolerance in wheat (*Triticum aestivum* L.) to lower leaf and spike temperatures under controlled conditions. Euphytica. 2011;**180**:181-194

[85] Barakat MN, Al-Doss AAA, Elshafei A, Moustafa KA. Identification of new microsatellite marker linked to the grain filling rate as indicator for heat tolerance genes in F_2 wheat population. Australian Journal of Crop Science. 2011;**2**:104-110

[86] Rajneesh P, Marion SR, Uttam K, Jai PS, Arun JK. QTL mapping of terminal heat tolerance inhexaploid wheat (*T. aestivum* L). Theoretical and Applied Genetics. 2012;**125**:561-575

[87] Hamid S, Julian DT, Iman L, Huwaida R, Chris B, Andy T, et al. A QTL on the short arm of wheat (*Triticum aestivum L.*) chromosome 3B affects the stability of grain weight in plants exposed to a brief heat shock early in grain filling. BMC Plant Biology. 2016;**16**(100):1-15

[88] Dion B, Matthew R, Daniel M, Ali I, Hayda K, Peter L, et al. Detection of two major grain yield QTLl in bread wheat (*Triticum aestivum* L.) under heat, drought and high yield potential environments. Theoretical and Applied Genetics. 2012;**125**(7):1473-1485

[89] Sadat S, Saeid KA, Bihamta MR, Sepideh T, Ghasem HS, Lotfali GAA. Marker assisted selection for heat tolerance in bread wheat. World Applied Sciences Journal. 2013;**21**(8):1181-1189

[90] Golabadi M, Arzani A, Mir MSAM. Identification of microsatellite markers

associated with grain protein content in durum wheat grown under drought stress at terminal growth stages. Cereal Research Communications. 2012;**40**(2):215-224

[91] Kirigwi M, Ginkel VG, Brown GBS, Gill GM, Paulsen AKF. Markers associated with a QTL for grain yield in wheat under drought. Molecular Breeding. 2007;**20**:401-413

[92] Pinto RS, Reynolds MP, Mathews KL, McIntyre CL, Olivares VJJ, Chapman SC. Heat and drought adaptive QTL in a wheat population designed to minimize confounding agronomic effects. Theoretical and Applied Genetics. 2010;**121**:1001-1021

[93] Sultan MD, Mia HL, Xingyi W, Guijun Y. Multiple near-isogenic lines targeting a QTLs hotspot of drought tolerance showed contrasting performance under post-anthesis water stress. Frontiers in Plant Science Journal. 2019;**10**:1-271

[94] Kordenaeej A, Nejad AAN, Shojaeian AA, Lelley T. Mapping QTLs for Yield and Yield Components under Drought Stress in Bread Wheat. 2007. Available from: https://www.researchgate.net/publication/291808129.

[95] Israr A, Niaz A, Habib A, Ullah I. Association mapping of root traits for drought tolerance in bread wheat, wheat improvement, management and utilization. 2017. Available from: https://www.intechopen.com/books/wheat-improvement-management-and-utilization/association-mapping-of-root-traits-for-drought-tolerance-in-bread-wheat

[96] Zhang J, Hao C, Ren Q, Chang X, Liu G. Ruilian Association mapping of dynamic developmental plant in common. Wheat Planta. 2011;**234**:891-902

[97] Jaiswal B, Prasad S, Rani R, Singh S, Kumar A, Kumar A, et al. Evaluation of wheat (*Triticum aestivum L.*) lines at reproductive stage for heat stress tolerance. International Journal of Current Microbiology and Applied Science. 2018;**7**:1350-1357

[98] Hafiz GMA, Sajad MLM, Azmat MA, Rizwan M, Maqsood RH, Khan SH. Selection criteria for drought tolerance bread wheat genotypes at seedling stage. Sustainability. 2019;**11**:1-7

[99] Hasan MA, Ahmed JU, Hossain T, Hossain MM, Ullah MA. Germination characters and seed reserve mobilization during germination of different wheat genotypes under variable temperature regimes. Journal of the National Science Foundation of Sri Lanka. 2004;**32**(3&4):97-107

[100] Zulkiffal M, Ahsan A, Javed A, Aziz R, Muhammad A, Saima G, et al. Appraisal of bread wheat *(Triticum aestivum L.)* genotypes under normal, drought and heat prone environments for morpho-physiological multiplicity and constancy. International Journal of Agriculture and Environmental Research. 2018;**4**(6):1298-1306

[101] Gulnaz S, Zulkiffal M, Sajjad M, Ahmed J, Musa M, Abdullah M, et al. Identifying Pakistani wheat (*Triticum Spp* L.) landraces as genetic resources for yield potential, heat tolerance and rust resistance. International Journal of Agriculture and Biology. 2018;**21**:520-526

[102] Mirza FQ, Rahmatullah Q, Humaira S. Effects of pre-anthesis drought, heat and their combination on the growth, yield and physiology of diverse wheat (*Triticum aestivum L.*) genotypes varying in sensitivity to heat and drought stress. Scientific Reports. 2019;**9**:6955

[103] Katarzyna K, Artur N, Jerzy L. Effect of drought and heat stresses on transpiration and photosynthesis of

wheat. In: Abstract. MACSUR Scientific Conference. Italy: University of Sassari; 2014

[104] Moaed A, Deshmukh PS, Sairam RK, Kushwaha SR, Singh TP. Protective role of antioxidant enzymes under high temperature stress. Plant Science. 2006;**171**:382-388

[105] Sami UK, Jalal U, Ali RG, Abdul Q, Hakim K. Heat tolerance evaluation of wheat (*Triticum aestivum L.*) genotypes based on some potential heat tolerance indicators. Journal of the Chemical Society of Pakistan. 2013;**35**(3):647-653

[106] Saeidi M, Ardalani S, Jalali HS, Ghobadi M, Abdoli M. Antioxidant enzyme responses and crop yield of wheat under drought stress and re-watering at vegetative growth period. Iranian Journal of Plant Physiology. 2017;**8**(1):2257-2267

[107] Mirza H, Kamrun N, Tasnim FB, Taufika IA, Masashi I, Hirosuke O, et al. Salicylic acid: An all-rounder in regulating abiotic stress responses in plants. Rijeka, Croatia: IntechOpen. DOI: 10.5772/intechopen.68213.2017. Available from: https://www.intechopen.com/books/phytohormones-signaling-mechanisms-and-crosstalk-in-plant-development-and-stress-responses/salicylic-acid-an-all-rounder-in-regulating-abiotic-stress-responses-in-plants

[108] Amarshettiwar SB, Berad PB. Biochemical and yield responses of wheat genotypes to normal and heat stress condition. Journal of Pharmacognosy and Phytochemistry. 2018;**7**(1):2663-2666

[109] Quarrie SA, Jones HG. Effects of abscisic acid and water stress on development and morphology of wheat. Journal of Experimental Botany. 1977;**28**(102):192-203

[110] Zhao Y, Tian X, Wang F, Zhang L, Xin M, Hu Z, et al. Characterization of wheat MBY genes responsive To High Temperatures. BMC Plant Biology. 2017;**17**:208

[111] Abu HMK, Ki HK, Kwang HS, Jong SC, Byung K, Hisashi T, et al. Abiotic stress responsive proteins of wheat grain determined using proteomics technique. Australian Journal of Crop Science. 2010;**4**(3):196-208

[112] Kaushal N, Bhandari K, Siddique KM, Nayyar H. Food crops face rising temperatures: An overview of responses, adaptive mechanisms, and approaches to improve heat tolerance. Cogent Food and Agriculture. 2016;**2**:1-42

[113] Oviedo AFP, Herz R, Rudorff BFT. Effect of water stress and planting density on the use of radiation and wheat crop productivity (*Triticum aestivum L.)*. Taubate Bioscience Review. 2001;**7**(1):23-33

[114] Labuschagne MT, Elago O, Koen E. Influence of extreme temperatures during grain filling on protein fractions, and its relationship to some quality characteristics in bread, biscuit, and durum wheat. Cereal Chemistry. 2009;**J86**:61-66

[115] Tsenov N, Atanasova D, Stoeva I, Tsenova E. Effects of drought on grain productivity and quality in winter bread wheat. Bulgarian Journal of Agricultural Science. 2015;**21**(3):592-598

[116] Balla K, Rakszegi MLZ, Bekes F, Bencze S, Veisz O. Quality of winter wheat in relation to heat and drought shock after anthesis. Czech Journal of Food Sciences. 2011;**29**(2):117-128

Chapter 2

Characterization of Selected Drought Tolerance Rice Landraces: A Case in Kerala, India

Kallingil Gopi Divya

Abstract

Rice is a staple food of more than half of the world's population. The successful cultivation of rice depends on a variety of climatic and soil conditions. There are lot of factors both biotic and abiotic, which affect the growth and yield of crops. Drought is one of the major abiotic stresses, which leads to drastic decline in the production of rice worldwide. In the present situation of severe climatic change, the scarcity of fresh water is diminishing at an alarming rate. Due to the sensitivity of rice crop and the enormous requirement for optimum yield, drought affects rice when compared to other food crops. Rice germplasm is endowed with scores of varieties and landraces that are reservoirs of genes which is capable of withstanding various abiotic stresses. These landraces can be used to tackle this abiotic stress and can fulfill the increasing demand of food.

Keywords: abiotic stress, drought, heat shock protein, landraces, proline

1. Introduction

Human population is exploding day by day. By 2050, the world population is expected to reach the mark of nine billion, and to feed this enormous population, the food production needs to be increased by more than 50% [1]. Rice is agronomical and nutritional, one among the most important staple food satisfying the need of the calorie intake of approximately half of the population. To meet the future demand for food, anticipated from the projected world population increase, there is an urgent need to take all necessary steps to enhance the productivity of rice. The in-depth understanding of plant responses to abiotic stress needs to be achieved as the climate prediction models have demonstrated the increased rate of abiotic factors like drought, floods, salinity, and soaring temperature during the crop growing periods [2, 3]. Of the 470 Mts of worldwide production of rice, 90% is contributed from the Asian countries. The uneven pattern of rainfall has led to a marked decrease in the yield of upland cultivation. The reduction in the amount of fresh water causes serious stress to the crop production since nearly 5000 L/kg of water is required [4].

The staple food for almost two-third of the world's population, the rice is both nutritional and has medicinal values. Rice is a major source of carbohydrates, which are broken down into glucose upon digestion. Thus, it becomes a rich source of energy. The presence of resistant starch adds to its nutritive value. Since, rice lacks sodium and cholesterol; it is greatly preferred as the staple food. Rice is gluten free and rich in vitamins like thiamine, niacin, iron, and riboflavin. The absence of preservatives and gluten makes it a preferable choice. In addition to the nutritional properties, there are some notable medicinal values for rice. The resistance starch encourages the growth of beneficial bacteria, which helps in keeping the bowel healthy. The high amount of insoluble fiber present in the whole grain is believed to be protective against many varieties of cancer. In addition to being the best food in dysenteric problems, it is also considered to be tonic, fattening, and aphrodisiac [5].

The IRRI 2010 have projected that rice requirement of 496 Mt in 2020 will increase up to 555 Mt in 2035 and will see an overall increase in the future. Global rice production in 2019 has been capped at 512 million tones [6]. Global rice consumption is recorded 494.5 million tons in 2019/20 [7]. Demand for rice is estimated to be 2000 million metric tons by 2030 [8], which has to be answered by immediate promising improvement in rice production in this era of drastic climate change and drought [9]. Of the total worldwide, rice production more than 50% of the area is rainfed and which accounts one quarter of total rice production [10]. World rice production is geographically concentrated in Western and Eastern Asia. Asia is the biggest rice producer, accounting for 90% of the world's production and consumption of rice. Asian farmers still account for 92% of the world's total rice production. Around 280 major rice varieties have been reported from India. There are thousands of strains of rice today, including those grown in wild and those-which are cultivated as crop.

Rice is grown under diverse conditions such as irrigated, rainfed lowland and flood prone ecosystems. It is the only crop that is grown in most fragile ecosystems and cultivated in all the agro-climatic zones below sea levels to the hilly areas. In rainfed ecosystems, drought is the major obstacle for rice production. As it is evident, drought seems to be the major obstacle in the rainfed ecosystem and causes severe loss in grain production. Pandey et al. have cited the example of various Indian states like Jharkhand, Orissa, and Chhattisgarh, which faces severe drought that had recorded a yield loss of 40% valued more than $650 million [11]. Rice production in Kerala is tapering down, yielding just 6.3 lakh tone last year. The state produced only about 15% of its requirement in 2008 compared to 45% in 1956. The work undertaken by the Kerala Agricultural University focusing on the landraces pointed out the fact that Palakkad District contributes around one-third of the total state production [12].

2. Drought, the severe abiotic factor affecting rice production

There are various factors affecting the growth development and yield of agriculture crops which can be classified as biotic and abiotic factors. According to the International Rice Research Institute (IRRI, 1998) projection, over the next two decades, India, Pakistan, the Philippines, and Vietnam will suffer, due to the sharp decrease in per capita water availability. Drought will be the most important abiotic factor which will lead to inter sectorial competition drastically affecting the need of agricultural water [13]. In Asia, approximately 34 million ha of shallow rain-fed lowland rice and 8 million ha of upland rice, totaling approximately one-third of the total Asian rice area are subjected to occasional or frequent drought stress [14]. Continuous drought and fresh-water scarcity has led to the increase of salinity level again giving rise to a new abiotic stress [15]. As paddy crop suffers from water stress at soil water contents even above field capacity drought affects the yield output. Farmers in most rainfed ecosystems, therefore, use the less risky traditional varieties and small amounts of fertilizers which often cause low production.

Drought is the stress situation that lowers plant water potential and turgor pressure and alters the execution of normal physiological process. The challenge of drought is even greater for crops such as rice when compared with other crops such as maize and wheat due to its relatively higher water needs [16]. Around 4000 L of water is required for the production of 1 k of rice which is about double the amount needed for other crops [17]. Since rice cultivars have been grown since time memorable under flood irrigation conditions, rice is very sensitive to deficiency in soil water level as compared to many other field crops; the production systems of rice are more sensitive to drought and possess relatively weak resistances [18]. The water deficit arising with the drought can be classified as the – severe water deficit: available soil moisture 40–50%; moderate water deficit: available soil moisture 50–60%; mild water deficit: available soil moisture 60–70%.

Sarvestani et al. has reported that as a result of reduced water level in the soil, the plant root stem uptakes reduced water leading to the yellowing of leaves due to the reduction in chlorophyll content [19]. The first noticeable response in respect to drought is the rolling of leaves which can help in reducing the internal water level [20]. Henderson et al. has also reported that leaf tip drying is also observed in drought affected plants [21]. The rice genotypes that possess high leaf rolling ability can produce high yield in comparison to other genotypes [22]. Bosco et al. have sustained the fact that the decrease in yield is a result of decrease in plant height, reduction in leaf area, and tillering ability [23]. Root characteristics, leaf temperature, flowering time, panicle exertion, maturity time, and spikelet fertility also affect the crop production [24]. An additional factor occurring due to drought stress resulting in yield reduction is the closure of stomata causing a sharp decrease in carbon dioxide intake directly reducing the photosynthesis rate [25]. The duration and severity of water scarcity determines the rate of reduction of life cycle and the extent of grain filling [26].

3. Drought tolerance in rice landraces

Erratic rainfall distribution is the most limiting factor of growing upland rice in India. Identification of relevant physiological stress tolerance mechanism and the genetic improvement of drought tolerance in crop plants need great attention [27]. Traditional varieties, wild relatives, native species, and modern cultivars together represent the wide genetic resources of the agriculture cops which form the basis of global food security. Genetic diversity provided farmers, plant physiologists, plant breeders, and biotechnologists with options to develop, through the natural selection, breeding and genetic manipulation, new crops, that are resistant to pests, diseases, and adapted to changing environments (abiotic stress) [10].

Landraces are generally called as the native varieties. They harbor a great genetic potential for crop improvement. As these are not subjected to subtle selection over many decades, they are endowed with tremendous genetic variability. This heterozygosity aids in the adaptation of landraces to wide agro-ecological niches and also possess unmatched qualitative traits and medicinal properties. This rich variability of complex quantitative traits still remains unexploited. Rice cultivation in the Kerala state of India situated in the southern Western Ghats dates back to 3000 B.C. [28]. There is immense genetic diversity in the germplasm of both wild and cultivated rice. The landraces present in the state are named according to morphological features, seed color, specific uses, growing conditions, etc. South Asian region is considered as the primary center [29], and Jeypore tract of Orissa is considered as the secondary center of rice and its wild relatives [30] in addition to the humid tropical coastal and mid-lands of Kerala with innumerable wild relative together with more than 600 landraces [31].

Landraces are also important genetic resources for resistant to pest and fungal diseases. "Velluthachira," "Bengle," and "Bhumanasam" are resistant to Rice gall midge; "Thadakan" is resistant to blast,"Buhjan" and "Laka" are resistant to brown plant hopper [32]. Owing to their specific domination in geographical niches, landraces have gene of resistance to abiotic stresses, which have not been widely used or incorporated into modern varieties. The South Indian Landraces "Norungan" and "Noortripathu" are now used as donors for drought tolerance [33].

The present scenario of predicted global food security needs to be tacked with the better understanding of natural selection, germplasm, responses of plant to abiotic stress, and improved yield under stress conditions. The identification of a traditional race, which can thrive well in stress prone area and genetic manipulation of plants that can maintain higher photosynthetic rates, better foliage growth and improved yield under stress conditions requires to be attended [34, 35]. There are a number of comprehensive reviews on drought response in plants [36–38]. Kamoshita et al. have reviewed the drought responses in upland rice [39]. Drought tolerance studies of the under-utilized heterozygous landraces which are bestowed with numerous beneficial properties are the need of the hour.

4. Screening of drought tolerance in rice

One of the major phenomena encountered in almost all rice growing environments is drought. The methods to mitigate are to cultivate either the genotypes which have the capability to strive in the water scarcity or to develop new cultivars which can withstand the drought stress. The factors, which might have operated to create intra-varietal differences in the cultivated rice of Kerala can be attributed to the diverse climatic and ecological conditions which would have led to selection of varieties. The investigation was carried out to find out the best landrace with yield in both stressed and non-stressed environment which can be suggested as the best cultivar for upland cultivation. The recent approaches by functional genomics and genomics assisted breeding of abiotic stress tolerance are helpful in generating valuable information for engineering stress tolerant plants for their use in sustainable agriculture.

Exploration and collection of different accessions of landraces from different districts of Kerala were carried out. After primary screening, the following germplasm were selected to study the effect of drought and the ability of landraces to withstand it (**Tables 1** and **2**).

The seeds of selected landraces varieties were kept for germination. The approximate period for germination was recorded to be 7 days. After all the seeds were germinated, they were transferred to drought condition by withholding the water supply. The withholding resulted in the appearance of wilting signs. These seedlings were screened for their regaining capacities after re-watering. After completion of 7 days of germination, the seedlings were transferred to drought by withholding the input of water. Neither of them could survive 4 days in drought and they wilted and died. In the second phase, drought period was reduced to 3 days. After 72 hours of drought, the seedlings did show signs of wilting but once re-watered, they started to regain. The regaining capacity of the landraces showed district variation. After normal growth for 7 days the seedlings were transferred to drought condition. The areas of the leaves were recorded prior to their transfer into drought with the help of a graph paper and recorded as L1. After 3 days of drought, the leaves showed signs of wilting and started to roll. At that time, the leaf area was again calculated with the help of a graph paper and recorded, L2. Induction of heat shock proteins was done and quantified together with the total carbohydrates and proline. The healthy grains of the selected landraces along with check varieties were

SN	Name of landraces
1.	Navara punja
2.	Navara
3.	Oarkazhama
4.	Vadakan
5.	Vellaryan
6.	Malakkaran
7.	Mundakan
8.	Kuthiru
9.	Kazhama

Table 1.
Landraces selected.

SN	Landraces	Plumule	Radicle
1	Navara punja	9.425	11.27
2	Navara	9.275	11.9
3	Orkazhama	9.675	15.6
4	Vadakan	9.65	10.02
5	Vellaryan	9.7	10.85
6	Malakaran	10.975	15.02
7	Mundakan	9.75	14.27
8	Kuthiru	9.77	12.27
9	Kazhama	8.82	13.27

Table 2.
Plumule and radicle length (cm).

germinated under controlled condition by soaking in double distilled water 72 h, after germination and 50 seeds were exposed to a temperature of 42°C at uniform time intervals. The amount of protein accumulated was correlated with the level of drought tolerance of different landraces and the control and was estimated by the method of Lowry et al. [40]. Accumulation of proline is seen under stress conditions which is used as an energy source for survival and growth [41]. Proline is an osmoregulatory molecule which allows the cell to balance the osmotic strength of its cytoplasm with that of its surroundings to prevent a net loss of water. In addition to functioning as osmotic balancing agents, proline interacts with crucial macromolecules of the cell to maintain their biological activity during stress. It has been suggested that proline may also function as a sink of energy and reducing power, hydroxyl radical scavenger, a compatible solute that protects enzymes and also as a means of reducing acidity. The proline content of the drought treated samples and control seeds were analyzed according to the procedure of Bates et al. [42].

4.1 Screening of root architecture

The root and shoot lengths of all the accessions studied before and after subjection to drought were recorded. From the data obtained its seen that with the induction of drought, there is a remarkable increase in the length of radicle, and

SN	Landraces	Plumule length	Radicle length
1	Orkazhama	>10	Replaced by fibrous roots
2	Malakaran	>9	>7.8.
3	Mundakan	>8.7	>7.5
4	Kazhama	>11	Replaced by fibrous roots
5	Kuthiru	>14	Replaced by fibrous roots

Table 3.
Root performance under drought (cm).

numerous root hairs make their appearance. This might be for the increasing of the absorption area. Kuthiru has the longest shoot while Kazhama has the shortest. With respect to roots, Oarkazhama has the longest roots followed by Malakkaran and Kuthiru. Vaddakan has the smallest, and deteriorated radicle was shown by Navara. After a period of 3 days, the seedlings were watered and their regaining capacity and performance is depicted in **Table 3**. Root system architecture is one of the most important contributors to drought resistance in crops [26]. A well-developed root system is a key to ensuring stable and high yields under drought [43], and the greater root length in deeper soil layers has been shown to increase yield by allowing more water extraction [44]. Supporting to this, the current investigation also revealed the formation of dense root system on subjection to drought. Since rice is characterized by a shallower and more fibrous root system, it has limited water extraction below 60 cm [45]. The lateral root of rice showed greater development under drought [46], which would accelerate drought stress in 20 cm deep soil adverse to rice production.

Since the landraces Njavara, Njavara punja Vadakan and Vellaryan did not show any marked drought tolerance and could not withstand the initial drought stress; they were not subjected to further studies.

4.2 Estimation of leaf area

The leaf area index that reduces due to the subjective stress was compared with normal ones grown under optimum conditions. All the accessions showed decrease in leaf area, and the maximum reduction was shown by the accession, "Malakkaran" and minimum by "Kazhama." There was a clear indication that the accessions tend to reduce the leaf area during drought condition (**Table 4**).

4.3 Estimation of heat shock proteins (HSPS)

The concentration of heat shock accumulated is then found out by the method of Lowry et al. The amount of heat shock protein accumulated by induction together with the amount induced by drought is summarized in **Table 5**.

SN	Landraces	Normal	Drought induced	% Reduction
1	Oarkazhama	17.0	12.5	26
2	Malakaran	18.9	11.7	38
3	Mundakan	19.4	12.3	36
4	Kuthiru	16.9	11.0	34
5	Kazhama	20.7	17.9	14

Table 4.
Leaf area index of drought induction (in cm).

SN	Landraces	Raw	Control	Drought induced
1	Orkazhama	139.41	169.41	181.17
2	Malakaran	116.47	110.0	199.41
3	Mundakan	104.7	173.5	194.70
4	Kuthiru	113.52	169.38	193.53
5	Kazhama	124.7	165.49	181.76

Table 5.
Heat shock proteins in rice.

The concentration of the heat shock protein increases when the seeds are subjected to drought. And the amount is the lowest in Oarkazhama and highest in Malakkaran in the case of raw seeds. Once drought is induced, it is seen that there is an increase in the Hsp's. These may help them to tolerate the advent of water stress. Plant growth regulators modulate plant response to unfavorable environment. Heat shock proteins (HSPs) are expressed at a very high rate when the cells are exposed to increased temperature, and its activity and expressions are co regulated.

The increasing global temperature and the reduction in the available water content for the crops subject them to survive in elevated temperatures. Park et al. reported that the hsp 90 genes and hsp 90 proteins are found in rice [47]. They are induced by elevated temperature stress but their levels are reasonably large even at low temperatures. These hsps provide the temperature affected rice to combat the degradative effect of heat. Yeh et al. reported that a recombinant rice of 16.9 kda heat shock protein can provide thermo protection in vitro [48]. Hsp 100 family has been directly implicated in the induction of thermo tolerance in microbial and animal cells. Jie Zou showed that rice small Hsp gene, shsps 17.7, the product which acts as molecular chaperons aid to determine the mechanism of acquisition of tolerance to drought stress by heat shock [49].

4.4 Estimation of proline

Following the procedure of Bates et al., the concentration of the basic amino acid proline, which has to play a great role in stress condition was determined. The results obtained are summarized below (**Table 6**).

The amount of proline in raw seeds is very less. The value increases only negligibly once the seedlings germinated. But the induction of drought stress remarkably contributes toward the increase of proline content which helps the seedlings to survive in stress period. The landraces Malakkaran showed the highest amount of proline in both raw seeds and also after introduction of drought. While Kuthiru and Kazhama were the landraces possessing the least amount. The landraces which had thrived well in the drought condition showed increase in the proline content.

Proline, a compatible solute and an amino acid, is involved in osmotic adjustment (OA) and protection of cells during dehydration [50]. Cell turgor is maintained due to osmotic adjustments, which allows cell enlargement and plant growth during water stress. It also enables stomata to remain partially open and CO_2 assimilation to continue at water potentials that would be otherwise inhibitory for CO_2 assimilation [51]. Proline can scavenge free radicals and reduce damage due to free radicles during drought stress. Growing body of evidence indicated that proline content increases during drought stress and proline accumulation is associated with improvement in drought tolerance in plants [50, 52].

SN	Landraces	Raw	Control	Drought induced
1	Orkazhama	0.23	0.56	0.86
2	Malakaran	0.28	0.65	0.92
3	Mundakan	0.13	0.49	0.81
6	Kuthiru	0.14	0.38	0.83
5	Kazhama	0.07	0.41	0.80

Table 6.
Proline content in rice.

From the physiological and biochemical results, it is quite clear that the increased higher accumulation of the stress amino acid proline renders Malakkaran followed by Kuthiru the ability to withstand the drought conditions. These two landraces showed more adaptability to drought stress. And if the particular seeds can withstand the severe drought condition, then they can be successfully used for cultivation purposes. From the study undertaken, it is obvious that the two landraces, Malakkaran and Kuthiru can be used for even the upland cultivation even in severe drought periods.

Drought contributes directly to an increase in the incidence and severity of poverty. It is therefore critical that we establish effective strategies to mitigate the effects of drought in order to ensure agricultural productivity and environmental sustainability. The use of landraces which can thrive well in the drastic water scarcity can result in optimum yield which would help in fulfilling the adequate food availability of the increasing human population. As the pressure on land and its resources are increasing with the population growth, there is a noticeable decrease in the low land areas suitable for rice cultivation. The practice of cultivating cultivars which have promising characteristics to tolerate the abiotic stress is the need of the hour.

5. Conclusion

The increased frequency of drought in arable region threatens rice production and demands the development of rice genotype capable of producing more from diminishing water resources. The development of rice cultivars with improved drought tolerance is thus an important element in reducing risk, increasing productivity, and alleviating poverty in communities dependent on rain-fed production. These landraces with the potential to withstand drought condition should be cultivated to harness their potential. The repository of hidden treasures of tolerant genes in the underutilized landraces present in the state of Kerala with maximum water use efficiency mitigate can prove to be the best in alleviating the drought stress.

Acknowledgements

I deeply acknowledge the support and suggestions provided by Dr. Maya C Nair, Head and Associate Professor of Botany, Government Victoria College, Palakkad, Kerala.

Conflict of interest

Nil

Author details

Kallingil Gopi Divya
Department of Pharmacognosy, Siddha Central Research Institute, Chennai, India

*Address all correspondence to: minnu.kg@gmail.com

References

[1] Alexandratos N, Bruinsma J. Interim Report World Agriculture Towards 2030/2050: The 2012 Revision. FAO; 2016

[2] Bates BC, Kundzewicz ZW, Wu S, Palutikof JP. Climate change and water. In: Technical Paper of the Intergovernmental Panel on Climate Change, IPCC Secretariat. Vol. 168. No. 1. Geneva, Switzerland: The American Midland Naturalist; 2008

[3] Mittler R, Blumwald E. Genetic engineering for modern agriculture: Challenges and perspectives. Annual Review of Plant Biology. 2010;**61**:443-462

[4] Fahad S, Adnan M, Noor M, Arif M, Alam M, Khan IA, et al. Major constraints for global rice production. In: Advances in Rice Research for Abiotic Stress Tolerance. Woodhead Publishing; 2019. pp. 1-22

[5] Divya KG. Identification and characterization of drought tolerance in traditional rice landraces in Kerala [MSc thesis]. Palakkad: Govenment Victoria College; 2009

[6] Crop prospects and food situation. Food and Agriculture Organization of United Nations Quatery Global Report; 2020

[7] Nathan Childs, RCS-19J, USDA, Economic Research Service; October 15, 2019

[8] Bruinsma J. World Agriculture: towards 2015/2030: Summary Report. Food and Agriculture Organization of the United Nations (FAO); 2002

[9] Bouman BA, Humphreys E, Tuong TP, Barker R. Rice and water. Advances in Agronomy. 2007;**92**:187-237

[10] McLean JL, Dawe DC, Hardy B, Hettel GP. RiceAlmanac. International Rice Research Institute, LosBa˜nos, Philippines; WARDA, Bouak'e, Côte d'Ivoire; CIAT, Cali, Colombia; and FAO, Rome; 2002. p. 253

[11] Pandey S, Bhandari HS, Hardy B. Economic Costs of Drought and Rice Farmers' Coping Mechanisms: A Cross-Country Comparative Analysis. Los Banos, Philipines: International Rice Research Institute; 2007. p. 203

[12] Kumari SL. Status Paper on Rice in Kerala; 2012

[13] IRRI. Sustaining Food Security Beyond the Year 2020. LosBaños, Philippines: A Global Partnership for Rice Research, IRRI Rolling Medium Term Plan 1991-2000; 1998

[14] Huke RE, Huke EH. Rice: Then and Now. Los Banos, Philipines: International Rice Research Institute; 1990

[15] De Datta SK, Mikkelsen DS. Potassium nutrition of rice. Potassium in Agriculture. 1985;**1**:665-699

[16] Todaka D, Shinozaki K, Yamaguchi-Shinozaki K. Recent advances in the dissection of drought-stress regulatory networks and strategies for development of drought-tolerant transgenic rice plants. Frontiers in Plant Science. 2015;**6**:84

[17] Bouman BA, Peng S, Castaneda AR, Visperas RM. Yield and water use of irrigated tropical aerobic rice systems. Agricultural Water Management. 2005;**74**(2):87-105

[18] O'Toole JC. Rice and water: The final frontier. In: The First International Conference on Rice for the Future Rockefeller Foundation. Bangkok, Thailand; 2004

[19] Sarvestani ZT, Pirdashti H, Sanavy SA, Balouchi H. Study of water stress effects in different growth stages on yield and yield components of different rice (Oryza sativa L.) cultivars. Pakistan Journal of Biological Sciences. 2008;**11**(10):1303-1309

[20] Sié M, Sanni K, Futakuchi K, Manneh B, Mandé S, Vodouhe RS, et al. Towards a rational use of African rice (*Oryza glaberrima* Steud.) for breeding in sub-Saharan Africa. Genes, Genome and Genomics. Global Science Books. 2012;**6**:1-7

[21] Henderson SA, Kamboonruang V, Copper M. Evaluation of a glasshouse screening method to select for drought resistance in rainfed lowland rice. In: International Rice Research Conference on Fragile Lives in Fragile Ecosystems, Los Banos, Laguna (Philippines), 13-17 Feb 1995. IRRI; 1995

[22] Fukai S, Cooper M. Field screening of adaptability in drought-prone rainfed lowland rice: ACIAR experience in Thailand and Laos. In: Saxena NP, O'Toole JC, editors. Field Screening for Drought Tolerance in Crop Plants with Emphasis on Rice. Proc. Int. Workshop Field Screening Drought Tolerance Rice, Patancheru, India. Patancheru, India: International Crops Research Institute for the SemiArid Tropics; 2002, 2000. pp. 61-62

[23] Bosco R, Lorieux M, Futakuchi K, Manney B, Baimey H, Ndjiondjop MN. Agro-morphological characterization of a population of introgression lines derived from crosses between IR 64 (*O. sativa indica*) and TOG 5681 (*O. glaberrima*) for drought tolerance. Plant Science. 2011;**183**:65-76

[24] Ndjiondjop MN, Cisse F, Girma G, Sow M, Bocco R, Djedatin G, et al. Morpho-agronomic and molecular characterization of Oryza glaberrima germplasm from Mali. African Journal of Biotechnology. 2010;**9**(44):7409-7417

[25] Flexas J, Bota J, Cifre J, Mariano Escalona J, Galmés J, Gulías J, et al. Understanding down-regulation of photosynthesis under water stress: Future prospects and searching for physiological tools for irrigation management. The Annals of Applied Biology. 2004;**144**(3):273-283

[26] Farooq M, Hussain M, Siddique KHM. Drought stress in wheat during flowering and grain-filling periods. Critical Reviews in Plant Sciences. 2014;**33**:331-349. DOI: 10.1080/07352689.2014.875291

[27] Theresa K, Adhikary S, Ulrichs C, Goswami A. Evaluation of drought tolerance in some rainfed upland rice cultivars. Indian Agriculture. 2003;**47**:259-263

[28] Manilal KS. Ethnobotany of rices of Malabar. In: Jain SK, editor. Glimpses of Indian Ethnobotany; 1990. pp. 297-307

[29] Chang TT. The origin, evolution, cultivation, dissemination, and diversification of Asian and African rices. Euphytica. 1976;**25**(1):425-441

[30] Sampath S. Wild rices of Orissa and their relationship to the cultivated varieties. Rice News Teller. 1958;**6**(3):17-20

[31] Latha M, Abraham Z, Nair RA, Mani S, Dutta M. Rice landraces of Kerala state of India: A documentation. International Journal of Biodiversity and Conservation. 2013;**5**(4):250-263

[32] Siddiq EA, Saxena S, Malik SS. Plant Genetic Resources: Food Grain Crops. New Delhi, India: Indian Society of Plant Genetic Resources, Narosa Publishing House; 2005. pp. 27-57

[33] Ram SG, Thiruvengadam V, Vinod KK. Genetic diversity among cultivars, landraces and wild relatives of rice as revealed by microsatellite

markers. Journal of Applied Genetics. 2007;**48**(4):337-345

[34] Condon AG, Richards RA, Rebetzke GJ, Farquhar GD. Breeding for high water-use efficiency. Journal of Experimental Botany. 2004;**55**(407):2447-2460

[35] Morison JIL, Baker NR, Mullineaux PM, Davies WJ. Improving water use in crop production. Philosophical Transactions of the Royal Society of London. Series B, Biological Sciences. 2008;**363**:639-658

[36] Ingram J, Bartels D. The molecular basis of dehydration tolerance in plants. Annual Review of Plant Physiology and Plant Molecular Biology. 1996;**47**:377-403

[37] Holmberg N, Bülow L. Improving stresstoleranceinplantsbygene transfer. Trends in Plant Science. 1998;**3**:61-66

[38] Farooq M, Basra SMA, Khalid M, Tabassum R, Mehmood T. Nutrient homeostasis, reserves metabolism and seedling vigor as affected by seed priming in coarse rice. Canadian Journal of Botany. 2006a;**84**:1196-1202

[39] Kamoshita A, Babu RC, Boopathi NM, Fukai S. Phenotypic and genotypic analysis of drought-resistance traits for development of rice cultivars adapted to rainfed environments. Field Crops Research. 2008;**109**:1-23

[40] Lowry OH, Rosebrough NJ, Farr AL, Randall RJ. Protein measurement with the Folin phenol reagent. The Journal of Biological Chemistry. 1951;**193**(1):265-275

[41] Kaur-Sawhney R, Tiburcio AF, Altabella T, Galston AW. Polyamines in plants: An overview. Cellular and Molecular Biology. 2003;**2**:1-2

[42] Bates LS, Waldren RP, Teare ID. Rapid determination of free proline for water-stress studies. Plant and Soil. 1973;**39**(1):205-207

[43] Dodd IC, Whalley WR, Ober ES, Parry MA. Genetic and management approaches to boost UK wheat yields by ameliorating water deficits. Journal of Experimental Botany. 2011;**62**:5241-5248. DOI: 10.1093/jxb/err242

[44] Manschadi AM, Christopher J, Devoil P, Hammer GL. The role of root architectural traits in adaptation of wheat to water-limited environments. Functional Plant Biology. 2006;**33**: 823-837. DOI: 10.1071/FP06055

[45] Fukai S, Inthapan P. Growth and yield of rice cultivars under sprinkler irrigation in South-Eastern Queensland. 1. Effects of sowing time. Aust. Journal of Experimental Agriculture. 1988;**28**:237-242. DOI: 10.1071/EA9880237

[46] Kameoka E, Suralta RR, Mitsuya S, Yamauchi A. Developmental plasticity of rice root system grown under mild drought stress condition with shallow soil depth; comparison between nodal and lateral roots. Plant Production Science. 2016;**19**:411-419. DOI: 10.1080/1343943X.2015.1128094

[47] Park M, Kang CY, Krishna P. *Brassica napus* Hsp90 can autophosphorylate and phosphorylate other protein substrates. Molecular and Cellular Biochemistry. 1998;**185**:33-38

[48] Yeh CH, Yeh KW, Wu SH, Chang PF, Chen YM, Lin CY. A recombinant rice 16.9-kDa heat shock protein can provide thermoprotection in vitro. Plant & Cell Physiology. 1995;**36**(7):1341-1348

[49] Zou J, Liu A, Chen X, Zhou X, Gao G, Wang W, et al. Expression analysis of nine rice heat shock protein genes under abiotic stresses and ABA treatment. Journal of Plant Physiology. 2009;**166**(8):851-861

[50] Zhang X, Ervin EH, Evanylo GK, Haering KC. Impact of biosolids on hormone metabolism in drought-stressed tall fescue. Crop Science. 2009;**49**:1893-1901

[51] Alves AAG, Setter TL. Abscisic acid accumulation and osmotic adjustment in cassava under water deficit. Environmental and Experimental Botany. 2004;**51**:259-279

[52] Seki M, Umezawa T, Urano K, Shinozaki K. Regulatory metabolic networks in drought stress responses. Current Opinion in Plant Biology. 2007;**10**:296-302

Chapter 3

The Response of Maize Physiology under Salinity Stress and Its Coping Strategies

Shazia Iqbal, Sajid Hussain, Muhammad Abdul Qayyaum, Muhammad Ashraf and Saifullah

Abstract

Maize is a cross-pollinated, polymorphic plant in nature. It is commonly a moderately salt-sensitive crop. Salinity stress is the main abiotic factor that arrests the physiological characteristics and plant growth of a maize plant. It causes the osmotic effect, associated with an increase in phytotoxic ions, oxidative stress by increased reactive oxygen species (ROS) production, and ionic effect in the cytosol. These salinity effects hinder the maize plant's physiological processes such as respiration, photosynthesis, transpiration, stomatal functioning, hormone regulation, and functioning, seed germination, and dormancy and water relation with plants and ultimately reduce the plant growth and yield. However, the physiology of maize subjected to salinity shows various responses that depend on the genetic responses and growth stages. Maize plant undergoes many physiological changes and adapts some mechanism internally to cope with salinity stress. Numerous mitigating strategies such as application of chemicals, application of plant growth-promoting rhizobacteria (PGPR), application of hormones, and use of genetic and molecular techniques are used to handle salinity. This chapter will cover the effect of salinity on maize growth, its physiology, and physiological adaptations of maize plants with management strategies.

Keywords: *Zea mays* L., salinity, physiology, genetic and molecular techniques, antioxidants

1. Introduction

Soils with an excessive amount of soluble salts or exchangeable sodium in the root zone are termed salt-affected soils. Owing to limited rainfall and high evapotranspiration demand, coupled with poor soil and water management practices, salt stress has become a serious threat to crop production in arid and semi-arid regions of the world [1, 2]. Although the general perception is that salinization only occurs in arid and semi-arid regions, no climatic zone is free from this problem [3]. More than 800 million hectares of land worldwide is affected by either salinity (397 million hectares) or sodicity (434 million hectares) [4–6].

Maize (*Zea mays* L.) is the third most important cereal crop after rice and wheat and is grown under a wide spectrum of soil and climatic conditions. It is an important C4 plant from the *Poaceae* family and is moderately sensitive to salt stress;

nonetheless, wide intraspecific genetic variation for salt resistance exists in maize. According to the biphasic model of salinity-induced growth reduction [7], osmotic stress during the first phase and ion toxicity during the second phase are responsible for reduced growth in cereals, specifically wheat. The same model for salinity-induced growth reduction in maize was confirmed by [8], but ion toxicity and the associated growth reduction can occur, to a small extent, in the first phase in maize. The sensitivity of maize to salinity is associated with higher accretion of Na^+ in the leaves. A saline level of more than 0.25 M NaCl damages maize plants and may stunt growth and cause severe wilting [9].

2. Salinity stress

Sodium is the main toxic ion interfering with potassium uptake and thus disturbs stomatal undulations causing severe water loss and necrosis in maize; a reduction in potassium content in the leaf symplast of maize has been reported under saline conditions. High osmotic stress due to low external water potential, ion toxicity by sodium and/or chloride, and imbalanced nutrition due to interference with the uptake and transport of essential nutrients are three potential effects of salt stress on crop growth. The latter may not have an immediate effect because plants have some nutrient reserves which can be remobilized [10, 11]. Osmotic stress is linked to ion accumulation in the soil solution, whereas nutritional imbalance and specific ion effects are connected to ion buildup, mainly sodium and chloride, to toxic levels which interferes with the availability of other essential elements such as calcium and potassium [12]. Toxic levels of sodium in plant organs damage biological membranes and subcellular organelles, reducing growth and causing abnormal development before plant mortality [13, 14]. Several physiological processes such as photosynthesis, respiration, starch metabolism, and nitrogen fixation are also affected under saline conditions, leading to losses in crop productivity.

Moreover, salt stress also induces oxidative damage to plant cells with overproduction of reactive oxygen species in maize [15]. The ability of plants to survive and produce harvestable yields under salt stress is called salt resistance. Salt resistance is a complex phenomenon, and plants manifest a variety of adaptations at subcellular, cellular, and organ levels such as stomatal regulation, ion homeostasis, hormonal balance, activation of the antioxidant defense system, osmotic adjustment, and maintenance of tissue water status to grow successfully under salinity [16–20]. An integrated approach encompassing conventional breeding together with marker-assisted selection, biotechnology, exogenous use of growth regulators/osmoprotectants, and nutrient management may be needed for successful maize cultivation on salt-affected soils [21–23].

3. The response of salinity stress on maize plant physiology

Salt stress affects the growth and development of maize; however, the response of plants varies with the degree of stress and crop growth stage. Short-term exposure of maize plants to salt stress influences plant growth, owing to osmotic stress in the first phase of salt stress without reaching toxic sodium concentrations.

3.1 Salinity stress response on seed germination

Seed germination is the most critical stage in a seedling establishment which determines the success of crop production on salt-affected soils. Generally, salt stress during germination delays the start, reduces the rate, and enhances the dispersion of germination events [23–25]. It is important to note that germination and early seedling growth are more sensitive to salinity than later developmental stages [26]. Salt stress influences seed germination primarily by sufficiently lowering the osmotic potential of the soil solution to retard water absorption by seeds, by causing sodium and/or chloride toxicity to the embryo or by altering protein synthesis. Hyper-osmotic stress and toxic effects of sodium and chloride ions on germinating seeds in a saline environment may delay or inhibit germination [25, 27]. However, in maize, it is sodium toxicity and not chloride toxicity that is the major problem in the second phase of salt stress.

3.2 Salinity stress response on vegetative, reproductive growth and grain yield

Although the root is the first organ exposed to salt stress, shoots are more sensitive to salt stress than roots [7]. Salinity promotes the suberization of the hypodermis and endodermis, and the Casparian strip develops closer to the root tip than in non-saline roots [28]. Salinity reduces shoot growth by suppressing leaf initiation and expansion, as well as internode growth, and by accelerating leaf abscission [29–31]. Salt stress rapidly reduces the leaf growth rate due to a reduction in the number of elongating cells and/or the rate of cell elongation. As a salt-sensitive crop, shoot growth in maize is strongly inhibited in the first phase of salt stress [32–34].

Salt stress may also displace calcium from plasma membrane binding sites, thus causing membrane leakiness as a primary cellular response to salt stress [35]. However, it is interesting to note that if salt stress influences the integrity of the plasma membrane, then the cell wall acidification process, which is partially dependent on adenosine triphosphate-driven outward pumping of protons across the intact plasma membrane, may also be affected [36]. Acidification of the apoplast is the major requirement for increasing cell wall extensibility, which controls extension growth [37]. In this regard, cell wall proteins such as expansions are of great interest. Expansions, wall-loosening enzymes located within the apoplast of the elongation zone of leaves [38], regulate cell elongation. The assimilate supply to growing tissues is not limited during the first phase of salt stress [39], suggesting that photosynthesis is not responsible for any growth reduction in maize during this phase. Salinity-induced growth reduction in maize is caused by suppressed leaf initiation, expansion, and internode growth and by increased leaf abscission. In maize, suppression of expansion growth by salinity is principally caused by reduced apoplastic acidification and activity of wall-loosening enzymes.

In salt-affected soils, excessive buildup of sodium and chloride ions in the rhizosphere leads to severe nutritional imbalances in maize due to strong interference of these ions with other essential mineral elements such as potassium, calcium, nitrogen, phosphorus, magnesium, iron, manganese, copper, and zinc [40, 41]. Generally, salt stress reduces the uptake of nitrogen, potassium, calcium, magnesium, and iron [42]. For maize, sodium is the principal toxic ion interfering with potassium uptake and transport, leading to disturbance in stomatal modulations and causing water loss and necrosis. Competition between potassium and sodium under salt stress severely reduces potassium content in both leaves and roots of maize [19] and reduces potassium content by up to 64% in the symplast of expanding tissues under salt stress. Moreover, salt stress not only reduces potassium uptake rates but, to a greater extent, disturbs potassium translocation from root to shoot tissues in maize, leading to lower potassium shoot contents than root contents. Reduced leaf expansion with reduced calcium contents in expanding shoot tissues in maize is due to reduced transport in a saline environment; some calcium is required to uphold cell membrane integrity for proper functioning [43]. The high values for sodium/potassium, sodium/calcium, and sodium/magnesium ratios in the total plant and apoplast and symplast of expanding tissues in maize confirm that impaired transport of potassium, calcium, and magnesium by sodium might upset plant metabolism, leading to reduced growth under saline conditions. Besides potassium and calcium, nitrogen uptake and translocation are severely inhibited under salt stress, leading to reduced nitrogen contents in different maize tissues [41, 44]. Higher buildup of sodium and chloride concentrations in different plant tissues is the principal reason for nutritional imbalances. Accumulation of high sodium and chloride ions, due to salinity, in the rhizosphere decreases plant uptake of nitrogen, potassium, calcium, magnesium, and iron and thus causes severe nutritional imbalances in maize.

Carbon fixation in maize is very sensitive to salt stress [45]. Reduced stomatal conductance, impaired activities of carbon fixation enzymes, reduced photosynthetic pigments, and destruction of photosynthetic apparatus are among the key factors limiting carbon fixation capacity of maize plants under salt stress [31, 46]. Total photosynthesis decreases due to inhibited leaf development and expansion as well as early leaf abscission, and as salt stress is prolonged, ion toxicity, membrane disruption, and complete stomatal closure become the prime factors responsible for photosynthetic inhibition. Salt stress affects stomatal conductance immediately due to perturbed water relations and shortly afterward due to the local synthesis of abscisic acid. Gas exchange analysis confirmed that reductions in net photosynthetic rates are connected with the limited availability of intercellular carbon dioxide due to reduced rates of transpiration and stomatal conductance in salt-treated maize plants.

Salt stress in maize, during the reproductive phase, decreases grain weight and number, resulting in substantial reductions in grain yield [47, 48]. Salinity-induced reductions in photosynthesis and sink limitations are the major causes of poor kernel settings and reduced grain number [49]. Salinity-induced reductions in assimilate translocation, from source to developing grains, are also responsible for poor grain setting and filling and ultimately grain yield [50].

4. Mechanisms of salt tolerance in maize

Maize plants undergo a variety of adaptations at subcellular, cellular, and organ levels to grow successfully under salinity. Salt tolerance is a complex phenomenon; maize plants manifest several adaptations such as stomatal regulation, changes in hormonal balance, activation of the antioxidant defense system, osmotic adjustment, maintenance of tissue water contents, and various mechanisms of toxic ion exclusion under salt stress.

Osmotic adjustment or osmoregulation is the key adaptation of plants at the cellular level to minimize the effects of salinity-induced drought stress, especially during the first phase of salt stress. Osmoregulation is primarily met with the accretion of organic and inorganic solutes under salinity and/or drought to lower water potential without lessening actual water contents [51]. Soluble sugars, sugar alcohols, proline, glycine betaine, organic acids, and trehalose are among the major osmolytes. Proline and glycine betaine are the major osmolytes responsible for osmoregulation in maize under salt stress. Physiologically, the exclusion of excessive salt is an adaptive trait of plants to acquire salt resistance. Accumulation of sodium in excessive amounts is highly toxic for maize growth due to its strong interference with potassium, leading to disturbed stomatal regulation. Therefore, the exclusion of excessive sodium or its compartmentation into vacuoles through tonoplast hydrogen/sodium antiporters driven by the proton gradient is an important adaptive strategy for plants under salt stress. Through this strategy, maize plants not only evade the cytosol from the toxic effects of excessive sodium and gain tissue resistance for sodium but also significantly lower the osmotic potential which contributes to osmoregulation. In root cells of maize, shifting sodium into vacuoles through the tonoplast appears to be a viable strategy to minimize sodium transport to developing shoots [16]. Absorption of excessive sodium from xylem by parenchyma cells in the xylem to limit sodium translocation to shoots is also reported in maize [52]. However, salt tolerance in maize is not linked to sodium content in shoots, but rather the ability of cells to shift excessive sodium in vacuoles to maintain low sodium concentrations in the cytoplasm seemed more important [53].

Salt tolerance in maize is also linked with higher potassium and lower sodium and chloride fluxes and cytoplasmic contents and their ability to rule out sodium and chloride from leaves to sustain a higher potassium/sodium ratio. Moreover, shifting sodium and chloride in the stems and/or leaf sheaths to lessen the buildup of toxic ions in more sensitive leaf blades is another adaptive strategy of maize plants in a saline environment [54].

Salinity-induced osmotic effects alter general metabolic processes and enzymatic activities, leading to over-generation of reactive oxygen species which causes oxidative stress in maize. Overproduction of reactive oxygen species is highly toxic and damages proteins, lipids, carbohydrates, and deoxyribonucleic acid. Photosystems I and II in chloroplasts and complex I, ubiquinone, and complex III of the electron transport chain in mitochondria are key sites for reactive oxygen species synthesis [55]. Plants have multigenic responses against salt stress that involve both osmotic and ionic homeostasis, as well as cell detoxification, which is primarily met by antioxidant defense mechanisms [56, 57]. The better leaf growth, leaf water content, and membrane stability index observed in salt-tolerant maize were associated with higher antioxidant activity with greater accumulation of polyphenols under saline conditions [19]. Catalase, ascorbate peroxidase, and guaiacol peroxidase enzymes in combination with superoxide dismutase have the greatest hydrogen peroxide scavenger activity in both leaves and roots of salt-stressed maize plants [15].

Plant growth and development is governed by the synthesis of hormones with small amounts sufficient to regulate plant growth. Auxins, gibberellins, cytokinins, ethylene, and abscisic acid are the most important phytohormones; among them, the first three are growth promoters, while the other two are growth retardants. Maize plants under salt stress make certain modifications to the synthesis of these growth substances. In a saline environment, root tips are the first to sense impaired water availability due to the osmotic effect, sending a signal to shoots to adjust whole plant metabolism [18]. Higher abscisic acid levels in salt-tolerant maize help to minimize water loss and may even regulate growth promotion. Leaf growth sensitivity decreases as abscisic acid levels increase under such conditions.

Maize plants facing salt stress undergo a variety of adaptive mechanisms at the molecular level to counteract the damaging effects of salinity stress. Of these, accumulation or inhibition of several proteins and the upregulation and downregulation of many gene transcripts are important [58]. Expression of antioxidant defense genes is triggered in maize to protect the cells from salinity-induced oxidative damage. In photosynthesizing shoots of maize, catalase activity increased due to the induction of mRNA accumulation in response to higher reactive oxygen species levels under salt stress. Likewise, the buildup of superoxide dismutase transcripts increased without any notable change in total superoxide dismutase enzymatic activity or isozyme profiles [9]. The alteration/adaptation in cell wall chemical composition may also contribute to salt resistance in maize, as a low accumulation of non-methylated uronic acid in leaf cell walls may contribute to salt resistance in the first phase of salt stress [59].

5. Management strategies

Remediation of salt affected areas with low cost, efficient, and adaptable methods is a challenging goal for scientists [11]. Different practices are used to improve growth and tolerance of crops in salt-affected areas.

5.1 Agronomic approaches (soil amendments)

For saline soil management, many chemicals and organic amendments are applied to combat the adverse effect of salinity in maize crops. Chemicals mostly applied to soil for maize crops include silicon, salicylic acid, potassium, phosphorus, gypsum, biochar, and boron, and many organic amendments are also applied. Silicon application and an increase in their availability reduce the changes caused by salinity in plants by altering the plant and soil factors [60]. Silicon application increases the photosynthetic apparatus efficiency of maize plants under salinity stress by improving and maintaining the continuity of the electron transport chain [61] Silicon is recognized as a resistance improver against salinity in the maize crop. Silicon application lessened both oxidative and osmotic stress in maize crops by improving the defensive machinery performance under salinity stress. Silicon also improved water-use efficiency. Silicon-treated maize plants showed better survival under saline conditions, and their biochemical and photosynthetic apparatus was better working than silicon non-treated plants [62]. The application of brackish water is also reported in maize plants to reclaim salt effects. Brackish water irrigation boosted K uptake and retarded Na uptake in some maize genotypes. Selection of tolerant genotypes for growing in salt affected areas would be a better reclamation method [63]. Boron is an important element for many biochemical and physiological reactions of plants [64]. Boron application alleviated the negative effect of sodium chloride-induced salinity in sweet corn. Boron improved potassium concentration and maintained membrane integrity [65]. Combined application of silicon and boron also proved effective in alleviating the salinity effect on maize crops. They both in combination enhanced maize plant physical and biological parameters and also increased total soluble sugars and proline content [66]. In saline conditions, sodium concentration increased that caused an imbalance in sodium to potassium ratio. Application of potassium maintained or lowered this ratio and alleviates the deleterious effects of sodium. Potassium application to maize crop grown in saline soil decreased sodium percentage and enhanced potassium percentage in maize grain and stalk as well as distinctly boosted the maize salt tolerance by decreasing the sodium to potassium ratio. The most significant effect was observed at higher potassium fertilizer application rates [67].

Combined application of potassium sulfate and diammonium phosphate on maize in saline soil for maize (*Zea mays* L.) showed that maize responded well to potassium and phosphorus fertilization. Salinity effects were amended by potassium and phosphorus fertilizer application and improved yield. The influence of potassium was great on grain yield compared to phosphorus. K affected yield-related parameters, and phosphorus showed substantial effects on sodium, potassium, magnesium, and sodium to potassium ratio. Potassium application decreased sodium concentration and ultimately decreased sodium to potassium ratio [68]. Foliar application of potassium chloride, boron, and thidiazuron was done on maize crops in saline stress. Thidiazuron and potassium application improved the physiological parameters of the crop. Thidiazuron proved more efficient in alleviating the adverse effects of salinity than potassium and boron. Potassium content, chlorophyll content, total carbohydrate protein percentage, and total soluble salt percentage were substantially improved by foliar application of thidiazuron; however, transpiration rate and proline content were decreased [69].

Flue gas desulfurization gypsum (FGDG) application can reduce sodium toxicity by replacing it with calcium at the cation exchange site and results in increased clay particle flocculation near the surface of the soil [70]. Furfural residue is rich in organic carbon and can increase the SOC content, reduce soil bulk density, and lower soil pH [71]. Flue gas desulfurization gypsum and furfural residue combined application reduced the yield gap of maize and recovered soil properties. Flue gas desulfurization gypsum and furfural residue increased the organic carbon (SOC) and calcium contents and decreased the upper soil layer pH and sodium content. Mineral nutrients like phosphorus, nitrogen, potassium, magnesium, and calcium accumulations in maize increased significantly, and sodium accumulation decreased in the flue gas desulfurization gypsum and furfural residue treatment compared with control [72].

Hydrogen peroxide as foliar spray effectively curtailed the effects induced by salinity because of increased antioxidant enzyme activities: ascorbate peroxidase, guaiacol peroxidase, superoxide dismutase, and the most responsive catalase [73].

Salicylic acid is an imperative secondary metabolite that is used in salinity management as it induces resistance against salinity in plants by regulating physiological processes through signaling. Maize plants exposed to sodium chloride induced salinity, reduced plant dry biomass, increased membrane permeability, and reduced nutrient availability, while those plants supplied with exogenous salicylic acid increased dry biomass, decreased membrane permeability and lipid peroxidation, and increased iron, zinc, copper, and manganese contents. Salicylic acid application further improved nutrient uptake by maize plants except for zinc in the saline condition. Salicylic acid reduced chloride and sodium accumulation considerably [22].

In another study, a maize crop dry weight and leaf area decreased by 51.43 and 53.18%, respectively, when irrigated with saline water, while salicylic acid foliar application at the rate of 200 ppm remedied the harmful salinity effects and improved whole plant dry weights and leaf area and improved proline and amino acid contents such as lysine, arginine, glutamic acid, and serine accumulation under saline stress conditions [74].

Organic amendments proved as an effective strategy for saline soil amelioration. Organic amendments improve soil chemical and physical properties. Solid waste, vermicompost, and cow dung influence soil salinity and alleviate its adverse effects on the growth of plants by changing the physico-chemical properties of soil. Solid waste, vermicompost, and cow dung reduced soil electrical conductivity as well as improved shoot and root length [75].

Compost and vermicompost application increased maize plant dry matter and plant height and reduced soil pH and electrical conductivity. Extractable phosphorus, total nitrogen, soil organic carbon, cation exchange capacity, and potassium, calcium, and magnesium concentrations were improved by the application of vermicompost and compost. Sodium concentration decreased because of its replacement by calcium ions and then leaching. This results in a decrease in soil salinity levels [76].

Biochar also improved physico-chemical properties of soil, including soil cation exchange capacity, pH, water holding capacity, surface area, and soil structure under abiotic stresses [77]. Biochar application improved potassium availability uptake and decreased sodium availability and uptake under salt stress [78, 79]. Biochar made by cow manure is a rich source of many plant nutrients which significantly increased nutrient uptake in maize crop. Cow manure biochar application improved net WUE, field-saturated hydraulic conductivity, and significantly increased Oslen-P, total N, pH, total C, exchangeable cations, and cation exchange capacity [80]. Compost manure and crop straw biochar and pyroligneous solution can improve maize productivity and combat salinity stress. Compost manure and crop straw biochar both increased nutrient statuses and decreased salinity by reducing chloride and sodium accumulation and increasing potassium concentration. Manures also increased plant performance, maize grain yield, and leaf area index, with a decrease in electrolyte leakage. Leaf bioactivity associated with osmotic stress was improved significantly [81]. It is concluded that exogenously applied organic matters such as plant residues, manure, a by-product of municipal or farming activities, etc. are an efficient and feasible way to mitigate the effects of salinity on plant growth and soil health. Organic amendments at optimal rates (>50 tons per hectare) can improve soil chemical like cation exchangeable capacity, pH, etc. and physical properties like permeability, soil structure, water holding capacity, etc., approving maize plant growth [82].

5.2 Application of hormones

Hormones govern many processes inside plants that regulate plant growth: auxins, gibberellins, and cytokinins are growth promoter hormones, while abscisic acid and ethylene are the growth retardants. Under salt stress conditions, growth-promoting hormones are applied exogenously to overcome the adverse effects of salinity on maize plant growth and development.

Cytokinin is a plant growth regulator that plays a vital role in cytokinin-dependent processes that regulate plant adaptation, growth, and development processes [83]. It is reported in recent research that cytokinins of developing maize seeds may come from both transport and local synthesis. Cytokinin fertilization at higher rates suggested parental control on plant metabolism [84]. Cytokinin and auxin application alone or in combination with maize plants reduced the deleterious effect of salinity on plant growth and increased physical parameters like stem diameter, plant height, ear length, row number per ear, and biological yield like grain yield and number at different concentrations. A single application of cytokinin played a role in improving kernel number per row, while a single application of auxin increased grain weight and better harvest index in saline condition [85].

Kinetin is one form of cytokinins and is known to boost the crop plant growth grown under saline conditions [86]. Kinetin and indoleacetic acid (auxin) applications as foliar spray overcame to adversative effects of sodium chloride induced stress on physiological parameters at the earlier stages of maize plants at a variable extent. Foliar combined application of both kinetin and indoleacetic acid substantially increased K^+ and Ca^{2+} concentration and reduced those of Na^+. Their application also increased essential inorganic nutrients and maintained membrane permeability and in result thwarted some salt-persuaded adversative effects [19]. Exogenous combined application of inorganic nutrients and indoleacetic acid improved phosphorus, calcium, and magnesium contents and decreased sodium concentration in maize plants grown in saline condition. Improvement in growth by indoleacetic acid and organic nutrient application is linked with an improved concentration of photosynthetic pigment, more leaf sodium to potassium ratio, rehabilitated activities of some antioxidant enzymes such as CAT and SOD, and reduced membrane permeability under salinity. Exogenous foliar application of indoleacetic acid additionally improved the CAT and SOD activities in salt-stressed maize plants, while increasing effect was not detected in activities of POX or PPO [87]. Previously, foliar application of indoleacetic acid enhanced the essential nutrient uptake along with a noteworthy decrease in sodium uptake that resulted in better growth and yield of maize plant under salt stress condition [88].

The combined application of sodium chloride-induced salinity and gibberellic acid on maize plant growth and nutritional status was studied. Salinity decreased chlorophyll content, total dry matter, and relative water content, whereas increased enzyme activities peroxidase polyphenol oxidase superoxide dismutase and proline accumulation. Gibberellic acid overcame the deleterious effects of sodium chloride-induced salinity stress on the above physiological characteristics to a variable extent. Gibberellic acid decreased enzyme activities and increased physiological parameters and macro- and micronutrient concentration. Foliar application of gibberellic acid counteracted some salinity adverse effects by the buildup of proline concentration which sustained membrane permeability [89]. A comparison between gibberellic acid and salicylic acid under the saline condition in maize plant showed that gibberellic acid was more efficient in resisting salinity effect on leaves than salicylic acid. Gibberellic acid also improved the nutrient status of plant except for copper and manganese [90].

5.3 Application of PGPR

Soil has an enormous microbial versatility that belongs to different groups of fungi, Archaea, and bacteria [91]. Microorganisms are used in agricultural fields, and they can lessen many abiotic stresses [92, 93]. Usually, bacteria are used for promoting plant growth and alleviating many abiotic stresses. These bacteria are usually termed as plant growth-promoting rhizobacteria (PGPR). PGPR is rhizospheric or endophytic bacteria that colonize the root either interiorly or exteriorly. Bacterial genera such as *Achromobacter*, *Azospirillum*, *Bacillus*, *Burkholderia*, *Enterobacter*, *Methylobacterium*, *Microbacterium*, *Paenibacillus*, *Pantoea*, *Pseudomonas*, *Rhizobium*, *Variovorax*, etc. provide tolerance to host plants against abiotic stresses [94, 95]. Stress tolerance is boosted by microbes by various mechanisms and production of indoleacetic acid, gibberellins, and many other elements. These elements improved the root growth and enhance nutrient content, thus improving the plant health under salt stress [95]. Bacteria that help plants in alleviating salt stress are called halotolerant or salt-tolerant or salt-loving bacteria. These halotolerant microbes have vital importance in the field of agriculture. In arid and semi-arid regions, they improve crop productivity [91]. Specific PGPR inoculations help to boost salt stress tolerance in plants by induced systemic tolerance (IST). Induced systemic tolerance changes many biochemical and functional characteristics. The PGPR improves salinity tolerance by either direct mechanism (indoleacetic acid (IAA) synthesis phosphate solubilization, nitrogen fixation, etc.) or indirect mechanism (exopolysaccharides (EPS), antioxidant defense, osmotic balance, and volatile organic compounds (VOCs)) and improves plant growth [96] (**Table 1**).

5.3.1 Osmotic adjustment

Osmotic adjustment is the maintenance of cell turgidity by increasing compatible solutes, vital for regular cell functioning. Compatible solutes decrease osmotic stress caused by salts [55]. PGPR produce and secrete compatible osmolytes to mitigate the harmful effect of salts and help plants improve their growth. Proline is the main osmolytes in reducing osmotic stress and produced by the hydrolysis of proteins in the plant. Under salt stress, glycine betaine and proline are usually produced and accumulated in plants. There is a dearth of organic osmolytes production such as trehalose in plants [112]. Under salinity, proline plays a multifunctional role like regulating cytosolic acidity, protein maintenance, ROS

PGPR strain	Mechanism	Improvement in crop	References
Pseudomonas syringae, *P. fluorescens*	ACC deaminase	Improved plant growth	[97–100]
Pseudomonas spp.	EPS		[101]
P. aeruginosa	IAA production, ACC deaminase, phosphate solubilization, and biofilm formation		[102]
Pseudomonas spp.	Osmotic regulation		[103]
Proteus penneri	EPS		[101]
Pantoea agglomerans, *Staphylococcus sciuri*, *Arthrobacter pascens*	Upregulation of aquaporin genes		[94, 104, 105]
Gracilibacillus, *Staphylococcus*, *Virgibacillus*, *Salinicoccus*, *Zhihengliuella*, *Brevibacterium*, *Oceanobacillus*, *Exiguobacterium*, *Arthrobacter*, and *Halomonas* spp.	Antioxidant enzyme phosphate solubilization, osmotic regulation and antioxidant enzymes IAA production, ACC deaminase, phosphate solubilization, and biofilm formation		[102]
Serratia liquefaciens KM4	Facilitated gas exchange, osmoregulation, antioxidant enzymes, nutrient uptake, and downregulation of ABA biosynthesis		[106]
Enterobacter aerogenes, *Enterobacter* spp.	ACC deaminase	Reduced ethylene production	[97, 98]
Azospirillum brasilense	Ion toxicity, NOR, and nitrogenase activity	Improved chlorophyll content	[100]
A. faecalis, *A. brasilense* strains Ab-V5 and Ab-V6	EPS, antioxidant enzymes, and proline contents	Improved nutrition	[101, 107]
Azotobacter chroococcum	Improved K/Na ratio, polyphenol content, and proline concentration		[108]
B. amyloliquefaciens	Soluble sugar content and antioxidant enzymes	Improved plant growth and photosynthetic rate	[109]
Bacillus spp.	Phosphate solubilization, osmotic regulation, and antioxidant enzymes		[104]
Bacillus aquimaris	Chlorophyll content, osmotic regulation, and antioxidant enzymes		[110]
Bacillus	IAA production, ACC deaminase, phosphate solubilization, and biofilm formation		[102]
Geobacillus sp.	Increased proline content		[111]
Rhizobium	Osmotic regulation	Increased chlorophyll and photosynthesis rate	[103]
Rhizobium tropici strain CIAT 899	Antioxidant enzymes and proline contents		[107]

Table 1.
PGPR and their mechanisms for salt tolerance.

scavenging decrease in peroxidation of lipids, etc. PGPR inoculation in plants showed improved proline levels under salt stress. *Arthrobacter pascens* inoculation produces more proline in corn plants [104]. *Pseudomonas* spp. improved growth of maize plant by production of proline that helps in osmotic adjustments [103]. *Azotobacter chroococcum* improved nutrition [108], *Geobacillus* sp. increased photosynthetic rate [111], and *Rhizobium* spp., *Rhizobium tropici* strain CIAT, *A. brasilense* strains Ab-V5 and Ab-V6 [107], and *A. faecalis* [101] enhanced chlorophyll content and photosynthetic rate by increased accumulation of proline and osmotic adjustments in maize plants.

5.3.2 Antioxidants

Plants normally produce reactive oxygen species during cellular metabolism in less quantity. However, under salinity stress conditions, increased production of reactive oxygen species occurs, which alters redox state, denatures membrane bound proteins, reduces fluidity of membrane, causes DNA damage, destroys enzymatic actions, changes formation of protein, and destroys cell homeostasis, which can damage the cell and finally cause cell death [113]. PGPR excrete many enzymatic antioxidants (ascorbate peroxidase (APX), catalase (CAT) dehydro-ascorbate reductase, glutathione reductase (GR), superoxide dismutase (SOD), non-enzymatic antioxidants, ascorbate, tocopherols, glutathione, and cysteine) [114]. *Staphylococcus sciuri* induction induces more antioxidant production in maize plants that helped in the degradation of reactive oxygen species and improved plant growth [94]. *A. faecalis* [101], *Serratia liquefaciens* KM4 [106], and *Bacillus* sp. [104] are reported to increased maize growth, nutrition, and photosynthetic rate by producing more antioxidative enzymes. *Azotobacter vinelandii*, *Pseudomonas fluorescens*, and *Pseudomonas putida* restored lipids and antioxidant enzymes peroxidase and catalase to semi-normal levels under saline condition [115].

5.3.3 Exopolysaccharides

PGPR produce exopolysaccharides (EPS), which are either homo- or heteropolysaccharides. These EPS bind to the cell surface like a capsule and make a biofilm [116]. Different microbes produce different types of polysaccharides, but some common monomers comprise glucose, galactose, and mannose. Uronic acids (fucose and rhamnose), amino sugars (N-acetylamino sugars), neutral sugars (galacturonic), pyruvate ketals, and ester-linked substituents are EPS constituents [117]. PGPR produce EPS and form hydrophilic biofilms under saline conditions and improve plant growth significantly [118]. EPS producing PGPR makes rhizosheaths around roots that help fight against salt stress by attaching Na^+ ions with EPS. Attachment of Na^+ ions to EPS decreases the toxicity of Na^+ and makes it inaccessible for plants [119]. *P. aeruginosa* improved plant growth because of more EPS content production. *Pseudomonas* spp. produced more EPS and increased root growth and nutrition in maize plants [101]. Many other PGPRs such as *Gracilibacillus*, *Salinicoccus*, *Staphylococcus*, *Zhihengliuella*, *Bacillus*, *Brevibacterium*, *Virgibacillus*, *Oceanobacillus*, *Arthrobacter*, *Exiguobacterium*, and *Halomonas* spp. are reported to improve maize growth by the formation of biofilm [102]. *B. amyloliquefaciens* improved plant growth by the accumulation of soluble sugar content [109].

5.3.4 Volatile organic carbons

Rhizobacteria that produce lipophilic fluids with high vapor pressures are called volatile organic compounds. They communicate by cell signaling between organisms to improve growth. The VOCs are species-specific and promote the biosynthesis of glycine betaine and choline. These osmolytes improve plant tolerance against osmotic stress. A high level of VOCs in plants is a sign of activated self-protective response against salt stress [120].

The VOCs produced by *Bacillus subtilis* triggered the gene of HKT1/K^+ transporter and inhibited sodium ion influx through roots and eliminated salt stress. It also encouraged the glycine betaine synthesis that decreased the uptake of Na^+ through roots and transported more nutrients toward shoot than during salt stress [120].

5.4 Seed priming

Poor crop stands because of low seed germination rate in salt-affected areas are a challenge for the lucrative production of a crop. Maize seed germination rate is affected by toxic effects of chloride and sodium ions [25]. Seed priming helps to recover maize germination rate in salt-affected areas. Seed priming is a pre-sowing treatment either with water or any chemical of interest that boosts seed performance with a quicker and harmonized germination under sub-optimal and optimal conditions [121]. This is a physiological treatment under salinity in which seeds are moderately hydrated and radicle does not emerge [122]. Priming treatments include hydropriming with water, osmopriming with salts or osmolytes, and hormonal priming with hormones. Partial hydration is enough for the physiological process occurrence that is typical of the first stages of imbibition (pre-germinative metabolism) [123]. Under saline conditions germination rate improved by soaking maize seeds in water priming with water under salinity-enhanced maize seedling vigor index, germination index, final germination percentage, and seedling length, showing its potential as a seed invigoration technique under salinity for better maize performance [23].

Priming of seeds with salt solution enables them to break their dormancy and escape from disease-causing agents and competent seeds of weeds [124]. Priming seeds with NaCl significantly enhanced maize plant growth. Fresh and dry weights of roots and shoots were increased. Under salt stress, seed priming lessened the inhibitory effect of salt stress on maize seedling growth [125]. Priming with NaCl also increased plant height and yield and induced early emergence, more germination rate, more shoot length and dry weight, and more leaf chlorophyll, area, and number [126]. Seed halopriming with calcium chloride, sodium chloride, and potassium chloride was effective in mitigating the salt adversities on maize seed germination. Calcium chloride priming was most operative. Calcium, sodium, and potassium concentrations improved significantly in all parts of germinating seed. Most of the calcium was reserved in mesocotyl and seed, thus limiting its transference to radicles and plumules.

Seed priming with NaCl and $CaCl_2$ had significant effects on germination rate, earlier growth, number of branches, cobs number, and yield. This increase in growth traits likely helps to reduce the competition for water and nutrients with associated improvements in seed yield. Sodium chloride seed priming increased shoot length, and calcium chloride seed priming increased root length. In vertisol soil, seed priming is preferred for improved crop yield and stand establishment, while in lithosol soils, seed priming is preferred for well germination of seed and increased cob number [124].

Other priming agents include thiamin, pyridoxine, and ascorbic acid, which not only improved the germination of pretreated seed but also improved seed growth and yield under salinity. Enhanced maize seedling biomass under saline conditions is reported by hormonal priming with chloro-ethyl-phosphonic acid, an ethylene releaser [127]. Salicylic acid application under saline conditions at the rate of 0.1 mM enhanced growth and development of plants [128]. Priming with 28-homo-brassinolide improved the antioxidative enzyme activities and lowered lipid peroxidation and increased concentration of protein, thus signifying that

28-homo-brassinolide can lessen oxidative stress in salt-affected maize plants [129]. Priming with hydrogen peroxide improved activities of catalase ascorbate peroxidase and guaiacol peroxidase and increased seed germination percentage, under salt stress condition in maize plants [73].

Seeds of maize hybrid FH-810 were soaked in water (hydropriming), calcium chloride (2.2%, osmopriming), *Moringa* leaf extracts (MLE 3.3%, osmopriming), and salicylic acid (SA, 50 mg L^{-1}, hormonal priming), each for 18 h. Plant length, biological yield, 1000-grain weight, and harvest index were improved by seed priming. However, osmopriming with MLE and hormonal priming were more effective in these parameters. Hormonal priming at seedling stage increased the leaf chlorophyll contents and decreased the electrical conductivity followed by osmopriming with CaCl2. Hormonal or osmopriming with MLE improved the yield performance at early planting primarily by increased crop growth, net assimilation rates, a leaf area index, and maintenance of green leaf area at maturity. Hormonal priming with SA and osmopriming with MLE were the most economical methods in enlightening early planted spring maize productivity by early seedling growth stimulation at low temperature [130].

5.5 Application of molecular and genetic approaches such as MAS, selection, and breeding approaches

Maize is a polymorphic plant because of its cross-pollinated nature and genetic variations for salinity resistance. It is commonly a moderately salt-sensitive crop, but some salinity tolerant genotypes also exist. Tolerance in these genotypes occurs because of higher potassium and lower chloride and sodium cytoplasmic contents. Mass screening of maize genotypes is done to identify and isolate salt-tolerant germplasm for breeding purposes and to develop better performing genotypes. Screenings for salt tolerance or resistance are usually done at the early growth stages of maize plants [21]. Many plant characteristics are identified as salt-tolerant traits. Acidification of cell wall because of better H^+-ATPase activity in the plasma membrane in salt-tolerant maize hybrid (SR 03) appeared as an important tolerance/resistance trait. Turgor, cell wall acidification, and osmotic adjustment, in newly established salt-resistant maize hybrids, are a salt-resistant trait [48]. More abscisic acid accumulation in salt-resistant genotypes plays a role in osmotic adjustments under saline condition [39]. Salt-tolerant genotypes usually had lower sodium accumulation and more potassium to sodium and calcium to sodium ratio. Sensitive genotypes had more sodium accumulation, suggesting that accumulation of sodium in shoots is a reliable screening parameter for salt tolerance/resistance in maize at early stages of growth [21]. However, higher sodium accumulation was observed in salt-tolerant Giza 2 roots than in salt-sensitive Trihybrid 321. Many other traits of maize plants such as growth rate, seedling weight, and photochemical efficiency should also be used for screening and breeding of salt-tolerant crops [131].

A proteomic approach is also used to recognize salt resistance-associated proteins in maize in breeding programs for markers to develop salt-tolerant/salt-resistant genotypes. The use of physiological and molecular markers to recognize salt-resistant genotypes of maize is a reliable approach [132]. Sodium and soluble organic solute accumulations in roots were associated with maize salt resistance. More soluble organic solute and sodium accumulation in maize salt-tolerant genotype roots (BR5033) than in salt-sensitive genotype (BR5011) was reported. Hence, soluble organic solute and sodium accumulations in roots can be used as physiological markers to screen and isolate salt-resistant maize genotypes [15]. More total separated proteins (>80%) in severe saline stress in maize genotypes and 45 and

31% increase in root and shoot proteins under mild salinity showed differential regulation of proteins [58].

Transferring one or more salt-resistant genes from one species to another to insert required quantitative and qualitative characteristics is stated as the transgenic approach. This practice is much quicker than conventional breeding practices, and it warrants wanted genes induction without the addition of excess genes from the donor organism [133]. Improvements and advances in biotechnology and functional genomics have made it feasible to identify and distinguish salinity-tolerant genes that help to develop salt-resistant plants by the use of transgenic tactics (27).

By using the Flippase recombination enzyme P/Flippase recognition target-based marker elimination system to eliminate the *als* gene [134]. Marker-free salt-tolerant transgenic maize is produced to improve the bio-safety of the environment. Under the saline condition, transgenic maize seed inserted with AtNHX1 gene and wild-type maize were planted. Wild-type maize plants withered, and upper leaves shriveled, whereas 56% of transgenic plants survived salinity up to the six-leaf stage. More grain yield, 1000-grain weight, was recorded in transgenic plants under saline condition than those under non-saline conditions. More potassium accumulation in root and shoot was observed in transgenic plants [134].

The sodium vacuolar compartmentation or cytoplasmic exclusion into the apoplasts through tonoplast sodium/hydrogen antiporters or plasma membrane is an adaptive mechanism to alleviate the adverse excess sodium effects in maize plants [26]. Under saline conditions, transgenic maize plants were better than wild-type plants because of higher hydrogen/sodium exchange rates in vesicles of tonoplast. Also, the efficient sodium vacuolar compartmentalization in cells of transgenic maize plants improved salt tolerance as well as the productivity of grain [134].

Salt stress boosted ZmNHX transcription which caused an increase in antiporters (sodium/hydrogen) of tonoplast in salt-resistant maize leaves by impounding sodium into vacuoles of the leaf to reduce sodium ion effects on the cytoplasm [135]. Transgenic maize plants with inserted sodium/hydrogen antiporter (OsNHX1) gene from *Oryza sativa* gave better yield than wild-type maize at 200 mM NaCl. Lower osmotic potential coupled with higher potassium and sodium contents in transgenic maize leaves was recorded under saline condition compared to wild maize [136].

The complementary DNA (cDNA) micro-array is an operative method for expression profile studies to assess differences and similarities under salinity stress in diverse patterns of expression. A cDNA macro-array with 190 maize expressed sequence tags persuaded by water stress was applied to cold stress, abscisic acid, and high salinity conditions. High salinity stress upregulated 41 sequence tags in roots and 36 sequence tags in leaves [137]. Quan et al. (2004) [138] introduced the betA gene encoding choline dehydrogenase (AtNHX1), which was inserted in maize line DH4866 from *Escherichia coli* to develop transgenic maize. This gene improved the biosynthesis of glycine betaine from choline under salinity and increased salt resistance in maize plants [139]. In conclusion, maize genotypes with externally inserted genes of betaine aldehyde dehydrogenase and vacuolar sodium/hydrogen antiporter, etc. performed better under salinity stress and can be used for inducing salt resistance in maize plants.

6. Conclusions

Salinity stress poses a serious threat to maize. It affects the plant physiology and reduces growth and yield. Salinity affects the maize crop at different growth stages. Seed germination is the stage that is affected adversely by salinity, and germination rate is reduced. At vegetative and reproductive stages, salinity affects photosynthesis, respiration, transpiration, stomatal and hormonal regulation, and water relation processes. These processes affect the growth pattern of plants and cause reduction in growth and yield. To mitigate the effects of salinity on maize crop, different management practices are used. Management by agronomic means, such as application of nutrients (through the application of biochar, compost, gypsum and nutrient fertilizers, etc.), either exogenously or as seed priming with different chemical and hormones, exogenous application of hormones, and growing of resistant cultivars, proved effective in reducing the adverse effects of salinity on maize crops. PGPR application mitigates the salinity stress by the production of different hormones, exopolypolysaccharides, or volatile organic compounds. Different genetic and molecular techniques are also used for inducing salinity tolerance by the insertion of tolerant genes in maize plants. For the future, more work on improved genetic and molecular techniques is needed.

Acknowledgements

Shazia Iqbal is thankful to Saline Agriculture Research Center, Institute of Soil and Environmental Sciences, University of Agriculture, Faisalabad, Pakistan, for the award of doctoral fellowships. Shazia Iqbal is giving special thanks to Dr. Sajid Hussain and Dr. Muhammad Ashraf for motivating him to write this chapter and providing guidance. The authors are also highly thankful to Muhammad Qayyaum, for contributing in the chapter write-up and providing them supporting material.

Conflict of interest

There is no conflict of interest among all the authors. All the authors revised and approved the chapter.

Author details

Shazia Iqbal[1], Sajid Hussain[2*], Muhammad Abdul Qayyaum[3], Muhammad Ashraf[4] and Saifullah[5]

1 Institute of Soil and Environmental Sciences, University of Agriculture, Faisalabad, Pakistan

2 State Key Laboratory of Rice Biology, China National Rice Research Institute, Hangzhou, Zhejiang, China

3 Department of Soil and Environmental Sciences, Faculty of Agricultural Sciences, Ghazi University, Dera Ghazi Khan, Punjab, Pakistan

4 Department of Soil Science, Bahauddin Zakariya University, Multan, Pakistan

5 Department of Environmental Health, College of Public Health, Imam Abdulrehman Bin Faisal University, Dammam, Saudi Arabia

*Address all correspondence to: hussainsajid@caas.cn

References

[1] Flowers TJ, Yeo A. Breeding for salinity resistance in crops. Where next? Australian Journal of Plant Physiology. 1995;**22**:875-884

[2] Munns R. Comparative physiology of salt and water stress. Plant, Cell and Environment. 2002;**25**:239-250

[3] Rengasamy P. World salinization with emphasis on Australia. Journal of Experimental Botany. 2006;**57**:1017-1023

[4] FAO. Global Network on Integrated Soil Management for Sustainable Use of Salt-Affected Soils. Rome, Italy: FAO Land and Plant Nutrition Management Service; 2005

[5] Munns R. Genes and salt tolerance: Bringing them together. New Phytologist. 2005;**167**:645-663

[6] Munns R, James RA, Läuchli A. Approaches to increasing the salt tolerance of wheat and other cereals. Journal of Experimental Botany. 2006; **57**:1025-1043

[7] Munns R, Sharp RE. Involvement of abscisic acid in controlling plant growth in soils of low water potential. Australian Journal of Plant Physiology. 1993;**20**:425-437

[8] Fortmeier R, Schubert S. Salt tolerance ofmaize (*Zea mays* L.): The role of sodium exclusion. Plant, Cell and Environment. 1995;**18**:1041-1047

[9] Menezes-Benavente L, Kernodle SP, Margis-Pinheiro M, Scandalios JG. Salt induced antioxidant metabolism defenses in maize (*Zea mays* L.) seedlings. Redox Report. 2004;**9**:29-36

[10] Flowers TJ, Flowers SA. Why does salinity pose such a difficult problem for plant breeders? Agricultural Water Management. 2005;**78**:15-24

[11] Munns R, Tester M. Mechanisms of salinity tolerance. Annual Review of Plant Biology. 2008;**59**:651-681

[12] Hussain M, Park HW, Farooq M, Jabran H, Lee DJ. Morphological and physiological basis of salt resistance in different rice genotypes. International Journal of Agriculture and Biology. 2013;**15**:113-118

[13] Davenport R, James RA, Zakrisson-Plogander A, Tester M, Munns R. Control of sodium transport in durum wheat. Journal of Plant Physiology. 2005;**137**:807-818

[14] Quintero JM, Fournier JM, Benlloch M. Na^+ accumulation in shoot is related to water transport in K^+ starved sunflower plants but not in plants with a normal K^+ status. Journal of Plant Physiology. 2007;**164**:60-67

[15] de Azevedo Neto AD, Prisco JT, Eneas J, de Abreu CEB, Gomes-Filho E. Effect of salt stress on antioxidative enzymes and lipid peroxidation in leaves and roots of salt-tolerant and salt sensitive maize varieties. Environmental and Experimental Botany. 2006;**56**:87-94

[16] Neubert AB, Zörb C, Schubert S. Expression of vacuolar Na^+ /H^+ antiporters (ZmNHX) and Na^+ exclusion in roots of maize (*Zea mays* L.) genotypes with improved salt resistance. In: Li CJ et al., editors. Plant Nutrition for Food Security, Human Health and Environmental Protection. Bejing, China: Tsinghua University Press; 2005. pp. 544-545

[17] Hichem H, Mounir D, Naceur EA. Differential responses of two maize (*Zea mays* L.) varieties to salt stress: Changes on polyphenols composition of foliage and oxidative damages. Industrial Crops and Products. 2009;**30**:144-151

[18] Schubert S. Advances in alleviating growth limitations of maize under salt

stress. In: The Proceedings of the International Plant Nutrition Colloquium XVI; Department of Plant Sciences, UC Davis; 2009

[19] Kaya C, Tuna AL, Okant AM. Effect of foliar applied kinetin and indole acetic acid on maize plants grown under saline conditions. Turkish Journal of Agriculture and Forestry. 2010;**34**:529-538

[20] Jafar MZ, Farooq M, Cheema MA, Afzal I, Basra SMA, Wahid MA, et al. Improving the performance of wheat by seed priming under saline conditions. Journal of Agronomy and Crop Science. 2012;**198**:38-45

[21] Eker S, Comertpay G, Konuskan O, Ulger AC, Ozturk L, Cakmak I. Effect of salinity stress on dry matter production and ion accumulation in hybrids maize varieties. Turkish Journal of Agriculture and Forestry. 2006;**30**:365-373

[22] Gunes A, Inal A, Alpaslam M, Erslan F, Bagsi EG, Cicek N. Salicylic acid induced changes on some physiological parameters symptomatic for oxidative stress and mineral nutrition inmaize (*Zea mays* L.) grown under salinity. Journal of Plant Physiology. 2007;**164**: 728-736

[23] Janmohammadi M, Dezfuli PM, Sharifzadeh F. Seed invigoration techniques to improve germination and early growth of inbred line of maize under salinity and drought stress. General and Applied Plant Physiology. 2008;**34**:215-226

[24] Ashraf M, Foolad MR. Pre-sowing seed treatment-a shotgun approach to improve germination growth and crop yield under saline and none-saline conditions. Advances in Agronomy. 2005;**88**:223-271

[25] Khaje-Hosseini M, Powell AA, Bingham IJ. The interaction between salinity stress and seed vigour during germination of soybean seeds. Seed Science and Technolog. 2003;**31**:715-725

[26] Goldsworthy. Calcium and salinity. Annals of Applied Biology. 1994;**4**:1-6

[27] Hasegawa PM, Bressan RA, Zhu JK, Bohnert HJ. Plant cellular and molecular response to high salinity. Annual Review of Plant Physiology and Plant Molecular Biology. 2000;**51**:463-499

[28] Shannon MC, Grieve CM, Francois LE. Whole plant response to salinity. In: Wilkinson RE, editor. Plant-Environment Interactions. New York: Dekker; 1994. pp. 199-244

[29] Rios-Gonzalez K, Erdei L, Lips SH. The activity of antioxidant enzymes in maize and sunflower seedlings as affected by salinity and different nitrogen sources. Plant Science. 2002; **162**:923-930

[30] Akram M, Ashraf MY, Ahmad R, Rafiq M, Iqbal AIJ. Allometry and yield components of maize (*Zea mays* L.) hybrids to various potassium levels under saline conditions. Archives of Biological Sciences. 2010;**62**:1053-1061

[31] Qu C, Liu C, Gong X, Li C, Hong M, Wang L, et al. Impairment of maize seedling photosynthesis caused by a combination of potassium deficiency and salt stress. Environmental and Experimental Botany. 2012;**75**:134-141

[32] Pitann B, Kranz T, Mühling KH. The apoplastic pH and its significance in adaptation to salinity in corn (*Zea mays* L.): Comparison of fluorescence microscopy and pH-sensitive microelectrodes. Plant Science. 2009; **176**:497-504

[33] El Sayed HESA. Influence of salinity stress on growth parameters, photosynthetic activity and cytological studies of *Zea mays*,L. plant using hydrogel polymer. Agriculture and

Biology Journal of North America. 2011; **2**:907-920

[34] Wakeel A, Sümer A, Hanstein S, Yan F, Schubert S. In vitro effect of Na^+/K^+ ratios on the hydrolytic and pumping activity of the plasma membrane H-ATPase from maize (*Zea mays* L.) and sugar beet (*Beta vulgaris* L.) shoot. Plant Physiology and Biochemistry. 2011;**49**: 341-345

[35] Cramer GR, Epstein E, Lauchli A. Kinetics of root elongation of maize in response to short term exposure to NaCl and elevated calcium concentration. Journal of Experimental Botany. 1988; **39**:1513-1522

[36] Spanswick RM. Electrogenic ion pumps. Annual Review of Plant Physiology. 1981;**32**:267-289

[37] Hager A. Role of the plasma membrane H^+-ATPase in auxin-induced elongation growth: Historical and new aspects. Journal of Plant Research. 2003; **116**:483-505

[38] Cosgrove DJ. Loosening of plant cell walls by expansins. Nature. 2000;**407**: 321-326

[39] De Costa W, Zorb C, Hartung W, Schubert S. Salt resistance is determined by osmotic adjustment and abscisic acid in newly developedmaize hybrids in the first phase of salt stress. Physiologia Plantarum. 2007;**131**:311-321

[40] Karimi G, Ghorbanli M, Heidari H, Khavarinejad RA, Assareh MH. The effects of NaCl on growth, water relations, osmolytes and ion content in Kochia prostrate. Biologia Plantarum. 2005;**49**:301-304

[41] Turan MA, Elkarim AHA, Taban N, Taban S. Effect of salt stress on growth and ion distribution and accumulation in shoot and root of maize plant. African Journal of Agricultural Research. 2010;**5**: 584-588

[42] Yasmeen A, Basra SMA, Farooq M, Rehman H, Hussain N, Athar HR. Exogenous application of moringa leaf extract modulates the antioxidant enzyme system to improve wheat performance under saline conditions. Plant Growth Regulation. 2013;**69**: 225-233

[43] Hu Y, Burucs Z, Tucher SV, Schmidhalter U. Short-term effects of drought and salinity on mineral nutrient distribution along growing leaves of maize seedlings. Environmental and Experimental Botany. 2007;**60**:268-275

[44] Gadalla AM, Hamdy A, Galal YGM, Aziz HAA, Mohamed MAA. Evaluation of maize growth under salinity stress and N application strategies using stable nitrogen isotope. African Crop Science Conference Proceedings. 2007;**8**:1553-1562

[45] Omoto E, Taniguchi M, Miyake H. Adaptation responses in C4 photosynthesis of maize under salinity. Journal of Plant Physiology. 2012;**169**: 469-477

[46] Gong XL, Liu C, Zhou M, Luo LY, Wang L, Wang Y, et al. Oxidative damages of maize seedlings caused by combined stress of potassium deficiency and salt stress. Plant and Soil. 2011;**340**: 443-452

[47] Abdullah Z, Khan MA, Flowers TJ. Causes of sterility in seed set of rice under salinity stress. Journal of Agronomy and Crop Science. 2001;**187**: 25-32

[48] Schubert S, Neubert A, Schierholt A, Sumer A, Zorb C. Development of salt resistant maize hybrids: The combination of physiological strategies using conventional breeding methods. Plant Science. 2009;**177**:196-202

[49] Hiyane R, Hiyane S, Tang AC, Boyer JS. Sucrose feeding reverses

shade-induced kernel losses in maize. Annals of Botany. 2010;**106**:395-403

[50] Lohaus G, Hussmann M, Pennewiss K, Schneider H, Zhu JJ, Sattelmacher B. Solute balance of a maize (*Zea mays* L.) source leaf as affected by salt treatment with special emphasis on phloem re-translocation and ion leaching. Journal of Experimental Botany. 2000;**51**:1721-1732

[51] Serraj R, Sinclair TR. Osmolyte accumulation: Can it really help increase crop yield under drought conditions? Plant, Cell and Environment. 2002;**25**: 333-341

[52] Yeo AR, Kramer D, Lauchli A, Gullasch J. Ion distribution in salt-stressed mature *Zea mays* roots in relation to ultrastructure and retention of sodium. Journal of Experimental Botany. 1977;**28**:17-29

[53] Alberico GL, Cramer GR. Is the salt tolerance of maize related to sodium exclusion? I. Preliminary screening of seven cultivars. Journal of Plant Nutrition. 1993;**16**:2289-2303

[54] Isla R, Aragues R. Yield and plant ion concentrations in maize (*Zea mays* L.) subject to diurnal and nocturnal saline sprinkler irrigations. Field Crops Research. 2010;**116**:175-183

[55] Gill SS, Tuteja N. Reactive oxygen species and antioxidant machinery in abiotic stress tolerance in crop plants. Plant Physiology and Biochemistry. 2010;**48**:909-930

[56] Zhu JK. Plant salt tolerance. Trends in Plant Science. 2001;**6**:66-71

[57] Mittler R. Oxidative stress, antioxidants and stress tolerance. Trends in Plant Science. 2002;**7**:405-410

[58] Zörb C, Schmitt S, Neeb A, Karl S, Linder M, Schubert S. The biochemical reaction of maize (*Zea mays* L.) to salt stress is characterized by a mitigation of symptoms and not by a specific adaptation. Plant Science. 2004;**167**:91-100

[59] Uddin MN, Hanstein S, Leubner R, Schubert S. Leaf Cell-wall components as influenced in the first phase of salt stress in three maize (*Zea mays* L.) hybrids differing in salt resistance. Journal of Agronomy and Crop Science. 2013;**199**:405-415

[60] Gattullo CE, Allegretta I, Medici L, et al. Silicon dynamics in the rhizosphere: Connections with iron mobilization. Journal of Plant Nutrition and Soil Science. 2016;**179**:409-417

[61] Khan WUD, Aziz T, Hussain I, et al. Silicon: A beneficial nutrient for maize crop to enhance photochemical efficiency of photosystem II under salt stress. Archives of Agronomy and Soil Science. 2016a;**63**:599-611

[62] Khan WUD, Aziz T, Maqsood MA, Farooq M, Abdullah Y, Ramzani PMA, et al. Silicon nutrition mitigates salinity stress in maize by modulating ion accumulation, photosynthesis, and antioxidants. Photosynthetica. 2018;**56** (4):1047-1058

[63] Ahsan RA, Mehdi SM, Rasheed MK, Tahir FA, Bhat B, Uallah R, et al. Evaluating the salinity tolerance of maize (*Zea mays* L.) genotype under brackish water application in Punjab-Pakistan. Life Science Journal. 2013;**10** (3):913-919

[64] Sezer S. Effect of boron fertilizer applications on the growth and B, N uptake of maize (*Zea mays* L.) under the different soils. Journal of Food, Agriculture and Environment. 2014;**12** (2):1323-1327

[65] Bastías EI, González-Moro MB, González-Murua C. *Zea mays* L. amylacea from the Lluta Valley (Arica-Chile) tolerates salinity stress when high

levels of boron are available. Plant and Soil. 2004;**267**:73-84

[66] Salim BBM. Effect of boron and silicon on alleviating salt stress in maize. Middle East Journal of Agriculture Research. 2014;**3**(4):1196-1204

[67] El-Dissoky RA, Ebtsam M, Morsy, El-Shazly MA. Beneficial effect of potassium fertilization and yeast strains on maize plants grown on salt affected soil. Journal of Soil Science and Agricultural Engineering. 2013;**4**(9): 827-842

[68] Hussain Z, Khattak RA, Fareed I, Irshad M, Mahmood Q. Interaction of phosphorus and potassium on maize (*Zea mays* L.) in saline-sodic soil. Journal of Agricultural Science. 2015;7(3):1916-9760

[69] El-Sharkawy HM, Shehata SA, Eisa SS, Kishk ET, Khafaga HS, Abd. El-Naby AS. Foliar application of thidiazuron, potassium chloride and boron with early cultivation date elevated growth and productivity of hybrid corn grown under adverse conditions. International Journal of Environment. 2017;**06**(2):31-41

[70] Mahmoodabadi M, Yazdanpanah N, Sinobas LR, Pazira E, Neshat A. Reclamation of calcareous saline sodic soil with di_erent amendments (I): Redistribution of soluble cations within the soil profile. Agricultural Water Management Journal. 2013;**120**:30-38

[71] Wang L, Sun X, Li S, Zhang T, Zhang W, Zhai P. Application of organic amendments to a coastal saline soil in North China: Effects on soil physical and chemical properties and tree growth. PLoS One. 2014;**9**:e89185

[72] Jishi Z, Jiang X, Xue Y, Li Z, Yu B, Xu L, et al. Closing yield gaps through soil improvement for maize production in coastal saline soil. Agronomy. 2019;**9**: 573

[73] Gondim FA, Gomes-Filho E, Costa JH, Alencar NLM, Prisco JT. Catalase plays a key role in salt stress acclimation induced by hydrogen peroxide pretreatment in maize. Plant Physiology and Biochemistry. 2012;**56**:62-71

[74] Hussein MM, Balbaa LK, Gaballah MS. Salicylic acid and salinity effects on growth of maize plants. Research Journal of Agricultural and Biological Sciences. 2007;**3**:321-328

[75] Monowara K, Shuvo AR, Salam TB, Rahman H, Tareq MBS, Tareq MS. Effect of organic amendments on soil salinity and the growth of maize (*Zea mays* L.). Plant Science. 2019;**6**(2):106-111

[76] Oo AN, Iwai CB, Saenjan P. Soil properties and maize growth in saline and nonsaline soils using cassava-industrial waste compost and vermicompost with or without earthworms. Land Degradation and Development. 2015;**26**(3):300-310

[77] Bamminger C, Poll C, Sixt C, Högy P, Wüst D, Kandeler E, et al. Short-term response of soil microorganisms to biochar addition in a temperate agroecosystem under soil warming. Agriculture, Ecosystems and Environment. 2016;**233**:308-317

[78] Drake JA, Cavagnaro TR, Cunningham SC, Jackson WR, Patti AF. Does biochar improve establishment of tree seedlings in saline sodic soils? Land Degradation and Development. 2016;**27**: 52-59

[79] Usman ARA, Al-Wabel MI, Abdulaziz AH, Mahmoud WA, El-Naggar AH, Ahmad M, et al. Conocarpus biochar induces changes in soil nutrient availability and tomato growth under saline irrigation. Pedosphere. 2016;**26**:27-38

[80] Uzoma KC, Inoue M, Andry H, Fujimaki H, Zahoor A, Nishihara E.

Effect of cow manure biochar on maize productivity under sandy soil condition. Soil Use and Management. 2011;**27**:205-212

[81] Lashari MS, Ye MSY, Ji H, Li L, Kibue GW, Lu H, et al. Biochar-manure compost in conjunction with pyroligneous solution alleviated salt stress and improved leaf bioactivity of maize in a saline soil from Central China: A 2-year field experiment. Journal of the Science of Food and Agriculture. 2015;**95**:1321-1327

[82] Rita L, Vitti C. Use of organic amendments to reclaim saline and sodic soils: A review. Arid Land Research and Management. 2019;**33**(1):1-21

[83] Lomin SN, Krivosheev DM, Steklov MY, Arkhipov DV, Osolodkin DI, Schmulling TR, et al. Plant membrane assays with cytokinin receptors underpin the unique role of free cytokinin bases as biologically active ligands. Journal of Experimental Botany. 2015;**66**(7):1851-1863

[84] Tomaz R, Mukesh J, Marina D, Prem C. Spatial and temporal profiles of cytokinin biosynthesis and accumulation in developing caryopses of maize. Annals of Botany. 2011;**107**(7): 1235-1245

[85] Davani D, Nabipour M, Dezfouli RH. Effects of different concentrations of cytokinin and auxin hormones on yield and yield components of grain maize (*Zea mays* L.) in salinity conditions. Journal of Plant Productions. 2017;**40**:169-180

[86] Salama FM, Awadalla AA. The effect of different kinetin application methods on some chlorophyll parameters of two crop plants grown under salinity stress. Phyton. 1987;**27**: 181-193

[87] Kaya C, Ashraf M, Dikilitas M, Tuna AL. Alleviation of salt stress-induced adverse effects on maize plants by exogenous application of indoleacetic acid (IAA) and inorganic nutrients—A field trial. Australian Journal of Crop Science. 2013;7(2):249-254

[88] Darra BL, Saxena SN. Role of IAA on the mineral composition of maize crop under various osmotic stressed conditions. Plant and Soil. 1973;**38**:657-661

[89] Tuna AL, Kaya C, Dikilitas HD. The combined effects of gibberellic acid and salinity on some antioxidant enzyme activities, plant growth parameters and nutritional status in maize plants. Environmental and Experimental Botany. 2008;**62**(1):1-9

[90] Rashad RT, Rashad AH. A comparison study on the effect of some growth regulators on the nutrients content of maize plant under salinity conditions. Annals of Agricultural Science. 2014;**59**(1):89-94

[91] Niu X, Song L, Xiao Y, Ge W. Drought-tolerant plant growth-promoting rhizobacteria associated with foxtail millet in a semi-arid agroecosystem and their potential in alleviating drought stress. Frontiers in Microbiology. 2018;**8**:2580

[92] Banaei-Asl F, Bandehagh A, Uliaei ED, Farajzadeh D, Sakata K, Mustafa G, et al. Proteomic analysis of canola root inoculated with bacteria under salt stress. Journal of Proteomics. 2015;**124**: 88-111

[93] Wang Q, Dodd IC, Belimov AA, Jiang F. Rhizosphere bacteria containing 1-aminocyclopropane-1-carboxylate deaminase increase growth and photosynthesis of pea plants under salt stress by limiting Na^+ accumulation. Functional Plant Biology. 2016;**43**(2): 161-172

[94] Akram MS, Shahid M, Tariq M, Azeem M, Javed MT, Saleem S, et al.

Deciphering *Staphylococcus sciuri* SAT-17 mediated anti-oxidative defense mechanisms and growth modulations in salt stressed maize (*Zea mays* L.). Frontiers in Microbiology. 2016;7

[95] Shahid M, Akram MS, Khan MA, Zubair M, Shah SM, Ismail M, et al. A phytobeneficial strain *Planomicrobium* sp. MSSA-10 triggered oxidative stress responsive mechanisms and regulated the growth of pea plants under induced saline environment. Journal of Applied Microbiology. 2018;**124**: 1566-1579

[96] Yang J, Kloepper JW, Ryu CM. Rhizosphere bacteria help plants tolerate abiotic stress. Trends in Plant Science. 2009;**14**(1):1-4

[97] Nadeem SM, Zahir ZA, Naveed M, Arshad M. Preliminary investigations on inducing salt tolerance in maize through inoculation with rhizobacteria containing ACC deaminase activity. Canadian Journal of Microbiology. 2007; **53**:1141-1149

[98] Nadeem SM, Zahir ZA, Naveed M, Arshad M. Rhizobacteria containing ACC-deaminase confer salt tolerance in maize grown on salt-affected fields. Canadian Journal of Microbiology. 2009;**55**(11):1302-1309

[99] Zafar-ul-Hye M, Farooq HM, Zahir ZA, Hussain M, Hussain A. Application of ACC-deaminase containing rhizobacteria with fertilizer improvesmaize production under drought and salinity stress. International Journal of Agriculture and Biology. 2014;**16**:591-596

[100] Zerrouk IZ, Benchabane M, Khelifi L, Yokawa K, Ludwig-Müller J, Baluska F. A Pseudomonas strain isolated from date-palm rhizospheres improves root growth and promotes root formation in maize exposed to salt and aluminum stress. Journal of Plant Physiology. 2016; **191**:111-119

[101] Naseem H, Bano A. Role of plant growth-promoting rhizobacteria and their exopolysaccharide in drought tolerance of maize. Journal of Plant Interactions. 2014;**9**:689-701

[102] Aslam F, Ali B. Halotolerant bacterial diversity associated with *Suaeda fruticosa* (L.) Forssk. Improved growth of maize under salinity stress. Agronomy. 2018;**8**:131

[103] Bano A, Fatima M. Salt tolerance in *Zea mays* L. following inoculation with rhizobium and pseudomonas. Biology and Fertility of Soils. 2009;**45**(4):405-413

[104] Ullah S, Bano A. Isolation of plant-growth-promoting rhizobacteria from rhizospheric soil of halophytes and their impact on maize (*Zea mays* L.) under induced soil salinity. Canadian Journal of Microbiology. 2015;**61**:307-313

[105] Gond SK, Torres MS, Bergen MS, Helsel Z, White JFJ. Induction of salt tolerance and up-regulation of aquaporin genes in tropical corn by rhizobacterium *Pantoea agglomerans*. Letters in Applied Microbiology. 2015; **60**:392-399

[106] El-Esawi MA, Alaraidh IA, Alsahli AA, Alzahrani SA, Ali HM, Alayafi AA, et al. Serratia liquefaciens KM4 improves salt stress tolerance in maize by regulating redox potential, ion homeostasis, leaf gas exchange and stress-related gene expression. International Journal of Molecular Sciences. 2018;**19**:3310

[107] Fukami J, de la Osa C, Ollero FJ, Megías M, Hungria M. Coinoculation of maize with *Azospirillum brasilense* and *Rhizobium tropici* as a strategy to mitigate salinity stress. Functional Plant Biology. 2018;**45**:328-339

[108] Rojas-Tapias D, Moreno-Galván A, Pardo-Díaz S, Obando M, Rivera D, Bonilla R. Effect of inoculation with

plant growth-promoting bacteria (PGPB) on amelioration of saline stress in maize (*Zea mays*). Applied Soil Ecology. 2012;**61**:264-272

[109] Chen L, Liu Y, Wu G, Veronican Njeri K, Shen Q, Zhang N, et al. Induced maize salt tolerance by rhizosphere inoculation of *Bacillus amyloliquefaciens* SQR9. Physiologia Plantarum. 2016;**158**: 34-44

[110] Li H, Jiang X. Inoculation with plant growth-promoting bacteria (PGPB) improves salt tolerance of maize seedling. Russian Journal of Plant Physiology. 2017;**64**(2):235-241

[111] Abdelkader AF, Esawy MA. Case study of a biological control: *Geobacillus caldoxylosilyticus* (IRD) contributes to alleviate salt stress in maize (*Zea mays* L.) plants. Acta Physiologiae Plantarum. 2011;**33**(6):2289

[112] Krasensky J, Jonak C. Drought, salt, and temperature stress-induced metabolic rearrangements and regulatory networks. Journal of Experimental Botany. 2012;**63**(4):1593-1608

[113] Halo BA, Khan AL, Waqas M, Al-Harras A, Hussain J, Ali L, et al. Endophytic bacteria (*Sphingomonas* sp. LK11) and gibberellin can improve *Solanum lycopersicum* growth and oxidative stress under salinity. Journal of Plant Interactions. 2015;**10**(1):117-125

[114] Kim K, Jang YJ, Lee SM, Oh BT, Chae JC, Lee KJ. Alleviation of salt stress by *Enterobacter* sp. EJ01 in tomato and *Arabidopsis* is accompanied by upregulation of conserved salinity responsive factors in plants. Molecular Cell. 2014;**37**(2):109

[115] Abd El-Ghany TM, Masrahi YS, Mohamed A, Abboud A, Alawlaqi MM, et al. Maize (*Zea mays* L.) growth and metabolic dynamics with plant growth-promoting rhizobacteria under salt stresses. Journal of Plant Pathology and Microbiology. 2015;**6**:305

[116] Upadhyay S, Singh J, Singh D. Exopolysaccharide-producing plant growth promoting rhizobacteria under salinity condition. Pedosphere. 2011;**21** (2):214-222

[117] Miller G, Suzuki N, Ciftci-Yilmaz S, Mittler R. Reactive oxygen species homeostasis and signalling during drought and salinity stresses. Plant, Cell and Environment. 2010;**33**(4):453-467

[118] Rossi F, De Philippis R. Role of cyanobacterial exopolysaccharides in phototrophic biofilms and in complex microbial mats. Life. 2015;**5**(2):1218-1238

[119] Zhang H, Kim MS, Sun Y, Dowd SE, Shi H, Paré PW. Soil bacteria confer plant salt tolerance by tissue-specific regulation of the sodium transporter HKT1. Molecular Plant-Microbe Interactions. 2008;**21**(6):737-744

[120] Timmusk S, El-Daim IAA, Copolovici L, Tanilas T, Kännast A, Behers L, et al. Drought-tolerance of wheat improved by rhizosphere bacteria from harsh environments: Enhanced biomass production and reduced emissions of stress volatiles. PLoS One. 2014;**9**(5):e96086

[121] Sivritepe N, Sivritepe HO, Eris A. The effects of NaCl priming on salt tolerance in melon seedlings grown under saline conditions. Scientia Horticulturae (Amsterdam). 2003;**97**: 229-237

[122] Ibrahim EA. Seed priming to alleviate salinity stress in germinating seeds. Journal of Plant Physiology. 2016; **192**:38-46

[123] Paparella S, Araujo SS, Rossi G, Wijayasinghe M, Carbonera D, Balestrazzi A. Seed priming: State of the art and new perspectives. Plant Cell Reports. 2015;**34**:1281-1293

[124] Gebreslassie GB, Qufa CA. Plant physiological stimulation by seeds salt

priming in maize (*Zea mays*): Prospect for salt tolerance. African Journal of Biotechnology. 2017;**16**(5):209-223

[125] Abraha B, Yohannes G. The role of seed priming in improving seedling growth of maize (*Zea mays* L.) under salt stress at field conditions. Agricultural Sciences. 2013;**4**(12):666-672

[126] Farahbakhsh H, Saiid MS. Effect of seed priming with NaCl on maize germination under different saline conditions. African Journal of Agricultural Research. 2011;**6**:6095-6099

[127] Carvalho RF, Piotto FA, Schmidt D, Peters LP, Monteiro CC, Azevedo RA. Seed priming with hormones does not alleviate induced oxidative stress in maize seedlings subjected to salt stress. Journal of Scientific Agriculture. 2011; **68**:598-602

[128] Khodary SEA. Effect of salicylic acid on the growth, photosynthesis and carbohydrate metabolism in salt-stressed maize plants. International Journal of Agriculture and Biology. 2004;**6**:5-8

[129] Arora N, Bhardwaj R, Sharma P, Arora HK. 28-Homobrassinolide alleviates oxidative stress in salt treated maize (*Zea mays* L.) plants. Brazilian Journal of Plant Physiology. 2008;**20**: 153-157

[130] Rehman H, Hassan I, Basra SMA, Afzal I, Farooq M, Wakeel A, et al. Seed priming improves early seedling vigor, growth and productivity of spring maize. Journal of Integrative Agriculture. 2015;**14**(9):1745-1754

[131] Giaveno CD, Ribeiro RV, Souza GM, de Oliveira RF. Screening of tropical maize for salt stress tolerance. Crop Breeding and Applied Biotechnology. 2007;7:304-313

[132] Muhammad F, Hussain M, Wakeel A, Kadambot H, Siddique M. Salt stress in maize: Effects, resistance mechanisms, and management. A review. Agronomy for Sustainable Development. 2015;**35**:461-481

[133] Gosal SS, Wani SH, Kang MS. Biotechnology and drought tolerance. Journal of Crop Improvement. 2009;**23**: 19-54

[134] Li B, Li N, Duan X, Wei A, Yang A, Zhang J. Generation of marker-free transgenic maize with improved salt tolerance using the FLP/FRT recombination system. Journal of Biotechnology. 2010;**145**:206-213

[135] Pitann B, Mohamed A-K, Neubert AB, Schubert S. Tonoplast Na^+/H^+ antiporters of newly developed maize (*Zea mays*) hybrids contribute to salt resistance during the second phase of salt stress. Journal of Plant Nutrition and Soil Science. 2013;**176**:148-156

[136] Chen M, Chen Q-J, Niu X-G, Zhang R, Lin H-Q, Xu C-Y, et al. Expression of OsNHX1 gene in maize confers salt tolerance and promotes plant growth in the field. Plant, Soil and Environment. 2007;**53**:490-498

[137] Zheng J, Zhao J, Zhang J, Fu J, Gou M, Dong Z, et al. Comparative expression profiles of maize genes from a water stress-specific cDNA macroarray in response to high salinity, cold or abscisic acid. Plant Science. 2006;**170**:1125-1132

[138] Quan R, Shang M, Zhang H, Zhao Y, Zhang J. Improved chilling tolerance by transformation with betaA gene for the enhancement of glycinebetaine synthesis in maize. Plant Science. 2004; **166**:141-149

[139] Yin XY, Yang AF, Zhang KW, Zhang JR. Production and analysis of transgenic maize with improved salt tolerance by the introduction of AtNHX1 gene. Acta Botanica Sinica. 2004;**46**:854-861

Chapter 4

Production and Salinity Tolerance of Fodder Beet (*Beta vulgaris* L. ssp. Maritima)

Sami Ullah Khan, Zulfiqar Ali Gurmani, Waseem Ahmed, Shahzad Ahmed and Alvina Gul

Abstract

Fodder beet (*Beta vulgaris* L. ssp. maritima) belongs to the Amaranthaceae family. It was introduced first in the Europe and then to USA in 1800 and is currently being grown under cool environmental conditions of the world. It can be cultivated at temperature ranging from 8°C to 25°C. Both shoots and roots of fodder beet can be used as a feed for livestock. In the face of changing climate, there is a dire need to find out climate-resilient crops in new niches that can fulfill the growing needs of farming communities. In this context, fodder beet could be a good option for growers having sizable marginal as well as salt-affected soils. The chapter discusses in detail the efficient salinity-tolerance mechanism of fodder beet that enables it to survive under moderate salinity. Selective ion uptake mechanism, efficient antioxidant defensive mechanism and osmoregulation by accumulation of compatible solutes enable it to thrive well under saline environment. Hence, fodder beet is a relatively salt-tolerant crop that can be successfully grown on normal, marginal as well as salt-affected soils to fulfill the fodder requirements of livestock in fodder-scarce times and salinity amelioration.

Keywords: fodder beet, salinity, compatible solutes, salt tolerant crop, livestock

1. Introduction

1.1 Origin, history, and adaptation

Fodder beet (*Beta vulgaris* L. ssp. maritima) is known to have been originated in Mesopotamia (Middle East) and ancient Greece in 500 BC chiefly used as animal fodder [1]. It belongs to the Amaranthaceae family, which consists of about 105 genera separated into 1400 species, mainly herbaceous dicotyledonous plants [2]. *It was introduced firstly in Europe and then to USA in 1800 and is currently grown under cool environmental conditions of the world, mainly Northern America and New Zealand at 600–1000 m altitudes in the tropics. It can be cultivated at a temperature ranging from* 8 to 25°C. However, frost can damage the seedlings below −3°C. Suitable soil pH for beet cultivation is greater than 6.5 but acid soils are not adequate for beet growth. The crop is relatively salt tolerant and can also be cultivated with brackish water. It is drought tolerant and could be grown successfully at the end of a dry summer when other crops cannot be grown [3]. Both shoots and roots of fodder beet can be used as a feed for livestock. Roots of fodder beet contain sugars mainly in the form of sucrose (up to 60%), low crude protein (approximately 10%) as well as neutral detergent fiber contents (approximately 12%). The shoots of the beet make up about one-third of the dry matter of the whole plant and are considered to possess high protein content of 11.4–15.8% [4].

1.2 Soil and climate

The crop can be successfully grown on friable, deep, and well-drained soil containing sufficient calcium contents. Usually well-drained sandy loam soils are good for fodder beet cultivation. For good beet growth, soil pH should be 5.8–7.8. It is suitable for cultivation on temperate areas of the globe having mild winters and moderate summer temperature. The average annual precipitation for its development should be 60–65 cm. The suitable temperature for garden beet growth is between 15.5 and 25°C. It is a biannual crop and heat can damage its growth during the second year of its growth mainly at the start of pollination and seed formation.

2. General insights into salinity stress

The process of soil salinity is natural and closely linked with the formation of landscape and soil. Nevertheless, human-induced practices can promote salinity processes and hence may cause long-term degradation of water and land resources. When a high concentration of sodium salt adversely affects the growth of plants then salinity becomes a land issue, but whenever it affects the uptake of water because of the high concentration of salts in the water it becomes a water issue. Salinization is a serious problem of irrigated arable lands across the world. According to assessments, around more than 6% area of the world is affected by salinization due to natural causes or faculty irrigation practices. This situation has rendered the soils unfit for agriculture production annually [5]. When the saturation extract electrical conductivity in the rhizosphere surpasses 4 dS m^{-1} at the temperature of 25°C then the soil is recognized as saline soil and these soils possess exchangeable sodium greater than 15%. At this electrical conductivity, the crop yield is reduced by most crop plants [6].

The chief cause of the salinization of water and land in semiarid and arid areas of the world is mainly excessive irrigation. Salt stress occurs as ions such as electrically charged atoms or compounds in the soil. Due to mineral weathering, these salts are released in the soil. However, they might accumulate due to irrigation water application or sometimes from low groundwater they may travel upward in the soil. Low precipitation is unable to leach down these ions from the soil profile; as a result, accumulation of salts occurs in the soil and causes salt stress problem [7]. Water-soluble salts are contained by all soil types. Plants uptake essential nutrients in the form of soluble salts but excessive accumulation of these nutrients in the soil intensely suppresses the plant growth. The saline area in the world is increasing continuously each year mainly because more areas are being covered under irrigation [8].

3. Fodder beet as fodder crop in the world

In the face of changing climate, there is a dire need to find out climate-resilient crops that can fulfill the growing needs of the farming community. Fodder beet is a halophyte that can successfully be grown on salt-affected lands throughout the world. It not only fulfills fodder requirements of the ruminants and other cattle but also proved helpful for effective utilization of salt-affected marginal lands where no other crop can be grown. In countries like Pakistan and India, it can successfully be grown from August to September to fulfill fodder needs during the peak winter months where no other fodder crop can survive. Fodder beet has the ability to tolerate salinity as compared to other fodder crops; hence it can be successfully grown on salt-affected soils across the Globe. The scope of crop production has been limited due to millions of hectares of marginal and salt-affected soils [9]. For grazing young stock and to fill the feed gaps in the late lactation, fodder beet could be good choice as a feed for lactating cows. Likewise, in coastal areas of many European countries, the fodder beet is cultivated as a fodder and forage crop [10]. At South Island in New Zealand, fodder beet cultivation has become a renewed interest, particularly because dairy herbs fodder beet is being used for winter feeding of the livestock. Due to the wide acceptance of fodder beet in New Zealand, it is being cultivated on about 6–10,000 ha of land annually [5, 11].

4. Adverse effects of salinity stress on plants

Due to soil salinization and increased use of irrigation water with poor-quality water, soil salinity has become one of the most brutal abiotic stresses that limit crop productivity in many sections of the world [12]. There are numerous adverse effects of salinity stress on plant growth and development. Two major threats of salt stress to plant growth are osmotic stress and ionic stress. Firstly, soil salinity represses the growth of plants due to osmotic stress followed by the toxicity of ions [13]. Salinity stress also causes oxidative stress in plants. Various metabolic and physiological processes are adversely affected by salt stress in plants. The prominent symptoms of salinity stress include a reduction in leaf area, leaf abscission, enhancement of leaf succulence and thickness, reduction in length of internodes, and shoot and root necrosis [14]. Soil salinity stress also results in reduced water absorption capacity of roots, with a concomitant increase in the rate of transpiration, which is facilitated by high salt accumulation due to osmotic stress in plants and soils. As a result, osmotic stress due to salinity causes numerous biochemical changes inside the plant cell such as nutrient inequity, disruption of membranes, reduced ability to quench reactive oxygen species (ROS), and decreased stomatal conductance and photosynthetic activity [15]. Salinity stress is also known as hyper ionic stress. When plants are exposed to high NaCl concentrated soils, then salinity stress causes Cl^- and Na^+ ions accumulation in plant tissue, which is considered as the main harmful effect of salinity stress. The ionic balance of the plant cell is disturbed and significant physiological disorders may take place due to the introduction of Na^+ and Cl^- ions into the cells. K^+ ions are a key element for crop productivity but uptake of these ions is inhibited because of excessive concentration of Na^+ ions inside the cells. Consequently, deficiency of K^+ ions inside the cells results in less productivity and ultimately death of the plants. Moreover, reduction in leaf area, dry and fresh weight of root and shoot is a common feature of salinity stress [16].

Enhanced production of reactive oxygen species (ROS) as a result of salinity stress causes the creation of superoxide, singlet oxygen, hydrogen peroxide, and hydroxyl radicals. These ROS result in various oxidative damage to cellular components such as lipids, proteins, and DNA and also interrupt important cellular functions in plants [17].

5. Mechanism of salinity tolerance in fodder beet

Various biochemical and physiological mechanisms are involved in surviving fodder beet plants under high salt concentration.

5.1 Salt tolerance and ion homeostasis

Under salt stress conditions, maintenance of ion homeostasis is important for normal growth. Under extreme salt concentration, the halophytic plants are unable to tolerate salts in their cytoplasm, so the surplus amount of salt is either translocated to the vacuoles or seized in older tissues, which are finally scarified to protect the plant from salt stress condition. NaCl is the major salt present in the saline soils. Na^+/H^+ antiporters transport Na^+ ions that have entered the cytoplasm to the vacuoles. Vacuolar type H^+-ATPase (V-ATPase) and vacuolar pyrophosphatase (V-PPase) are two kinds of H^+ pumps that exist in the vacuolar membranes. The activity of these H^+ is upregulated under salt stress to mitigate the effects of salinity on plants [18]. Fodder beet plants develop efficient methods to keep low level of ion concentration in the cytoplasm. A significant role is performed by membranes and their related components for ion concentration maintenance within the cytosol during the stress period by regulation of ion transport and uptake. Different channels and the carrier proteins, symporters, and antiporters carried out the phenomenon of ion transport. Maintenance of cellular Na^+/K^+ balance is very essential for plant survival under salinity stress. During salt stress, due to enhanced Na^+ concentration in the soil, competition occurs between K^+ and Na^+ ion for the transporters because both these elements have the same transport mechanism, which reduces the uptake of K^+ [19].

5.2 Compatible solute accumulation and osmotic protection

The compatible solutes can be defined as a group of organic compounds that are chemically diverse and these are polar, uncharged, and naturally soluble. At high concentration, they do not hinder cellular metabolism. Polyols, proline, glycine betaine, and sugar are the main compatible solutes [26–28]. Arginine, cysteine, and methionine amino acids constitute about 55% of the total free amino acids and exposure of salinity stress decreases the concentration of these amino acids while the concentration of proline increases under salinity stress conditions [20]. Increased proline concentration in fodder beet helps the crop to cope with salinity stress and accumulation of proline is an eminent feature for salinity stress mitigation. It was also observed in some previous studies that higher proline accumulation in olive plants increased salt tolerance by improving photosynthetic activity, antioxidative enzymatic activity, and plant growth and helped to maintain suitable water balance in the cells under salt stress conditions [21]. During recovery from stress, proline accumulation in fodder beet served as an organic nitrogen. Glycine betaine plays a vital role in the mitigation of stress in the fodder beet by raising the cell osmolarity during salinity stress. Glycine betaine helps in protein stabilization, provides protection to the cell through osmotic adjustment, guards the chlorophyll against stress injuries as well as reduces reactive oxygen species. Salinity stress in fodder increased the accumulation of soluble sugars. These sugars serve as a source of carbon storage, provide osmoprotection, and help in the scavenging of reactive oxygen species [22].

5.3 Role of antioxidant enzymes in salinity tolerance

Salinity stress in plants may cause overproduction or disruption of electron transport chains (ETCs) in subcellular organelles such as chloroplasts and mitochondria. In this scenario, molecular oxygen or $^1O_2$ acts as an electron acceptor, causing the accumulation of reactive oxygen species (ROS). This singlet oxygen ($^1O_2$), the superoxide radical, the hydroxyl radical (OH^-), and hydrogen peroxide (H_2O_2) are all strongly reactive compounds and hence can cause damage to the cell integrity. Upregulation of antioxidant defensive mechanisms in fodder beet plants plays a vital role in the detoxification of ROS, which are otherwise triggered under salinity stress. The activity of antioxidant enzymes is positively correlated with salinity tolerance. Three specific traits help the plants to better adapt under salinity stress environment mainly through ion exclusion and tissue tolerance ability. Thus antioxidant enzymes contribute in maintaining tissue turgidity and in the mechanism of salinity tolerance [12].

5.4 Roles of polyamines in salinity tolerance

In abiotic stresses, the polyamines play an important role such as salt stress and stress tolerance in plants is correlated with an increase in the level of polyamines. Endogenous polyamines level in fodder beet and other salt-tolerant plants increases with exposure to salinity stress. Polyamines play a positive role in salt stress by maintaining the membrane integrity; regulating the genes expression for solutes synthesis, which are osmotically active; reducing reactive oxygen species (ROS); and most importantly controlling the Na^+ and Cl^- ion accumulation [23].

5.5 Hormonal regulation of salinity tolerance

The increased concentration of abscisic acid (ABA) can reduce the impact of salinity stress on assimilates translocation and photosynthesis. The positive association between salinity tolerance and ABA accumulation is attributed to the K^+, Ca^{2+} accumulation, and accumulation of sugars, and proline in root vacuole, which restrict the uptake of Na^+ and Cl^- [24]. The compounds like brassinosteroids (BRs) and salicylic acid (SA) have hormonal properties and paly a role in plant responses to abiotic stresses. The application of these compounds improves salt tolerance in plants by regulation of various physiological and biochemical processes [24].

6. Mechanism against salinity-induced oxidative stress

Salinity stress results in a continuous increment in cellular membrane injury and a reduction in relative water content. Further, increased ion leakage of cellular membranes due to salinity stress results in malfunctioning of cellular membranes. It has been observed that plasma membrane deteriorates owing to the salt ions action. Cell membrane stability and maintenance of suitable relative water content are significantly reduced by salinity stress [25]. The primary site of salt injury is the plasma membrane because salt stress causes changes in plasma membrane permeability and the lipid composition of membranes and also alters the activities of membrane-bound enzymes. That is why plasma membrane permeability is an effective selection criterion for salinity stress in fodder beet and other plants. Alteration in plasma membrane permeability occurs significantly in salt-sensitive crops but in the case of fodder beet is less affected under salinity stress. The inherited and induced protection of membranes in fodder beet and some other salt-tolerant crops helps in the maintenance of cell membrane permeability and stability of the plasma membrane. Sustained composition of lipids and protein and accumulation of several protecting agents under salt stress in salt-tolerant crops help to retain and stabilize plasma membrane integrity. In salt-tolerant crops such as fodder beet, some specific protein and lipids are induced under salt stress and contribute to the maintenance of cellular membrane function and structure. Cellular membrane stabilization and protection are also achieved by proline, glycine betaine, and polyamines and these are known as protecting agents of the cellular membrane. It has been proved that salt stress correlates with plasma membrane permeability and this feature of plasma membrane is a useful character for selecting salt-tolerant crop genotypes [26, 27]. An important adaptive mechanism of fodder beet plants and other halophytes under salinity stress is the expression of stress proteins, which helps in the maintenance of cell membrane integrity, topology, and native configuration. Under wheat plants exposed to salinity stress, protein content and molecular weight of the protein were found to decrease, which ultimately affected the activities of different proteins. This change in protein activities suggests that only some proteins are directly participating in salinity tolerance [28].

7. Osmotic adjustment under salinity stress

To reduce cell water potential, fodder beet and other halophytes accumulate inorganic ions in their vacuoles because the consumption of energy from synthesizing organic compounds is far less than absorbing inorganic ions [29]. Under salt stress, the main inorganic osmolyte in the vacuole is Na^+ ion. In many plants, salt stress inhibits the accumulation of Mg^{2+} and Ca^{2+}. But fodder beet crops can accumulate Ca^{2+} and Mg^{2+} ions under salinity stress and hence contribute to better osmotic adjustment. To maintain various enzymatic processes, it is essential to maintain low Na^+ ion and high K^+ ions in the vacuoles. Under salinity stress, absorption of K^+ is inhibited while the absorption of Na^+ is increased in many halophytes. But in case of fodder beet, the accumulation of both Na^+ and K^+ ions increases under salt stress. This phenomenon proves that fodder beet plants may have a distinctive pathway for absorption of Na^+ independent of K^+ pathway [30].

When plants are exposed to salinity, their primary reaction is osmotic stress. To alleviate osmotic imbalances due to salt stress, osmotic adjustment is very essential for the maintenance of cell turgor [31]. It encompasses cellular accumulation of solute in response to a reduction in the water potential of the environment. Fodder beet plants have a high osmotic adjustment capacity, as reflected by the organic and inorganic osmolyte accumulation in salinity stress [32]. Earlier in the chapter, it has been emphasized that accumulation of glycine betaine, proline, free amino acids, and choline occurs in fodder beet leaves when the concentration of NaCl is increased in the growth medium. Under normal growth conditions, high level of glycine betaine in young leaves of fodder is detected because glycine betaine is primarily synthesized in the old leaves and then translocated to the young leaves; that is why young leaves of fodder contain a high accumulation of glycine betaine. It is important to point out here that a glycine betaine plays a key role in fodder beet exposed to salt stress [33]. Similarly, proline accumulation was found to occur mainly to facilitate osmotic adjustment and salinity stress mitigation in many halophytes and fodder beet plants. It has been proved that proline concentration in shoots of fodder beet and other salt-tolerant plants was higher than in salt-sensitive plants [34]. High proline concentration in salt-tolerant genotypes of fodder is induced by cellular demand for membrane stabilization and osmotic adjustment. But the contribution of proline for osmotic adjustment and salinity stress mitigation is small as compared to the contribution of glycine betaine. The presence of inorganic salt ions in fodder beet and other halophytes also plays an essential role in an osmotic adjustment under salinity stress. In an earlier research investigation, it was proposed that high levels of ions such as Cl^-, K^+, and N^+ in shoots of fodder beet seedlings played a role in salinity stress mitigation and effective osmotic adjustment during salinity stress [35].

8. Biochemical indicators of salinity stress

Adaptive mechanisms that are utilized by plants to survive under salinity stress and metabolic sites which damages due to salt stress are not well understood. Due to this, no well-defined salinity stress indicator is accessible to help plant breeders for the improvement of tolerance to salinity of main crops. In recent times plant breeders have effectively enhanced salt stress forbearance of some crops such as fodder beet using seed yield or vigor of the plant as the key selection criteria but in order to have more convenient and practicable selection crop must possess distinctive indicators of salt stress at cellular, tissue or whole plant level [36].

Some of the biochemical indicators of salinity stress are discussed below:

8.1 Biochemical markers

As already discussed, osmotic adjustment to mitigate salt stress can be accomplished by accumulation of high levels of inorganic ions or low-molecular weight organic solutes. Both of these play a key role in salt stress tolerance in higher plants. The compatible osmolytes that are found in fodder beet and higher plants are organic acids, sugars with low molecular weight, polyols, and nitrogen-containing compounds [37].

8.2 Soluble sugars

In glycophytes in saline situation, sugars contribution is up to 50% of total osmotic potential. Despite a significant decrease in the net CO_2 assimilation rate the soluble carbohydrate accumulation has been reported widely in plants under salt stress. It has been found that salt-tolerant crops such as fodder beet accumulate high levels of soluble sugars under salt stress conditions. It is evident that considerable variations in the soluble sugars accumulation in response to salinity stress exist at both intraspecific and interspecific levels and even among all genotypes of different salt-tolerant plants [38].

8.3 Soluble proteins

In fodder beet and other salt-tolerant crops, proteins that accumulate under salinity stress may provide a storage form of nitrogen, which is reutilized when stress is over and may also play a part in osmotic adjustment. When salt-tolerant plants such as fodder are exposed to salt stress, the accumulation of soluble proteins is increased to play a role in mitigating the adverse effect of salinity. The soluble proteins are the essential molecular markers for betterment of salt tolerance by the means of genetic engineering techniques but the use of soluble proteins as biochemical indicator depends on the nature of plant cultivar or species [39].

8.4 Amino acids and amides

In fodder beets and higher salt-tolerant plants, the accumulation of amino acids has been reported under salt stress. In salt-tolerant plants, glutamine and asparagine amides have also reported to accumulate under salt stress. It has been reported that total free amino acids tend to be higher in leaves of fodder and salt-tolerant lines of sunflower than in salt-sensitive lines of fodder beet, sunflower, safflower and *Lens culinaris* [40]. For example, proline is accumulated at a higher level in fodder beet under salt stress.

8.5 Polyamines and polyols

Accumulation of polyamines can also be used as an indicator of salinity stress. In several studies, it has been reported that the accumulation of polyamines increased when plants were exposed to salinity stress [41].

Polyols are also thought to play a role in salt tolerance in salt-tolerant plants. Polyols accumulate in the cytoplasm of salt-tolerant plants to overcome the osmotic disturbances, which occurred due to high levels of inorganic ions that are compartmentalized in vacuoles. Polyols also play a part in oxygen radical scavenging. Polyols accumulation has been reported in several studies in response to salt stress in many higher plants; thus it can also be used as a biochemical indicator of salt stress [42].

8.6 ATPases

One of the important factors responsible for salt tolerance of plants is the regulation of ion transport. A significant role is played by membrane proteins in selective distribution of ions with the cell or whole plant. Salinity tolerance in plants is linked with low accumulation and uptake of Na^+ ions. ATPases can be used as a biochemical indicator of salinity stress because it has been reported that the activity of ATPases increases in roots, leaves, and cells of tested plants under induced salinity stress. It was found in wheat and fodder beet that activity of ATPases increased in salt-tolerant genotypes as compared to salt-sensitive genotypes under induced salinity stress [43].

9. Antioxidants and ROS-scavenging

Plants are protected from oxidative damages by antioxidant defense machinery. Several enzymatic antioxidant defense systems are possessed by plants such as super oxide dismutase, peroxidases, glutathione reductase, catalases, ascorbate peroxidase, dehydroascorbate reductases, monodehydroascorbate, glutathione peroxidase, glutathione-S-transferase, guaiacol peroxidase, ascorbic acid, glutathione, phenolic compounds, alkaloids, α-tocopherols, and non-protein amino acids, which help to control the negative effects of uncontrolled oxidation as well as provide protection to plant cells from oxidative damages caused via scavenging of ROS. The ROS also effect the gene expression of many genes and thus control many processes like abiotic stress (salinity) response, programmed cell death, growth, pathogen defense, cell cycle, systemic signaling, and development [44].

ROS are recognized as the main cause of cellular damage under biotic and abiotic stresses. During aerobic metabolism when electrons from the electron transport chains in chloroplast and mitochondria are leaked and react with oxygen in the nonappearance of other acceptors the active oxygen species such as hydrogen peroxide, hydroxy radical, super oxide, and singlet oxygen are produced [45]. Nonetheless, by superoxide dismutase (SOD), plants can eliminate super oxide, which catalyzes the dismutation of super oxide into H_20_2 and O_2 and is essential in the prevention of metal ions reduction and hence the synthesis of hydroxyl radicals. An ascorbate peroxide, which is located in the thylakoid membrane, can also eliminate hydrogen peroxide [46].

It has been reported that the production of ROS is increased in plants in response to different abiotic stresses such as salinity stress, drought, high- and low-temperature stress, water-logging stress, light stress, etc. [47]. One of the key limiting factors in crop production is oxidative stress. Due to the production of ROS under salinity stress, the plants come under oxidative stress. It has been reported that ROS, which is generated during metabolic processes, results in damage to cellular functions, which finally lead to senescence, disease, and ultimately cell death. As discussed earlier, efficient defense systems of plants scavenge ROS by antioxidant enzymes. Several attempts have been made by researchers to lessen the oxidative damage under the salinity stress by the management of enzymes that can scavenge ROS by technology used for gene transfer [48].

In a comparison of the antioxidant production mechanism in salt-sensitive and salt-tolerant plants, it was found that peroxidase activity increases while a decline was noted in SOD activity [49].

10. Salinity tolerance improvement in fodder beet

When comparing with other fodder and forage crops fodder beet is a fodder crop with salt tolerance ability as it can be successfully grown on salt-affected lands. The most serious and important threats to crop productivity worldwide are drought and salinity [50]. Estimates show that excessive and regular irrigation results in the salinization, which leads to the desertion of 107 hectares of arable land annually. Moreover, 0.25–0.5 M ha of agricultural land is lost yearly in semiarid and arid areas because of the salinity problem worldwide [51, 52]. Salt stress causes a reduction in field crop production of most crops [53]. It has been reported that salinity greatly influenced the growth attributes of fodder beet genotypes. Fodder beet has the greater ability to thrive best under salinity stress with the highest biomass production than other fodder crops. Overall, fodder beet plants grew successfully under moderate salinity up to 200 mM saline soil [54].

Fodder beet is a more salt-tolerant crop and can be grown on saline soil than other forage and fodder crops. Fodder beet is used as animal feed in many European countries as well as in Egypt. The roots and leaves can be fed to animals in both fresh and silage form [55]. On saline barren lands, high economic yield production can be achieved by growing fodder beet as a fodder crop [56]. All parts of fodder beet such as tuberous roots and aboveground leaves are utilized as animal feed alone or in combination in Europe and numerous other countries of the world [57]. There is a dire need to identify mutants to develop high biomass-producing, high-protein fodder beet plants with the ability to grow not only on normal soils but also on salt-affected soils in the world.

11. Future perspectives

Fodder beet is a potential high-biomass fodder crop that can be introduced with guaranteed success to fulfill the fodder needs of small ruminants as well as lactating cows and buffaloes. There is dire need to promote its seed production in the northern parts of Pakistan like Swat, Naran, and Kaghan to provide seed to the local growers. This will help to reduce reliance on imported seeds on one hand and promote its cultivation during fodder-scarce months, which is a limiting factor for the livestock industry in the country. Fodder beet can also be used for effective utilization of sizable salt-affected soils in Pakistan, which otherwise remain barren or could not be used for any crop production. It will help the local growers to improve their socioeconomic conditions.

Fodder beet can be cultivated and promoted as a potential fodder crop in Pakistan and other countries of South Asia along with the conventional crops such as Oat (*Avena fatua*), Barley (*Hordeum vulgare* L.), Alfalfa (*Medicago sativa* L.) etc. The cropping season of fodder beet in Pakistan also matches with the conventional fodder shortage period for livestock.

In the future, the main area of research should be to develop local fodder beet varieties adapted under local agroecological conditions with the ability to produce high fresh biomass on normal as well as saline environmental conditions.

12. Conclusion

Fodder beet crops can thrive under moderate salinity due to an efficient salinity tolerance mechanism. Generally, salt stress reduces the shoot and root growth of fodder beet plants. The ability of fodder plants to survive under salinity stress depends on the stage of crop growth, the intensity of salinity stress, and duration of salinity. Fodder beet being a halophytic plant possess the ability to selectively uptake beneficial ions like calcium and potassium and reduce uptake of toxic and harmful ions like Na^+ and Cl^-. Moreover, the efficient antioxidant defensive mechanism makes it able to thrive under the saline environment by deleting reactive oxygen species generated in the chloroplast and mitochondria. The enhanced concentration of compatible solutes such as polyols, proline, glycine betaine, and soluble sugars in fodder beet under abiotic stresses makes it suitable to grow under abiotic stresses especially under saline environments. Thus, it can be concluded that fodder beet is a relatively salt-tolerant crop, which can be successfully grown on normal, marginal as well as under salt-affected soils to fulfill the fodder requirements of livestock in fodder-scarce times.

Conflicts of interest

The authors declare no conflicts of interest.

Disclaimer

We hereby declare that the book chapter does not have any material which has been accepted to publish any journal or publisher, and also has no copy of any material in previously published, except where due permission and reference is made in the text.

Author details

Sami Ullah Khan[1*], Zulfiqar Ali Gurmani[3], Waseem Ahmed[2], Shahzad Ahmed[1] and Alvina Gul[4]

1 Department of Agronomy, Faculty of Basic and Applied Sciences, The University of Haripur, Pakistan

2 Department of Horticulture, Faculty of Basic and Applied Sciences, The University of Haripur, Pakistan

3 Maize, Millet and Fodder Research Program, Crop Sciences Institute, National Agricultural Research Centre, Islamabad, Pakistan

4 Department of Plant Biotechnology, Atta-ur-Rahman School of Applied Biosciences, National University of Sciences and Technology, Islamabad, Pakistan

*Address all correspondence to: samiullah@uoh.edu.pk

References

[1] Henry K. Fodder beet. In: Root and Tuber Crops. A Handbook of Plant Breeding. New York, NY: Springer; 2010. pp. 221-243

[2] Watson L, Dallwitz MJ. The Families of Flowering Plants: Descriptions, Illustrations, Identification, and Information Retrieval. University of New Orleans; 1999

[3] Oyen LPA. *Beta vulgaris* L. record from Protabase. In: Grubben GJH, Denton OA, editors. Plant Resources of Tropical Africa. Wageningen, Netherlands; 1999. Available from: http://www.prota4u.org

[4] Matthew C, Nelson NJ, Ferguson D, Xie Y. Fodder beet revisited. Agronomy New Zealand Journal. 2011;**41**:39-48

[5] Maathuis FJ, Sanders D. Sodium uptake in Arabidopsis roots is regulated by cyclic nucleotides. Journal of Plant Physiology. 2001;**127**(4):1617-1625

[6] Jamil A, Riaz S, Ashraf M, Foolad MR. Gene expression profiling of plants under salt stress. Published in Critical Reviews in Plant Sciences. May 2011;**30**(5):435-458

[7] Blaylock AD. Soil Salinity, Salt Tolerance, and Growth Potential of Horticultural and Landscape Plants. University of Wyoming, Cooperative Extension Service, Department of Plant, Soil, and Insect Sciences, College of Agriculture; 1994

[8] Patel BB, Dave RS. Studies on infiltration of saline-alkali soils of several parts of Mehsana and Patan districts of North Gujarat. Journal of Applied Technology in Environmental Sanitation. 2011;**1**(1):87-92

[9] Wang Q, Wu C, Xie B, Liu Y, Cui J, Chen G, et al. Model analyzing the antioxidant responses of leaves and roots of switchgrass to NaCl-salinity stress. Journal of Plant Physiology and Biochemistry. 2012;**58**:288-296

[10] Roy SJ, Negrão S, Tester M. Salt resistant crop plants. Current Opinion in Biotechnology Journal. 2014;**26**:115-124

[11] Chakwizira E, Maley S, George M, Hubber R, Morton J, Stafford A. Effects of potassium, sodium and chloride fertilisers on yield and mineral composition of fodder beet. In: Proceedings of the 5th Australasian Dairy Science Symposium, Melbourne, Australia, 13-15 November 2012. 2012. pp. 431-434

[12] Gupta KJ, Stoimenova M, Kaiser WM. In higher plants, only root mitochondria, but not leaf mitochondria reduce nitrite to NO, in vitro and in situ. Journal of Experimental Botany. 2005;**56**(420):2601-2609

[13] Rozema J, Flowers T. Crops for a salinized world. Science. 2008;**322**(5907):1478-1480

[14] Parida AK, Das AB, Mohanty P. Investigations on the antioxidative defense responses to NaCl stress in a mangrove, *Bruguiera parviflora*: Differential regulations of isoforms of some antioxidative enzymes. Plant Growth Regulation Journal. 2004;**42**(3):213-226

[15] Munns R. Genes and salt tolerance: Bringing them together. New Phytologist Journal. 2005;**167**(3):645-663

[16] James RA, Blake C, Byrt CS, Munns R. Major genes for Na^+ exclusion, Nax1 and Nax2 (wheat HKT1; 4 and HKT1; 5), decrease Na^+ accumulation in bread wheat leaves under saline and waterlogged conditions. Journal of Experimental Botany. 2011;**62**(8):2939-2947

[17] Ahmad P, Nabi G, Jeleel CA, Umar S. Free radical production, oxidative damage and antioxidant defense mechanisms in plants under abiotic stress. In: Oxidative Stress: Role of Antioxidants in Plants. New Delhi: Studium Press; 2011. pp. 19-53

[18] Hasegawa PM. Sodium (Na^+) homeostasis and salt tolerance of plants. Journal of Environmental and Experimental Botany. 2013;**92**:19-31

[19] Sairam RK, Tyagi A. Physiology and molecular biology of salinity stress tolerance in plants. Current Science Journal. 2004:407-421

[20] Hoque MA, Banu MN, Okuma E, Amako K, Nakamura Y, Shimoishi Y, et al. Exogenous proline and glycinebetaine increase NaCl-induced ascorbate-glutathione cycle enzyme activities, and proline improves salt tolerance more than glycinebetaine in tobacco bright yellow-2 suspension-cultured cells. Journal of Plant Physiology. 2007;**164**(11):1457-1468

[21] Khan MA, Ungar IA, Showalter AM. Effects of sodium chloride treatments on growth and ion accumulation of the halophyte *Haloxylon recurvum*. Communications in Soil Science and Plant Analysis. 2000;**31**(17-18):2763-2774

[22] Saxena SC, Kaur H, Verma P, Petla BP, Andugula VR, Majee M. Osmoprotectants: Potential for crop improvement under adverse conditions. In: Plant Acclimation to Environmental Stress. New York, NY: Springer; 2013. pp. 197-232

[23] El-Shintinawy F, El-Shourbagy MN. Alleviation of changes in protein metabolism in NaCl-stressed wheat seedlings by thiamine. Biologia Plantarum Journal. 2001;**44**(4):541-545

[24] Ben Ahmed C, Ben Rouina B, Sensoy S, Boukhriss M, Ben Abdullah F. Exogenous proline effects on photosynthetic performance and antioxidant defense system of young olive tree. Journal of Agricultural and Food Chemistry. 2010;**58**(7):4216-4222

[25] Chaum S, Kirdmanee C. Effect of glycinebetaine on proline, water use, and photosynthetic efficiencies, and growth of rice seedlings under salt stress. Turkish Journal of Agriculture and Forestry. 2010;**34**(6):517-527

[26] Yiu JC, Juang LD, Fang DY, Liu CW, Wu SJ. Exogenous putrescine reduces flooding-induced oxidative damage by increasing the antioxidant properties of Welsh onion. Scientia Horticulturae Journal. 2009;**120**(3):306-314

[27] Gurmani AR, Bano A, Khan SU, Din J, Zhang JL. Alleviation of salt stress by seed treatment with abscisic acid (ABA), 6-benzylaminopurine (BA) and chlormequat chloride (CCC) optimizes ion and organic matter accumulation and increases yield of rice ('*Oryza sativa*' L.). Australian Journal of Crop Science. 2011;**5**(10):1278

[28] Ashraf M, Akram NA, Arteca RN, Foolad MR. The physiological, biochemical and molecular roles of brassinosteroids and salicylic acid in plant processes and salt tolerance. Critical Reviews in Plant Sciences Journal. 2010;**29**(3):162-190

[29] Jamil M, Ashraf M, Rehman S, Ahmad M, Rha ES. Salinity induced changes in cell membrane stability, protein and RNA contents. African Journal of Biotechnology. 2012;**11**(24):6476-6483

[30] Mansour MM. Plasma membrane permeability as an indicator of salt tolerance in plants. Biologia Plantarum Journal. 2013;**57**(1):1-0

[31] Ashraf M, Ali Q. Relative membrane permeability and activities of some antioxidant enzymes as the

key determinants of salt tolerance in canola (*Brassica napus* L.). Journal of Environmental and Experimental Botany. 2008;**63**(1-3):266-273

[32] Wahid A, Perveen M, Gelani S, Basra SM. Pretreatment of seed with H_2O_2 improves salt tolerance of wheat seedlings by alleviation of oxidative damage and expression of stress proteins. Journal of Plant Physiology. 2007;**164**(3):283-294

[33] Munns R. Comparative physiology of salt and water stress. Plant, Cell & Environment Journal. 2002;**25**(2):239-250

[34] Yang C, Shi D, Wang D. Comparative effects of salt and alkali stresses on growth, osmotic adjustment and ionic balance of an alkali-resistant halophyte *Suaeda glauca* (Bge.). Plant Growth Regulation Journal. 2008;**56**(2):179

[35] Liang W, Ma X, Wan P, Liu L. Plant salt-tolerance mechanism: A review. Biochemical and Biophysical Research Communications. 2018;**495**(1):286-291

[36] Wu GQ, Feng RJ, Liang N, Yuan HJ, Sun WB. Sodium chloride stimulates growth and alleviates sorbitol-induced osmotic stress in sugar beet seedlings. Plant growth regulation. 2015 Jan 1;**75**(1):307-16

[37] Waditee R, Bhuiyan NH, Hirata E, Hibino T, Tanaka Y, Shikata M, et al. Metabolic engineering for betaine accumulation in microbes and plants. Journal of Biological Chemistry. 2007;**282**(47):34185-34193

[38] Ghoulam C, Foursy A, Fares K. Effects of salt stress on growth, inorganic ions and proline accumulation in relation to osmotic adjustment in five sugar beet cultivars. Journal of Environmental and Experimental Botany. 2002;**47**(1):39-50

[39] Per TS, Khan NA, Reddy PS, Masood A, Hasanuzzaman M, Khan MI, et al. Approaches in modulating proline metabolism in plants for salt and drought stress tolerance: Phytohormones, mineral nutrients and transgenics. Journal of Plant Physiology and Biochemistry. 2017;**115**:126-140

[40] Ashraf MP, Harris PJ. Potential biochemical indicators of salinity tolerance in plants. Plant Science Journal. 2004;**166**(1):3-16

[41] Greenway H, Munns R. Mechanisms of salt tolerance in nonhalophytes. Annual Review of Plant Physiology. 1980;**31**(1):149-190

[42] Murakeözy ÉP, Nagy Z, Duhazé C, Bouchereau A, Tuba Z. Seasonal changes in the levels of compatible osmolytes in three halophytic species of inland saline vegetation in Hungary. Journal of Plant Physiology. 2003;**160**(4):395-401

[43] Ali G, Srivastava PS, Iqbal M. Proline accumulation, protein pattern and photosynthesis in *Bacopa monniera* regenerants grown under NaCl stress. Biologia Plantarum Journal. 1999;**42**(1):89-95

[44] Johnson CB. Physiological Processes Limiting Plant Productivity. Elsevier; 2013

[45] Abebe T, Guenzi AC, Martin B, Cushman JC. Tolerance of mannitol-accumulating transgenic wheat to water stress and salinity. Plant Physiology. 2003;**131**(4):1748-1755

[46] DuPont FM. Salt-induced changes in ion transport: Regulation of primary pumps and secondary transporters. In: Transport and Receptor Proteins of Plant Membranes. Boston, MA: Springer; 1992. pp. 91-100

[47] Gill SS, Tuteja N. Reactive oxygen species and antioxidant machinery in abiotic stress tolerance in crop plants.

Plant Physiology and Biochemistry. 2010;**48**(12):909-930

[48] Mittova V, Volokita M, Guy M, Tal M. Activities of SOD and the ascorbate-glutathione cycle enzymes in subcellular compartments in leaves and roots of the cultivated tomato and its wild salt-tolerant relative *Lycopersicon pennellii*. Physiologia Plantarum Journal. 2000;**110**(1):42-51

[49] Joseph B, Jini D. Insight into the role of antioxidant enzymes for salt tolerance in plants. International Journal of Botany. 2010;**6**(4):456-464

[50] Guo J, Ling H, Wu Q, Xu L, Que Y. The choice of reference genes for assessing gene expression in sugarcane under salinity and drought stresses. Scientific Reports. 2014;**4**(1):1-0

[51] Peng YL, Gao ZW, Gao Y, Liu GF, Sheng LX, Wang DL. Eco-physiological characteristics of alfalfa seedlings in response to various mixed salt-alkaline stresses. Journal of Integrative Plant Biology. 2008;**1**:29-39

[52] Qadir M, Quillérou E, Nangia V, Murtaza G, Singh M, Thomas RJ, et al. Economics of salt-induced land degradation and restoration. Natural Resources Forum. 2014;**38**(4):282-295

[53] Hussain MI, Lyra DA, Farooq M, Nikoloudakis N, Khalid N. Salt and drought stresses in safflower: A review. Agronomy for Sustainable Development Journal. 2016;**36**(1):4

[54] Ali A, Khan SU, Qayyum A, Billah M, Ahmed W, Malik S. Silicon and thiourea mediated stimulation of salt tolerance varying between three fodder beet (*Beta vulgaris* L.) genotypes. Journal of Applied Ecology and Environmental Research. 2019;**17**(5):10781-10791

[55] Sakr HO, Awad HA, Seadh SE, Abido WA. Influence of irrigation withholding and potassium levels on forage yields and its quality of fodder beet. Journal of Crop Science. 2014;**5**(1):116

[56] Abdallah EF, Yassen AA. Fodder beet productivity under fertilization treatments and water augmentation. Australian Journal of Basic and Applied Sciences. 2008;**2**(2):282-287

[57] El-Sarag EI. Response of fodder beet cultivars to water stress and nitrogen fertilization in semi-arid regions. American-Eurasian Journal of Agricultural & Environmental Sciences. 2013;**13**:1168-1175

Chapter 5

Sustainability and Determinate of Farmers' Mitigation Strategies to Greenhouse Gases Emission: A Case in Rice Agric-Food System of Nigeria

Nnaemaka Success Esiobu, Chinedu Gilbert Onubuogu, Sylvarlene Munachim Njoku and Blessing Chidinma Nwachukwu

Abstract

Sustainable production refers to the production that meets the needs of the present, without compromising the ability of future generations to meet their own needs. At global level and mainly across Nigeria, rice fields are considered as one of the most important sources of atmospheric concentration of two greenhouse gases, mainly anthropogenic methane (CH_4) and nitrous oxide (N_2O) emissions. These greenhouse gases (GHGs) are produced under anaerobic conditions, and their production has negative environmental and health implications. Additionally, the growing demand for rice across Nigeria exceeds supply, resulting in a rice deficit. To overcome this challenge, rice production should be increased, with so much regard to less GHG emission. Moving forward, understanding the determinate of farmers' mitigation strategies to GHGs will definitely enhance effort made for farmers to continue to mitigate easily over-time. Incidentally, empirical study on the present discourse is relatively scanty, isolated, and devoid of in-depth and quantitative analyses. Most empirical studies did not pay close attention to the determinants of rice farmers' decisions to mitigation options to GHGs. Studies on mitigation of GHGs at a farm or household level should rigorously examine the socioeconomic characteristics that influence farmers' decisions to practice GHG mitigation or not. These create a gap in research and make it extremely difficult if not impossible for the governments/interest groups to know the method they can adopt in helping farmers mitigate the negative impact of GHG emission in rice production. It was against this backdrop that this study was systematically undertaken.

Keywords: rice, greenhouse gases (GHGs), mitigation strategies, sustainability and multinomial model, Nigeria

1. Introduction

Sustainable production refers to the production that meets the needs of the present, without compromising the ability of future generations to meet their own needs [1]. For an agricultural production to be sustainable, it must produce food with regard not only to the environment (to ensure production can continue on an indefinite basis) but also to generating sufficient production to meet the demand and producing an adequate return for farmers to support their standard of living of those yet unborn. Therefore, rice (*Oryza* spp.), which is the second-largest most consumed cereal (after wheat), shapes the lives of millions of households globally [2]. More than half of the worlds' population depends on rice for about 80% of its food calorie requirements [3, 4]. It has become a staple food in Nigeria such that every household, both the rich and the poor, consumes a great quantity. A combination of various factors seems to have triggered the structural increase in rice consumption over the years with consumption broadening across all socioeconomic classes, including the poor [5]. The rising demand could be as a result of increasing population growth and income level coupled with the ease of its preparation and storage. Currently, due to the present government objective on diversification of the economy, rice is grown in almost 36 states in Nigeria including Federal Capital Territory (FCT) under diverse production systems and agroclimatic conditions. Additionally, the growing demand for rice across sub-Saharan Africa and particularly in Nigeria exceeds supply, resulting in a rice deficit. In the same way, Nigeria is the continent's leading consumer of rice, one of the largest producers of rice in Africa, and simultaneously one of the largest rice importers in the world. Incidentally, rice field is a significant anthropogenic source of methane (CH_4) and nitrous oxide (N_2O), two important greenhouse gases (GHGs). Methane, which accounts for 20–30% of the global warming effect, is second only to carbon dioxide (CO_2) as the most significant GHG [6]. Methane from rice fields represents about 10% of non-CO_2 emissions from agriculture [7] and about 89% of the global warming potential (GWP) from rice [8]. The current understanding of the determinate of farmers' mitigation strategies to GHG emission in rice agric-food system in Nigeria has not much been empirically documented. Additionally, to the best of our knowledge, no study has systematically modeled farmers' mitigation strategies for GHG emission using multinomial logit regression. The multinomial logit model is an extension of the binary logit model for modeling categorical dependent variables with more than two categories. The dependent variable is assumed to follow a multinomial distribution, a generalization of the binomial distribution. This creates a gap in knowledge and makes it absolutely difficult if not impossible for researchers, the government, and policy-makers to know the method they can adopt in assisting the farmers increase their production, their standard of living and livelihood in a cleaner environment. Despite the importance attached to understanding rice production under a cleaner environment, it is somewhat surprising that little or nothing is known about farmers' socioeconomic characteristic; farmers' mitigation strategies for GHG emissions; how farmers' socioeconomic characteristic influences their mitigation strategies; and the barrier they encounter in mitigating GHGs in the area. Empirical evidence remains largely scanty, isolated, and devoid of in-depth and quantitative analysis. It was against these backdrops that it became increasingly pertinent that the study was systematically and logically undertaken.

2. Methodology

The study was carried out in Imo State, Nigeria. Imo State is located in the eastern zone of Nigeria. The state lies between latitude 4°45′N and 7°15′N and longitude 6°50′E and 7°25′E [9]. It is bounded on the east by Abia State, on the west by the river Niger and Delta State, and on the north by Anambra State, while Rivers State lies to the south. Imo State covers an area of about 5067.20 km^2, with a population of

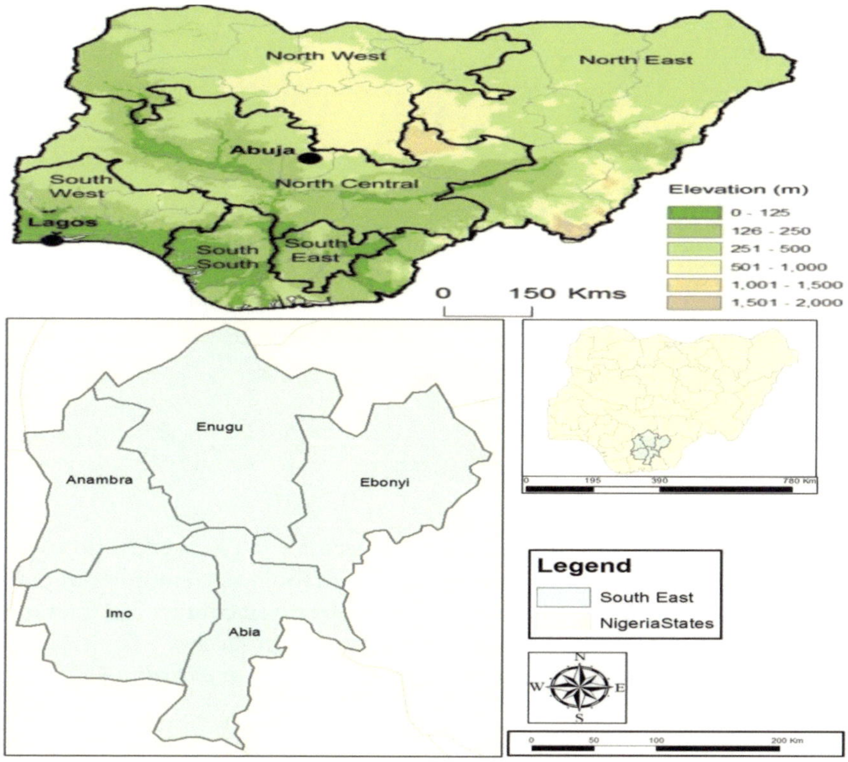

Figure 1.
Map of Nigeria showing the study area.

3,934,899 [10, 11] and population density of about 725 km^2 [12]. The state has three agricultural zones namely Orlu, Owerri, and Okigwe (**Figure 1**). The state has an average annual temperature of 28°C, an average annual relative humidity of 80%, average annual rainfall of 1800–2500 mm, and an altitude of about 100 m above sea level [12]. It experiences two major seasons: dry and rainy seasons. The state has fertile and well-drained soil suitable for rice farming and a good proportion of the population are essentially farmers. A multistage and purposive random method was used in the selection of respondents. Purposive sampling method was used to select respondents who are predominantly rice farmers. The sample size comprised 120 rice farms. A well-structured questionnaire was the main tool for data collection. Data collected were analyzed using descriptive statistical tools and a multinomial logit model. The model is given below:

If p_{ij} is the probability of y_i falling in category $j, j = 1,2,\ldots,J$, then

$$\ln\left(\frac{p_{ij}}{p_{iJ}}\right) = \alpha_j + \beta_j X_i, j = 1,2,\ldots,J-1 \tag{1}$$

leading to

$$p_{ij} = \frac{e^{\alpha_j + \beta_j X_i}}{1 + \sum_{k=1}^{J-1} e^{\alpha_k + \beta_k X_i}}, j = 1,\ldots,J-1 \tag{2}$$

and

$$p_{iJ} = \frac{1}{1+\sum_{k=1}^{J-1} e^{\alpha_k+\beta_k X_i}} \tag{3}$$

where P = response probability (J = 0, 1, 2, 3,...7); Y = mitigation category, J = 1, 2,...,8; 1 = alternate wetting and drying of rice (AWD); 2 = system of rice intensification (SRI); 3 = changing tillage operations (CTO); 4 = Nitrogen Fertilizer Management (NFM); 5 = residue management (RM); 6 = aerobic rice varieties (ARC); 7 = no mitigation strategies.

The explanatory variables are as follows:

$$Y = f(X_1, X_2, X_3, X_4, X_5, X_6, X_7, X_8, X_9, X_{10}, X_{11}) + ei \tag{4}$$

where X_1 = age (years); X_2 = sex (male = 1, female = 0); X_3 = educational level (years); X_4 = farming experience (years); X_5 = household size (number of persons); X_6 = farm income (N); X_7 = farm size (ha); X_8 = extension contact (contact = 1, no-contact = 0); X_9 = access to farm credit (access = 1, no-access = 0); X_{10} = access to GHG emission information (access = 1, no-access = 0); e_i = error term.

3. Results and discussion

3.1 Socioeconomic characteristics of rice farmers

Table 1 reveals that majority (59.17%) fell within the age range of 41–50 years. The mean age was 45.00 years. This shows that farmers in the area are vibrant, young, and still within the active age. Rice farming is so strenuous. The implication is that younger farmers are more likely to practice more and modern mitigation strategies in GHG emission faster than the older ones. Young farmers are more likely to know about new mitigation strategies to avert GHG emission with the willingness to bear risk. **Table 1** also reveals that majority (75.85%) of the farmers were males. The finding implies that both sexes are involved in rice farming but males are more in number in the area. This is true as male farmers have been found to be relatively more efficient than women [13].

Entries in **Table 1** also show that greater proportion (53.33%) had secondary school education. The main education level is 12 years, which is equivalent to secondary school education. The finding implies that approximately 95.00% of the farmers had formal education, which is expected to increase their level of understanding on the effect of GHG emissions in rice farms and various mitigation strategies to practice in thwarting the negative effect. Result in **Table 1** shows that majority (84.17%) were married. The finding implies that rice farming is an enterprise of married individuals who are expected to be responsible according to societal standard. Married farmers have more likelihood of adapting to climate change easily than their unmarried counterparts since they have access to labor. Result of farming experience is shown in **Table 1** and it shows that about 27.50% of the farmers had a farming experience ranging from 11 to 15 years. The mean year of experience in farming was 15.00 years. This shows that the farmers were quite experienced in rice farming and may have been adapting to several mitigation strategies for GHG emissions in the area. It is expected that farmers with more experience are more

	Frequency	Percentage	Mean (X)
Age (years)			
21–30	5	4.17	
31–40	11	9.16	
41–50	71	59.17	
51–60	30	25.00	
61–70	3	2.50	
Total	120	100.0	45.00
Sex			
Male	91	75.83	
Female	29	24.16	
Total	120	100.0	
Educational level (years)			
No formal education	6	5.00	
Primary	41	34.17	
Secondary	64	53.33	
Tertiary	9	7.50	
Total	120	100.0	12 years equivalent to secondary education
Marital status			
Single	8	6.67	
Married	101	84.17	
Divorced	4	3.33	
Widowed	7	5.83	
Total	120	100.0	
Farming experience (years)			
1–5	38	63.33	
6–10	9	15.00	
11–15	5	8.33	
16–20	8	13.33	
21–25	9	7.50	
Total	120	100.0	23
Household size (number of persons)			
1–2	2	1.67	
3–4	5	4.17	
5–6	11	9.17	
7–8	29	24.17	
9–10	51	42.50	
11–12	13	10.83	
13–14	9	6.67	
Total	120	100.0	9.00

	Frequency	Percentage	Mean (X)
Extension contact			
Contact (yes)	31	25.83	
No contact (no)	89	74.17	
Total	120	100.0	
Access to credit			
Access	46	76.67	
No access	14	23.33	
Total	120	100.0	
Access to GHG information			
Access	107	89.17	
No access	13	10.83	
Total	120	100.0	
Farm size (ha)			
0.1–0.99	27	22.50	
1.0–2.50	83	69.17	
2.60–3.00	10	8.33	2.28
Total	120	100	
Annual farm income (N)			
100,001–200,000	21	17.50	
200,001–300,000	25	20.83	
300,001–400,000	65	54.17	
400,001–500,000	9	7.50	
Total	120	100.0	400,790.00 (1034.40 USD)

Source: field survey data, 2020.

Table 1.
Socioeconomic characteristics of rice farmers.

likely to accept innovations and new mitigation strategies for GHG emissions than inexperienced farmers. The number of years of farming helps to cushion the effects of GHG emissions, since GHG emissions is yearly recurring decimal during rice farming. Results in **Table 1** also show that majority (74.17%) of the farmers had no contact with extension agents. The implication is that majority of the farmers may not have the opportunity of learning new mitigation options in GHG emissions and consequently exposing their rice farming to incidence of CH_4 and N_2O impact in the area. It becomes clear that there is a need for the government to strengthen the Agricultural Development Programme (ADP) to facilitate timely extension contacts with farmers in the area. The provision of information and guidance to farmers in any farming season would increase mitigation of GHG emissions and improvement in their faming enterprise in a cleaner environment. Entries in **Table 1** reveal that about 42.50% had a household size ranging from 9 to 10. The mean household size was found to be 9.00 persons. The result shows that farmers had large households. The implication is that they could draw farm labor from their households for the practice of various mitigation strategies for GHG emissions in rice farming. **Table 1** shows that majority (89.17%) of the farmers have access to GHG emission information. This implies that farmers in the study area have access to GHG emissions information, which enhances their easy mitigation to multiple choices in GHG emissions. It is expected that farmers who have access to GHG emissions information will be more aware of effect of GHG emissions and practice better mitigation measures than farmers with no access to information. **Table 1** reveals that majority (69.17%) of the farmers had farm size of between 2.00 and 2.50 ha. The finding implies that the farmers in the area are mainly smallholder farmers operating on less than or equal to 2.50 ha of farmland. This could be as a result of land tenure system or increasing population prevalent in the area. Additionally, the small farm size is not even contiguous plot but rather small plots scattered in different areas of the community. It is expected that farmers with large farm size will practice more GHG strategies than those with lesser farmland in the area. More so, larger farm size enhances the probability of households choosing multiple and better measures to mitigate GHG emission than of households with smaller farm size. Finally, **Table 1** indicates that majority (54.14%) had an average annual farm income of between N300,001 and N400,000. The mean annual farm income was N400,790.00 while monthly farm income was estimated to be N33,399.167. The finding implies that the farmers have a relatively low farm income despite the larger household size, which they recorded. The implication of the findings is that farmers may not have the much needed financial capacity to mitigate GHG emission. This is true as some mitigation strategies for GHG emission are costly. Hence, farmers may have several GHG emission strategies they want to practice but limited fund will continue to hinder them.

3.2 Farmers' GHG emission mitigation strategies in rice farming

The result in **Figure 2** reveals farmers' GHG emission mitigation strategies in rice farming in the area. Similarly, it is very possible that the various mitigation strategies used by the rice farmers to reduce the negative impacts of GHG emission in their farming activities could be profit driven rather than GHG emission driven. In strengthening the above assertion, the study of [14] reported that the action of farmers in reducing the negative impact of climate change over time has basically been climate change driven; hence, the study assumed that the rice farmers' various mitigation measures are therefore GHG emission driven. The result reveals that about 98.10% of the farmers identified alternate wetting and drying of rice (AWD) as one of their several mitigation strategies for climate change. AWD is a method of reducing 30.00% of water in rice farms to influence GHG emission reduction by 48%. The AWD process influences rice production, CH_4 and N_2O emissions from rice systems. The finding is supported by the study of [15] who found that single or

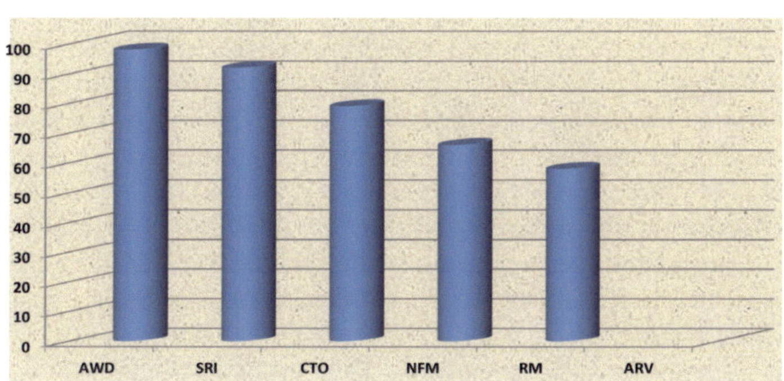

Figure 2.
GHG emission mitigation strategies of rice farmers in the study area. Keys: AWD: alternate wetting and drying of rice; SRI: system of rice intensification; CTO: changing tillage operations; NFM: Nitrogen Fertilizer Management; RM: residue management; ARV: aerobic rice varieties.

multiple drainage management during a rice-growing season (e.g., AWD) reduces CH_4 emissions by 48–93% compared to those observed under continuous flooding systems. Approximately, 92.00% identified system of rice intensification (SRI). The SRI is a holistic approach for sustainable rice cultivation. It involved planting a single seedling with more space between them rather than by the handful and bunched closely together. It also involves watering intermittently and allowing for dry spells rather than using continuous flooding and using organic input. The study of [16] confirmed a similar finding as one of the strategies used by rice farmers in GHG mitigation. Additionally, about 79.00% of the farmers practiced changing tillage operations (CTO). The study of [17, 18] concluded that biomass incorporation under conventional tillage is the main cause of the higher CH_4 emissions, implying that rice production systems where residue incorporation is excluded (no-till) may contribute to mitigation of GHG emissions. Similarly, the finding agrees with the study of Ahmad et al. [1] who also reported significant reductions in CH_4 emissions (21–60%) from no-till compared to tilled fields. In the same vein, Nitrogen Fertilizer Management (NFM) was identified by 66.00% of the farmers. The application of nitrogen (N) fertilizer to agricultural soils increases productivity and may also influence GHG emissions from rice systems. The finding of [19] found that N fertilizer-induced N_2O emissions were reported to be 0.21% under continuous flooding and 0.40% under alternate wetting and drying (AWD) rice production systems. In the same meta-analysis, an effect of fertilizer type was reported, with N_2O emissions shown to increase by 24% and CH_4 emissions to decrease by 40% when urea was replaced by ammonium sulfate. Others (58.00 and 35.00%) identified residue management (RM) and aerobic rice varieties (ARC), respectively. The incorporation of rice residues contributes toward long-term nutrient cycling but may, due to high C/N ratios, cause short-term N immobilization and thus affect N availability for subsequent crops [19]. Meanwhile, aerobic rice varieties (ARV) is a production system in which especially developed "aerobic rice" varieties are grown in well-drained, non-puddled, and non-saturated soils [20]. With a good management, the system aims for yields of at least 4–6 tons per ha. Therefore, the finding became clear that farmers are noticing changes in rice field and have started practicing several strategies to thwart the negative effect of GHG emission in their rice farming.

3.3 Determinants of rice farmers' mitigation strategies for GHG emission

Table 2 shows determinants of rice farmers' mitigation strategies for GHG emission. The estimation of the multinomial logit model for this study was undertaken by normalizing one category, which is normally referred to as the "reference or base category." In this analysis, the last category (no mitigation strategies) is the reference category. The model was run and tested for the validity of the independence of the irrelevant alternatives (IIA) assumption by using the Hausman test for IIA. The test accepted the null hypothesis of independence of the mitigation strategies for GHG emission, suggesting that the multinomial logit specification is appropriate and a good fit to model farmers' mitigation strategies for GHG emission. Results reveal a likelihood ratio chi-square (χ^2) value of 0.9770 implying that 97.70% of variation in the model for the mitigation strategies was explained by the explanatory variables while the remaining 2.30% was accounted for by stochastic error. The model was also statistically significant at 1% ($P < 0.00001$), suggesting that the models have strong explanatory power. This indicates that all the models had good fit to the model. The significance of this likelihood ratio statistics test indicates that rice farmers' socioeconomic characteristics significantly influence the use of mitigation strategies for GHG emission in the area. Consequently, the interpretation and discussion of the multinomial logit result indicate the following:

Explanatory variables	AWD	SRI	CTO	NFM	RM	ARV
Age (X_1)	-1.0079e-03 (-3.11)***	0.00085 (4.02)***	-0.021 (-3.10)***	0.004 (3.84)**	0.0093 (3.38)***	-0.0098 (-3.92)***
Sex (X_2)	-0.00015 (-0.11)	0.0006 (0.76)	0.234 (1.17)*	-0.155 (-0.12)	-0.23 (-0.05)	0.14 (0.87)
Educational level (X_3)	4.20e-06 (1.08)*	0.00009 (0.63)	0.008 (0.96)	0.012 (-0.68)	-0.02 (-1.64)*	-0.009 (-0.91)
Farming experience (X_4)	-4.96e-06 (-0.76)	-0.00005 (-0.51)	0.011 (1.35)*	0.0015 (1.01)	-0.011 (-0.52)	-0.007 (-0.63)
Household size (X_5)	-0.000042 (-0.25)	0.0004 (0.14)	0.003 (0.35)	0.017 (0.12)	-0.009 (-0.19)	-0.001 (-0.29)
Farm income (X_6)	1.39e-08 (2.16)**	3.79e-09 (1.94)*	7.54e-09 (1.09)	3.02e-06 (1.63)*	2.66e-06 (1.50)*	2.74e-06 (0.69)
Farm size (X_7)	-0.00046 (-0.68)	-0.0006 (-1.46)	-0.07 (-0.88)	-0.112 (-0.98)	-0.03 (-0.59)	-0.12 (-1.45)
Extension contact (X_8)	0.0051 (3.21)***	0.006 (5.04)***	0.013 (4.85)***	0.054 (5.10)***	0.08 (4.69)***	0.23 (4.97)***
Access to farm credit (X_9)	0.027 (4.04)***	-0.00098 (-1.63)	-0.134 (-1.60)	0.161 (1.84)*	0.11 (0.95)	0.08 (0.95)
Access to GHG emission information (X_{10})	4.37e-06 (0.37)	0.179 (5.01)***	-0.169 (-0.13)	-0.023 (-0.25)	0.04 (0.54)	-0.04 (-0.21)
Pseudo R^2			0.5919			
Likelihood Chi square			97.70***			
Sample size (n)			120			
Reference/base category			No mitigation strategies			

Output of STATA; values in parenthesis are Z values.
****Significant at 1% level.*
***Significant at 5% level.*
**Significant at 10% level.*
Field survey, 2020.
Keys: AWD: alternate wetting and drying of rice; SRI: system of rice intensification; CTO: changing tillage operations; NFM: Nitrogen Fertilizer Management; RM: residue management; ARV: aerobic rice varieties.

Table 2.
Estimated multinomial logit model of the determinants of rice farmers' mitigation strategies for GHG emission.

Age (X_1): age of the rice farmers significantly influences mitigation of GHG emission. Age of the farmers was positively related across the practice of alternate wetting and drying of rice (AWD); system of rice intensification (SRI); Nitrogen Fertilizer Management (NFM); residue management (RM); and aerobic rice varieties (ARV). This reason could be because the options have been practiced for a long period of time and are well known by older farmers than their younger counterparts. On the other hand, age of the farmers had a negative influence on the probability of uptake of CTO. The result shows that a unit increase in the age of the farmers decreases the likelihood of taking up CTO by 0.21 (2.10%). This could be because CTO may require more physical strength and energy to practice in rice farming of which older farmers may not have the capacity to do. The result is consistent with the findings of [14] who noted that the older farmers become more risk averse and practice less strategies, particularly those requiring more energy over time.

Sex (X_2): the result indicated that female-headed households practiced efficiently and more mitigation strategies for GHG emission than their male counterparts. On the other hand, male-headed households were more readily resilient to GHG emission than their female counterparts by practicing SRI and CTO. The finding tallies with the study of [21] who asserted that females are more involved in rural agriculture. This is true as women use it to support their families nutritionally and income-wise while the male households usually migrate to urban cities in search of nonagricultural jobs. Additionally, it is also expected that females will understand perceived effect of GHG emission in rice farming and practice modern mitigation strategies than their male counterpart.

Educational level (X_3): education of the farmers was positively related across all the mitigation strategies for GHG emission. This result is in line with the *a priori* expectation of the model. The finding is in line with the study of [22] who asserted that exposure to higher education of the farmer increases the probability of choosing different sustainable farming methods. The probable reason could be due to the fact that educated farmers have more knowledge of GHG emission and are already aware of various techniques and management practices that could be employed to mitigate the emissions easily. Additionally, the study of [23] also confirmed the importance of education on choice of mitigation strategies for GHG emission.

Farming experience (X_4): farming experience had a positive and significant relationship across all the mitigation strategies for GHG emissions modeled. This implies that increase in years of experience increases the probability of uptake of AWD, SRI, CTO, NFM, RM, and ARV. Highly experienced farmers are likely to have more information and knowledge on GHG emission than their counterpart with limited years of experience. In addition, experience exposes farmers to various GHG emission strategies they could employ in the face of anticipated environmental situations. The findings support [24] who asserted that farmers with more experience would be more efficient, have better knowledge of climatic conditions and market situation, and are, thus, expected to run a more efficient and profitable enterprise.

Household size (X_5): household size of farmers increased the likelihood of using CTO, RM, and SRI practices by 0.001(1.00%). This indicates that household size increases the probability of uptake of these mitigation measures to climate change because such options require additional labor from the farmers, which is usually provided by his/her household members. On the other hand, household size of farmers decreased the likelihood of practicing ARV and NFM by 0.0001 (0.1%). This is because, as the hectare of farmland cultivated by each farmer reduces, the labor needed by such farmers also reduces. The finding tallies with the study of [5] who reported that large household size is associated with a higher labor endowment, which would enable the household to accomplish various agricultural tasks especially at the peak seasons and ensure ease of adaptation to climate change. The finding is also supported by the result of [14] who opined that large household size has shown to provide cheap and available source of labor for farmers in adapting easily to climate change.

Farm income (X_6): the income of farmers had a positive and significant influence on the likelihood of practicing all the mitigation measures identified. Higher income farmers are less risk averse and have more access to information, a lower discount rate, a longer term planning horizon, and are wealthier than low-income farmers. Additionally, with more financial and other resources at their disposal, farmers are able to change their management practices in response to changing climatic, GHG emissions and other factors and are better able to make use of all the available information they might have on changing conditions, both climatic and other socioeconomic factors. The result shows that a unit increase in the income of the farmers increased the likelihood of adopting the practice of AWD, SRI, CTO, NFM, RM, and ARV. The study of [15] reported that farmers with higher farm income will make better decision, use necessary productive inputs, and realize huge yield/output than their counterparts who have low farm income. Additionally, the study of [14] also reported that adaptation options to climate change are costly.

Farm size (X_7): farmers' land area cultivated was negatively related to mitigation strategies for GHG emissions in the area. The negative relationship between farmers' mitigation strategies for GHG emissions and farm size is inconsistent with the study carried out by [25] but in line with [14] who reported that the probable reason could be due to the fact that adaptation/mitigation measures are plot-specific. It is expected that farmers with large farm size will practice more mitigation strategies for GHG emissions than those with lesser farmland in the area. More so, larger farm size enhances the probability of household choosing multiple and better mitigation strategies for GHG emissions than households with smaller farm size. This means that it is not the size of the farm but the specific characteristics of the farm that dictate the need for specific adaptation mitigation strategies for GHG emissions in rice production.

Extension contact (X_8): extension contact had a positive and significant influence across all the mitigation strategies for GHG emissions modeled. The finding shows that a unit increase in the number of extension visits to the farmers increased the likelihood of AWD by 0.006 (0.6%), SRI by 0.013 (1.3%), CTO by 0.054 (5.4%), NFM by 0.08 (8.00%), RM by 0.0051 (5.1%), and ARV by 0.23 (23.00%). Contact with extension agents, which denotes access to information, had a positive effect across all adaption measures indicating that extension contact increases the likelihood of mitigating GHG emissions in rice farm easily. Access to extension services significantly increased the probability of taking up AWD, SRI, CTO, NFM, RM, and ARV. Extension services provide an important source of information on GHG emissions as well as agricultural production and management practices. Farmers who have significant extension contacts have better chances to be aware of changing climatic conditions and also of the various management practices that they can use to adapt to changes in climatic conditions. The findings are in line with the study [26] which argued that extension contact enhances farmers' production and promotes their knowledge on modern farming methods.

Access to farm credit (X_9): results showed that farmers' access to credit significantly increased the probability of uptake of AWD, SRI, CTO, NFM, RM, and ARV. Inadequate fund is one of the main constraints in adjusting to climate change [14]. Despite the various mitigation strategies farmers could be aware of and willing to practice, inadequate fund to purchase the necessary inputs and other associated equipment remains one of the significant barriers to mitigation strategies for GHG emissions in rice production.

Access to GHG emission information (X_{10}): this depicts the level of awareness of GHG emissions significantly increased the probability of uptake of all the mitigation strategies identified. Farmers who have access to GHG emissions and climate information are more aware of changes in climatic conditions and have higher chances of taking adaptive measures in response to observed changes. It is an important precondition for farmers to take up mitigation strategies. Information on climate variables like temperature amount, relative humidity, rainfall amount, and sunshine duration has really helped farmers in the area on the time to plant a particular breed of rice. Farmers' access to information on GHG emissions is likely to enhance their probability to understand GHG emissions and climate change impact and hence enable them take up better mitigation strategies to increase their farm yield and income.

3.4 Rice farmers' barrier to mitigation of GHG emission

The findings in **Figure 3** show rice farmers' barrier to mitigation of GHG emission in the area. The finding reveals that about 98.30% of the farmers identified inadequate information. This could be attributed to dearth in research on GHG emission and mitigation strategies as well as lack of information on GHG and climatic variables which should always be disseminated by Nigerian Meteorological Agency (NiMET) and agricultural extension agents. This constraint left the farmers unable to get the much needed information on climate change and GHG emission. In the present information age, inadequate information could pose serious challenges to the farmers' coping strategies as they may not be aware of recent developments regarding GHG emission, mitigation strategies, and the necessary readjustments. Poor information on mitigation strategies for GHG emission in rice farming may result in food insecurity and unsustainable production over time. About 94.75% identified inadequate fund. Inadequate fund left most of the rice farmers unable to get necessary resources in mitigating GHG emission in the area. This could be attributed to high cost of mitigation options. Inadequate fund hinders farmers from getting the necessary resources and technologies that assist to efficiently mitigate GHG emission. The result shares view with the study of [14] who argued that adaptation options are costly and hence farmers need adequate fund to adapt. Going forward, poor extension contact, high cost of inputs, poor access to farm credit, limited availability of farmland were identified by 87.50, 83.33, 82.50, and 75.74% of the rice farmers, respectively. High cost of farm inputs could also be attributed to inadequate fund. With limited fund, the acquisition of necessary facilities will be difficult. They may not only be costly, but may also appear scarce for poor farmers. In addition, the farmers may not also have the necessary facilities for current information like radio and television to obtain weather forecasts. Poor access to credit could be linked to lack of information or awareness of the presence of loan facilities, high collateral requirements, and location of banks in urban areas, which are far from the rural areas where farmers live. Limited farmland could be attributed to land tenure system or increasing population prevalent in the area. High population pressures compel farmers to intensively farm over a small plot of land and make them unable to practice several GHG mitigation strategies that will improve their farm yield and income. It becomes clear that this constraint is responsible for poor production of rice and GHG emission mitigation in the area. Curbing this barrier will be vital in promoting not just local mitigation strategies but global strategies of GHG emission in the area and perhaps beyond.

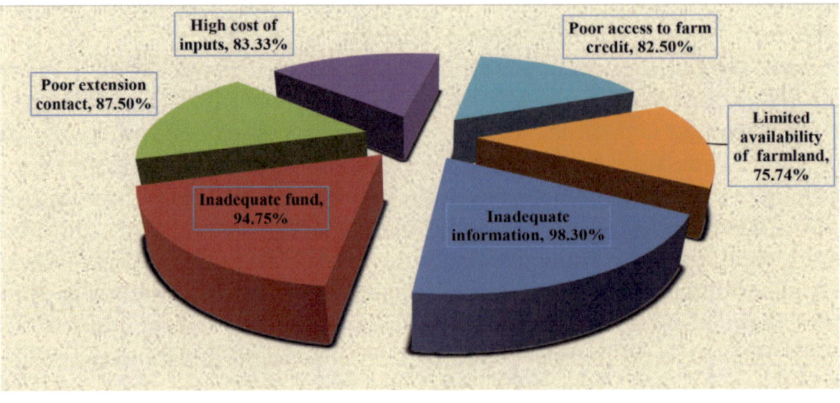

Figure 3.
Rice farmers' barrier to mitigation of GHG emission.

4. Conclusion

Conclusively, the study was logically guided by describing the socioeconomic characteristics of the rice farmers; identifying and describing the mitigation strategies for GHGs used by rice farmers and constraints in mitigating GHGs in rice farming. A multistage and purposive random method was used in the selection of respondents. Purposive sampling method was used to select respondents who are predominantly rice farmers. The sample size comprised 120 rice farms. A well-structured questionnaire was the main tool for data collection. Data collected were analyzed using descriptive statistical tools and a multinomial logit model. The result shows that the mean age was 45.00 years. Greater proportions (75.83%) were male. Majority (84.17%) were married with an average household size of nine persons. The mean educational level and farming experiences were 12 years (equivalent to secondary school education) and 23.00 years, respectively. Average farm size and annual farm income were 2.28 ha and N400,790.00 (1027.67 USD), respectively. The result confirmed the incidence of GHG emission in rice farm in the area. Interestingly, farmers are becoming increasingly aware and have started practicing several mitigation strategies. The major GHG mitigation strategies the farmers practice were alternate wetting and drying of rice (AWD) (98.10%) and the system of rice intensification (SRI) (92.00%) among various strategies they practiced simultaneously. Estimated multinomial logit model revealed that household size (X_5), farm size (X_7), and education (X_9) significantly influence their choice of GHG mitigation strategies at 1% level of probability. Regrettably, farmers complained of inadequate fund (98.33%). It was therefore recommended that farmers should form a stable cooperative to access fund, information and government support effectively. In the same way, the study confirmed the incidence of GHG emission in rice farm the area. Interestingly, farmers are becoming increasingly aware and are noticing the GHG emission. The farmers have started practicing several mitigation strategies to thwart the negative effect of GHG emission while remaining sustainable. The major GHG mitigation strategies of farmers were alternate wetting and drying of rice (98.10%) and the system of rice intensification (92.00%) among various strategies they practice simultaneously. The study also looked at the determinants of rice farmers' use of various mitigation options for GHG emission using a multinomial logit model. The model permits the analysis of decisions across dichotomous categories, allowing the determination of choice probabilities for different categories. Multinomial logit results confirmed that access to credit, extension services, farming experience, education, access to climate change information, and farm size were some of the significant determinants of farm-level mitigation options. The main barrier to the mitigation of GHG emission was lack of information on appropriate mitigation option, which could be attributed to dearth in research on GHG emission as well as poor information dissemination on the part of extension agents in the study area.

4.1 Recommendations

The following recommendations were made based on the major research observations and findings of the study.

i. Effective agricultural policies and programs should focus on how to intensify awareness on GHG emission in rice farm as well as its mitigation strategies. This should be done through strengthened agricultural extension delivery.

ii. Since education and farmland were found to significantly increase mitigation, investment strategies should also focus on expansion of farmers' farmland and improvement of their education as this would affect their mitigation of GHG emission positively.

iii. The government must also design policies in such a way that farmers have access to affordable credit as well as subsidized agricultural inputs in order to increase their ability and flexibility to change production strategies in response to the forecasted climatic conditions.

iv. The government or interested organization should endeavor to build weather stations in all local government areas in Nigeria to reduce the incidence of poor climate record keeping and to provide mid-term forecast of weather and other climatic variables.

v. Ultimately, incorporating local knowledge into GHG emission concerns should not be done at the expense of modern/western scientific knowledge. Local knowledge should complement rather than compete with global modern practices in counteracting the negative impact of GHG emission in the area and beyond.

Acknowledgments

Special thanks to the local rice farmers in the study area who provided the data for the study. Additionally, many thanks to our volunteer field enumerator who helped in visiting the sampled farmers in their remote rice farms for evidence-based data collection. Thanks to all those involved in data entry, data cleaning, data coding, and analysis. We cannot thank you all enough.

Author details

Nnaemaka Success Esiobu[1*], Chinedu Gilbert Onubuogu[1],
Sylvarlene Munachim Njoku[2] and Blessing Chidinma Nwachukwu[3]

1 Department of Agricultural Economics, Extension and Rural Development, Imo State University, Owerri, Nigeria

2 Department of Nutrition and Dietetics, Imo State University, Owerri, Nigeria

3 Food Security and Safety Niche, Faculty of Natural and Agricultural Sciences, North-West University, South Africa

*Address all correspondence to: esiobunnaemekasuccess@gmail.com

References

[1] Ahmad S, Li C, Dai G, Zhan M, Wang J, Pan S, et al. Greenhouse gas emission from direct seeding paddy field under different rice tillage systems in central China. Soil and Tillage Research. 2009;**106**:54-61

[2] Akande T. (2020). An Overview of The Nigerian Rice Economy; Agriculture and Rural Development Department The Nigerian Institute of Social and Economic Research (NISER). 2020. Available from: https://unep.ch/etb/etp/events/Agriculture/nigeria.pdf

[3] Ariyo OC, Ariyo MO, Okelola OE, Aasa OS, Awotide OG, Aaron AJ, et al. Assessment of the role of mass media in the dissemination of agricultural technologies among farmers in Kaduna North local government area of Kaduna State, Nigeria. Journal of Biology, Agriculture and Healthcare. 2013;**3**(6):19-28

[4] Food and Agricultural Organisation (FAO). The State of Food and Agriculture; 2010-2019. Rome: FAO; 2020. p. 2020

[5] Ojo OT, Ogundeji AA, Babu SC, Alimi T. Estimating financing gaps in rice production in Southwestern Nigeria. 2020;**9**:12. DOI: 10.1186/s40008-020-0190-y

[6] Intergovernmental Panel on Climate Change (IPCC) (2019) Climate Change 2019: Mitigation of Climate Change. Contribution of Working Group III to the Fifth Assessment Report of the Intergovernmental Panel on Climate Change; The Intergovernmental Panel on Climate Change (IPCC) presented at the UN Climate Change Conference (COP 25) in Madrid on 2-13 December 2019; Available from: https://www.ipcc.ch/2019/

[7] Ngonidzashe C, Laura A, Maria K, Sandra L, Fernando C, Manabu I, et al. Sustainable and low greenhouse gas emitting rice production in Latin America and the Caribbean: A review on the transition from ideality to reality. Sustainability. 2018;**10**:671

[8] Zhang CL, Chao L, Jun Z, Aixing D, Kees J, Weijian Z. Acclimation of methane emissions from rice paddy fields to straw addition. Science Advances. 2019;**5**(1):eaau9038

[9] Nigerian Meteorological Agency (NiMET). Drought and Flood Monitoring in South East Bulletin. 2018. Available from: www.nimet.gov.ng [Accessed: 25 October 2018]

[10] National Bureau of Statistics (NBS). National Bureau of Statistics Official Gazette (FGP 71/52007/2,500(OL24)): Legal Notice on Publication of the Details of the Breakdown of the National and State Provisional Totals, 2006 Census. 2007. Available from: www.nigerianstat.gov.ng [Accessed: 28 February 2016]

[11] Nigeria Population Commission (NPC). Nigeria Population Commission, Nigeria Federal Government Initiative of individual head count by gender. Spread, State by State. MOFINEWS. 2006;**6**(3):01-32

[12] Imo State Agricultural Development Programme (Imo-ADP). Work Programme. Owerri, Imo State, Nigeria: Imo ADP; 2018

[13] Esiobu NS. Relative efficiencies of resource use among cassava farmers in Imo State, Nigeria [M.Sc. thesis]. Owerri, Nigeria: Department of Agricultural Economics, Extension and Rural Development, Faculty of Agriculture and Veterinary Medicine; Imo State University; 2018

[14] Esiobu NS, Onubuogu GC. Trend, perceptions and adaptation options of arable crop farmers to climate change

in Imo State, Nigeria; a multinomial logit approach. World Science Journal. 2014;**5**(9):12-24

[15] Munonye JO, Esiobu NS. Sustainability and agribusiness development in Nigeria. Journal of Sustainable Development; An International Peer-reviewed Journal. 2017;**27**:40-44

[16] Bharali A, Baruah KK, Gogoi N. Potential option for mitigating methane emission from tropical paddy rice through selection of suitable rice varieties. Crop & Pasture Science. 2017;**68**:421-433

[17] Bayer C, Costa FS, Pedroso GM, Zschornack T, Camargo ES, Lima MA, et al. Yield-scaled greenhouse gas emissions from flood irrigated rice under long term conventional tillage and no-till systems in a humid subtropical climate. Field Crops Research. 2014;**162**:60-69

[18] Bayer C, Zschornack T, Pedroso GM, da Rosa CM, Camargo ES, Boeni M, et al. A seven-year study on the effects of fall soil tillage on yield-scaled greenhouse gas emission from flood irrigated rice in a humid subtropical climate. Soil and Tillage Research. 2015;**145**:118-125

[19] Linquist BA, Adviento-Borbe MA, Pittelkow CM, van Kessel C, van Groenigen KJ. Fertilizer management practices and greenhouse gas emissions from rice systems: A quantitative review and analysis. Field Crops Research. 2012;**135**:10-21

[20] Yamano T, Arouna A, Labarta RA, Huelgas ZM. Adoption and impacts of international rice research technologies. Global Food Security. 2016;**8**:1-8

[21] Kughur P, Gyanden O, Omaku S, Isa M. Assessment of input needs of women vegetable farmers in Gwer-East local government area of Benue State, Nigeria. 2018;**2**:20-30. DOI: 10.31058/j.as.2018.23013

[22] Esiobu NS, Onubuogu GC. Determinant of risk-smart options among farming households in agricultural risk management in Imo State, Nigeria; (a multinomial logit model approach). Journal of Environment Protection and Sustainable Development. 2018;**3**(10):20-39

[23] Kushal KB. Analysis of greenhouse gas (methane and nitrous oxide) emission and global warming potential from rice fields: With reference to biological mitigation of climate change. Journal of Earth Sciences & Environmental Studies. 2018;**3**(2):395-407

[24] Onubuogu GC, Esiobu NS. Determinants of allocative (pricing) efficiency of cassava farms in Imo State, Nigeria. Journal of Agriculture and Food Sciences. 2020;**17**(2):86-99. Available from: https://www.ajol.info/index.php/jafs/article/view/194543

[25] Nhemachena C, Hassan R. Micro-Level Analysis of Farmers' Adaptation to Climate Change in Southern Africa. International Food Policy Research Institute (IFPRI) Discussion Paper No. 00714. Washington, D.C.: Environment and Production Technology Division, IFPRI; 2007

[26] Nwaiwu JC. Farmers' adoption of some arable crop soil conservation practices in Imo State [unpublished PhD thesis]. Owerri, Nigeria: Department of Agricultural Economics, Extension and Rural Development, Imo State University; 2015

Section 2

Consequences and Mitigation Strategies of Biotic and Abiotic Stress

Chapter 6

Consequences and Mitigation Strategies of Biotic and Abiotic Stress in Rice (*Oryza sativa* L.)

Shandrea Stallworth, Brooklyn Schumaker, Mary Gracen Fuller and Te-Ming Tseng

Abstract

Rice (*Oryza sativa*) is the staple food for more than 3.5 billion people worldwide. Yield levels in Asia have tripled and are expected to increase by 70% over the next 30 years due to population growth. In the US, Arkansas accounts for more than 50% of rice production. Over the last 68 years, rice production has continued to grow in Mississippi, placing it in fourth place after Arkansas, Louisiana, and California. Due to increasing rice acreage, regionally and worldwide, the need to develop abiotic stress tolerant rice has increased. Unfortunately, current rice breeding programs lack genetic diversity, and many traits have been lost through the domestication of cultivated rice. Currently, stressors stemming from the continued effects of climate change continue to impact rice. This chapter highlights current research that strives to discover abiotic and biotic stress tolerant rice. This chapter calls for directed research in genetics and genomics to address the need to discover biotic and abiotic stress tolerant traits. While many genes have been uncovered to arm rice against these stresses, decreased genetic variability in current rice traits presents a small gene pool for discovery.

Keywords: rice, *Oryza sativa*, abiotic, biotic, stress tolerance

1. Introduction

Rice, *Oryza sativa*, is a cultivated, food staple feeding more than one-half of the world's population [1]. Rice is regarded as one of the world's most important crops and is grown in more than one hundred countries producing more than 700 million tons annually [2, 3]. Asia currently accounts for more than 90% of rice that is grown and consumed [4]. In southern China alone, rice consumption is almost 50% higher than the global average due to a diet heavily rooted in rice [5]. It is predicted that rice yield must increase by 1% annually to continue to feed the growing population [6]. To meet this expectation, the development of high-yielding, stress-tolerant rice cultivars is necessary [7].

Rice is a tropical and sub-tropical plant that requires temperatures ranging from 20 to 40°C with flooded conditions, and is highly influenced by solar radiation [8]. It is an annual grass with a life cycle ranging from 105 to 145 days from germination to maturity contingent on various types of environmental contributions [9]. Rice domestication is estimated to have started more than 9000 years ago via wild rice from China [10]. There are two species of cultivated rice (*O. sativa* and *O. glaberrima*) originating from Asia and Africa, respectively [11].

Globally, the Asian cultivar is grown on a large scale while the African cultivar is confined to West Africa [11]. It is a diploid species with an AA genome that can be subjected to traditional hybridization and selection [12]. In addition to two distinct species of rice, the crop can further be divided into two different varieties, *indica* and *japonica* [13]. In this review, knowledge of major abiotic and biotic associated with rice are presented to highlight a need for genomic-focused studies. As climate change continues to be a major contributor to stress in rice production fields, understanding current research strategies in rice stress is necessary.

2. Consequences and mitigation strategies of biotic stresses with specific focus on disease

2.1 Rice blast disease

Rice blast disease is caused by the fungal pathogen *Magnaporthe oryzae* (*M. oryzae*) formerly named *M. grisea*, which is a hemibiotrophic filamentous ascomycete that infects rice and causes yield losses worldwide [14, 15]. This is one of the most economically damaging grass fungi causing annual losses of up to 10% in global rice production [16]. Specific areas that reported devastating losses to blast include India, Japan, South Korea, and Indonesia ranging from 20 to 70% yield losses [17]. This fungal pathogen is so damaging because it can infect the crop at any growth stage and in any tissue above or below ground [18]. Symptoms of rice blast in leaf tissue include diamond-shaped tan lesions with a dark brown edge appearing 5–7 days after infection [19]. Economic losses are caused when the panicle of the plant, where the desired seed is exhibits symptoms of infection and lack of filling [14]. The infection can be located on the collar of the plant causing the tissue to rot leading to the entire panicle collapsing from lack of adequate structure [14].

Similar to most fungal pathogens, the conidia of *M. oryzae* are an essential part of the pathogens cycle [15]. The fungi attach to the plant tissue where they form structures called appressoria, a primary hypha, after germination. Bulbous invasive hypha structures at the end of the germ tubes of the conidia cause a buildup of turgor pressure allowing the pathogen to invade the outer plant tissue and settle in its host [15, 18]. After penetration, infected hypha migrates through the rice leaf cells. Molecules called effectors are released by the pathogen that competes for binding to the plant proteins, thereby disrupting natural processes [20]. These effectors are sometimes translocated to the cytosol of the host cell, where they alter the host's natural immunity responses [15]. Symptoms of blast infection manifest as streaks of dead leaf tissue on which the conidia are produced by the thousand and released for the spread of further infections [21]. *M. oryzae* favors the same humid growing conditions as rice and is spread through the air making it very hard to control when it has been introduced to a field. The pathogen can circulate through multiple lifecycles in one growing season making it dangerous to the crop at any stage [14].

In recent years, research efforts have focused on understanding the process of conidiation, appressoria formation, and responses in the host, rice, to the infection [20]. The effectors that are secreted by the pathogen into the intercellular spaces of the host interrupt natural plant processes and prey on a variety of host proteins, such as important parts during pathogen-associated molecular pattern (PAMP)-triggered immunity (PTI), vesicle trafficking, effector-triggered immunity (ETI), autophagy, sugar transport, chloroplast and mitochondrial functions phytoalexin production, and more [20]. Pathogen response in plants often imitate normal development processes for instance response to fungal gibberellic acid is very similar to the plants' natural response to its own produced gibberellin and auxins [22]. Gene expression regulates plant development and stress response and therefore, it can be stated that the proteins involved in these processes are regulated by a specific set of genes [22]. Study's find similar genes respond to environmental cues and stresses [22].

Understanding innate plant immunity is crucial for the advancing knowledge of plant stress response and the mechanisms it encompasses. Resistant rice cultivars are grown and developed to help control disease outbreaks and infections that can wipe out entire fields [23]. Natural immunity in plants stems from the recognition of PAMPS by the host sensors on the cell surface or pattern recognition receptors [23]. Pathogen-associated triggered immunity is the first kind of plant immunity [23]. Plant receptor protein kinases (RPKs) recognize microbial molecules like lipopolysaccharides (LPS), chitin, peptides, double-stranded RNA, as well as microbial DNA which then activate a mitogen-activated protein kinase (MAPK) element. This process is one of the earliest signaling actions post-plant sensing of the invading pathogen [23, 24]. The signaling channels of MAPKs regulate the production and function of a multitude of enzymes, transcription factors (TFs), hormones, antimicrobial chemicals and peptides, that all play critical roles in resistance to bacterial and fungal pathogens [24]. This first responder component of immunity is vital if a host is to survive an infection. Immune responses in rice have been found to be triggered by sulfated peptide Ax21, chitin, flagellin peptides and LPS [23]. Another level of plant immunity is effector-triggered immunity (ETI) which is a response to a wide variety of microbial molecules known as effectors, which are secreted from the fungus into the host during infection. Lastly, systemic acquired resistance (SAR) is another defense mechanism in plants [23].

Sensors in the rice membrane, as well as Nucleotide binding site/Leucine rice repeat (NBS-LRR) proteins, are needed for immune response. These NBS-LRRs job is to recognize the effectors secreted by the fungi. Four small protein effectors produced by the rice blast fungus *M. oryzae* have been characterized as *AvrPita*, *AvrPiz-t*, *AvrPik/km/kp*, and *AvrPia*, all having distinct structures. Recognition by these NBS-LRR proteins in rice depends on direct, decoy or bait models [23]. Thirteen NBS-LRR proteins that cause resistance to *M. oryzae* have been characterized in rice. Pita is the most studied rice NBS-LRR protein and interestingly has only one amino acid difference between susceptible and resistant alleles [23]. This protein triggers programmed cell death after it binds to a specific effector protein, *AvrPita*, which is hypothesized to keep the fungus from migrating to adjacent cells. This binding is an example of a recognition model of effector identification [23].

Chitin is a well-known product of PAMP that signals defense responses in plants, both monocots, and dicots [25]. This polymer of N-acetyl-D-glucosamine, is part of the fungal cell wall but is not found in plants although they do possess chitin degrading enzymes. The enzymes in plants can degrade the fungal cell wall, are also able to recognize when the fungal cell wall is releasing these chitin particles, and respond during infection [26]. Recognition of chitin by these enzymes activates the plants' defense pathway [26]. How chitin and its fragments, chitin oligosaccharides or N-acetylchitooligosaccharides, are able to notice a harmful pathogen and trigger defense is a topic of research [25]. Proteins in collaboration with receptor-like kinases, which serve as chitin elicitor binding proteins, subsequently bind chitin and play a critical role in chitin signaling in rice, thus activating intracellular events. Rice contains a large number of genes regulated by chitin. Many of these regulated genes are defense-related genes, such as those encoding pathogenesis-related proteins, TFs, and disease resistance proteins [26]. Studies found that in cultured rice cells in the lab, the recognition of chitin elicitor induces a series of defense responses including the activation of MAPKs, ROS production, defense gene expression, phytoalexin production and the accumulation of an important signal molecule in plant response to stress, phosphatidic acid (PA) [23].

The ability to uptake nutrients can affect plant and immunity in response to pathogen introduction. Some of the genes that respond to mycorrhiza colonization may be involved in the uptake of phosphate in rice [27]. *M. oryzae* possessed mechanisms by which it can suppress rice host immunity by regulating the K^+ channel 7. Potassium is important in many plat function including enzyme activation, cellular homeostasis, membrane transport, osmoregulation and immunoreaction [20]. Transporters and channels are responsible for the uptake and translocation of K. Often K fertilizers are applied in production systems and have been noted to reduce the occurrence of disease but it was unclear how. It is thought that high K concentrations in rice support immune response and inhibit the growth of the fungus [20]. Excess nitrogen applied as a fertilizer drastically increases rice susceptibility to *M. oryzae* [19].

After effector recognition and signal initiation, plant defense is activated causing responses such as cell wall reinforcement, accumulation of antimicrobial secondary metabolites, and expression of PR proteins [19, 23]. Several classes of antifungal metabolites are known and are non-essential for basic plant metabolism. Large numbers of terpenoid compounds are present in rice and serve in reducing pathogen toxicity [28]. Terpenoids have a plethora of key jobs including plant hormone functions, electron carriers, vitamins, pigments and membrane components, and importantly plant-pathogen interaction. Rice leaves produce momilactones A and B upon the introduction of *M. oryzae* infection and are widely studied for their anti-fungal activity during attack [28]. Also present in the rice leaves is oryzalexin A–D, which is classified as phytoalexin. Furthermore, Flavonoids belong to a large class of phytoanticipan and phytoalexin phenolic metabolites that are synthesized from phenylalanine in the shikimate pathway play part in plant resistance and defense. These flavonoids are known to directly inhibit the growth and germination of *M. oryzae* [28].

The cell wall in plant cells is armor against many pathogens that not only acts as a barrier, but it also produced chemicals to fight off pathogens that enter the cell. These chemicals include reactive oxygen species (ROS) and phenolics [29]. ROS production serves many functions in eukaryotic cells, including those in cellular defense. The generation of ROS is regarded as one of the first responses to fungal invasion [21]. In *M. oryzae*, intracellular ROS is critical to its pathogenicity in rice at the seedling stage. Interruption of the ROS production in the fungus causes it to lose toxicity in the host [21]. *M. oryzae* hyphae seem to initiate quick production of H_2O_2 in the host rice cells at the penetration sites. Small GTPase Rac complexes regulate accumulation of ROS generated through NADPH oxidases. Highly lethal strains of *M. oryzae* have been noted to repress the production of ROS thus suppressing immune responses. Plants that have high ROS accumulation tend to also have crosslinking of cell well proteins that develop tick cell walls [30].

Plant proteins that are explicitly stimulated during pathogen invasion are referred to as Pathogenesis-related proteins (PRs) [23]. The accumulation of these proteins plays an essential role in active plant defense response [23]. Rice has several groups of PR genes that have been found to be triggered after species-specific infections [23]. Ubiquitin-proteasome system is used by plants to regulate protein production and usage for growth as well as abiotic and biotic stress response. E3s are a class of ligases that are common among the main types of ubiquitin-proteasome enzymes that play a role in pathogen response and interaction. About 1500 E3s are encoded in the rice genome, some of which are key to immunity in rice against fungal pathogens [29]. Some E3s are thought to play a vital role in cell wall reinforcement, specifically after fungal infection. Pathogens try to interfere with the PTI pathway by producing proteins and sending them into the cell. Hyphae infected with *M. oryzae* release proteins that can direct the natural response process of the host cell defense in its favor [21]. Recently it has been found that the MoAP1 protein in the conidia of the fungus is highly expressed during the invasive stage of its lifespan [21]. Some of these proteins inhibit PTI by interfering with the E3 ligase activity, which normally acts as a positive regulator in immune response, or degrading them altogether. In immune plants, some of the defense E3 ligases in rice have the ability to degrade the proteins produces by the fungus [29].

Pathogens can create and mimic plant auxins that interrupt the pants pathways and repressed defense. Rice is often treated with a growth hormone called brassinolide (BL) to confer resistance to *M. oryzae* by reinforcing its natural hormone defenses [23]. Pathogenesis-related proteins and their corresponding receptor kinases in this case BAK1 induce the signal to initial PTI. Hormones like abscisic acid (ABA), jasmonic acid (JA) have been used as well to trigger gene defense expression against *M. oryzae* [23].

About 60 genes for rice blast resistance have been found but this number includes allelic resistance; therefore, only five genes have been extensively used by breeders and shown to be reliable over the past few decades [17]. Molecular markers have made it possible to tag resistance genes [17]. Fungal pathogens have to break through physical surface barriers that serve as the plant's line of defensive as well as the antimicrobial chemicals supplied by the host to survive [15]. Once it has penetrated the host immunity is stimulated, and defenses are up. Sensors in the host recognize the pathogenic microbes and the NBS-LRR proteins read the effectors that are projected into the cell. *M. oryzae* secretes about 740 different proteins during an invasion; therefore, in rice, its primary means of recognition are the NBS-LRR proteins [15, 23]. The signals these initial receptors trigger initiate MAPK activity and transcription factors, which then activate PR expression. Defense responses in the host attempt to reinforce the cell wall and produce secondary metabolites to compact that microbial invasion [23]. In-Depth knowledge of the mechanisms of resistance and host response to fungal pathogen invasion will help further agricultural development of blast resistant crops. Enhanced management practices for rice blast can also be achieved with an understanding of pathogen interactions and stress responses.

2.2 Rice sheath blight disease

Regarded as one of the most critical diseases in cultivated rice, rice sheath blight (*Rhizoctonia solani* Kuhn) is a widely distributed soil-borne disease prevalent in most rice-growing areas [31]. Within the U.S., rice sheath blight has increased due to increased crop rotation of rice with soybean (*Glycine max*) as aerial blight in soybean is caused by *R. solani* [32]. Globally, the widespread cultivation of semi-dwarf, high-yielding rice cultivars acclimated to high rates of nitrogen fertilizer has contributed to 50% yield reduction in susceptible cultivars [31]. The primary source of rice sheath blight inoculum is attributed to the formation of lesions near the waterline due to germinating sclerotia [33]. While effective fungicides are available to manage rice sheath blight, they are not considered a long-term solution due to health and environmental concerns [34].

Rice sheath blight (RSB) has the proven ability to survive from one crop season to another via sclerotia, plant debris, and weed hosts that may have dropped in the field during harvest [35]. Infection can occur at any growth stage from seedling to flowering due to different inoculum sources making it more prominent and common than other rice diseases [35]. Rice inoculation by the RSB pathogen results in the production of enzymes that lead to callus breaking, degradation of sheath cells and organelles, cell wall cracking, and mitochondrial damage [36]. In addition to the production of degrading enzymes, toxin production can lead to visual symptoms on rice leaves, seedling wilting, reduced radical growth [35].

To decrease chemical control usage of fungicides against *R. solani*, researchers have turned to nanoparticle treatments. In an in vivo and in vitro study to assess the impact of silver nanoparticles on *R. solani*, the research found that increased inhibition of sclerotia formation (92%) and mycelia growth (85%) was observed when particles were applied at a concentration of 50 ppm [37]. Similar results were seen under microscopic observations of hyphae exposed to silver nanoparticles resulting in decreased sclerotial germination of 12% with just 7 ppm of silver nanoparticle-containing medium [38]. In a study done to observe the impact of silver nanoparticles on detached rice leaves infected with varying inoculations of *R. solani*, lesion lengths were significantly lower when leaves were treated with a nutrient broth containing silver nanoparticles compared to those that directly applied the nanoparticles [39].

3. Consequences and mitigation strategies of abiotic stresses

3.1 Cold stress

Although rice originates from tropical and sub-tropical areas, erratic climate change has led to cold-sensitive rice cultivars. While cold damage can occur at any growth stage, chilling injury at the early seedling stage can lead to slow growth, delayed crop maturity, poor establishment, and decreased yield [40]. In a universal screening method developed by Shirasawa et al. [41], rice plants are maintained in a cold deep-water irrigated pool during the entire booting stage, and the completion, spikelet fertility is used to determine cold tolerance in the population [42]. To evaluate cold tolerance, the parameters of germination percent, germination index, root, shoot, and seedling length, and seedling vigor are usually observed [43].

When evaluating rice genotypes for cold stress tolerance at the seedling stage, Rahul et al. [44] identified three rice cultivars with a decreased reduction in seedling vigor to two cold environments, 8 and 13°C, after germinating for 28 days. Evaluation of cold temperatures at the reproductive stage has pointed to spikelet sterility (90%), reduced numbers of spikelets (41%), and a decrease in panicle number per tiller (43%) when the duration of cold is greater than a day [45].

3.2 Drought stress

Drought has been identified as a significant problem in rain-fed lowland and upland rice [46]. Drought stress severely impacts rice production and affects approximately 23 million hectares of rice annually [47]. To produce 1 kg of rice seeds, 3000–5000 L of water is required making rice one of the least water-efficient crops produced [48]. Drought stress has been tied to rice yield variability due to climate variability with approximately 32% of rice harvesting regions experiencing climate variability due to global climate change [49]. As a result of drought stress, yield reduction can be 100% dependent upon the growth stage of the crop [50].

Drought patterns can be unpredictable, and traits are complex to understand; response mechanisms make it difficult to identify components tied directly to drought stress tolerance [51]. At the onset of drought stress, rice plants can respond by slowing down or stopping their growth, which is perceived as a survival technique [52]. Physically, drought stress can result in poor root development, reduced leaf traits such as shape and epicuticular wax formation that can lead to direct affects to the leaf canopy, and decreased stem nutrient reserves [53]. Morphologically, drought stress can cause a reduction in germination, plant height, biomass, number of tillers, and leaf number and size [51].

To decrease the impact of drought stress in cultivated rice fields, researchers have attempted to introgress drought tolerant traits into high yielding cultivars using recombinant inbred lines (RILs) to detect quantitative trait locus (QTL) associated with plant protection traits [54]. Researchers used locally adapted *indica* rice lines to uncover these QTLs, but none were located in relation to yield [54]. Additional studies using RILs to detect drought tolerant traits have seen little success with one study reporting a 50% reduction in biomass, 39% reduction in grain yield, and 28% reduction in straw yield [55]. Although research continues to evolve, there are currently no commercially available, high-yielding rice cultivars available generating a direct need for improved research.

3.3 Heat stress

Although rice originated from tropical and subtropical regions, extremely high temperatures (above 38°C) can cause significant yield loss during the grain-filling stage [56]. While acclimated to higher temperatures, rice is most sensitive to heat stress during anthesis [57]. Heat stressed rice grains are subjected to decreased grain plumpness, starch content, and protein accumulation that decrease yield and grain quality [58]. In a study from 1992 to 2003 to access the impact of climate change on rice production, researchers used weather data for the International Rice Research Institute to find that rice yield declined by 10% for each 1°C increase in temperature during the dry season [59]. High temperatures during reproductive stages such as before or during anthesis have resulted in decreased grain filling represented by a negative screening index [60]. When evaluating day and night temperature fluctuations, daytime temperatures above 33°C have proven to disrupt pollen tube formation, while nighttime temperatures above 29°C have led to seed sterility and decreased grain yield [57]. High temperature stress at the flowering stage has also proven to be injurious to rice. Temperatures above 35°C have led to decreased yield (60–90%), low grain quality, increased pollen and spikelet sterility (100%), resulting in an overall low harvest index [61].

Although heat tolerance is a multivariate trait, one *aus indica* variety "Nagina22" or N22 has been characterized as a heat tolerant cultivar [62]. While this rice variety has some physically undesirable traits, it does contain some morphophysiological traits such as early maturity, increased regeneration and recovery processes, and variability in the accumulation and mobilization of carbohydrates [63]. Many studies have been conducted in the development of recombinant inbred lines for quantitative trait loci (QTL) mapping for candidate genes with one study uncovering 11 QTLs that control young seedling tolerance to heat stress [64].

3.4 Salinity stress

The rising global population has put pressure on agricultural production systems, and it is becoming more common for unsuitable land, such as saline soils, to be used for agricultural production [65]. When soil or land is said to be salt-affected, it means that the soil is characterized by a high accumulation or concentration of soluble salts, such as NaCl. For soil to be officially defined as salt-affected, the electrical conductivity of the saturated paste extract (ECe) is measured to be at least 4 dS/m. This is the lowest ECe value that has been observed to affect crop growth and yield [66]. It is expected for global warming and rising sea levels to increase the number of arable hectares degraded by salt waters, and to increase the concentration of salt in already salt-affected areas [67]. Saline soils are a problem for agricultural crops around the world. In Pakistan, around 40,000 hectares/year of cultivable land are reduced in quality due to the effects of increased saline concentration [68, 69]. Additionally, it is globally estimated that 800 million hectares are degraded by salt. Out of the 1500 hectares used for agriculture without irrigation systems in place, it is estimated that approximately 32 million hectares are salt-affected. Out of the 230 million hectares used for agriculture with irrigation systems in place, it is estimated that approximately 45 million hectares have accumulated deleterious salt concentrations [66].

Soils can acquire salt ions through natural and anthropogenic means. Oceanic winds and rains are capable of depositing salt ions onto land [66]. The use of brackish watering systems and other unsuitable irrigation systems further contribute to saline soils [65]. Land clearing also contributes to the salinization of lands because it leads to the water table rising and salt concentrating in the rhizosphere's root zone [66]. The deleterious effects of salty environments are exacerbated by dry or arid climates. Saline environments are a huge stressor for plants, affecting their overall growth and fecundity [70].

Rice (*Oryza* ssp.) is a crop grown around the world, and makes up the diet of 3.5 billion people, especially in developing and poverty-stricken regions [67, 71]. There are approximately 22 species of *Oryza*, but only two are cultivated: *Oryza glaberrima* and *Oryza sativa* [71]. Nearly 75% of cultivated rice is grown on 85–90 million irrigated hectares out of the 230 million hectares utilized for agricultural production worldwide. From 1995 to 2000, rice demand exceeded rice production, and prices skyrocketed globally [71]. As climate change progresses and seawater rises, many rice fields are increasingly threatened by salt stress [67, 69, 71].

Stress in a plant is understood to be when the plant's energy levels are decreased and when energy or resources within the plant are allocated towards defense rather than biomass accumulation, reproduction, or maintenance [65]. The most evident effects of saline stress are inhibited growth and reduced photosynthetic rates [70]. Many plants will exhibit no growth because the energy gained through photosynthesis and cellular respiration is equivalent to the energy required to tolerate the salt stress. When the energy required to tolerate the salt stress exceeds the energy gained through photosynthesis and cellular respiration, the plant may exhibit injury or death [65]. When rice (*Oryza sativa*) is grown in the presence of salt, cell wall elasticity is reduced due to changes in metabolic pathways. The altered osmotic gradients and the interruption of photosynthetic activities result in imbalanced nutrient levels. These processes ultimately reduce plant growth, chlorophyll concentration, and overall leaf area [69]. Sterile panicles and panicles with reduced seed have been observed in rice plants grown in environments with a high saline concentration [72]. Reactive oxygen species (ROS) have been observed to act as signaling agents in response to salt stress; however, ROS are also capable of damaging cellular structures, such as the lipid membrane or enzyme activity, if they are present in high enough concentrations [65]. Although all the mechanisms have not been identified or fully understood, it is clear that saline environments cause stress in rice plants that reduce overall growth and fecundity.

Plants that grow in saline environments exhibit marked differences, both physiologically and biochemically, from plants that grow in non-saline environments [72]. Salt inhibits plant growth and development through the alteration of osmotic gradients, the accumulation of ions leading to ion toxicity, and the interruption of pathways related to photosynthesis and nutrition [69]. Salt concentrations present outside of plant tissues, such as outside of roots, make it difficult for the plant to uptake water and other nutrients [66]. Salinity injures plants through inhibition of osmotic properties [70]. It was originally thought that rice crops suffered injury in saline environments as a result of increased chloride uptake due to hypothesized osmotic adjustments. In a research review by Gregoria et al. [73] studies conducted by Clarkson and Hanson indicated that the damage is a result of increased Na^+ uptake, which reduces yield because of an imbalance in Na-K within the plant's tissues. Osmotic responses to salinity concentrations outside of the plant tissues are the most rapid salt stress response and are primarily exhibited through a reduction in new shoot growth. New shoot growth is one of the first noticeable inhibitions because a reduction in leaf tissue/root ratio allows for decreased water required for the plant. This allows the plant to better control soil moisture when salt concentrations are high [66].

While salt concentrations outside of the plant cause stress and osmotic imbalances, the accumulation of salt and ions within the plant tissues also causes major biochemical and physiological problems [66]. The accumulation of Na^+ and Cl^- ions in the plant tissues, a mechanism of osmotic adjustment, can confer toxicity [68]. This saline stress response is slower than the initial osmotic effect previously discussed and eventually is illustrated in the discoloration and injury observed in older leaves [66]. It can result in other reduced growth parameters including shoot dry weight, root dry weight, and tillering capacity [68]. This phenomenon is observed in more mature plant tissues because mature tissues do not filter or dilute nutrients or ions, as new growth tissues do [66]. Furthermore, when experiencing salt stress, plants will close their stomata due to decreased water potential [61]. Closed stomata limit the amount of carbon dioxide that can be assimilated. This ultimately interrupts photosynthetic pathways, reducing energy assimilation and growth rates. Photosynthetic machinery has also been observed to be damaged as a result of extended stomata closure. This phenomenon also changes source-to-sink carbohydrate allocation in the plant [70].

When experiencing drought stress, plants maintain diverse biochemical and physiological mechanisms for coping. First, a plant must perceive a stressful environment and induce stress-related defense mechanisms to protect itself [74]. Salinity tolerance can be achieved through different mechanisms, such as the accumulation of compensatory osmolytes, which help to maintain stable osmotic environments within the cell or the overexpression of stress defense genes [70, 74]. To tolerate salty situations, many plants will make osmotic adjustments, which serve to sustain turgor pressure [65]. The two mechanisms involved in plant osmotic adjustments are ion exclusion and tissue tolerance. Ion exclusion is when the plant removes Na^+ and Cl^- ions from leaves and rely on organic solutes, such as proline and glycine betaine (GB) as osmoprotectants [65, 70, 75]. The osmoprotectants serve to regulate osmotic gradients and processes to continue water uptake in the presence of soluble salt ions [75]. Parenchyma cells located in the root xylem act as 'gatekeepers' in the exclusion of Na^+ and Cl^-. Tissue tolerance is when the plant acquires Na^+ and Cl within compartmentalized plant tissues [65]. In both mechanisms of osmotic adjustments, the plant accumulates osmolytes within cells. This lowers the osmotic potential of the cell to mimic the low water potential of the substrate. This mechanism helps to maintain stable osmotic environments within the cells, conferring tolerance of environmental stressors, such as drought or salinity, to the plant. Plant species that are able to tolerate higher salt concentrations have been observed to maintain high concentrations of Na^+ and Cl^- ions within plant tissues at all times. Plant species that are more sensitive to saline environments have been observed to maintain lower Na^+ concentrations within plant leaves and are hypothesized to rely more heavily on the ion exclusion mechanism [65].

It is unclear if osmotic adjustments are the only mechanisms a plant has to tolerate stressful environments, and the overall tolerant effect is likely a result of multiple mechanisms working synergistically within the plant [70]. For example, when plants experience stressful environmental conditions, such as high salt concentrations or drought, stress-related genes may be induced for defense purposes. Some of the proteins observed to function in this manner are chaperone proteins, ion channels, transporters, antioxidation proteins, detoxification proteins, and osmotic adjustment proteins. The gene family NAC is a highly conserved group that is only observed in plants [74], which over 100 NAC genes described in the rice (*Oryza sativa*) genome [76]. While this gene family codes for a large array of proteins with various functions, a few genes have been identified in environmental stress response pathways. In 2015, Hakim et al. [77] identified SNAC1 and determined that this gene is expressed in response to abiotic stressors, including salt stress. Further analysis revealed the role of SNAC1 in regulating stomatal guard cells on the upper and lower sides of leaves. In transgenic plants designed to overexpress the SNAC1 gene, increased closure of stomata was observed; however, photosynthetic rates were not reduced. It was further determined that when the SNAC1 gene is overexpressed, no fitness costs are incurred [74]. Additional studies have identified the expression of NAC genes, OsNAC6 and SNAC2, as enhancing abiotic stress tolerance, including cold, salt, and drought tolerance [76].

As populations and sea levels rise, salt-affected soils will continue to pose challenges in agricultural production around the world [65]. Rice is an important crop, feeding approximately 3.5 billion people around the world [71]. It is understood that high concentrations of saline reduce photosynthesis, growth, and yield in rice through the interruption of osmotic gradients [66, 70]. Some plants and varieties of rice have been observed to tolerate salt stress through two osmotic adjustments: ion exclusion and tissue tolerance. Ion exclusion involves the accumulation of organic solutes, such as proline and glycine betaine (GB), to maintain osmotic gradients and uptake water from substrates with high concentrations of soluble salts [65, 70, 75]. Ongoing research has identified several genes within the NAC gene family that regulate abiotic stress response factors. These genes are SNAC1, SNAC2, and OsNAC6 [74, 76]. Transgenic rice crops have been developed to overexpress SNAC1, SNAC2, and OsNAC6 to better understand their role in abiotic stress defense, and it was determined that the overexpression of these genes incurs no fitness cost to the transgenic plant [76].

3.5 Submergence stress

Due to the negative impact of climate change, a majority of rice-producing countries have witnessed a steady decline in the performance of culturally selected rice cultivars [78]. The largest difficulties are due to rice's inability to adapt to multiple abiotic stresses such as flooding, drought, and soil salinity [79, 80]. Flooding causes the most negative impact because of difficulties in water accumulation due to increased rainwater, increased river discharges, and unexpected tidal movements [80]. In rice, flash flooding caused by extremely heavy rainfall can lead to crops being completely submerged for 10–15 days or longer in some countries causing them a substantial reduction in yield [81].

To counteract yield reduction, researchers set out to characterize and discover rice cultivars that demonstrated tolerance to complete submergence. Research completed at the Huntra Rice Experiment Station in Thailand showed only 6% of 3156 screened rice cultivars having a level of tolerance after 10 days of complete submergence, while IRRI only found 2% of 18,115 screened rice cultivars having submergence tolerance [82]. Comparisons were made to rice cultivar FR13A that was released in 1940, and showed increased tolerance to submergence serving as the model standard for complete submergence tolerance in rice [78]. While FR13A survives complete submergence, it is still characterized as a low-yielding cultivar [83]. After additional studies, researchers were able to discover and clone the gene associated with submergence tolerance in FR13A, SUB1. This gene was able to be introgressed into high-yielding cultivars through marker-assisted backcrossing creating mega-varieties such as Swarna and IR64 [78]. These mega-varieties have proven to be very useful in affected areas, but increased tolerance is necessary to provide relief in low-lying areas [78].

In a study conducted by Iftekharuddaula et al. [84], it was observed that a mega-variety, BR11, introgressed with the SUB1 QTL produced BR11-Sub1 that was 99.8% identical to BR11 in yield, yield-component parameters, and grain properties. This is important, as BR11 is grown on more than 40% of the rice acreage in rain-fed lowland rice (RLR) of Bangladesh [85]. BR11-Sub1 demonstrated submergence tolerance after complete submergence stress of 21 days. This was comparable to the tolerant donor, IR40931-33-1-3-2, and the tolerant check, FR13A [84]. In both studies, Iftekharuddaula et al. [84] and Gonzaga et al. [78], submergence levels varied from 0.7 to 1.7 m, respectively demonstrating submergence tolerance in RLR areas that receive an average of 0.3 m of water or more [86].

4. Conclusions

As the demand for rice yield continues to increase over the next 20 years, research must continue to develop abiotic and biotic stress-tolerant rice cultivars. The impacts of climate change can be seen as research focuses on cold, drought, heat, and submergence stress. Climate change can lead to decreases in rice yields up to 100% due to fluctuating temperatures, and unexpected instances of flash flooding. There is also a need to identify a more diverse slate of traits that can arm rice in overcoming stress due to climate change. In rice, research has shifted towards studies focused on screening weedy rice (*Oryza sativa*) for abiotic stress tolerance. Recent discoveries have uncovered cold, heat, drought, and submergence tolerant weedy rice that could lead to traits that can be introgressed into the cultivated rice germplasm. This chapter focused on strategies used to discover biotic and abiotic stress tolerant rice germplasms to protect rice against climate change and disease resistance.

Author details

Shandrea Stallworth, Brooklyn Schumaker, Mary Gracen Fuller
and Te-Ming Tseng*
Mississippi State University, Mississippi State, MS, USA

*Address all correspondence to: t.tseng@msstate.edu

References

[1] Bhattacharjee P, Singhal RS, Kulkarni PR. Basmati rice: A review. International Journal of Food Science and Technology. 2002;**37**(1):1-2

[2] Londo JP, Chiang YC, Hung KH, Chiang TY, Schaal BA. Phylogeography of Asian wild rice, *Oryza rufipogon*, reveals multiple independent domestications of cultivated rice, *Oryza sativa*. Proceedings of the National Academy of Sciences. 2006;**103**(25):9578-9583

[3] Rice productivity—Ricepedia [Internet]. Available from: http://ricepedia.org/rice-as-a-crop/rice-productivity [Accessed: 8 December 2019]

[4] Mitchell PL, Hardy B. Redesigning Rice Photosynthesis to Increase Yield. Studies in Plant Science. IRRI Elsevier Science B.V.; 2000;**7**

[5] Hu Y, Cheng H, Tao S. The challenges and solutions for cadmium-contaminated rice in China: A critical review. Environment International. 2016;**92**:515-532

[6] Normile D. Reinventing rice to feed the world. Science. 2010;**321**:330-333

[7] Huang M, Jiang P, Shan S, Gao W, Ma G, Zou Y, et al. Higher yields of hybrid rice do not depend on nitrogen fertilization under moderate to high soil fertility conditions. Rice. 2017;**10**(1):43

[8] Singh R, Srivastava M, Shukla A. Environmental sustainability of bioethanol production from rice straw in India: A review. Renewable and Sustainable Energy Reviews. 2016;**54**:202-216

[9] Moldenhauer K, Counce P, Hardke J. Rice growth and development. In: Hardke J, editor. Arkansas Rice Production Handbook. Little Rock, Arkansas: University of Arkansas Division of Agriculture Cooperative Extension Service; 2001. p. 192

[10] Harlan JR, de Wet JM. Toward a rational classification of cultivated plants. Taxon. 1971;**1**:509-517

[11] Khush GS. Origin, dispersal, cultivation and variation of rice. Plant Molecular Biology. 1997;**35**(1-2):25-34

[12] Multani DS, Jena KK, Brar DS, de los Reyes BG, Angeles ER, Khush GS. Development of monosomic alien addition lines and introgression of genes from Oryza australiensis Domin. To cultivated rice *O. sativa* L. Theoretical and Applied Genetics. 1994;**88**(1):102-109

[13] Kato S. On the affinity of rice varieties as shown by fertility of hybrid plants. Scientific Bulletin of the Faculty of Agriculture Kyushu University. 1928;**3**:132-147

[14] Sanchez T, LaForest J, Harmon CL. *Magnaporthe oryzea* [Internet]. Available from: https://wiki.bugwood.org/Magnaporthe_oryzae [Accessed: 8 December 2019]

[15] Koeck M, Hardham AR, Dodds PN. The role of effectors of biotrophic and hemibiotrophic fungi in infection. Cellular Microbiology. 2011;**13**(12):18649-11857

[16] Skamnioti P, Gurr SJ. Against the grain: Safeguarding rice from rice blast disease. Trends in Biotechnology. 2009;**27**(3):141-115

[17] Khush GS, Jena K. Current status and future prospects for research on blast resistance in rice (*Oryza sativa* L.). In: Wang GL, Valent B, editors. Advances in Genetics, Genomics and Control of Rice Blast Disease. Dordrecht: Springer; 2009

[18] Raman V, Simon SA, Romag A, et al. Physiological stressors and invasive plant infections alter the small RNA transcriptome of the rice blast fungus, *Magnaporthe oryzae*. BMC Genomics. 2013;**14**:326. DOI: 10.1186/1471-2164-14-326

[19] Ribot C, Hirsch J, Balzergue S, et al. Susceptibility of rice to the blast fungus, *Magnaporthe grisea*. The Journal of Plant Physiology. 2008:114-124

[20] Shi X, Long Y, He F, Zhang C, Wang R, Zhang T, et al. The fungal pathogen *Magnaporthe oryzae* suppresses innate immunity by modulating a host potassium channel. PLoS Pathogens. 2018;**14**(1):e1006878

[21] Guo M, Chen Y, Du Y, Dong Y, Guo W, et al. Correction: The bZIP transcription factor MoAP1 mediates the oxidative stress response and is critical for pathogenicity of the Rice blast fungus *Magnaporthe oryzae*. PLoS Pathogens. 2019;**15**(11):e1008196

[22] Cooper B, Clarke JD, Budworth P, Kreps J, Hutchison D, Park S, et al. A network of rice genes associated with stress response and seed development. Proceedings of the National Academy of Sciences of the United States of America. 2003;**100**(8):4945-4950

[23] Chen X, Ronald PC. Innate immunity in rice. Trends in Plant Science. 2011;**16**(8):451-459

[24] Taj G, Giri P, Tasleem M, Kumar A. MAPK signaling cascades and transcriptional reprogramming in plant–pathogen interactions. In: Gaur R, Sharma P, editors. Approaches to Plant Stress and their Management. New Delhi: Springer; 2014

[25] Shimizu T, Nakano T, Takamizawa D, Desaki Y, Ishii-Minami N, Nishizawa Y, et al. Two LysM receptor molecules, CEBiP and OsCERK1, cooperatively regulate chitin elicitor signaling in rice. The Plant Journal. 2010;**64**:204-214

[26] Wan J, Zhang X, Stacey G. Chitin signaling and plant disease resistance. Plant Signaling & Behavior. 2008;**3**(10):831-833

[27] Güimil S, Chang HS, Zhu T, Sesma A, Osbourn A, Roux C, et al. Comparative transcriptomics of rice reveals an ancient pattern of response to microbial colonization. Proceedings of the National Academy of Sciences of the United States of America. 2005;**102**(22):8066-8070

[28] Du Fall LA, Solomon PS. Role of cereal secondary metabolites involved in mediating the outcome of plant-pathogen interactions. Metabolites. 2011;**1**:64-78

[29] Ning Y, Wang R, Shi X, Zhou X, Wang GL. A layered defense strategy mediated by rice E3 ubiquitin ligases against diverse pathogens. Molecular Plant. 2016;**9**:1096-1098

[30] Li W, Zhong S, Li G, et al. Rice RING protein OsBBI1 with E3 ligase activity confers broad-spectrum resistance against *Magnaporthe oryzae* by modifying the cell wall defense. Cell Research. 2011;**21**:835-848. DOI: 10.1038/cr.2011.4

[31] Prasad B, Eizenga GC. Rice sheath blight disease resistance identified in Oryza spp. accessions. Plant Disease. 2008;**92**(11):1503-1509

[32] Lee FN, Rush MC. Rice sheath blight: A major rice disease. Plant Disease. 1983;**67**:829-832

[33] Ou SH. Rice Diseases. Kew, England: Commonwealth Mycology Institute; 1985. pp. 256-268

[34] Radja Commare R, Nandakumar R, Kandan A, Suresh S, Bharathi M,

Raguchander T, et al. Pseudomonas fluorescens based bio-formulation for the management of sheath blight disease and leaffolder insect in rice. Crop Protection. 2002;**21**(8):671-677

[35] Kumar KV, Reddy MS, Kloepper JW, Lawrence KS, Groth DE, Miller ME. Sheath blight disease of rice (*Oryza sativa* L.)—An overview. Biosciences, Biotechnology Research Asia. 2016;**6**(2):465-480

[36] Zhang H, Chen XJ, Tong YH, Ji ZL, Xu JY. Damage of cell wall degrading enzymes produced by *Rhizoctonia solani* to rice tissue and cells. Journal of Yangzhou University(Agricultural and Life Science Edition). 2005;**26**(4):83

[37] Nejad MS, Bonjar GH, Khatami M, Amini A, Aghighi S. In vitro and in vivo antifungal properties of silver nanoparticles against *Rhizoctonia solani*, a common agent of rice sheath blight disease. IET Nanobiotechnology. 2016;**11**(3):236-240

[38] Min JS, Kim KS, Kim SW, Jung JH, Lamsal K, Kim SB, et al. Effects of colloidal silver nanoparticles on sclerotium-forming phytopathogenic fungi. The Plant Pathology Journal. 2009;**25**(4):376-380

[39] Chiranjeevi N, Kumar PA, Jayalakshmi RS, Prasad KH, Prasad TN. Bio efficacy of biogenic silver nanoparticles against rice sheath blight causing pathogen *Rhizoctonia solani* Kuhn. International Journal of Current Microbiology and Applied Sciences. 2018;**7**(7):4148-4160

[40] Zhao J, Zhang S, Yang T, Zeng Z, Huang Z, Liu Q, et al. Global transcriptional profiling of a cold-tolerant rice variety under moderate cold stress reveals different cold stress response mechanisms. Physiologia Plantarum. 2015;**154**(3):381-394

[41] Shirasawa S, Endo T, Nakagomi K, Yamaguchi M, Nishio T. Delimitation of a QTL region controlling cold tolerance at booting stage of a cultivar, 'Lijiangxintuanheigu', in rice, *Oryza sativa* L. Theoretical and Applied Genetics. 2012;**124**(5):937-946

[42] Sun J, Yang L, Wang J, Liu H, Zheng H, Xie D, et al. Identification of a cold-tolerant locus in rice (*Oryza sativa* L.) using bulked segregant analysis with a next-generation sequencing strategy. Rice. 2018;**11**(1):24

[43] Cruz RP, Milach SC. Cold tolerance at the germination stage of rice: Methods of evaluation and characterization of genotypes. Scientia Agricola. 2004;**61**(1):1-8

[44] Rahul NS, Bhadru D, Sreedhar M, Vanisri S. Screening of cold tolerant Rice genotypes for seedling traits under low temperature regimes. International Journal of Current Microbiology and Applied Sciences. 2017;**6**(12):4074-4081

[45] Jacobs BC, Pearson CJ. Growth, development and yield of rice in response to cold temperature. Journal of Agronomy and Crop Science. 1999;**182**(2):79-88

[46] Fukai S, Cooper M. Development of drought-resistant cultivars using physiomorphological traits in rice. Field Crops Research. 1995;**40**(2):67-86

[47] Serraj R, McNally KL, Slamet-Loedin I, Kohli A, Haefele SM, Atlin G, et al. Drought resistance improvement in rice: An integrated genetic and resource management strategy. Plant Production Science. 2011;**14**(1):1-4

[48] Sigh AK, Choudhury BU, Bouman BAM. Water-wise rice production. In: Bouman BAM, Hengsdijk H, Hardy B, Bindraban PS, Tuong TP, Ladha JK, editors. Proceedings

of the International Workshop on Water-wise Rice Production. Los Banos, Philippines: International Rice Research Institute; 2002. pp. 237-248

[49] Ray DK, Gerber JS, MacDonald GK, West PC. Climate variation explains a third of global crop yield variability. Nature Communications. 2015;**6**:5989

[50] Oladosu Y, Rafii MY, Samuel C, Fatai A, Magaji U, Kareem I, et al. Drought resistance in rice from conventional to molecular breeding: A review. International Journal of Molecular Sciences. 2019;**20**(14):3519

[51] Pandey V, Shukla A. Acclimation and tolerance strategies of rice under drought stress. Rice Science. 2015;**22**(4):147-161

[52] Zhu JK. Salt and drought stress signal transduction in plants. Annual Review of Plant Biology. 2002;**53**(1):247-273

[53] Blum A. Drought resistance–is it really a complex trait? Functional Plant Biology. 2011;**38**(10):753-757

[54] Gomez SM, Boopathi NM, Kumar SS, Ramasubramanian T, Chengsong Z, Jeyaprakash P, et al. Molecular mapping and location of QTLs for drought-resistance traits in indica rice (*Oryza sativa* L.) lines adapted to target environments. Acta Physiologiae Plantarum. 2010;**32**(2):355-364

[55] Prince SJ, Beena R, Gomez SM, et al. Mapping consistent rice (*Oryza sativa* L.) yield QTLs under drought stress in target rainfed environments. Rice. 2015;**8**(25):1-13

[56] Morita S, Yonemaru JI, Takanashi JI. Grain growth and endosperm cell size under high night temperatures in rice (*Oryza sativa* L.). Annals of Botany. 2005;**95**(4):695-701

[57] Prasad PV, Boote KJ, Allen LH Jr, Sheehy JE, Thomas JM. Species, ecotype and cultivar differences in spikelet fertility and harvest index of rice in response to high temperature stress. Field Crops Research. 2006;**95**(2-3):398-411

[58] Lin CJ, Li CY, Lin SK, Yang FH, Huang JJ, Liu YH, et al. Influence of high temperature during grain filling on the accumulation of storage proteins and grain quality in rice (*Oryza sativa* L.). Journal of Agricultural and Food Chemistry. 2010;**58**(19):10545-10552

[59] Peng S, Huang J, Sheehy JE, Laza RC, Visperas RM, Zhong X, et al. Rice yields decline with higher night temperature from global warming. Proceedings of the National Academy of Sciences. 2004;**101**(27):9971-9975

[60] Mackill DJ, Coffman WR, Rutger JN. Pollen shedding and combining ability for high temperature tolerance in rice. Crop Science. 1982;**22**(4):730-733

[61] Ye C, Tenorio FA, Argayoso MA, Laza MA, Koh HJ, Redoña ED, et al. Identifying and confirming quantitative trait loci associated with heat tolerance at flowering stage in different rice populations. BMC Genetics. 2015;**16**(1):41

[62] Bahuguna RN, Jha J, Pal M, Shah D, Lawas LM, Khetarpal S, et al. Physiological and biochemical characterization of NERICA-L-44: A novel source of heat tolerance at the vegetative and reproductive stages in rice. Physiologia Plantarum. 2015;**154**(4):543-559

[63] Gorantla M, Babu PR, Reddy Lachagari VB, Reddy AM, Wusirika R, Bennetzen JL, et al. Identification of stress-responsive genes in an indica rice (*Oryza sativa* L.) using ESTs generated from drought-stressed seedlings.

Journal of Experimental Botany. 2006;**58**(2):253-265

[64] Kilasi NL, Singh J, Vallejos CE, Ye C, Jagadish K, Kusolwa P, et al. Heat stress tolerance in rice (*Oryza sativa* L.): Identification of quantitative trait loci and candidate genes for seedling growth under heat stress. Frontiers in Plant Science. 2018;**9**:1578

[65] Munns R, Gilliham M. Salinity tolerance of crops–what is the cost? New Phytologist. 2015;**208**(3):668-673

[66] Munns R, Tester M. Mechanisms of salinity tolerance. Annual Review of Plant Biology. 2008;**59**:651-681

[67] Hakim MA, Juraimi AS, Hanafi MM, Selamat A, Ismail MR, Karim SR. Studies on seed germination and growth in weed species of rice field under salinity stress. Journal of Environmental Biology. 2011;**32**(5):529

[68] Yeo AR, Flowers SA, Rao G, Welfare K, Senanayake N, Flowers TJ. Silicon reduces sodium uptake in rice (*Oryza sativa* L.) in saline conditions and this is accounted for by a reduction in the transpirational bypass flow. Plant, Cell & Environment. 1999;**22**(5):559-565

[69] Ali Y, Aslam Z, Ashraf MY, Tahir GR. Effect of salinity on chlorophyll concentration, leaf area, yield and yield components of rice genotypes grown under saline environment. International Journal of Environmental Science and Technology. 2004;**1**(3):221-225

[70] Pattanagul W, Thitisaksakul M. Effect of salinity stress on growth and carbohydrate metabolism in three rice (*Oryza sativa* L.) cultivars differing in salinity tolerance. CSIR. Indian Journal of Experimental Biology. 2008;**46**(10):736-742

[71] Seck PA, Diagne A, Mohanty S, Wopereis MC. Crops that feed the world 7: Rice. Food Security. 2012;**4**(1):7-24

[72] Abdullah ZKMA, Khan MA, Flowers TJ. Causes of sterility in seed set of rice under salinity stress. Journal of Agronomy and Crop Science. 2001;**187**(1):25-32

[73] Gregoria GB, Senadhira D, Mendoza RD. Screening rice for salinity tolerance. IRRI Discussion Series No. 22. Manila, Philippines: International Rice Research Institute. 1997; (No. 2169-2019-1605)

[74] Hu H, Dai M, Yao J, Xiao B, Li X, Zhang Q, et al. Overexpressing a NAM, ATAF, and CUC (NAC) transcription factor enhances drought resistance and salt tolerance in rice. Proceedings of the National Academy of Sciences. 2006;**103**(35):12987-12992

[75] Moghaieb RE, Saneoka H, Fujita K. Effect of salinity on osmotic adjustment, glycinebetaine accumulation and the betaine aldehyde dehydrogenase gene expression in two halophytic plants, *Salicornia europaea* and *Suaeda maritima*. Plant Science. 2004;**166**(5):1345-1349

[76] Zheng X, Chen B, Lu G, Han B. Overexpression of a NAC transcription factor enhances rice drought and salt tolerance. Biochemical and Biophysical Research Communications. 2009;**379**(4):985-989

[77] Hakim MA, Juraimi AS, Hanafi MM, Rafii MY, Ismail MR, Karim SR, et al. Integration of herbicides with manual weeding for controlling the weeds in rice under saline environment. Journal of Environmental Biology. 2015;**36**(6):1311

[78] Gonzaga ZJC, Carandang J, Sanchez DL, Mackill DJ, Septiningsih EM. Mapping additional QTLs from FR13A to increase

submergence tolerance in rice beyond SUB1. Euphytica [Internet]. 2016;**209**(3):627-636. Available from: http://link.springer.com/10.1007/s10681-016-1636-z [Accessed: 15 September 2017]

[79] Mortimer M. Weedy rice: Approaches to ecological appraisal and implications for research priorities. Wild and Weedy Rice in Rice Ecosystems in Asia—A Review. Manila, Philippines: International Rice Research Institute; 2000:97-105

[80] Singh A, Septiningsih EM, Balyan HS, Singh NK, Rai V. Genetics, physiological mechanisms and breeding of flood-tolerant rice (*Oryza sativa* L.). Plant and Cell Physiology [Internet]. 2017;**58**:185-197. Available from: https://academic.oup.com/pcp/article-lookup/doi/10.1093/pcp/pcw206 [Accessed: 15 September 2017]

[81] Zeigler RS, Puckridge DW. Improving sustainable productivity in rice-based rainfed lowland systems of South and Southeast Asia. GeoJournal. 1995;**35**(3):307-324

[82] Jackson MB, Ram PC. Physiological and molecular basis of susceptibility and tolerance of rice plants to complete submergence. Annals of Botany. 2003;**91**(2):227-241

[83] Mackill DJ, Coffman WR, Garrity DP. Rainfed Lowland Rice Improvement. Manila, Philippines: International Rice Research Institute; 1996:242. ISBN: 971-22-0071-x

[84] Iftekharuddaula KM, Newaz MA, Salam MA, Ahmed HU, Mahbub MA, Septiningsih EM, et al. Rapid and high-precision marker assisted backcrossing to introgress the SUB1 QTL into BR11, the rainfed lowland rice mega variety of Bangladesh. Euphytica. 2011;**178**(1):83-97

[85] BRRI (Bangladesh Rice research Institute). BRRI Annual Internal Review 2004-2005. Gazipur, Bangladesh: Agriculture Economics Division; 2007

[86] Ismail AM, Singh US, Singh S, Dar MH, Mackill DJ. The contribution of submergence-tolerant (Sub1) rice varieties to food security in flood-prone rainfed lowland areas in Asia. Field Crops Research. 2013;**152**:83-93

Chapter 7

Consequences and Mitigation Strategies of Heat Stress for Sustainability of Soybean (*Glycine max* L. Merr.) Production under the Changing Climate

Ayman EL Sabagh, Akbar Hossain, Mohammad Sohidul Islam, Muhammad Aamir Iqbal, Shah Fahad, Disna Ratnasekera, Faraz Azeem, Allah Wasaya, Oksana Sytar, Narendra Kumar, Analía Llanes, Murat Erman, Mustafa Ceritoğlu, Huseyin Arslan, Doğan Arslan, Sajjad Hussain, Muhammad Mubeen, Muhammad Ikram, Ram Swaroop Meena, Hany Gharib, Ejaz Waraich, Wajid Nasim, Liyun Liu and Hirofumi Saneoka

Abstract

Increasing ambient temperature is a major climatic factor that negatively affects plant growth and development, and causes significant losses in soybean crop yield worldwide. Thus, high temperatures (HT) result in less seed germination, which leads to pathogenic infection, and decreases the economic yield of soybean. In addition, the efficiency of photosynthesis and transpiration of plants are affected by high temperatures, which have negative impact on the physio-biochemical process in the plant system, finally deteriorate the yield and quality of the affected crop. However, plants have several mechanisms of specific cellular detection of HT stress that help in the transduction of signals, producing the activation of transcription factors and genes to counteract the harmful effects caused by the stressful condition. Among the contributors to help the plant in re-establishing cellular homeostasis are the applications of organic stimulants (antioxidants, osmoprotectants, and hormones), which enhance the productivity and quality of soybean against HT stress. In this chapter, we summarized the physiological and biochemical mechanisms of soybean plants at various growth stages under HT. Furthermore, it also depicts the mitigation strategies to overcome the adverse effects of HT on soybean using exogenous applications of bioregulators. These studies intend to increase the understanding of exogenous biochemical compounds that could reduce the adverse effects of HT on the growth, yield, and quality of soybean.

Keywords: *Glycine max* L., osmoprotectants, crop productivity, heat stress, mitigation strategies

1. Introduction

Soybean is one of the key source of food energy for humans, it has principal economic value for the high-quality oil and protein, and it is grown about 6% arable lands across the globe [1, 2]. Being the members of the Leguminosae (Fabaceae) family, soybean seeds are predominantly rich in proteins and essential-fatty acids [3]. Presently, it is also dignified as a prospective plant for the production of biodiesel [4].

Adverse environmental conditions such as increasing ambient temperature, water deficit, salinity, among others, are expected as a part of the phenomenon called global climate change and these are the great threat in agriculture. Heat stress is a foremost unfavorable weather factor of climate change, which has a negative impact on crop production [5, 6]. An increase in air temperatures modifies photosynthetic rates by affecting photosystems of plants which decreases the growth and development of plant, resulted in the reduction of crop yield [7]. At the physio-biochemical level, HT tempts to denature protein, increases lipid fluidity in cells membrane, over production of ROS, ultimately inhibits the role of the photosynthetic apparatus [8]. Besides, a variety of mechanisms are developed in plants that allow them to survive with HT stress including fluctuations in leaf positioning, alteration of membrane lipid configuration, stimulation of antioxidant defense, buildup of osmolites, hormonal regulation, and quick ripening [9, 10].

Environmental stress negatively influences the growth, yield, and quality of plants and there have been efforts to improve genotypes for higher stress tolerance [11–13]. However, HT stress limits the growth and yield of soybean by changing the different physiological and biochemical processes of plants. Several antioxidants, such as glycinebetaine (GB) and proline (Pro), act as compatible solutes or osmoprotectants which can be used to mitigate the hostile impacts of HT stress [14–16]. Osmoprotectants can influence plant growth through various ways via the rooting-medium, foliar spray, and pre-sowing seeds treatment. It is reported that the application of Pro alleviates the unfavorable effects of environmental stresses [17]. GB applications enhance plant tolerance under stressful environments [18]. Considering the above discussion, the chapter aims to clarify the physiological and biochemical responses of soybean during various growth stages under HT stress conditions and to evaluate the exogenous application of different compounds for the mitigation of antagonistic effects of HT stress on soybean and exploiting the yield.

2. The consequences of heat stress on the productivity of soybean

High-temperature (HT) stress has been directly linked to a decrease in photosynthetic efficiency and finally decreased crop yield [19]. High temperature induces a limited supply of water and nutrition, which influences the leaf expansion, internodes elongation, motivates the flower bud abortion in plants [20]. Heat-associated damage to the reproductive part of different crops is the major reason for yield loss worldwide [21, 22]. HT stress during flowering has a destructive effect on legume seed yield, mainly due to loss of seed number. A series of biochemical mechanisms comprising the accumulation of HT shock proteins, metabolites, antioxidants, and hormones are proposed to play a key role in regulating legume seed set in response to HT stress [23]. A diverse set of antioxidant metabolites, including tocopherols, flavonoids, phenylpropanoids, and ascorbate precursors, were found to be enriched in the seed of the heat-tolerant genotype [24]. Studies in soybean plants showed that the stomatal conductance or non-stomatal factors under HT stress are associated with the low photosynthetic rate [25, 26] also concluded that HT stress increased the production of reactive oxygen species (ROS) which results in premature leaf senescence and lower leaf photosynthesis. However, the specific mechanisms causing lower photosynthesis under HT stress in soybeans are still not clearly understood.

2.1 The adverse effect of heat stress on germination and seedling establishment

Complete, rapid, and uniform germination is essential for having a good green area and crop growth rate for better radiation utilization and higher yield. The percentage of germination and other traits related to germination are severely influenced by abiotic stress [3]. Imposing long-term high-temperature stress during crop growth life cycle delays seed emergence, grain vigor, and reduces dry matter accumulation [27]. Germination and early seedling development in soybean is highly sensitive to HT stress. During the early germination process of soybean, high temperature significantly reduces the rate of imbibition, the ability of embryo tissue to expand, and mitochondrial respiration. Thus, temperature stress causes harmful effects to plant metabolism by disrupting their cellular homeostasis. Exposure of plants to high-temperature above the range of optimal levels can cause disturbance to the overall life cycle of the plant. HT stress can generate oxidative stress by accumulating the ROS [28]. Many physiological processes (such as photosynthesis, respiration) in surviving cells are sensitive to temperature stress [29]. Interestingly, exposure to low temperature during the seedling stage substantially extends the vegetative growth rate and increases the number of axillary branches, the rate of dry weight per plant and pod setting. Seed vigor is also reduced due to exposure of plants at the seedling stage. Germination is declined, as the number of days at 33/28°C (day/night temperatures) during seed development increased. Seed vigor determined by measured axis dry weight is also reduced [30]. Previous researches have demonstrated that the high-temperature stress during the seed filling period reduced the germination and vigor in soybean seed [30]. Increased temperatures have a strong negative effect on seed germination potential and result in a decrease in seed viability and poor germination [31].

2.2 The adverse effect of heat stress on growth and development of soybean

The growth performance of crops is adversely affected by high-temperature stress. Unfavorable environmental conditions (temperature and rainfall variability) during the reproductive growth stage can reduce the seed yield of soybean [32]. Disturbance induced by high-temperature stress in various crops reduces crop growth and development and severely reduces the physiological growth attributes [33, 34]. It has been reported that temperature and photoperiod predominantly affect the vegetative growth and development of soybean plants among other environmental variables. Reproductive growth periods of soybean are more sensitive to high temperatures than vegetative growth periods [35]. Environmental conditions, particularly day-time temperature have a direct effect on photosynthesis and transpiration, consequently affecting soybean yield. Therefore, plant reproductive organs are more vulnerable to changes in short episodes of high temperatures prior to and during the early flowering stage [36].

It is known that the roots of plants play an important role in the establishment of symbiotic associations with different microorganisms [37]. Genome-wide transcriptomic and proteomic studies on isolated root hairs of soybean plants (a single, epidermal cell type) as compared to stripped roots under HT stress showed global changes in their transcriptional and proteomic profiles. A diversity of proteins was determined whose expression changed after 3 h of HT stress application. Most such proteins were supposed to play a significant role in thermo-tolerance, post-transcriptional regulation and in the remodeling of chromatin [5].

The negative effects of HT stress are also observed in photosynthesis, transpiration, stomatal conductance, and yield. Thereby, a significant reduction was observed in dry matter accumulation, crop phenology (grain-filling duration), crop growth rate and relative growth rate as well as yield contributing characters (grains per plant and grain weight) under HT and water stress [38]. Similarly, stress condition causes a reduction in chlorophylls (Chl a and b) and carotenoids contents as well as the Chl 'a/b' and carotenoid/Chl 'a+b' ratios in the leaves that leads to decrease in the final yield [39].

In addition, a decrease in photosynthesis at HT stress can be mediated through anatomical and structural changes in the cell and cell organelles, particularly the chloroplast and mitochondria. For example, leaves of soybean under HT stress are characterized by a higher carbon isotope ratio and increased content of leaf reducing sugars [40]. Furthermore, temperature stress is the main reason for reactive oxygen species (ROS) production, such as hydrogen peroxide and hydroxyl radicals that cause severe damage to cellular membranes, and antioxidant activity resulting in decreased crop growth rate [41]. HT stress destroys the chlorophyll pigments and also declines the photosynthesizing efficiency that may produce ROS and ultimately negatively affects plant growth [42]. Photosynthesis metabolisms like mitochondrial membrane and catabolism of carbon present in the stroma are usually influenced by temperature stress [43]. Thus, reduction in photosynthesizing efficiency during high-temperature stress reduces crop growth which reduces crop yield [44]. Recently, several studies also found that stress conditions resulted in a decrease in relative leaf water content, membrane stability index and an increase in lipid peroxidation level and catalase and peroxidase activities [10]. Taken together, these studies demonstrate that several compounds and processes are contributing to reduce the growth and development of soybean plants under HT stress.

2.3 The adverse effect of heat stress on the yield of soybean

HT stress can significantly modify the seed development and decreases seed yield in legumes [45–47]. HT stress has a negative impact on the process of seed filling and ultimately influences the seed yield. Collectively, these adverse effects eventually decrease assimilate production and mobilization to developing seeds in various crops [48]. Exposure to HT stress during pod and seed filling stages results in a substantial decrease in the economic yield of crop plants by the reduction in seed weight. The decline in seed weight and seed number due to high temperatures has been reported in several crops including legumes [49]. High-temperature stress speeds up the rate of seed filling by reducing the duration of this stage and therefore reduces the yield potential [50–52]. The time of seed filling was reduced in pea, soybean and white lupin, resulting in smaller grains. High temperature during seed filling stimulates leaf senescence and reduces reduction in seed size is related to structural and functional reasons.

The yield and yield attributed traits have been significantly reduced by the photosynthetic capacity, which impacts seed development and reduces growth and yield traits in grain legumes [47, 52]. Accordingly, the environmental stresses [53]. Further, it is observed that water stress for a short period during the grain development stage decreases grain size and grain weight which ultimately affects the final grain yield. The seed yield reduction of soybean due to water deficit stress was recently reported [12]. Further, reduction of growth, yield, and attributing traits of various crops has been well-documented [54–56].

Flower initiation was reduced by temperature > 32°C and seed formation was delayed at 40–30°C [57]. The yield reduction of about 27% was measured when soybean plants were exposed to temperature at 35°C for 10 h during the day. Hence, it is essential to protect crop yield from higher and more frequent episodes of extremely higher temperatures both in current and future climates [58]. Physiologically, the high-temperature stress during reproductive development may have affected flower abortion, sequent sink site, and later pod abscission resulting in a decreased number of seeds per plant [59]. These results indicate that branch seed yield of determinate soybean is dependent on the vegetative growth of the branch that occurs during the flowering and pod formation stages [59]. Less information is known about the effects of temperature stress on soybean branch growth and branch seed yield or how temperature stress affects the distribution of seed yield between the main stem and branches [60].

Temperature exceeds about 35°C caused high-temperature stress. HT stress declined the plant development and grains in pods that ultimately decreased the biomass accumulation [61–64]. High-temperature stress produces less sterile pollen grains which decreased the grain formation. Temperature range about 29.4°C reduces pods quantity while when temperature range exceeds about 37.2°C strictly stops production of pods that ultimately reduces biomass production of various crops (**Figure 1**).

2.4 The adverse effect of HT stress on seed quality of soybean

High-quality seed production is a major obstacle to the expansion of soybean production to new areas of the tropic. Tropical conditions with high relative humidity and temperature are not conducive to seed growth and production of soybean. Such conditions do not support harvestable moisture levels for soybean growth with the final aim to get the high-quality seed. Modeling soybean yields based on carbon assimilation alone underestimated yield loss with high-intensity heat-wave and overestimated yield loss with low-intensity heat-wave, thus supporting the influence of direct HT stress on reproductive processes in determining yield [65]. The uniformity of seed development within the crop is a major factor that depends on

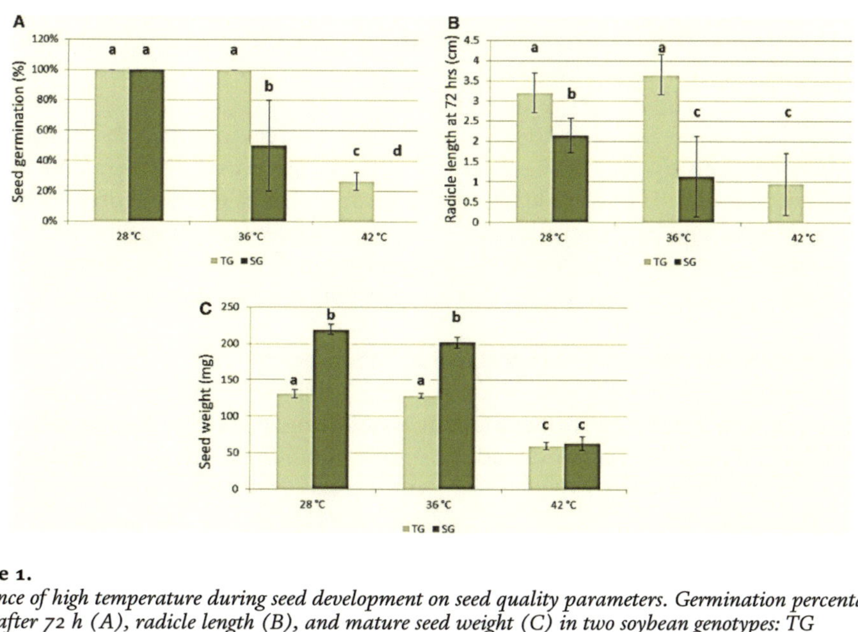

Figure 1.
Influence of high temperature during seed development on seed quality parameters. Germination percentage at 25°C after 72 h (A), radicle length (B), and mature seed weight (C) in two soybean genotypes: TG (high-temperature tolerant) and SG (high-temperature sensitive).

production practices and growing conditions. During the growth of field crops, maximum seed quality is generally regarded to be attained at physiological maturity, i.e., at the end of seed filling. Seed quality is, however, sensitive to temperature during the seed-filling period because high temperature differentially affects the various processes involved in seed filling and seed composition. In addition, high-temperature stress reduced the size of seeds and their milling quality [59]. **Figure 1** shows the effect of HT stress imposed during seed development on various seed quality characteristics like germination, radicle length, and mature seed weight.

The composition of soybean seed depends on many factors, including genotype, growing season, geographic location, and agronomic practices. The fatty acid composition of soybean oils is not constant. The fatty acid composition of soybean oils varies depending on mainly temperature and genetic factors. Environmental conditions play a decisive role in oil content and fatty acid development [66]. Temperature is the primary factor that contributes to seed filling which is the most critical growth stage in soybean. Oil content in developing seeds begins to accumulate at 15–20 days after flowering. Jung et al. [67] reported that the composition of oleic acid was positively influenced by increasing temperature, whereas the proportions of linoleic and linolenic acid were reduced. Severe water stress or high temperature resulted in higher C16:0 but lower C18:0. Genotypes differed in their responses to temperature and water stress [68].

HT and drought stress hinders the accumulation of various seed constituents, primarily starch and proteins [52, 69], through inhibiting the enzymatic processes of synthesis of starch [70] and proteins [71]. At a biochemical level, high temperature induces protein denaturation, increases membrane lipid fluidity and ROS production, and inhibits the function of photosynthetic apparatus [8, 72].

During the growth period of plants, seed formation is an important growth stage that includes the assembling of several compounds of leaves into the seed during the chemical formation of several organic compounds like starch, lipid, glucose, etc. [73, 74]. Grain formation is a very sensitive growth phase that is severely affected by high-temperature stress. Plant yield is severely declined when plants are directly exposed to high-temperature stress during seed formation stage and it ultimately reduces the seed weight and biological yield and quality of seeds. The reason for this decrease is that plants are unable to stand their growth under temperature stress circumstances; therefore, minimum photosynthetic efficiency was observed during the whole growth lifecycle. Thus, the assimilation of various seed constituents like protein, lipid, starch and carbohydrates, etc., get affected due to disturbance in enzymatic activity under high-temperature stress conditions [70, 71, 75]. Protein assimilation was decreased due to high-temperature stress in seeds [76], since there was a close relationship was noticed among leaf nitrogen concentration and seed protein contents [77]. HT stress leads to results decrease in gluten protein concentration and lactic acid concentration. Seed protein concentration is totally dependent on sedimenting amino acids while high-temperature stress decreases these sedimenting amino acids due to which seed protein contents were reduced [78].

2.5 Heat stress effects on nitrogen fixation in soybean

The understanding of environmental stress on nitrogen fixation is intensely required for growing soybean under adverse environmental conditions. High root temperatures strongly affect the bacterial infection and N_2 fixation in several legume species, including soybean. Indeed, temperature affects the root hair infection, bacteroid differentiation, nodule structure, and the functioning of the legume root nodule.

Several studies have shown that Rhizobium, a Gram negative N-fixing soil bacterium, has a positive impact on legumes.

Several environmental conditions are critical factors which can have detrimental effects on the steps involved in Rhizobium-legume symbiosis as infection process, nodules development and function, resulting in low nitrogen fixation and crop yield [79]. Under stress conditions, the aerobic bacteria have shown their ability to use nitrogen oxides as terminal electron acceptors which can help them to survive and grow during periods of anoxia. This may present a great advantage for the survival of rhizobia in soil [80]. High temperature is one of the main factors influencing symbiotic nitrogen fixation [81].

Nitrogen fixation is often especially inhibited by temperature extremes which have less effect on plant growth. High soil temperature is one of the critical factors that can prevent the development of a nitrogen-fixing association between the two symbiotic partners especially in arid and semi-arid regions [80]. High temperature can induce an inhibiting effect on bacterial adherence to root hairs, on bacteroid differentiation, on nodule structure and on legume root nodule's functioning [80]. High soil temperatures will delay nodulation or restrict it to the subsurface region. A better understanding of nodule activity physiological responses to extrinsic stress factors is very important to improve productivity by harnessing the biological nitrogen fixation process.

3. Strategies to mitigate heat stress on soybean for the sustainability of soybean production

3.1 By using "Stay-Green" genotypes or delay leaf senescence

Delayed senescence or Stay-Green (SG) genotypes constitute an important source of germplasm for the genetic improvement of plants to mitigate HT stress. These genotypes are of agronomic interest because their green leaves and photosynthesis capacity are maintained for a longer time after anthesis as compared to standard genotypes [82]. These plants are tolerant to biotic and abiotic stresses showing delayed leaf senescence under stress and improved yield production [83–85].

It has been reported that HT stress induces the leaf senescence by a decline in the Chl content of leaves, due to accelerated Chl degradation. Proteins encoded by the so-called "Stay-Green Rice" (SGR) genes may function as positive or negative regulators of Chl degradation during senescence [86]. For example, soybean plants have two SGR genes called D1 and D2, which encode GmSGR1 and GmSGR2, respectively [87]. Studies of these genes demonstrated that the leaves of d1 d2 double mutants exhibited a stronger "Stay-Green" phenotype than leaves of d1 mutants. These results indicate that the two GmSGRs have redundant functions and suggesting that SGR and SGRL could act in Chl catabolism during vegetative growth [87].

The utilization of SG trait in breeding programs results in important genetic progress for high grain yield and tolerance to HT stress. Thus, there is a need to increase the knowledge of the SG potentiality to increase grain yield under high-temperature conditions in soybean and to explore genotypes of SG ability in leaves to sustain seed filling in breeding programs.

3.2 By enhancing the production or exogenous application of antioxidants

HT stress triggers sudden and abrupt changes at the time of pollination as well as grain-filling stage which leads to early maturity along with deteriorating appropriate development of grains [88, 89]. Recently, global warming has multiplied the incidence of environmental stress leading to a serious decline in crop yield [90]. One of the ways to deal with adverse effects of HT stress may involve exploring some molecules that have the potential to protect the plants from the harmful effects of high temperature [91]. Previous studies report that exogenous proline application improves the tolerance against different types of abiotic stresses such as osmotic stress, but not in HT stress. HT stress often leads to excess accumulation of ROS such as superoxide radical (O_2^-) and hydrogen peroxide (H_2O_2), causing oxidative damage to DNA, proteins, and lipids and thus reproductive failure [23, 92]. Different types of antioxidants produced endogenously or applied have the potential to impart HT tolerance to crop plants under varying agro-climatic conditions. Antioxidants may improve HT tolerance through improvement of gaseous exchange and modulating metabolic activities of the plants along with reducing the generation of reactive oxygen species. Antioxidants enable plants to cope with oxidative burst and prevent damage to chloroplast [93].

Crop failure owing to HT stress becomes evident if the temperature gets increased even by 3–6°C during vegetative or reproductive growth stages of field crops. Liang et al. [94] reported that exogenous application of melatonin improved the activity of antioxidants especially catalase (CAT), superoxide dismutase (SOD), and peroxidase (POD) and thus enabling plants to cope with HT stress. It was inferred that glutathione reductase (GR), monodehydroascorbate reductase (MDHAR), ascorbate peroxidase (APX), and dehydroascorbate reductase (DHAR) imparted HT resistance owing to the regulation of AsA-GSH cycle. Plants can accumulate proline to cope with HT stress while antioxidants effectively enhance the production of melatonin in tea, wheat, cherry, tomato, and kiwi leaves by triggering proline biosynthesis pathway [95]. Antioxidants, such as glutathione (GSH), ascorbic acid (AsA), and proline play essential roles in protecting plants from oxidative damage by scavenging ROS and thus enhance HT tolerance of legumes. For example, the application of exogenous GSH enhanced mung bean seedling tolerance of short-term high-temperature stress (42°C) by modulating the antioxidant and glyoxalase systems [8, 23]. Under abiotic stresses, such as heat, drought, and salinity, plants often over-produce different types of compatible organic solutes, among which proline and glycine betaine are important in stress tolerance of plants by acting as osmoprotectants and ROS scavengers [17]. Thus, the exogenous application of antioxidants offers tremendous potential to enable crop plants to cope with HT stress especially at the reproductive growth stage because abrupt changes at the grain-filling stage drastically reduce grain development as well as its quality. However, further in-depth field and in-vitro investigations are direly needed to explore underlying plant mechanisms for the production of antioxidants and their ameliorative effect on plants subjected to HT stress.

3.3 Other compatible solutes as a means of heat stress defensive mechanism

HT stress causes the plant to gradually wilt at the vegetative growth stage while its incidence at reproductive stages severely hampers grain formation. One of the mechanisms to cope with HT stress is the synthesis of compatible solutes for regulating water content. Most of the solutes improve water retention by modulating cellular water potential and thus referred to as compatible solutes or osmoprotectants. The extensively studied osmolytes include betaine, trehalose, glycine, proline, and mannitol. However, proline is one the most effective compatible solute and it may be ranked at the top among osmoprotectants in plants [96, 97].

Several studies reported that proline plays a regulatory role in the activity and function of the enzymes in plant cells and in their participation in the development of metabolic responses to environmental factors [98]. Thus, proline can be a promising signaling molecule to take HT stress in the plant [88]. Similarly, these mechanisms are promoting photosynthesis, maintaining enzymatic activity, and scavenging ROS. Earlier studies noticed that the exogenous application of proline regulates the uptake of mineral nutrients in plants subjected to water deficit conditions [99] and it is one of the osmotic protection mechanisms in the plant under water [100]. However, the proposed functions of accumulated proline are osmoregulation, maintenance of membrane, and protein stability under water stress conditions [101]. Enhancement of proline concentration in whole plant organs is considered to be correlated with HT and water stress tolerance. The accumulation of proline to mitigate the negative effect on plant growth and development under HT stress was reported in chickpea [102, 103] and sorghum [104]. Much attention has been paid to define the role of proline in stress environment tolerance as a compatible osmolyte. However, little attention has been given to its role in affecting the uptake and accumulation of inorganic nutrients in plants [105].

Proline may enhance HT tolerance of chickpea through alleviating the inhibition of HT stress on key enzymes in carbon and oxidative metabolism in seedlings [106]. Therefore, it is speculated that proline and its transportation might regulate the response of legume reproduction to HT stress, which should be further testified by more direct evidence [23]. There are many defense mechanisms in plants such as osmoregulation, ion homeostasis, antioxidant and hormonal systems which induce HT stress tolerance in plants. Many plants in dry habitats are known to accumulate organic solutes such as GB [107]. GB is known to serve as compatible osmolytes, macromolecules protections, and also as scavengers of ROS under stressful environments [17]. In a stressful environment, plants store multiple groups of compatible solutes such as sugars, free amino acids like GB polyols to survive [108]. GB is a member of quaternary ammonium compounds that are pre-dominant in higher plants subjected to HT and water stress conditions. In [17, 109], the positive effects of exogenous application of GB on plant growth and final crop yield of soybean under water stress are reported. Wang et al. [110] reported that the application of GB increased the osmotic adjustment in plants for water stress tolerance by improving the anti-oxidative defense system including anti-oxidative enzymes in wheat. Although the exact mechanism is still unclear, it has been suggested that GB can mitigate HT stress via a number of different mechanisms. One of them is the protection of photosynthetic machinery [111].

Most studies with GB have focused on its physiological role and biosynthetic pathway, with little interest in its effect on the anti-oxidative defense system. GB, as one of the compatible solutes, which plays an important role in stress environment by osmotic adjustment in plants [112, 113], through protecting the proteins by maintaining the structure of enzymes such as Rubisco1996, protecting the membrane structure, protection of cytoplasm and chloroplasts [114], protection of photosynthetic mechanism [115], and by functioning as oxygen radical sweeper.

3.4 Production of stress defensive phytohormones

Plants have developed a variety of adaptations that allow them to cope with HT stress. Some of these responses include changes in leaf orientation, modification of membrane lipid composition, activation of anti-oxidative mechanisms, accumulation of osmolites, and hormonal regulation [8]. HT stress often leads to excess accumulation of ROS such as superoxide radical (O_2^-) and hydrogen peroxide (H_2O_2), causing oxidative damage to DNA, proteins, and lipids, and thus reproductive failure [23, 92].

Hormones are chemical messengers that control plants growth and development in response to adverse environmental conditions. The application of mineral fertilizers harmfully influences the environment, so eco-friendly agro-technologies are required, to improve crop production [2]. Small fluctuations of hormone contents alter the cellular dynamics and, hence, they have a central role in regulating plant growth responses to abiotic stress [116]. Moreover, hormones play vital roles in plant reproduction under both normal and HT stress conditions. In general, auxins (AX), gibberellins (GA), and cytokinins (CK) positively regulate plant reproductive tolerance to HT stress [117]. Ethylene (ET) may play a negative role in legume reproduction under HT stress. HT treatments in soybean plants increases the rate of ET production along with induction of oxidative damage, which triggers flower abscission and decreased pod set percentage [40]. A few studies have been conducted on the role of hormones in the HT tolerance of legume reproduction to date [23].

ABA and GA play several roles in the regulation of seed dormancy and germination. The metabolism and signaling by both hormones, ABA and GA, are modified during the development, dormancy and germination of seeds and the establishment of plants [118]. Recently, Shuai and his group, in 2017, demonstrated that applications of AX on soybean seeds represses the germination by increasing of ABA biosynthesis, while impairing the GA biogenesis, and finally decreasing GA1/ABA and GA4/ABA ratios. Accordingly, treatments of fluridone (ABA biosynthesis inhibitor) on seeds reversed the delayed-germination phenotype associated to AX applications, while treatments of Paclobutrazol (GA biosynthesis inhibitor) inhibited the germination of soybean seeds [119]. However, changes in hormones contents and signaling in soybean seed germination under HT stress remain unclear.

Ethylene, a gaseous phytohormone, affects seed germination, plant development and fruit production under abiotic stress [120, 121]. It is well-known that this hormone (provides tolerance to HT stress [122, 123]). It triggers the expression of certain genes essential for stress tolerance adaptation by influencing different osmolytes, which can protect the plants under stressful conditions [124–126]. Further researches are needed to identify the effects of ET and other hormones on seed germination under adverse environmental conditions. Recently, the indoleamine molecule (melatonin) has been proposed as a new plant hormone [127]. Melatonin is involved in several physiological processes in plants playing as an antioxidant molecule and triggers antioxidant responses in plants under abiotic stress [128]. Therefore, applications of this molecule in plants are being evaluated by numerous researchers. Wei et al. [129] studied the effect of melatonin on soybean growth and development. Applications of this molecule in seeds promoted the leaf size and height of soybean plants. In addition, melatonin increased the pod number and seed number, but not 100-seed weight. Under salinity and drought stress, melatonin applications showed an improvement of tolerance in soybean plants [129]. Similarly, melatonin applications could increase the soybean tolerance to HT stress; the evidence indicates that adverse environmental conditions can increase the melatonin content in plants as a protective response [127].

3.5 Biotechnological strategies to improve heat stress tolerance in soybean plants

The development of genotypes with tolerance to HT stress and agronomic practices avoiding the detrimental effects of high temperatures are required to sustain and increase the production of soybean plants. Therefore, scientists are looking for strategies to enhance soybean productivity to manage the necessity of feeding an increasing population. Among the strategies for improving soybean tolerance under challenging environments, numerous technologies are contributing to this purpose. Omics is one of the most emerging technologies which allows for studying the global metabolomic, transcriptomic and/or genomic responses of soybeans to HT stress for developing metabolomic markers, utilizing metabolic pathways, and assisting soybean breeding programs. Recently, Das et al. [130] performed a soybean metabolomic study of leaves and they determined differential abundances of various primary and secondary metabolites in response to HT stress. Metabolites for several processes, such as glycolysis, the tricarboxylic acid cycle, the pentose phosphate pathway, and amino acid metabolism, peptide metabolism, and purine and pyrimidine biosynthesis, were found to be affected by HT stress. Thus, soybean metabolomic profiling demonstrated that carbohydrate and nitrogen compounds are of prime significance under high-temperature conditions [130]. These results provide useful information for the development of tolerant soybean varieties to HT stress varieties. Similarly, seed metabolites were analyzed in several soybean genotypes with differential tolerance to high temperatures [24]. A total of 275 metabolites were identified. Antioxidant metabolites, such as tocopherols, flavonoids, phenylpropanoids, and ascorbate precursors were found to be enriched in seeds of the heat-tolerant soybean genotype. These metabolites in the tolerant genotype could be responsible, at least in part, for the greater tolerance to high temperatures during seed development. Moreover, studies of transcriptomic in soybean plants grown at high-temperature conditions were performed. For example, Xu and his group, in 2019, used a high-throughput RNA-Seq profiling technique to study the molecular mechanisms in the reproductive stage soybean in response to heat. They demonstrated that a total of 633 annotated genes were differentially expressed in heat-stressed soybeans, in which 417 genes were up-regulated and 216 were down-regulated. These genes encode for compounds related to flowering, oxidative stress, protein and mRNA folding and degradation, protective molecule synthesis, and hormonal biosynthesis and signaling [131]. Besides these, the transcriptomic analysis was performed on soybean seeds in response to abiotic stresses. Gene expression analysis revealed 49, 148, and 1576 differentially expressed genes in the soybean seed coat in response to drought, elevated ozone, and high temperatures, respectively [132]. The expressed genes in the seeds under high temperate were involved in DNA replication and several metabolic processes, suggesting that the timing of events that are important for cell division and development of seed were altered in a stressful growth environment.

Taken together, these studies show that soybeans plants employ diverse pathways and complex mechanisms to cope with high-temperature conditions. However, some of the identified genes and pathways could be used to improve HT tolerance in soybeans via either molecular breeding methods or genetic engineering.

4. Concluding remarks

In conclusion, this review clarified the numerous physiological and biochemical responses of different growth stages of soybean plants under HT stress. Therefore, HT stress has an adverse effect on growth, physiology, yield, and quality of soybean. However, applications of several compounds have a direct role in supporting enzymes, proteins, aminoacids, and lipids involved in protecting systems that participate in reducing HT stress in plants. Application of antioxidants, osmoprotectants, and phytohormones may improve the HT tolerance in soybean plants through different mechanisms. Accordingly, the antioxidant protection activity of several compounds, such as antioxidants, compatible solutes, and hormones against HT stress is powerful and can solve the seasonal HT stress problem to a greater extent and also provide the technical knowledge for sustainable development in agriculture. Emerging "omics" intervention, including genomics, epigenomics, transcriptomics, proteomics, and metabolomics could greatly improve our current understanding of the intricate gene networks and signaling cascades involved in the role of these compounds applied to minimize the harmful effects of HT stress on soybean plants.

Conflicts of interest

The authors declare no conflicts of interest.

Disclosure statement

Authors declare that no conflict of interest could arise.

Author details

Ayman EL Sabagh[1,13]*, Akbar Hossain[2]*, Mohammad Sohidul Islam[3], Muhammad Aamir Iqbal[4], Shah Fahad[5], Disna Ratnasekera[6], Faraz Azeem[7], Allah Wasaya[8], Oksana Sytar[9,10], Narendra Kumar[11], Analía Llanes[12], Murat Erman[13], Mustafa Ceritoğlu[13], Huseyin Arslan[13], Doğan Arslan[13], Sajjad Hussain[14], Muhammad Mubeen[14], Muhammad Ikram[15], Ram Swaroop Meena[16], Hany Gharib[1], Ejaz Waraich[17], Wajid Nasim[15], Liyun Liu[18] and Hirofumi Saneoka[18]

1 Department of Agronomy, Faculty of Agriculture, Kafrelsheikh University, Kafrelsheikh, Egypt

2 Bangladesh Wheat and Maize Research Institute, Dinajpur, Bangladesh

3 Department of Agronomy, Hajee Mohammad Danesh Science and Technology University, Dinajpur, Bangladesh

4 Department of Agronomy, University of Poonch Rawalakot Azadkashmir, Pakistan

5 Agriculture Department, The University of Swabi, Khyber Paktunkhwa, Pakistan

6 Department of Agricultural Biology, Faculty of Agriculture, University of Ruhuna, Matara, Sri Lanka

7 Department of Botany, Ch. Charan Singh University Campus, Meerut, Uttar Pradesh, India

8 College of Agriculture, BZU, Bahadur Sub-Campus Layyah, Pakistan

9 Department of Plant Biology, Educational and Scientific Center "Institute of Biology and Medicine", Kiev National University of Taras Shevchenko, Kyiv, Ukraine

10 Department of Plant Physiology, Slovak University of Agriculture in Nitra, Nitra, Slovakia

11 Division of Crop Production, ICAR-Indian Institute of Pulses Research, Kanpur, Uttar Pradesh, India

12 Plant Physiology Laboratory, Universidad Nacional de Río Cuarto (UNRC)-Instituto Nacional de Investigaciones Agrobiotecnológicas (INIAB), Consejo Nacional de Investigaciones Científicas y Técnicas (CONICET), Río Cuarto, Argentina

13 Department of Field Crops, Faculty of Agriculture, Siirt University, Turkey

14 Department of Environmental Sciences, COMSATS University Islamabad, Vehari, Pakistan

15 Department of Agronomy, Faculty of Agricultural Sciences and Technology, Bahauddin Zakariya University (BZU) Multan, Pakistan

16 Department of Agronomy, Institute of Agricultural Sciences, Banaras Hindu University, Varanasi, UP, India

17 Department of Agronomy, Faculty of Agriculture, University of Poonch Rawalakot (AJK), Pakistan

18 Graduate School of Integrated Sciences for Life, Hiroshima University, Higashi-Hiroshima, Japan

*Address all correspondence to: aymanelsabagh@gmail.com and akbarhossainwrc@gmail.com

References

[1] Hartman GL, West ED, Herman TK. Crops that feed the World 2. Soybean-worldwide production, use, and constraints caused by pathogens and pests. Food Security. 2011;**3**:5-17

[2] EL Sabagh A, Hossain A, Islam MS, Barutçular C, Ratnasekera D, Kumar N, et al. Sustainable soybean production and abiotic stress management in saline environments: A critical review. Australian Journal of Crop Science. 2019;**13**:228-236

[3] EL Sabagh A, Omar AE, Saneoka H, Barutçular C. Physiological performance of soybean germination and seedling growth under salinity stress Soyada. Dicle University Journal of Institute of Natural and Applied Science. 2015;**4**: 6-15

[4] Meena RS, Gogaoi N, Kumar S. Alarming issues on agricultural crop production and environmental stresses. Journal of Cleaner Production. 2017;**142**: 3357-3359

[5] Valdés-López O, Batek J, Gomez-Hernandez N, Nguyen CT, Isidra-Arellano MC, Zhang N, et al. Soybean roots grown under heat stress show global changes in their transcriptional and proteomic profiles. Frontiers in Plant Science. 2016;7:517. DOI: 10.3389/fpls.2016.00517

[6] Raza A, Razzaq A, Mehmood SS, Zou X, Zhang X, Lv Y, et al. Impact of climate change on crops adaptation and strategies to tackle its outcome: A review. Plants. 2019;**8**:34. DOI: 10.3390/plants8020034

[7] Mathur S, Jajoo A. Effects of heat stress on growth and crop yield of wheat (*Triticum astivum*). In: Physiological Mechanisms and Adaptation Strategies in Plants Under Changing Environment. New York, NY: Springer; 2014. pp. 163-191

[8] Hasanuzzaman M, Nahar K, Fujita M. Extreme temperature responses, oxidative stress and antioxidant defense in plants. Abiotic Stress-Plant Responses and Applications in Agriculture. 2013;**13**:169-205

[9] Fahad S, Ullah A, Ali U, Ali E, Saud S, Rehman K, et al. 7 drought tolerance in plants role of phytohormones and scavenging system of ROS. In: Hasanuzzaman M, Fujita M, Oku H, Islam MT, editors. Plant Tolerance to Environmental Stress: Role of Phytoprotectants. Boca Raton: CRC Press; 2019. p. 10

[10] Saleem MH, Fahad S, Khan SU, Din M, Ullah A, EL Sabagh A, et al. Copper-induced oxidative stress, initiation of antioxidants and phytoremediation potential of flax (*Linum usitatissimum* L.) seedlings grown under the mixing of two different soils of China. Environmental Science and Pollution Research. 2019;**1**:5211-5221. DOI: 10.1007/s11356-019-07264-7

[11] Islam MS, Akhter MM, EL Sabagh A, Liu LY, Nguyen NT, Ueda A, et al. Comparative studies on growth and physiological responses to saline and alkaline stresses of foxtail millet (*Setaria italic* L.) and Proso millet (*Panicum miliaceum* L.). Australian Journal of Crop Science. 2011;**5**:1269

[12] EL Sabagh A, Omara A, Saneokab H, Islamc MS. Roles of compost fertilizer on nitrogen fixation in soybean (*Glycine max* L.) under water deficit conditions. Agricultural Advances. 2016;**5**:340-344. DOI: 10.14196/aa.v5i7.2326

[13] Kovar M, Brestic M, Sytar O, Barek V, Hauptvogel P, Zivcak M. Evaluation of hyperspectral reflectance parameters to assess the leaf water content in soybean. Water. 2019;**11**:443

[14] Hadiarto T, Tran LS. Progress studies of drought-responsive genes in

rice. Plant Cell Reports. 2011;**30**: 297-310. DOI: 10.1007/s00299-010-0956-z

[15] Thapa GD, Dey M, Sahoo L, Panda SK. An insight into the drought stress induced alterations in plants. Biologia Plantarum. 2011;**55**:603-613

[16] Dadhich RK, Reager ML, Kansoti BC, Meena RS. Efficacy of growth substances on mustard (*Brassica juncea* L.) under hyper arid environmental condition of Rajasthan. The Ecoscan. 2014;**8**:269-272

[17] Ashraf MF, Foolad MR. Roles of glycine betaine and proline in improving plant abiotic stress resistance. Environmental and Experimental Botany. 2007;**59**:206-216

[18] Meena H, Meena RS, Rajput BS, Kumar S. Response of bio-regulators to morphology and yield of clusterbean [*Cyamopsis tetragonoloba* (L.) Taub.] under different sowing environments. Journal of Applied and Natural Science. 2016;**8**:715-718

[19] Mathur S, Allakhverdiev SI, Jajoo A. Analysis of high temperature stress on the dynamics of antenna size and reducing side heterogeneity of Photosystem II in wheat leaves (*Triticum aestivum*). Biochimica et Biophysica Acta (BBA)-Bioenergetics. 2011;**1807**:22-29

[20] Young LW, Wilen RW, Bonham-Smith PC. High temperature stress of *Brassica napus* during flowering reduces micro-and megagametophyte fertility, induces fruit abortion, and disrupts seed production. Journal of Experimental Botany. 2004;**55**:485-495

[21] Suzuki N, Miller G, Sejima H, Harper J, Mittler R. Enhanced seed production under prolonged heat stress conditions in *Arabidopsis thaliana* plants deficient in cytosolic ascorbate peroxidase 2. Journal of Experimental Botany. 2013;**64**:253-263

[22] Fahad S, Hussain S, Saud S, Khan F, Hassan S, Nasim W, et al. Exogenously applied plant growth regulators affect heat-stressed rice pollens. Journal of Agronomy and Crop Science. 2016;**202**: 139-150

[23] Liu Y, Li J, Zhu Y, Jones A, Rose RJ, Song Y. Heat stress in legume seed setting: Effects, causes, and future prospects. Frontiers in Plant Science. 2019;**10**:938. DOI: 10.3389/fpls.2019.00938

[24] Chebrolu KK, Fritschi FB, Ye S, Krishnan HB, Smith JR, Gillman JD. Impact of heat stress during seed development on soybean seed metabolome. Metabolomics. 2016;**12**:28. DOI: 10.1007/s11306-015-0941-1

[25] Djanaguiraman M, Boyle DL, Welti R, Jagadish SV, Prasad PV. Decreased photosynthetic rate under high temperature in wheat is due to lipid desaturation, oxidation, acylation, and damage of organelles. BMC Plant Biology. 2018;**18**:55. DOI: 10.1186/s12870-018-1263-z

[26] Djanaguiraman M, Prasad PV. Ethylene production under high temperature stress causes premature leaf senescence in soybean. Functional Plant Biology. 2010;**37**:1071-1084

[27] Wahid A, Gelani S, Ashraf M, Foolad MR. Heat tolerance in plants: An overview. Environmental and Experimental Botany. 2007;**61**:199-223

[28] Essemine J, Ammar S, Bouzid S. Impact of heat stress on germination and growth in higher plants: Physiological, biochemical and molecular repercussions and mechanisms of defence. Journal of Biological Sciences. 2010:565-572

[29] Suzuki N, Mittler R. Reactive oxygen species and temperature stresses: A delicate balance between signaling and destruction. Physiologia

Plantarum. 2006;**126**:45-51. DOI: 10.1111/j.0031-9317.2005.00582.x

[30] Egli DB, TeKrony DM, Heitholt JJ, Rupe J. Air temperature during seed filling and soybean seed germination and vigor. Crop Science. 2005;**45**: 1329-1335

[31] Fahad S, Bajwa AA, Nazir U, Anjum SA, Farooq A, Zohaib A, et al. Crop production under drought and heat stress: Plant responses and management options. Frontiers in Plant Science. 2017;**8**:1147

[32] Thanacharoenchanaphas K, Rugchati O. Simulation of climate variability for assessing impacts on yield and genetic change of Thai soybean. Genetics. 2011;**21**:4-5

[33] Naz N, Durrani F, Shah Z, Khan NA, Ullah I. Influence of heat stress on growth and physiological activities of potato (*Solanum tuberosum* L.). Phyton: International Journal of Experimental Botany. 2018;**87**:225-230

[34] Setiyono TD, Weiss A, Specht J, Bastidas AM, Cassman KG, Dobermann A. Understanding and modeling the effect of temperature and daylength on soybean phenology under high-yield conditions. Field Crops Research. 2007;**100**:257-271. DOI: 10.1016/j.fcr.2006. 07.011

[35] Reddy KR, Kakani VG. Screening capsicum species of different origins for high temperance tolerance by in vitro pollen germination and pollen tube length. Scientia Horticulturae. 2007;**112**: 130-135

[36] Puteh AB, ThuZar M, Mondal MM, Abdullah AP, Halim MR. Soybean [*Glycine max* (L.) Merrill] seed yield response to high temperature stress during reproductive growth stages. Australian Journal of Crop Science. 2013;7:1472-1479

[37] Braga RM, Dourado MN, Araújo WL. Microbial interactions: ecology in a molecular perspective. Brazilian Journal of Microbiology. 2016; **47**:86-98

[38] Ghassemi-Golezani K, Ghanehpoor S, Dabbagh M-NA. Effects of water limitation on growth and grain filling of faba bean cultivars. Journal of Food, Agriculture and Environment. 2009;7:442-447

[39] El-Tayeb MA. Differential response of two Vicia faba cultivars to drought: Growth, pigments, lipid peroxidation, organic solutes, catalase and peroxidase activity. Acta Agronomica Hungarica. 2006;**54**:25-37

[40] Djanaguiraman M, Prasad PV, Boyle DL, Schapaugh WT. High-temperature stress and soybean leaves: Leaf anatomy and photosynthesis. Crop Science. 2011;**51**:2125-2131

[41] Yin H, Chen Q, Yi M. Effects of short-term heat stress on oxidative damage and responses of antioxidant system in *Lilium longiflorum*. Plant Growth Regulation. 2008;**4**:45-54

[42] Camejo D, Jiménez A, Alarcón JJ, Torres W, Gómez JM, Sevilla F. Changes in photosynthetic parameters and antioxidant activities following heat-shock treatment in tomato plants. Functional Plant Biology. 2006;**33**: 177-187

[43] Wise RR, Olson AJ, Schrader SM, Sharkey TD. Electron transport is the functional limitation of photosynthesis in field-grown Pima cotton plants at high temperature. Plant, Cell & Environment. 2004;**27**:717-724

[44] Xu S, Li J, Zhang X, Wei H, Cui L. Effects of heat acclimation pretreatment on changes of membrane lipid peroxidation, antioxidant metabolites, and ultrastructure of chloroplasts in two cool-season turfgrass species under heat

stress. Environmental and Experimental Botany. 2006;**56**:274-285

[45] Bhandari K, Siddique KH, Turner NC, Kaur J, Singh S, Agrawal SK, et al. Heat stress at reproductive stage disrupts leaf carbohydrate metabolism, impairs reproductive function, and severely reduces seed yield in lentil. Journal of Crop Improvement. 2016; **30**(2):118-151

[46] Sharma L, Priya M, Bindumadhava H, Nair RM, Nayyar H. Influence of high temperature stress on growth, phenology and yield performance of mungbean [*Vigna radiata* (L.) Wilczek] under managed growth conditions. Scientia Horticulturae. 2016;**213**:379-391. DOI: 10.1016/j.scienta.2016.10.033

[47] Sita K, Sehgal A, Kumar J, Kumar S, Singh S, Siddique KH, et al. Identification of high-temperature tolerant lentil (*Lens culinaris* Medik.) genotypes through leaf and pollen traits. Frontiers in Plant Science. 2017;**8**:744. DOI: 10.3389/fpls.2017.00744

[48] Zare M, Nejad MG, Bazrafshan F. Influence of drought stress on some traits in five mung bean (*Vigna radiata* (L.) R. Wilczek) genotypes. International Journal of Agronomy and Plant Production. 2012;**3**:234-240

[49] Devasirvatham V, Tan DK, Trethowan RM, Gaur PM, Mallikarjuna N. Impact of high temperature on the reproductive stage of chickpea. In: Food Security from Sustainable Agriculture Proceedings of the 15th Australian Society of Agronomy Conference; 2010. pp. 15-18

[50] Prasad PV, Pisipati SR, Momčilović I, Ristic Z. Independent and combined effects of high temperature and drought stress during grain filling on plant yield and chloroplast EF-Tu expression in spring wheat. Journal of Agronomy and Crop Science. 2011;**197**: 430-441. DOI: 10.1111/j.1439-037X.2011.00477.x

[51] Kaushal N, Bhandari K, Siddique KH, Nayyar H. Food crops face rising temperatures: An overview of responses, adaptive mechanisms, and approaches to improve heat tolerance. Cogent Food & Agriculture. 2016;**2**: 1134380. DOI: 10.1080/23311932.2015.1134380

[52] Farooq M, Nadeem F, Gogoi N, Ullah A, Alghamdi SS, Nayyar H, et al. Heat stress in grain legumes during reproductive and grain-filling phases. Crop & Pasture Science. 2017;**68**: 985-1005

[53] EL Sabagh A, Islam MS, Ueda A, Saneoka H, Barutçular C. Increasing reproductive stage tolerance to salinity stress in soybean. The International Journal of Agriculture and Crop Sciences. 2015;**8**:738-745

[54] EL Sabagh A, Sorour S, Ueda A, Saneoka H. Evaluation of salinity stress effects on seed yield and quality of three soybean cultivars. Azarian Journal of Agriculture. 2015;**2**:138-141

[55] EL Sabagh A, Sorour S, Ragab A, Saneoka H, Islam MS. The effect of exogenous application of proline and glycine betaineon the nodule activity of soybean under saline condition. Journal of Agriculture Biotechnology. 2017;**2**: 01-05

[56] EL Sabagh A, Abdelaal KA, Barutcular C. Impact of antioxidants supplementation on growth, yield and quality traits of canola (*Brassica napus* L.) under irrigation intervals in North Nile Delta of Egypt. Journal of Experimental Biology and Agricultural Sciences. 2017;**5**:163-172

[57] Thomas JM, Boote KJ, Allen LH, Gallo-Meagher M, Davis JM. Elevated temperature and carbon dioxide effects on soybean seed composition and

transcript abundance. Crop Science. 2003;**43**:1548-1557. DOI: 10.2135/cropsci2003.1548

[58] Salem MA, Kakani VG, Koti S, Reddy KR. Pollen-based screening of soybean genotypes for high temperatures. Crop Science. 2007;**47**: 219-231. DOI: 10.2135/cropsci2006.07.0443

[59] Thuzar M. The effects of temperature stress on the quality and yield of soya bean [(*Glycine max* L.) Merrill.]. The Journal of Agricultural Science. 2010;**2**:172-179

[60] Frederick JR, Camp CR, Bauer PJ. Drought-stress effects on branch and mainstem seed yield and yield components of determinate soybean. Crop Science. 2001;**41**:759-763

[61] Tubiello FN, Soussana JF, Howden SM. Crop and pasture response to climate change. Proceedings of the National Academy of Sciences. 2007; **104**:19686-19690

[62] Canci H, Toker C. Evaluation of yield criteria for drought and heat resistance in chickpea (*Cicer arietinum* L.). Journal of Agronomy and Crop Science. 2009;**195**:47-54

[63] Canci H, Toker C. Evaluation of annual wild *Cicer* species for drought and heat resistance under field conditions. Genetic Resources and Crop Evolution. 2009;**56**:1. DOI: 10.1007/s10722-008-9335-9

[64] Wheeler T, Von Braun J. Climate change impacts on global food security. Science. 2013;**341**:508-513

[65] Thomey ML, Slattery RA, Köhler IH, Bernacchi CJ, Ort DR. Yield response of field-grown soybean exposed to heat waves under current and elevated [CO_2]. Global Change Biology. 2019;**25**:4352-4368

[66] Bellaloui N, Bruns HA, Abbas HK, Mengistu A, Fisher DK, Reddy KN. Agricultural practices altered soybean seed protein, oil, fatty acids, sugars, and minerals in the Midsouth USA. Frontiers in Plant Science. 2015;**6**:31. DOI: 10.3389/fpls.2015.00031

[67] Jung G, Lee J, Kim Y, Kim D, Hwang T, Lee K, et al. Effect of planting date, temperature on plant growth, isoflavone content, and fatty acid composition of soybean. Korean Journal of Crop Science/Hanguk Jakmul Hakhoe Chi. 2012;**57**:373-383

[68] Gulluoglu L, Bakal H, EL Sabagh A, Arioglu H. Soybean managing for maximize production: Plant population density effects on seed yield and some agronomical traits in main cropped soybean production. Journal of Experimental Biology and Agricultural Sciences. 2017;**5**:31-37

[69] Farooq M, Gogoi N, Barthakur S, Baroowa B, Bharadwaj N, Alghamdi SS, et al. Drought stress in grain legumes during reproduction and grain filling. Journal of Agronomy and Crop Science. 2017;**203**:81-102. DOI: 10.1093/jxb/err139

[70] Ahmadi A, Baker DA. The effect of water stress on grain filling processes in wheat. The Journal of Agricultural Science. 2001;**136**:257-269

[71] Triboï E, Martre P, Triboï-Blondel AM. Environmentally-induced changes in protein composition in developing grains of wheat are related to changes in total protein content. Journal of Experimental Botany. 2003;**54**: 1731-1742

[72] Qu AL, Ding YF, Jiang Q, Zhu C. Molecular mechanisms of the plant heat stress response. Biochemical and Biophysical Research Communications. 2013;**432**:203-207. DOI: 10.1016/j.Bbrc.2013.01.104

[73] Barnabás B, Jäger K, Fehér A. The effect of drought and heat stress on reproductive processes in cereals. Plant, Cell & Environment. 2008;**31**:11-38

[74] Awasthi R, Kaushal N, Vadez V, Turner NC, Berger J, Siddique KH, et al. Individual and combined effects of transient drought and heat stress on carbon assimilation and seed filling in chickpea. Functional Plant Biology. 2014;**41**:1148-1167

[75] Behboudian MH, Ma Q, Turner NC, Palta JA. Reactions of chickpea to water stress: Yield and seed composition. Journal of the Science of Food and Agriculture. 2001;**81**:1288-2891

[76] Lizana XC, Calderini DF. Yield and grain quality of wheat in response to increased temperatures at key periods for grain number and grain weight determination: Considerations for the climatic change scenarios of Chile. The Journal of Agricultural Science. 2013; **151**:209-221

[77] Iqbal M, Raja NI, Yasmeen F, Hussain M, Ejaz M, Shah MA. Impacts of heat stress on wheat: A critical review. Advances in Crop Science and Technology. 2017;**5**:251-259

[78] Dias AS, Bagulho AS, Lidon FC. Ultrastructure and biochemical traits of bread and durum wheat grains under heat stress. Brazilian Journal of Plant Physiology. 2008;**20**:323-333

[79] Lebrazi S, Benbrahim KF. Environmental stress conditions affecting the N_2 fixing rhizobium-legume symbiosis and adaptation mechanisms. African Journal of Microbiological Research. 2014;**8**: 4053-4061

[80] Abd-Alla MH, Issa AA, Ohyama T. Impact of harsh environmental conditions on nodule formation and dinitrogen fixation of legumes. Advances in Biology and Ecology of Nitrogen Fixation. 2014;**29**:9

[81] Keerio MI. Nitrogenase activity of soybean root nodules inhibited after heat stress. OnLine Journal of Biological Sciences. 2001;**1**:297-300

[82] Thomas H, Ougham H. The stay-green trait. Journal of Experimental Botany. 2014;**65**:3889-3900

[83] Peleg Z, Blumwald E. Hormone balance and abiotic stress tolerance in crop plants. Current Opinion in Plant Biology. 2011;**14**:290-295

[84] Reguera M, Peleg Z, Abdel-Tawab YM, Tumimbang EB, Delatorre CA, Blumwald E. Stress-induced cytokinin synthesis increases drought tolerance through the coordinated regulation of carbon and nitrogen assimilation in rice. Plant Physiology. 2013;**163**:1609-1622

[85] Luche HD, Silva JA, Maia LC, Oliveira AC. Stay-green: A potentiality in plant breeding. Ciência Rural. 2015; **45**:1755-1760

[86] Sakuraba Y, Piao W, Lim JH, Han SH, Kim YS, An G, et al. Rice ONAC106 inhibits leaf senescence and increases salt tolerance and tiller angle. Plant & Cell Physiology. 2015;**56**(12): 2325-2339

[87] Nakano M, Yamada T, Masuda Y, Sato Y, Kobayashi H, Ueda H, et al. A green-cotyledon/stay-green mutant exemplifies the ancient whole-genome duplications in soybean. Plant & Cell Physiology. 2014;**55**:1763-1771

[88] Iqbal N, Fatma M, Khan NA, Umar S. Regulatory role of proline in heat stress tolerance: Modulation by salicylic acid. In: Plant Signaling Molecules. Kidlington, United Kingdom: Woodhead Publishing; 2019. pp. 437-448. DOI: 10.1016/B978-0-12-816451-8.00027-7

[89] Ahmed JU, Hassan MA. Evaluation of seedling proline content of wheat genotypes in relation to heat tolerance. Bangladesh Journal of Botany. 2011;**40**: 17-22

[90] Kumar S, Meena RS, Lal R, Yadav GS, Mitran T, Meena BL, et al. Role of legumes in soil carbon sequestration. In: Legumes for Soil Health and Sustainable Management. Singapore: Springer; 2018. pp. 109-138

[91] Sharma L, Priya M, Kaushal N, Bhandhari K, Chaudhary S, Dhankher OP, et al. Plant growth-regulating molecules as thermoprotectants: Functional relevance and prospects for improving heat tolerance in food crops. Journal of Experimental Botany. 2020;**71**:569-594

[92] Oktyabrsky ON, Smirnova GV. Redox regulation of cellular functions. Biochemistry (Moscow). 2007;**72**: 132-145

[93] Bonnefont-Rousselot D, Collin F, Jore D, Gardès-Albert M. Reaction mechanism of melatonin oxidation by reactive oxygen species in vitro. Journal of Pineal Research. 2011;**50**:328-335

[94] Liang D, Gao F, Ni Z, Lin L, Deng Q, Tang Y, et al. Melatonin improves heat tolerance in kiwifruit seedlings through promoting antioxidant enzymatic activity and glutathione S-transferase transcription. Molecules. 2018;**23**:584. DOI: 10.3390/molecules23030584

[95] Sarropoulou V, Dimassi-Theriou K, Therios I, Koukourikou-Petridou M. Melatonin enhances root regeneration, photosynthetic pigments, biomass, total carbohydrates and poline content in the cherry rootstock PHL-C (*Prunus avium* × *Prunus cerasus*). Plant Physiology and Biochemistry. 2012;**61**: 162-168

[96] Lehmann S, Funck D, Szabados L, Rentsch D. Proline metabolism and transport in plant development. Amino Acids. 2010;**39**:949-962

[97] Siddique A, Kandpal G, Kumar P. Proline accumulation and its defensive role under diverse stress condition in plants: An overview. Journal of Pure and Applied Microbiology. 2018;**12**: 1655-1659

[98] Öztürk L, Demir Y. In vivo and in vitro protective role of proline. Plant Growth Regulation. 2002;**38**:2592-2564. DOI: 10.1023/A:1021579713832

[99] Jaleel CA, Gopi R, Manivannan P, Panneerselvam R. Responses of antioxidant defense system of *Catharanthus roseus* (L.) G. Don. to paclobutrazol treatment under salinity. Acta Physiologiae Plantarum. 2007;**29**: 205-209

[100] Yamada M, Morishita H, Urano K, Shiozaki N, Yamaguchi-Shinozaki K, Shinozaki K, et al. Effects of free proline accumulation in petunias under drought stress. Journal of Experimental Botany. 2005;**56**:1975-1981. DOI: 10.1093/jxb/eri195

[101] Taiz L, Zeiger E. Plant Physiology. 4th ed. Sunderland, MA, USA: Sinauer Associates, Inc.; 2006

[102] Chakraborty U, Tongden C. Evaluation of heat acclimation and salicylic acid treatments as potent inducers of thermotolerance in *Cicer arietinum* L. Current Science. 2005;**89**: 384-389

[103] Lv WT, Lin B, Zhang M, Hua XJ. Proline accumulation is inhibitory to Arabidopsis seedlings during heat stress. Plant Physiology. 2011;**156**:1921-1933

[104] Gosavi GU, Jadhav AS, Kale AA, Gadakh SR, Pawar BD, Chimote VP. Effect of heat stress on proline, chlorophyll content, heat shock proteins and antioxidant enzyme activity in sorghum (*Sorghum bicolor*) at seedlings

stage. Indian Journal of Biotechnology. 2014;**13**:356-363

[105] Khedr AH, Abbas MA, Wahid AA, Quick WP, Abogadallah GM. Proline induces the expression of salt-stress-responsive proteins and may improve the adaptation of *Pancratium maritimum* L. to salt-stress. Journal of Experimental Botany. 2003;**54**:2553-2562. DOI: 10.1093/ jxb/erg277

[106] Kaushal N, Gupta K, Bhandhari K, Kumar S, Thakur P, Nayyar H. Proline induces heat tolerance in chickpea (*Cicer arietinum* L.) plants by protecting vital enzymes of carbon and antioxidative metabolism. Physiology and Molecular Biology of Plants. 2011;**17**:203-213. DOI: 10.1007/s12298- 011-0078-2

[107] Khan MA, Gul B. Halophyte seed germination. In: Ecophysiology of High Salinity Tolerant Plants. Dordrecht: Springer; 2006. pp. 11-30

[108] Hoque MA, Banu MN, Okuma E, Amako K, Nakamura Y, Shimoishi Y, et al. Exogenous proline and glycinebetaine increase NaCl-induced ascorbate–glutathione cycle enzyme activities, and proline improves salt tolerance more than glycinebetaine in tobacco bright yellow-2 suspension-cultured cells. Journal of Plant Physiology. 2007;**164**:1457-1468

[109] Chaitanya KV, Rasineni GK, Reddy AR. Biochemical responses to drought stress in mulberry (*Morus alba* L.): Evaluation of proline, glycine betaine and abscisic acid accumulation in five cultivars. Acta Physiologiae Plantarum. 2009;**31**:437-443

[110] Wang GP, Hui Z, Li F, Zhao MR, Zhang J, Wang W. Improvement of heat and drought photosynthetic tolerance in wheat by over accumulation of glycinebetaine. Plant Biotechnology Reports. 2010;**4**:213-222. DOI: 10.1007/ s11816-010-0139-y

[111] Chen TH, Murata N. Glycinebetaine protects plants against abiotic stress: Mechanisms and biotechnological applications. Plant, Cell & Environment. 2011;**34**:1-20

[112] Meena RS, Yadav RS. Yield and profitability of groundnut (*Arachis hypogaea* L.) as influenced by sowing dates and nutrient levels with different varieties. Legume Research. 2015;**38**(6): 791-797

[113] Meena RS, Dhakal Y, Bohra JS, Singh SP, Singh MK, Sanodiya P. Influence of bioinorganic combinations on yield, quality and economics of mungbean. American Journal of Experimental Agriculture. 2015;**8**(3): 159-166

[114] Rahman MS, Miyake H, Takeoka Y. Effects of exogenous glycinebetaine on growth and ultrastructure of salt-stressed rice seedlings (*Oryza sativa* L.). Plant Production Science. 2002;**5**:33-44

[115] Sakamoto A, Murata N. The role of glycine betaine in the protection of plants from stress: Clues from transgenic plants. Plant, Cell & Environment. 2002;**25**:163-171

[116] Singh A, Meena RS. Response of bioregulators and irrigation on plant height of Indian mustard (*Brassica juncea* L.). Journal of Oilseed Brassica. 2020;**11**(1):9-14

[117] Ozga JA, Kaur H, Savada RP, Reinecke DM. Hormonal regulation of reproductive growth under normal and heat-stress conditions in legume and other model crop species. Journal of Experimental Botany. 2017;**68**: 1885-1894. DOI: 10.1093/ jxb/erw464

[118] Shu K, Qi Y, Chen F, Meng Y, Luo X, Shuai H, et al. Salt stress represses soybean seed germination by negatively regulating GA biosynthesis while positively mediating ABA biosynthesis. Frontiers in Plant Science. 2017;**8**:1372. DOI: 10.3389/ fpls.2017.01372

[119] Shuai H, Meng Y, Luo X, Chen F, Zhou W, Dai Y, et al. Exogenous auxin represses soybean seed germination through decreasing the gibberellin/abscisic acid (GA/ABA) ratio. Scientific Reports. 2017;7:12620. DOI: 10.1038/s41598-017-13093-w

[120] Abeles FB, Morgan PW, Saltveit Jr ME. Ethylene in plant biology. 2nd edition. United States: Elsevier Science Publishing Co. Inc.; 1992. p. 414. Available from: https://doi.org/10.1016/C2009-0-03226-7

[121] Meena RS, Lal R, Yadav GS. Long term impacts of topsoil depth and amendments on soil physical and hydrological properties of an Alfisol in Central Ohio, USA. Geoderma. 2020; **363**:1141164

[122] Meena RS, Kumar S, Datta R, Lal R, Vijayakumar V, Britnicky M, et al. Impact of agrochemicals on soil microbiota and management: A review. Land. 2020;**9**:34. DOI: 10.1016/j.geoderma.2019.114164

[123] Cao WH, Liu J, He XJ, Mu RL, Zhou HL, Chen SY, et al. Modulation of ethylene responses affects plant salt-stress responses. Plant Physiology. 2007; **143**:707-719

[124] Hattori T, Mitsuya S, Fujiwara T, Jagendorf AT, Takabe T. Tissue specificity of glycinebetaine synthesis in barley. Plant Science. 2009;**176**:112-118. DOI: 10.1016/j.plantsci.2008.10.003

[125] Iqbal N, Umar S, Khan NA. Nitrogen availability regulates proline and ethylene production and alleviates salinity stress in mustard (*Brassica juncea*). Journal of Plant Physiology. 2015;**178**:84-91

[126] Cui M, Lin Y, Zu Y, Efferth T, Li D, Tang Z. Ethylene increases accumulation of compatible solutes and decreases oxidative stress to improve plant tolerance to water stress in *Arabidopsis*. Journal of Plant Biology. 2015;**58**:193-201

[127] Arnao MB, Hernández-Ruiz J. Melatonin as a chemical substance or as phytomelatonin rich-extracts for use as plant protector and/or biostimulant in accordance with EC legislation. Agronomy. 2019;**10**:570. DOI: 10.3390/agronomy9100570

[128] Zhang N, Sun Q, Zhang H, Cao Y, Weeda S, Ren S, et al. Roles of melatonin in abiotic stress resistance in plants. Journal of Experimental Botany. 2015;**66**:647-656

[129] Wei W, Li QT, Chu YN, Reiter RJ, Yu XM, Zhu DH, et al. Melatonin enhances plant growth and abiotic stress tolerance in soybean plants. Journal of Experimental Botany. 2015;**66**:695-707

[130] Das A, Rushton PJ, Rohila JS. Metabolomic profiling of soybeans (*Glycine max* L.) reveals the importance of sugar and nitrogen metabolism under drought and heat stress. Plants. 2017;**6**: 21. DOI: 10.3390/plants6020021

[131] Xu C, Xia Z, Huang Z, Xia C, Huang J, Zha M, et al. Understanding the physiological and transcriptional mechanism of reproductive stage soybean in response to heat stress. Crop Breeding, Genetics and Genomics. 2019; **27**(1):2. DOI: 10.20900/cbgg20200004

[132] Meena RS, Kumar V, Yadav GS, Mitran T. Response and interaction of *Bradyrhizobium japonicum* and *Arbuscular mycorrhizal* fungi in the soybean rhizosphere: A review. Plant Growth Regulators. 2018;**84**:207-223

Chapter 8

Wheat (*Triticum aestivum* L.) in the Rice-Wheat Systems of South Asia Is Influenced by Terminal Heat Stress at Late Sown Condition: A Case in Bangladesh

Akbar Hossain, Mst. Tanjina Islam and M. Tofazzal Islam

Abstract

Wheat plays an important role in attaining food and nutritional security in Bangladesh after rice. The demand of wheat has been increasing every year at the rate of 13% due to rapid changes in dietary habits, socio-economic upliftment, enhancement of per capita income, etc. Bangladesh Wheat and Maize Research Institute (BWMRI) has already released 34 high yielding, disease-resistant, and abiotic stress-tolerant wheat varieties, and improved management practices to the farmers. Although all the released varieties have climatic yield potential as high as 6.0 t ha^{-1} with the attainable average yield is 4.0–4.5 t ha^{-1}, the national average yield in farmers' field is only 3.49 t ha^{-1}; it is specified that there is a huge yield gap existing among potential, attainable and actual yields. One of the most important reasons for this yield gap of wheat is the terminal high temperature stress (HS) in late sowing wheat. Generally, farmers in Bangladesh are sowing wheat lately due to delay in sowing monsoon rice and subsequent late harvest of the rice; as a result, late sown wheat faces terminal HS at reproductive stage. The chapter highlighted the consequences of terminal HS on wheat and potential approaches to mitigate the stress in Bangladesh.

Keywords: heat stress, late sowing, wheat, food security, South-Asia

1. Introduction

Wheat is one of the world leading cereals after rice and maize. One cup of whole wheat grain contains 33% protein, 29% carbohydrate and 5% fat. Currently, about 65, 17 and 12% of the wheat crop is used for food, animal feed and industrial applications, respectively [1, 2]. The world demand for cereals is increasing day by day. It is projected that the demand of cereals will be increased by 60% by 2050 in the developing world as compared to the demand in 2000. The demand of wheat will be increased by 26% by the same period of time [3–5].

In Bangladesh, wheat is an important cereal food crop next to rice. It is playing an important role in attaining national food and nutritional security [6, 7]. During 2018–2019, 1.15 million tons of wheat was produced from 0.33 million ha that can meet only 20% of the national requirement [8, 9]. On the other hand, the demand of wheat has been increasing every year at the rate of 13% due to rapid changes in dietary habits, socio-economic upliftment, enhancement of per capita income, the rapid growth of fast-food restaurants, the establishment of branded bakery and biscuit industries, etc. Due to the decrease in the wheat cultivation area by 15% in 2018–2019 than the previous year, wheat production also reduced to about 12%. There is a significant increase in national average wheat productivity (3.49 t ha^{-1}), which was possible through the development and dissemination of high yielding, disease-resistant and stress-tolerant wheat varieties, and also introduction of improved management practices to the farmers field [10]. Although all the released varieties have climatic yield potential as high as 6.0 t ha^{-1} with the attainable average yield is 4.0–4.5 t ha^{-1}. Bangladesh Wheat and Maize Research Institute (BWMRI) (formerly Wheat Research Center of Bangladesh Agricultural Research Institute) has already released 34 wheat varieties, which have climatic yield potential as high as 6.0 t ha^{-1} [11, 12] with the attainable average yield is 4.0–4.5 t ha^{-1} [10], while national average yield (farmers' field yield) is only 3.49 t ha^{-1}. Practically, a huge yield gap exists among potential, attainable and actual yields. Although the yield gap of wheat between the potential and national average was linked to many factors, the most important one is terminal high-temperature stress at late sowing conditions [13–17].

Wheat in Bangladesh is sowing after monsoon rice as a result of delays in sowing monsoon rice causing the late harvest of rice [18]. Therefore, late sown wheat faces two major stresses, (i) low temperature stress at germination to seedling stages; (ii) heat stress (HS) at the reproductive stage particularly during grain development [6, 16]. For proper growth and development of the existing Bangladeshi wheat varieties, optimal temperatures range is between 12 and 25°C, while temperatures below 10°C or above 30°C alter phenology, growth, and development and finally reduce the yield (reported by scientists of BWMRI: [6, 14–17]). In the meantime, it is noticed that global climate change will have a major impact on crop production in Bangladesh particularly winter crops including wheat [19]. The Organization for Economic Cooperation and Development [20] estimated a rise in temperature of 1.4°C by 2050 and 2.4°C by 2100 in Bangladesh. Likewise, CIMMYT-ICARDA projected that 20–30% of wheat yield losses will occur by 2050 in developing countries as a result of an assumed temperature increase of 2–3°C on a global scale [21].

Therefore, it is imperative to minimize the gaps between potential and attainable yields to reduce the import of wheat to ensure improved food security. This chapter presented a brief overview of research on phenology, growth, and yield of wheat as influenced by late sown terminal HS. The potential approaches to find out the genotypes and improvement of management practices for sustainable wheat production under the late sowing condition of Bangladesh are also discussed.

2. Phenology, growth, yield and yield-attributes of wheat are influenced by late sown heat stress

2.1 Effect of heat stress on the phenology of late sowing wheat

Temperature is a modifying factor in all stages of wheat development including germination, tillering, booting, ear emergence, anthesis, and maturity as it can influence the rate of water supply and other substances necessary for growth. However, the influence of temperature varies with plant species, variety and phenological stages of wheat plant [22]. Hossain et al. observed that phenology of three wheat varieties such as 'Gourab', 'BARI Gom 25' and 'BARI Gom 26' is influenced late sown (LS) HS condition (sown on Dec. 27th) [23–25]. They observed that days to the booting of all three varieties were reduced by 4–14% when sown at late. While the required days to first visible awn and days to heading were reduced by 14–25% due to the late sown HS condition. Similarly, Rahman et al. investigated the phenological variation of 10 wheat genotypes viz., 'Gen/3/Gov', 'PB 81/PVN', 'Fang 60', 'Kanchan', 'Sari 82', 'HI 977', 'HAR 424', 'PF 70354', 'Opata' and 'Fyn/Pvn' under optimum (Nov. 17) and late sowing HS condition (Dec. 21th) [26]. They observed that the number of days to anthesis, maturity as well as grain filling period of all wheat genotypes were reduced by 8.90, 12.80 and 20.10%, respectively, when sown at late sown HS condition. Since genotype 'Gen/3/Gov' was found the best entry for late sown with rationally high yield, moderate grain size and growth period. Alam et al. also observed the phenological variation of four wheat genotypes viz., 'BARI Gom 26', 'BAW 1051', 'BAW 1120' and 'BAW 1141' under four sowing conditions of Bangladesh dates of sowing viz., 30th Nov, 15th Dec, 30th Dec, and 14th Jan [27]. It was investigated that the duration of booting, heading, anthesis and physiological maturity stages was reduced significantly when sown at 30th Dec and 14th Jan, which lead to decrease the grain yield of all wheat genotypes. Developmental stages of eight modern wheat varieties viz., 'Sourav', 'Gourab', 'Shatabdi', 'Sufi', 'Bijoy', 'Prodip', 'BARI Gom 25' and 'BARI Gom 26' were investigated under eight sowing conditions viz., 8th Nov., 15th Nov., 22th Nov., 29th Nov., 6th Dec., 13th Dec., 20th Dec. and 27th Dec. Days to physiological maturity was decreased significantly from early to late sowing. While all the variety sown on 8th Nov. need maximum days for physiological maturity as compared to late sowing (27th Dec.). Among the variety 'Shatabdi' took the highest days to reach physiological maturity in all sowing condition, which closely followed by 'Sourav', 'Bijoy', 'BARI Gom 26', 'Prodip', 'BARI Gom 25' and minimum days need for variety 'Sufi' and 'Gourab'. It was concluded that in very early sowing (Nov. 08th), variety 'Sourav' (yield reduction 20.47%) is recommended, followed by 'BARI Gom 25' (yield reduction 27.91%). On the other hand, in very late sowing (Dec. 27th), 'Sufi' is the best (yield reduction 8.60%), followed by 'Bijoy' (yield reduction 11.05%) [15]. To find out the heat-tolerant wheat variety, twenty wheat genotypes 'Shatabdi', 'Prodip', BARI Gom 26, 'E-6', 'E-8', 'E-10', 'E-14', 'E-19', 'E-36', 'E-37', 'E-40', 'E-42', 'E-60', 'E-61', 'E-65', 'E-67', 'E-68', 'E-69', 'E-71' and 'E72' were evaluated under optimum (15 Nov.), late (25 Dec.) and very late sowing HS conditions (15th Jan.) of Bangladesh. Under the late and very late sowing HS conditions, days to heading and maturity of all genotypes were reduced significantly as a result of high-temperature stress [17]. Likewise, Nahar et al. conducted a field experiment at the Sher-e-Bangla Agricultural University, Dhaka-Bangladesh to study the effect of HS on the phenology of five modern wheat varieties viz., 'Sourav', 'Pradip', 'Sufi', 'Shatabdi' and 'Bijoy' grown under optimum (sown at Nov. 30th) and post-anthesis HS environment (sown at 30th Dec.) [28]. High-temperature stress at late sowing conditions significantly affected the days required to germination, booting, anthesis, and also maturity. Among these tested varieties, days to booting was shortened by 2.88% in 'Pradip', 3% in 'Sourav', 9.79% in 'Shatabdi', 12.65% in 'Bijoy' and 14.71% in 'Sufi' due to the late sowing HS. Whereas, the days to anthesis was reduced by 10.83% in 'Pradip', 11.59% in 'Sourav', 11.83% in 'Shatabdi', 12.74% in 'Bijoy' and 12.95% in 'Sufi' due to the late sowing HS. Ultimately, days to maturity were reduced by 8.8% in 'Bijoy', 13.26% in 'Shatabdi', 14.24% in 'Pradip', 15.13% in 'Sufi' and 15.61% in 'Sourav', as a result of high-temperature stress. To know the phenological variation of existing wheat varieties, six wheat cultivars viz., 'Gourab', 'Sourav', 'Kanchan' and 'Shatabdi', 'Sonora' and 'Kalyansona' were sown in two environmental conditions viz., Nov. 30th and Dec. 30th at Hajee Mohammad Danesh Science and Technology University, Dinajpur, Bangladesh farm [29]. Plants of optimum sowing took the higher heat units than the late sowing condition. At the earlier phenological stages, whereas the pheno-thermal indices decreased with late sowing compared to optimum sowing but increased at the later stages. Among the tested genotypes, 'Gourab', 'Sourav', 'Kanchan' and 'Shatabdi' were established as heat-tolerant cultivars that 'exhibited better performance in phenology', growing degree day, helio-thermal unit and finally used heat more efficiently, whereas, cultivars 'Sonora' and 'Kalyansona' were found heat sensitive. Wheat is generally sown at an extremely late season in South-Asia (including, India, Pakistan, and Bangladesh) that face severe high-temperature stress during the reproductive stage particularly at the grain-filling stage. It ultimately reduces the productivity of wheat as a result of the shortened life span [30]. Earlier heading during HS is helpful to preserve green leaves for a long time at the anthesis stage and produce desirable grain yield of wheat [31]. Delayed sowing under the environmental condition of Pakistan reduces the duration of each development phase of wheat as a result of the high-temperature stage also reported by Riaz-ud-Din et al. [32]. Wheat crops are affected by late sown HS at the post-anthesis period under the condition of Bangladesh that lead to reduction in grain filling period which ultimately drastically reduce the grain yield [33].

2.2 Effect of heat stress on growth and development of late sowing wheat

The plant population per unit area is an important growth parameter to get maximum yield. The growth and development of plants also depend on the initial plant population, genotypes and the environmental condition [16, 24]. Seed germination rate is one of the most important phases of wheat which affects the growth and development. It ultimately lead to grain yield (GY) and quality of wheat [34, 35]. Furthermore, the interaction between the seedbed environment and seed quality plays an important role in seedling emergence and their initial establishment. Al-Karaki et al. stated that the combined effect of high atmospheric temperature ranges between 27 and 33°C and water stress ranges between −3 and −0.9 MPa were the most precarious factors that reduce the rate of germination and finally GY of wheat [36]. Water shortage reduced the seed germination rate, resulting in unequal seedling emergence, which eventually declined the GY and quality of wheat [37].

Plant population density (plant/m^2) of wheat genotypes was higher in early sowing followed by optimum and late sowing due to the available soil moisture [25]. They also reported that from early to late sowing plants/m^2 decreased by 12–70%. Yadav and Raghuvamshi conducted a field research in the clay loam soil of Madhya Pradesh in India and revealed that desirable plants per unit area gave the maximum yield when sown at 1st week of Dec. than the 1st week of Jan. [38]. However, the wheat that was sown on the 1st week of Nov. under the environmental condition of Peshawar-Pakistan resulted in the maximum seedlings per unit area, which lead to performing higher yield [39]. Wajid et al. reported that early sowing (1st week of Nov.) increased plant population from 200 to 300 plants/m^2 and 400 plants/m^2 than late sowing (25th Dec.) [40]. Similarly, Mumtaz et al. observed that wheat sown on 11th Nov. showed maximum average germination m^{-2} than crop sown on 21st Dec. sowing [41].

Generally, tillers of wheat grow from the axils of the main shoot leaves. While the potential tillers are varied from variety to variety and also on the growing environmental conditions [23]. Samre et al. identified that the maximum tillers m^{-2} (92.0) obtained when the crop was sown on 5th Nov. in loamy sand soil of Punjab [42]. Similarly, 15th Nov. crop sown recorded the highest average tillers $plant^{-1}$ (2.0) in Akola [43]. Singh and Pal found that delay in sowing by one month and two months from normal date of sowing (25th Nov.) reduced the number of tillers plant^{-1} by 0 and 18%, respectively at Indian Agriculture Research Institute (IARI), New Delhi-India [44]. Hameed et al. [45] conducted an experiment at Malakandar Research Farms NWFP Agricultural University Peshwar, Pakistan to study the effect of planting dates of wheat variety Fakhare Sarhad in three sowing time namely 25th October, 10th Nov. and 25th Nov. and concluded that tillers m^{-2} was significantly affected by different planting dates significantly. They also observed that early sowing (25th October) had significantly maximum tillers m^{-2}. Subhani showed that tillers m^{-2} less reduction (15.38%) under the late sowing HS condition (Jan.) [46]. Six selected wheat genotypes such as 'V1-Aari-11', 'V2-Aas-11', 'V3-Meraj-08', 'V4-Millat-11', 'V5-Punjab-11', and 'V6-Seher-06' were evaluated at six different sowing condition with 10 days intervals of Bahawalpur, Pakistan and found that wheat sown on 11th Nov. performed the best result with respect to tillers m^{-2} than those of others [41]. Thiry et al. reported that the crop sown on 1st Dec. produced the maximum number of fertile tillers m^{-2} (327.66), while the significantly minimum number of fertile tillers m^{-2} (189.55) was obtained when the crop was sown on Dec. 30th [47]. Wajid et al. observed that the average tillers m^{-2} were 320, 316 and 255 on 10th Nov., 25th Nov. and 10th Dec. sowings, respectively [40]. Suleiman et al. reported that optimum sowing increased (141) productive tillers m^{-2} than late sowing (121) [48]. Anwar et al. stated that the early sowing recorded more tillers m^{-2} than late sowing [49]. Shah et al. found that the highest number of productive tillers m^{-2} (292.67) were produced when the crop was sown on 1st Nov. which was at par with 16th Nov. sowing (282) while the lowest number of 31.25 productive tiller m^{-2} [39]. Hossain et al. observed that early sowing produced the highest number of productive tiller plant^{-1} (8) than late sowing (3) [23]. Upadhyay et al. stated that the effective numbers of tillers were significantly superior for the crop sown on 20th Nov. followed by 10th Dec. sowing, whereas the number of tillers was significantly lowest for the crop sown on 30th October [50]. The effective number of tillers decreased with early and delayed sowings. Sharma et al. recorded more dry matter on the 25th Nov. sown crop as compared to 25th Dec. sown crop (84.3 q ha^{-1}) at Hisar, Haryana-India [51]. In wheat irrespective of wheat cultivars, maximum dry weight was observed under normal sown conditions when compared to that of the late sown condition at IARI, New Delhi-India [52].

Eight elite spring wheat cultivars were evaluated under three HS conditions (early, late and very late HS conditions) of Bangladesh and all types of HS negatively influenced the growth and development, finally drastically reduced the yield and quality of wheat [16]. Although flag leaf area of genotypes 'Prodip' and 'Sufi', dry matter partitioning of flag leaf in 'Prodip', 'BARI Gom 26' and 'Shatabdi'; above-ground dry matter partitioning in 'Shatabdi' and 'BARI Gom 26'; seedling emergence in 'Sufi' and 'BARI Gom 26'; tiller production in 'Sufi' and 'BARI Gom 26' did not differ significantly. For selection of wheat genotypes suitable for late sowing and also to use future breeding program for development of wheat tolerant variety, ten spring wheat genotypes such as 'Gen/3/Gov', 'Pb 81/Pvn', 'Fang 60', 'Kanchan', 'Sari 82', 'Hi 977', 'Har 424', 'Pf 70354', 'Opata' and 'Fyn/Pvn' were evaluated in the optimum (sown on 17th Nov.) and late (sown on 21th Dec.) sowing conditions. After experimentation, it was noticed that the growth and development of all genotypes were reduced significantly at late sowing conditions as a result of high-temperature stress. While the Genotype 'Gen/3/Gov' was found the best entry for late planting with reasonably higher biomass and yield [26]. Khichar and Niwas found that delayed in sowing after 20th Nov. resulted in a decrease in biological and GY [53]. Donaldson et al. reported that early sowing resulted in higher biomass yield at optimum environmental condition is the result of a more number of tillers [54]. Singh and Uma [55] observed that wheat sowing on Nov. 22th in the environmental condition of India enhanced the biomass yield significantly then sown on late (Dec. 30th) and very late (Jan. 11th) sowing. Shivani et al. found that dry matter accumulation decreased with delay in sowing from timely (21st Nov.) to very late (7th Jan.) on sandy loam soil of Jharkhand, India [56]. Kulhari et al. reported that biological yield was found to be maximum (96.61 q ha^{-1}) in crop sown on 1st Nov. and minimum (82.7 q ha^{-1}) in 1st Dec. sown crop on sandy loam soil of Jaipur, India [57]. Significantly higher dry matter production was recorded from crop sown on 15th Nov. as compared to that of 30th Nov. and 15th Dec. sowings [58]. Jat et al. demonstrated that dry-matter accumulation is higher on 20th Nov. sowing and minimum on 23rd Dec. sowing [59]. Wajid et al. observed that early (10th Nov. or 25th Nov.) sowings significantly increased final biomass than the late (10 Dec.) sowing ranged from 11.15 to 12.7 t ha^{-1} among different sowing dates [40]. Said et al. carried an experiment Agricultural Research Institute, Tarnab, Peshawar-Pakistan to investigate the effects of various sowing dates and seeding rates on the yield and yield components of wheat (*Triticum aestivum* L.) [60]. Among four planting dates viz. 1st Nov. 15th Nov. 1st Dec. and 15th Dec. sowing, the maximum biomass yield was produced from 1st to 15th Nov. followed by late sowing biomass yield (15th Dec.).

2.3 Effect of heat stress on yield and yield attributes of late sowing wheat

The potential productivity of the most cereals depends on the number of productive tillers/spikes/ears/heads per unit area of land, which depends on the genotype and their growing conditions [61]. Wheat sown at 15th Nov. under the environmental condition of India produced the maximum number of productive tillers per unit of area than wheat sown at extremely late [62]; but in loamy sandy soil of Punjab-India wheat sown on 25th October produced the highest number of ears which lead to produce the maximum yield [63]. Gill also observed that wheat sown on 31st October in Punjab-India recorded the maximum (500.7) number of effective tillers m^{-2} [64]. In the environmental condition of Delhi-India, the highest spikes $plant^{-1}$ was recorded on 18 Nov. sowing while wheat sown on late (11th Dec.) produced the lowest number of spikes $plant^{-1}$ [52]. Likewise, Natu et al. observed that wheat sown on 27th Nov. at IARI, New Delhi-India produced the maximum ears per pot than crop sown on 28th Dec [65]. At Haryana-India, Khichar and Niwas reported that 20th Nov. sowing produced the maximum number of ears $plant^{-1}$ than that of 20th Dec. sowing [66]. On sandy loam soil of Jharkhand-India, 21st Nov. sowing wheat produced the maximum effective tillers m^{-2} as compared to late and extremely late sown crops [56]. In Madhya Pradesh-India, Yadav and Raghuvamshi observed that environmental condition in the 1st week of Dec. is the best for the production of the maximum heads m^{-2} of wheat, however, wheat was sown at the 1st week of Jan. produced the minimum heads m^{-2}, which lead to decrease the final grain weight of wheat [38]. Pandey et al. demonstrated that 23rd Nov. sowing in Pusa-India is the best for the production of maximum yield through the production of the significantly maximum number of spikes m^{-2} than crop was sown on 21st Dec. and 4th Jan. [67]. Under the environmental condition of Pakistan, Khan et al. found that 10th Nov. sowing produced the maximum number of spikes per unit area than sown on 10th January due to the terminal HS under late sown condition [68]. In Egypt, the 25th Nov. sowing produced the highest number of tillers, while wheat sown on 25th December produced the lowest number of productive tillers [69]. Sowing on 1st Nov. resulted in the greatest mean productive tillers m^{-2} (297.7) in Peshawar-Pakistan [39].

Spike length of wheat generally depends on the genetic makeup of a genotype [70]. Variation in spike length due to the inherent genetic makeup among genotypes was also confirmed by Shah et al. [39], Pandey et al. [67, 71]. Since Spike length of a genotype may be changed due to the growing environment as confirmed by Haider et al. [72], who reported that early sown plants (Nov. 15th) had the longest spike than late sown plants (Dec. 5th), due to early sowing wheat got the favorable condition which lead to produce the longest spike, while the late sowing wheat faced the high-temperature stress that limited for development of spike. These findings also confirmed by Hossain et al. [16, 23, 24], who also observed that early and late sown HS was shortened the length of spikes of wheat genotypes under the environmental condition of Bangladesh, which lead to decrease the GY of wheat.

Several earlier findings confirmed that the reproductive stage of wheat is the most sensitive than the vegetative stage [15, 16, 30]. While spikelet spike^{-1} is the most important trait in the reproductive stage, which is sensitive to the fluctuation of temperature (high and low temperature) [73]. An experiment was conducted in Jebel Marra-Sudan with three wheat cultivars (viz., 'Debira', 'El Nelein' and 'Donki') sown at four environmental conditions (i.e., early July, mid-July, early August, and mid-August) and found that early sowing (July) produced the maximum number of spikelets spike^{-1} compared to late sowing due to the favorable environment at early sowing condition [74]. Hossain et al. observed the maximum number of spikelet spike^{-1} under the optimum sowing condition of Bangladesh, it is due to a more favorable environmental condition than early and late sowing [25].

Grains spike^{-1} is a very important parameter contributing to GY since it depends on the spike length and also genetic make-up. However, similar to other yield traits environmental factors also influence the grains spike^{-1} [15]. Grains per spike or per unit area may be increased by sinking the size or number of competing organs such as the peduncle and number of sterile tillers during spike growth; while decrease due to the increasing temperature at the reproductive stage, which force to premature death in more distal and basal florets; as a result, grain number is decreased severely [75–77]. Wajid et al. observed that the 10 Nov. sowing significantly enhanced the number of grains spike^{-1}, however, the 10 Dec. sowings produced the lowest number of grains due to high temperature during the flowering stage decreases grain set due to lower fertilization caused by pollen sterility and/or ovule abortion [40]. Said et al. conducted a field research at the Agricultural Research Institute of Tarnab, Peshawar-Pakistan to detect the effects of sowing dates (i.e., 1st Nov., 15th Nov., 1st Dec. and 15th Dec.) on the yield and yield components of wheat [60]. They observed that grains spike^{-1} were found the maximum from the sowing dates 1st to 15th Nov. sowing while the minimum number of grains spike^{-1} was recorded from the late sowing (15th Dec.). Khan et al. and Qasim et al. also confirmed that optimum sowing is the best for the production of the maximum number of grains than late sowing [68, 78]. Under the environmental condition of Maharashtra-India, Jadhav and Karanjikar observed a higher number of grains ear^{-1} under the early Nov. sowing, while grains ear^{-1} were reduced when sown at late Nov. [79]. In Madhya Pradesh-India, wheat seed sown on 1st week of Dec. recorded the more number of grains head^{-1} compared to early (Nov.) and late sowing (Jan.) [38]. On loamy sandy soil of Haryana-India delay in sowing from 15th Nov. to 25th Dec. significantly reduced the grains ear^{-1} from 42.9 to 37.1.

Delayed sowing shortens the duration of each developmental phase, which ultimately reduces the grain-filling period and lowers grain weight (GW) [15, 17, 75]. A wheat crop is sown late had statistically smaller grains than the crop sown earlier, however, when the crop was sown at late, 1000-GW was decreased significantly, due to a decrease in individual GW. Due to the favorable environmental at the grain-filling stage of optimum sown wheat, individual grain weight was higher which lead to produce the 1000-GW [23–25]. It was also observed that high temperature (soil, air) and deficit soil moisture (drought) stress in early sowing and low-temperature stress in late sowing condition of southern Astrakhan-Russia reduced individual grain weight of spring wheat, which eventually affected 1000-GW [25]. An experiment was conducted at the Agricultural Research Institute of Tarnab, Peshawar-Pakistan to investigate the effects of various sowing dates (1st Nov., 15th Nov., 1st Dec. and 15th Dec.) on 1000-GW of wheat. They observed that the maximum 1000-GW was recorded from 1st to 15th Nov. sowing, while the lowest grain weight was recorded at the late sowing (15th Dec.) treatment. A similar observation was carried out at Agricultural Research Station, Varanasi, Uttar Pradesh-India, to study the effect of sowing dates (20th Nov. and 23rd Dec.) on the 1000-GW of wheat. Delay sowing (23th Dec.) reduced the grain size of wheat which ultimately affected the 1000-GW [59]. Tahir et al. reported that the crop sown on 1st Dec. produced significantly heavier grains (35.13 g) and decreased in late sowing due to the result of temperature fluctuation [80]. Natu et al. reported that test weight was found to be higher in the 27th Nov. sown crop compared to that of 28th Dec. sown crop and the decrease in test weight due to delayed sowing by one month ranged from 10.1 to 13.76% IARI, New Delhi-India [65].

The HS, singly or in combination with drought, is the biggest constraint during anthesis and grain-filling stages in many cereal crops of temperate regions. HS reduced the grain-filling period with a reduction in kernel growth leading to losses in kernel density and weight by up to 7% in spring wheat [81]. Excess radiation and high temperatures are the most limiting factors affecting plant growth and finally crop yield in tropical environments [22]. Growth, yield and yield-related components of tomato varieties were affected by water stress while a heat-sensitive variety was more affected than a heat-tolerant variety [82]. Eight wheat genotypes (viz., 'Sourav', 'Gourab', 'Shatabdi', 'Sufi', 'Bijoy', 'Prodip', 'BAW 1059' and 'BAW 1064') were evaluated under eight sowing conditions (Nov. 8th, Nov. 15th, Nov. 22th, Nov. 29th, Dec. 6th, Dec. 13th, Dec .20th and Dec.27th) to observe the grain weight of wheat [14]. It revealed that all genotypes sown on Nov. 15 to Dec. 6 performed better for a good yield followed by Dec. 20th for considerable yield. In the extremely late sowing condition (up to Dec. 27th) genotype BAW1059 gave good yield, whereas varieties 'Sourav', 'Shatabdi', 'Bijoy' and 'BAW 1064' may be sown in 2nd week of Nov for desirable yield. Kabir et al. conducted a field experiment in the research field of BWMRI, Dinajpur-Bangladesh to evaluate the yield stability of 8 wheat genotypes ('Sourav', 'Gourab', 'Shatabdi', 'Sufi, 'Bijoy', 'Prodip', 'BAW 1059' and 'BAW 1064') at eight different date of sowing (Nov. 8th, Nov.15th, Nov. 22th, Nov. 29th, Dec. 6th, Dec. 13th, Dec. 20th' and Dec. 27th) [83]. It revealed that among the varieties, 'Sourav' was found the most stable variety for GY, whereas the Nov. 29th sowing was the most optimum time for planting of wheat crop. In the same location, another field research was conducted to find out the suitable variety for optimum and late sown HS condition [15]. After two years of observation, they observed that wheat sown on Nov. 22th to Dec. 20th produced significantly higher yield as compared to Nov. 08th, 15th and Dec. 27th. Among the genotypes, 'Shatabdi' performed the best followed by 'BARI Gom 26', 'Sourav', 'Prodip', 'Bijoy', 'Gourab', 'Sufi', whereas 'BARI Gom 25' was found the highly susceptible genotype against late sown HS. An experiment was conducted at the Wheat Research Institute of Faisalabad-Pakistan to study the effect of late sowing HS on GY of ten wheat genotypes (viz., 'Inqilab-91', 'AS-2002', 'GA-2002', 'Manthar', 'Ufaq-2002', '00125', '00055', '01180', '00183' and '99022'). All wheat genotypes were evaluated under two sowing conditions: Nov. (optimum sowing) and Jan. (late sowing) and observed that wheat is sown on late reduced about 53.75% yield reduction, due to the adverse effect of high-temperature stress at late planting conditions [32]. Seven wheat genotypes viz., 'DBW 88', 'DBW 17', 'SD2967', 'BPW 621-50', 'HD 3086', 'WH 1105' and 'PBW 550' were sown on three dates of sowing i.e., normal (22nd

Nov.), late (30th Dec.) and very late (11th Jan.) at College Research Farm, Bichpuri, Agra-India and noticed that the normal date of sowing (Nov. 22th) enhanced significantly the grain and straw yield of wheat over late (Dec. 30th) and very late (Jan.11th) sowing dates [55]. Three wheat varieties viz., 'VL-616', 'UP-1109' and 'UP-2572' were evaluated under three sowing viz., 30th Oct., 20th Nov. and 10th Dec. sowing conditions at Ranichauri Campus, V.C.S.G. Uttarakhand University of Horticulture and Forestry, Bharsar-India and revealed that variety 'UP-2572' performed the best when sown at 20th Nov., whereas all genotypes produced the significantly lower grain at late sown HS condition [50].

Harvest index (HI %) is defined as the grain as a percentage of total plant biomass. The genetic improvement of the GY in wheat is closely linked with HI % [84]. Sandhu et al. found that the harvest index was the highest in normal sowing but decreased gradually with delayed sowing [63]. Gupta et al. also stated that crop sown at normal date recorded the highest HI (%) compared to that of late sowing [52]. Dixit and Gupta reported that higher HI (%) was obtained when the crop was sown on normal sowing than the late sowing [85]. Sowing on 16th Nov. (optimum sowing) resulted in the greatest mean of HI (%) (35.7%) than late sowing in Peshawar-Pakistan as reported by Shah et al. [39]. Natu et al. conducted an experiment at IARI, New Delhi-India and revealed that HI (%) of wheat was decreased when sown at late (28th Dec.) as compared to normal sowing (27th Nov.) [65]. A similar observation was also confirmed by Kaur et al., who found that wheat sown at delay from 15th Nov. to 25th Dec. in Haryana-India decreased the HI (%) from 38.2 to 37.7% . In Pusa-India, Pandey et al. found that the crop sown on 23rd Nov. recorded maximum HI (%) (41.0%) and it decreased significantly with the subsequent delay in sowing [67].

2.4 Heat stress effect on the quality of wheat grains

The most important baking qualities in wheat grains are protein, wet gluten and Zeleny sedimentation value [86], which decisive the grain quality of wheat [87]. Wheat grain anticipated for baking purposes should meet not less than 11.5% dry matter, at least 25% wet gluten content and more than 30 cm^3 sedimentation value [88–90]. The baking quality values depend on the sowing date and cultivar and on the interaction of these factors [86]. Generally, a delay in sowing of spring wheat causes a significant reduction in yield, but protein content in the grain is increased [86, 91]. Sułek et al. and Wenda-Piesik et al. observed that wheat sown in the season of autumn gives a lower protein content, including gluten, and poorer sedimentation, while it endorses a higher bulk-density and a higher-falling number [92, 93]. Although the protein content in wheat grain flour increases significantly in bread wheat as a result of HS [94–96], but the quality of grain mainly gluten quality is adversely influenced through decreasing the grain-filling period [97]. The HS at the grain-filling stage of wheat reduced the period of grain-filling [98], enhanced cell death, and forced to early maturity [99], resulting in significant changes in physico-biochemical properties of wheat grain such as the composition of protein and starch granules. Sial et al. observed that yield and quality of wheat grains are influenced by high-temperature stress due to delay sowing [100]. Sial and co-workers reported that when wheat was planted delay, the development of plant organs and transfer from source to sink were remarkably affected, which was reflected by overall shortening of plant height, reduction in number of internodes, days to heading, days to maturity and grain filling period and ultimately in the reduction of yield and quality. Since grain protein content was higher when sown at late, possibly due to low grain weight. Hakim et al. evaluated a total of 20 spring wheat genotypes under optimum, late and very late sowing conditions of Dinajpur-Bangladesh to evaluate the variation in GY, protein and starch content as a result of late sown HS [17]. They observed that due to a decrease in the reproductive stage, the yield of wheat was decreased drastically; although the protein was increased, but starch content was decreased. Among the genotypes, 'E-65' and 'E-60' had the highest and lowest protein content (15.5% and 12%) under HS condition. In the case of starch content, genotype 'Prodip' and 'E-37' had the highest, while 'E-14' and 'E-72' had the lowest content (64.8% vs. 62.9%), respectively in all sowing conditions [17].

3. Thermal unit indices may be used efficiently to find out the wheat genotypes suitable to grow under heat stress condition

Thermal unit indices (TUIs) viz., helio-thermal units (HTU), growing degree days (GDD), heat use efficiency (HUE) and pheno-thermal index (PTI) have a resilient association with the phenology, growth, and yield of crops and can be efficiently used to select suitable crop cultivars for a specific environmental conditions [101, 102]. These TUIs are related to crop growth and dry matter production which could effectively be used for prediction of growth, phenology, and yield of crops based on weather parameters such as temperature and sunshine hours [101]. Based on their research findings, Akhter et al. opined that TUIs may be useful for specific crop varieties including wheat to articulate the recommendation of exact weather parameters such as temperature and sunshine hours for a specific area on the basis of their requirement [103]. Among TUIs, GDD is an indispensable parameter that could be useful to recognize the hostile effect of temperature and also the intervention of the timing of diverse biological processes responsible for plant growth and development [104]. Phenological variation of crops is associated with environmental variables that could be enlightened on the basis of GDD [105]. Warthinhton and Hatchinson noticed that growth and developing stages of crops may be predicted more precisely through GDD rather than the calendar days (CDs) [106]. An experiment was conducted by Ram et al. under different crop growing environments of central Punjab of India to investigate the accumulation of heat unit requirement for a specific wheat variety and also their yield variation in relation to TUIs [107]. They found that the wheat sown on October 25 took maximum CDs, GDD, PTU and HTU for 75% heading and maturity, while these parameters were reduced significantly when sown at delayed. Among the tested varieties 'PBW 621', 'PBW 343', 'DBW 17' and 'WH 542' took the highest CDs, GDD, HTU, PTU and PTI for earing and maturity when sown at optimum sowing condition. Al-Karaki reported that GY was negatively correlated with GDD to maturity, while positively correlated with GDD to heading [108]. They also observed that increasing GDD to heading resulted in higher GY, while increasing grain filling duration had little effect. Likewise, Amrawat et al. tested different wheat genotypes under diverse sowing conditions and revealed that phenology, growth and yield of wheat are liked with several TUIs viz., GDD, HTU, HUE and PTI. They found that wheat sown on optimum condition (5th Nov.) needed the highest CDs, GDD, PTI, HTU to reach different phenological stages, while reduced significantly when sown at late [109]. Among the tested wheat cultivars, MP-1203 took the maximum CDs, GDD, PTU, and HTU to reach maturity.

4. Stress tolerance indices could be used effectively for the selection of wheat genotypes suitable to grow under heat stress condition

Heat stress in wheat can be alleviated by two ways, by developing and practicing upgraded heat management approaches, or by using and developing heat-tolerant cultivars [30, 33, 110]. However, modern breeding approaches may be effectively used to develop and detect the wheat genotypes which are suitable for HS condition in the era of climate change [111]. Although, development of wheat varieties tolerant to diverse stress is time-consuming and also needs sophisticated knowledge [112]. Therefore, scientists suggested several stress tolerance indices (STIs) which are linked with agronomic parameters [16, 113]. These STIs are stress susceptibility index (SSI) [114], geometric mean productivity (GMP) [115], mean productivity (MP) [116], stress tolerance (TOL) [117] and STI [118] that can be useful for documentation of stress tolerance crops cultivars with high-yield ability. Among STIs, genotypes with higher values for each of MP [47, 119], GMP [118, 120], yield index (YI) [121, 122], yield stability index (YSI) [119, 123] and relative performance (RP%) [23, 33] suggest higher stress tolerance compared to lower tolerance for lower values of these indices. Whereas, the lower value of SSI of a crop cultivar indicates that the cultivar is tolerant to stress, while a higher SSI value indicates relatively greater sensitivity to a given stress [110, 124–126]. Likewise, SSI value, the lower value of TOL of a genotype is relatively more stable under stress conditions [23, 33, 110]. Wheat genotypes with the higher value of both SSI and TOL were found highly susceptible to HS also observed by Sharma et al. [127] and Singh et al. [128]. There is a significant correlation (positive and negative) between STIs and GY, which can be used to screen the wheat genotypes which are tolerant to HS [110, 129]. The correlations between GY of wheat and each of MP, GMP, and GY were positive [130]. Similarly, Malekshahi et al. and Khalili et al. also observed a significantly and positively correlated with GY and each of GMP, MP, YI, and YSI under HS conditions [131–133]. Eight wheat genotypes were evaluated under three HS conditions (early, late and very late) of Dinajpur-Bangladesh. Considering on the SSI and RP (%), genotypes, 'Sufi' was found highly tolerant to HS, followed by 'BARI Gom26' and 'Shatabdi' [16]. Hossain and Teixeira da Silva evaluated selected heat-tolerant wheat genotypes on the basis of lower reduction in life span, SSI and higher RP (%) values [33]. They reported that wheat variety 'BARI Gom 25' (RP value 79%; SSI value 0.7) was the best performing variety followed by 'BARI Gom 26' (RP value 74%; SSI value 0.9) under HS condition, while variety 'Gourab' (RP value 61%; SSI value 1.3) was found the highly sensitive to HS.

5. International collaboration for selecting wheat genotypes suitable for heat stress condition

Bangladesh Wheat and Maize Research Institute (BWMRI) is entrusted to the research works for the improvement of wheat in Bangladesh. This report contains the results of research activities conducted during 2018–2019 across the country. A major thrust is given for development high yielding wheat varieties with resistance/tolerance to ranges of abiotic (heat, drought, salinity) and biotic (diseases, such as wheat blast) stresses and fitting well to the existing cropping systems. Variations and recombination are being created through hybridization every year at BWMRI to generate new genetic stocks and select climate-resilient and disease-resistant varieties. Moreover, Marker-Assisted Breeding has been introduced this year to bring new momentum in the variety improvement program. The CIMMYT has been closely working with the Bangladesh wheat programs and has been played a vital role in popularizing wheat cultivation in Bangladesh from the start. The CIMMYT provided enormous elite wheat germplasm from which promising types could be selected to suit the Bangladesh environment. This collaboration is a key factor in quickly turning Bangladesh into a wheat-growing country. In addition to collaboration with CIMMYT, other national and international organizations such as BARC,

DAE, SCA, BADC, BSMRAU, BAU, and international organizations CIMMYT, BGRI, Cornell University, KARLO-Kenya, USDA-ARS, KSU, and the donor agencies like University of Queensland, CSIRO, ACIAR, USAID, NATP, and KGF for extending their cooperation and support for development of wheat varieties which are suitable to grow against abiotic (heat, drought, salinity) and biotic (diseases such as wheat blast) stresses in the conditions of Bangladesh.

6. Conclusion

Several climate models already predicted that mean temperatures across the globe will continue to increase by 1–4°C by 2099. Where in Asia, the temperature will increase by 1.09–1.18°C from 2010 to 2039, by 1.89–3.07°C from 2040 to 2069, and by 2.82–5.33°C by 2070 to 2099. Since, scientists already warned that if mean temperatures will rise by 3–4°C, the productivity of temperature-sensitive crops will be decreased by 15–35% in Asia and the Middle East. In South Asia, it is expected that due to changing climate, the yield of wheat, maize, and rice will be declined as much as 30% in South Asia by 2099 and 20% in East and South-East Asia. Among the countries in South-Asia, Bangladesh is the 8th largest world population and the 13th highest world population density. While, as a deltaic country, it has heterogeneous geophysics and there is a great variation in climate with extreme events. Since crop productivity including wheat, maize and rice face a great threaten due to the extreme events of abiotic stresses. In the meantime, several findings estimated that wheat production in Bangladesh might be fallen by 32% by 2050 due to an increase in temperature. Since wheat is the second most important food grain after rice. Major limitations for wheat production in Bangladesh are late sowing wheat as wheat is sowing after monsoon rice. When monsoon rice is sown delay, wheat is also sown delay as a result of the late harvest of the rice. When wheat is sown at ate sown, late sown wheat face two stresses, initially low-temperature stress at germination to seedling stages, while heat stress (HS) in the reproductive stage, particularly during grain development ultimately reduced the yield of wheat. Scientists of BWMRI confirmed that optimal temperatures range is between 12 and 25°C for all existing wheat varieties of Bangladesh, while temperatures below 10°C or above 30°C limit the productivity of wheat. In the chapter we reviewed the consequences of heat stress on the growth and development of wheat and also discussed the potential approaches to mitigate the adverse effect of the late sowing heat stress.

Conflicts of interest

The authors declare no conflicts of interest.

Disclaimer

We hereby declare that the book chapter does not have any material which has been accepted to publish any journal or publisher, and also has no copy of any material in previously published, except where due permission and reference is made in the text.

Author details

Akbar Hossain[1*], Mst. Tanjina Islam[2] and M. Tofazzal Islam[3]

1 Bangladesh Wheat and Maize Research Institute, Dinajpur, Bangladesh

2 Department of Agronomy, Hajee Mohammad Danesh Science and Technology University, Dinajpur, Bangladesh

3 Institute of Biotechnology and Genetic Engineering (IBGE), Bangabandhu Sheikh Mujibur Rahman Agricultural University, Gazipur, Bangladesh

*Address all correspondence to: akbarhossainwrc@gmail.com

References

[1] Jones JM, Peña RJ, Korczak R, Braun HJ. CIMMYT series on carbohydrates, wheat, grains, and health: Carbohydrates, grains, and wheat in nutrition and health: An overview. Part I. Role of carbohydrates in health. Cereal Foods World. 2015;**60**(5):224-233. Available from: https://repository.cimmyt.org/xmlui/bitstream/handle/10883/19130/59027.pdf?sequence=1&isAllowed=y

[2] Bader Ul Ain H, Saeed F, Ahmad N, Imran A, Niaz B, Afzaal M, et al. Functional and health-endorsing properties of wheat and barley cell wall's non-starch polysaccharides. International Journal of Food Properties. 2018;**21**(1):1463-1480

[3] Hubert B, Rosegrant M, Van Boekel MA, Ortiz R. The future of food: Scenarios for 2050. Crop Science. 2010;**50**:S33-S50

[4] CIMMYT (International Maize and Wheat Improvement Center). Wheat: CGIAR Research Program (CRP) led by CIMMYT. 2013. Available from: http://wheat.org/who-we-are/about-us

[5] FAO (Food and Agricultural Organization). FAO Statistical Pocket Book. Rome, Italy: Food and Agricultural Organization; 2015. p. 28

[6] Hossain A, Teixeira da Silva JA. Wheat production in Bangladesh: Its future in the light of global warming. AoB Plants. 2013;**5**:pls042. DOI: 10.1093/aobpla/pls042

[7] Timsina J, Wolf J, Guilpart N, Van Bussel LG, Grassini P, Van Wart J, et al. Can Bangladesh produce enough cereals to meet future demand? Agricultural Systems. 2018;**163**:36-44. DOI: 10.1016/j.agsy.2016.11.003

[8] BBS (Bangladesh Bureau of Statistics). Statistical Year Book of Bangladesh. Dhaka: Statistics Division, Ministry of Finance and Planning, Government of Peoples Republic of Bangladesh; 2016. Available from: http://203.112.218.65/WebTestApplication/userfiles/Image/AgricultureWing/Wheat-16.pdf

[9] BWMRI (Bangladesh Wheat and Maize Research Institute). BWMRI Annual Report 2019-20. Dinajpur, Bangladesh; 2019. p. 337

[10] Barma NCD, Hossain A, Hakim MA, Mottaleb KA, Alam MA, Reza MMA, et al. Progress and challenges of wheat production in the era of climate change: A Bangladesh perspective. In: Hasanuzzaman M, Nahar K, Hossain M, editors. Wheat Production in Changing Environments. Singapore: Springer; 2019. DOI: 10.1007/978-981-13-6883-7_24

[11] Timsina J, Singh U, Badaruddin M, Meisner C. Cultivar, nitrogen, and moisture effects on a rice-wheat sequence: Experimentation and simulation. Agronomy Journal. 1998;**90**:119-130

[12] Timsina J, Jat ML, Majumdar K. Rice-maize systems of South Asia: Current status, future prospects and research priorities for nutrient management. Plant and Soil. 2010;**335**:65-82

[13] Ahmed SM, Meisner CA. Wheat Research and Development in Bangladesh. Mexico: Bangladesh Australian Wheat Improvement Project and CIMMYT; 1996. p. 162

[14] Hossain A, Sarker MAZ, Saifuzzaman M, Akhter MM, Mandal MSN. Effect of sowing dates on yield of wheat varieties and lines developed since 1998. Bangladesh Journal of Progressive Science Technology. 2009;7(1):5-8

[15] Hossain A, Sarker MAZ, Hakim MA, Lozovskaya MV, Zvolinsky VP. Effect of temperature on yield and some agronomic characters of spring wheat (*Triticum aestivum* L.) genotypes. International Journal Agricultural Research Innovation and Technology. 2011;**1**(1):44-54

[16] Hossain A, Sarker MA, Saifuzzaman M, Teixeira da Silva JA, Lozovskaya MV, Akhter MM. Evaluation of growth, yield, relative performance and heat susceptibility of eight wheat (*Triticum aestivum* L.) genotypes grown under heat stress. International Journal of Plant Production. 2013;7(3):615-636

[17] Hakim MA, Hossain A, da Silva JA, Zvolinsky VP, Khan MM. Yield, protein and starch content of 20 wheat (*Triticum aestivum* L.) genotypes exposed to high temperature under late sowing conditions. Journal of Scientific Research. 2012;**4**(2):477-489

[18] Badruddin M, Saunders DA, Siddique AB, Hossain MA, Ahmed MO, Rahman MM, et al. Determining yield constraints for wheat production in Bangladesh. In: Saunders DA, Hettel GP, editors. Wheat in Heat Stressed Environments; Irrigated, Dry Areas and Rice-Wheat Farming Systems. CIMMYT: Mexico; 1994. pp. 265-271

[19] IPCC. In: Core Writing Team, Pachauri RK, Reisinger A, editors. Climate Change (2007): Synthesis Report, Contribution of Working Groups I, II and III to the Fourth Assessment Report of the Intergovernmental Panel on Climate Change. Geneva, Switzerland: IPCC; 2007. p. 104

[20] OECD (Organization of Economic Cooperation and Development). Rising Food Prices: Causes and Consequences. 2003. p. 9

[21] CIMMYT-ICARDA. WHEAT-Global Alliance for Improving Food Security and the Livelihoods of the Resources-Poor in the Developing World. Proposal submitted by CIMMYT and ICARDA to the CGIAR consortium board, in collaboration with Bioversity, ICRISAT, IFPRI, ILRI, IRRI, IWMI, 86 NARS Institute, 13 Regional and International Organizations, 71 Universities and Advance Research Institutes, 15 Private Sector Organizations, 14 NGOs and Farmers Cooperatives and 20 Host Countries. 2011. p. 197. Available from: www.cimmyt.org/.../503-wheatglobal-alliance-for-improving-food

[22] Wahid A, Gelani S, Ashraf M, Foolad MR. Heat tolerance in plants: An overview. Environmental and Experimental Botany. 2007;**61**:199-233. DOI: 10.1016/j.envexpbot.2007.05.011

[23] Hossain A, Teixeira da Silva JA, Lozovskaya MV, Zvolinsky VP, Mukhortov VI. In South-Eastern Russia: Yield, relative performance and heat susceptibility index. Journal of Plant Breeding and Crop Science. 2012;**4**(11):184-196. DOI: 10.5897/JPBCS12.047

[24] Hossain A, Lozovskaya MV, Zvolinsky VP, Teixeira da Silva JA. Effect of soil and climatic conditions on phenology of spring wheat varieties in northern Bangladesh. Journal of Fundamental and Applied Sciences. 2012;**2**(39):78-86. Available from: http://inter.aspu.ru/sections/195.html

[25] Hossain A, Teixeira da Silva JA, Lozovskaya MV, Zvolinsky VP. The effect of high temperature stress on the phenology, growth and yield of five wheat (*Triticum aestivum* L.) varieties. Asian and Australasian Journal of Plant Science and Biotechnology. 2012;**6**(1):14-23

[26] Rahman MM, Hossain A, Hakim MA, Kabir MR, Shaha MMR. Performance of wheat genotypes under optimum and late sowing condition. International Journal of Sustainable Crop Production. 2009;**4**(6):34-39

[27] Alam MN, Akhter MM, Hossain MM, Mahbubul SM. Phenological changes of different wheat genotypes (*Triticum aestivum* L.) in high temperature imposed by late seeding. Journal of Biodiversity and Environmental Sciences. 2013;**3**(8):83-93

[28] Nahar K, Ahamed KU, Fujita M. Phenological variation and its relation with yield in several wheat (*Triticum aestivum* L.) cultivars under normal and late sowing mediated heat stress condition. Notulae Scientia Biologicae. 2010;**2**(3):51-56

[29] Sikder S. Accumulated heat unit and phenology of wheat cultivars as influenced by late sowing heat stress condition. Journal of Agriculture & Rural Development. 2009:59-64

[30] Farooq M, Bramley H, Palta JA, Siddique KH. Heat stress in wheat during reproductive and grain-filling phases. Critical Reviews in Plant Sciences. 2011;**30**(6):491-507

[31] Tewolde H, Fernandez CJ, Erickson CA. Wheat cultivars adapted to post-heading high temperature stress. Journal of Agronomy and Crop Science. 2006;**192**(2):111-120. DOI: 10.1111/j.1439-037X.2006.00189.x

[32] Riaz-ud-Din MS, Ahmad N, Hussain M, Rehman AU. Effect of temperature on development and grain formation in spring wheat. Pakistan Journal of Botany. 2010;**42**(2):899-906

[33] Hossain A, Teixeira da Silva JA. Phenology, growth and yield of three wheat (*Triticum aestivum* L.) varieties as affected by high temperature stress. Notulae Scientia Biologicae. 2012;**4**(3):97-109

[34] Almansouri M, Kinet JM, Lutts S. Effect of salt and osmotic stresses on germination in durum wheat (*Triticum durum* Desf.). Plant and Soil. 2001;**231**(2):243-254

[35] Khajeh-Hosseini M, Powell AA, Bingham IJ. The interaction between salinity stress and seed vigour during germination of soybean seeds. Seed Science and Technology. 2003;**31**(3):715-725

[36] Al-Karaki G, Al-Ajimi A, Othman Y. Seed germination and early root growth of three barley cultivars as affected by temperature and water stress. American-Eurasian Journal Agriculture and Environment Science. 2007;**2**(2):112-117

[37] Hampson CR, Simpson GM. Effects of temperature, salt, and osmotic potential on early growth of wheat (*Triticum aestivum*). I. Germination. Canadian Journal of Botany. 1990;**68**(3):524-528

[38] Yadav RP, Raghuvamshi NK. Performance of new wheat (*Triticum aestivum* L.) genotypes under late sown conditions. Agronomy Digest. 2006;**6&7**:9

[39] Shah WA, Bakht J, Ullah T, Khan AW, Zubair M, Khakwani AW. Effect of sowing dates on the yield and yield components of different wheat varieties. Journal of Agronomy. 2006;**5**(1):106-110

[40] Wajid AF, Hussain AB, Ahmad AS, Goheer AR, Ibrahim MU, Mussaddique M. Effect of sowing date and plant population on biomass, grain yield and yield components of wheat. International Journal of Agriculture and Biology. 2004;**6**:1003-1005

[41] Mumtaz MZ, Aslam M, Nasrullah HM, Akhtar M, Ali B. Effect of various sowing dates on growth, yield and yield components of different wheat genotypes. American-Eurasian Journal of Agriculture and Environmental Science. 2015;**15**(11):2230-2234

[42] Samre JS, Dhillon SS, Kahlon PS. Response of wheat varieties to dates of

sowing. Indian Journal of Agronomy. 1989;**34**(3):286-289

[43] Patil KS, Durge DV, Phadnawis BN, Shivankar RS, Rathod TH. Effect of sowing dates on biomass production of wheat cultivars. Annals of Plant Physiology. 2000;**14**(2):115-119

[44] Singh S, Pal M. Growth, yield and phenological response of wheat cultivars to delayed sowing. Indian Journal of Plant Physiology. 2003;**8**(3):277-286

[45] Hameed E, Shah WA, Shad AA, Bakht J, Muhammad T. Effect of different planting dates, seed rate and nitrogen levels on wheat. Asian Journal of Plant Sciences. 2003;**2**(6):467-474

[46] Subhani R. Effect of temperature on development and grain formation in spring wheat. Pakistan Journal of Botany. 2010;**42**(2):899-906

[47] Thiry AA, Chavez Dulanto PN, Reynolds MP, Davies WJ. How can we improve crop genotypes to increase stress resilience and productivity in a future climate? A new crop screening method based on productivity and resistance to abiotic stress. Journal of Experimental Botany. 2016;**67**(19):5593-5603. DOI: 10.1093/jxb/erw330

[48] Suleiman AA, Nganya JF, Ashraf MA. Effect of cultivar and sowing date on growth and yield of wheat (*Triticum aestivum* L.) in Khartoum, Sudan. Journal of Forest Products and Industries. 2014;**3**(4):198-203

[49] Anwar J, Khan SB, Rasul IJ, Zulkiffal MU, Hussain M. Effect of sowing dates on yield and yield components in wheat using stability analysis. International Journal of Agriculture and Biology. 2007;**9**(1):129-132

[50] Upadhyay RG, Rajeev R, Negi PS. Influence of sowing dates and varieties on productivity of wheat under mid Himalayan region of Uttarakhand. International Journal of Tropical Agriculture. 2015;**33**(2(Part IV)):1905-1909

[51] Karambir S, Ram N, Mahender S. Effect of sowing time on radiation use efficiency of wheat cultivars. Journal of Agrometeorology. 2000;**2**(2):166-169

[52] Gupta NK, Shukla DS, Pande PC. Interaction of yield determining parameters in late sown wheat genotypes. Indian Journal of Plant Physiology. 2002;**7**(3):264-269

[53] Khichar ML, Niwas R. Microclimatic profiles under different sowing environments in wheat. Journal of Agrometeorology. 2006;**8**(2):201-209

[54] Donaldson E, Schillinger WF, Dofing SM. Straw production and grain yield relationship in winter wheat. Crop Science. 2001;**41**:100-106

[55] Singh P, Uma. Effect of sowing dates on yield contributing characters and yield of some new wheat genotypes under irrigated conditions. Journal of Multidisciplinary Advance Research. 2015;**4**(1):32-35

[56] Verma UN, Kumar S, Pal SK, Thakur R. Growth analysis of wheat (*Triticum aestivum*) cultivars under different seeding dates and irrigation levels in Jharkhand. Indian Journal of Agronomy. 2003;**48**(4):282-286

[57] Kulhari SC, Sharma SL, Kantwa SR. Effect of varieties, sowing dates and nitrogen levels on yield nutrient uptake and quality of durum wheat (*Triticum deurum* DESF). Annals of Agricultural Research. 2003;**24**(2):332-336

[58] Kumar S, Kadian VS, Singh RC, Malik RK. Effect of planting date on performance of wheat (*Triticum aestivum*) genotypes. Indian

Journal of Agricultural Sciences. 2005;**75**(2):103-105

[59] Jat LK, Singh SK, Latare AM, Singh RS, Patel CB. Effect of dates of sowing and fertilizer on growth and yield of wheat (*Triticum aestivum*) in an Inceptisol of Varanasi. Indian Journal of Agronomy. 2013;**58**(4):611-614

[60] Said A, Gul H, Saeed B, Haleema B, Badshah NL, Parveen L. Response of wheat to different planting dates and seeding rates for yield and yield components. ARPN Journal of Agriculture and Biological Science. 2012;**2**(7):138-140

[61] Rebetzke GJ, Bonnett DG, Reynolds MP. Awns reduce grain number to increase grain size and harvestable yield in irrigated and rainfed spring wheat. Journal of Experimental Botany. 2016;**67**(9):573-2586

[62] Singh B. Response of wheat (*Triticum aestivum* L.) varieties to different sowing dates and growth regulator [Doctoral dissertation]. Ludhiana: Punjab Agricultural University; 2016. Available from: https://pdfs.semanticscholar.org/23d4/cf9fdc7a7ff8a90f6e01139e1f9efc5ecc2e.pdf

[63] Sandhu IS, Sharma AR, Sur HS. Yield performance and heat unit requirement of wheat (*Triticum aestivum*) varieties as affected by sowing dates under rainfed conditions. Indian Journal of Agricultural Sciences. 1999;**69**(3):175-179

[64] Gill DS. Agro physiological traits for screening heat tolerant lines of wheat (*Triticum aestivum* L.) under late sown conditions. Indian Journal of Agricultural Research. 2009;**43**(3):211-214

[65] Natu PS, Sumesh KV, Lohot VD, Ghildiyal M. High temperature effect on grain growth in wheat cultivars: An evaluation of responses. Indian Journal of Plant Physiology. 2006;**11**(3):239-245

[66] Khichar ML, Niwas R. Thermal effect on growth and yield of wheat under different sowing environments and planting systems. Indian Journal of Agricultural Research. 2007;**41**(2):92-96

[67] Pandey IB, Pandey RK, Dwivedi DK, Singh RS. Phenology, heat unit requirement and yield of wheat (*Triticum aestivum*) varieties under different crop-growing environment. Indian Journal of Agricultural Sciences. 2010;**80**(2):136-140

[68] Khan MA, Sabir W, Ahmad N, Rehman AR. Selection criterion for high yielding wheat genotypes under normal and heat stress conditions. SAARC Journal of Agriculture. 2007;**5**(2):101-110

[69] Abdel-Nour NA, Fateh HS. Influence of sowing date and nitrogen fertilization on yield and its components in some bread wheat genotypes. Egyptian Journal of Agricultural Research. 2011;**89**(4):1413-1433

[70] Ali Y, Babar MA, Javed A, Philippe M, Zahid L. Genetic variability, association and diversity studies in wheat (*Triticum aestivum* L.) germplasm. Pakistan Journal of Botany. 2008;**40**(5):2087-2097

[71] Pandey BK, Verma NK, Ram Y. Response of wheat (*Triticum aestivum* L. Emend feorypaol) varieties to fertilizer levels in Bundel Khand region of Uttar Pradesh. Agricultural Science Digest. 2007;**27**(2):134-135

[72] Haider SA, Alam MZ, Alam MF, Paul NK. Influence of different dates on the phenology and accumulated heat units in wheat. Journal of Biological Sciences. 2003;**3**(10):932-939

[73] Sangtarash MH. Responses of different wheat genotypes to drought

stress applied at different growth stages. Pakistan Journal of Biological Sciences. 2010;**13**:114-119

[74] Ahmed MF, Ahmed ASH, Burhan HO, Ahmed FE. Effects of sowing date on growth and yield of wheat at different elevations in Jebel Marra highlands under Rain-fed Conditions. In: Proceedings of the ARC; 2006

[75] Ugarte C, Calderini DF, Slafer GA. Grain weight and grain number responsiveness to pre-anthesis temperature in wheat, barley and triticale. Field Crops Research. 2007;**100**(2-3):240-248

[76] Kaur V, Behl RK. Grain yield in wheat as affected by short periods of high temperature, drought and their interaction during pre-and post-anthesis stages. Cereal Research Communications. 2010;**38**(4):514-520

[77] Dreccer MF, Wockner KB, Palta JA, McIntyre CL, Borgognone MG, Bourgault M, et al. More fertile florets and grains per spike can be achieved at higher temperature in wheat lines with high spike biomass and sugar content at booting. Functional Plant Biology. 2014;**41**(5):482-495

[78] Qasim M, Qamer M, Alam M, Alam M. Sowing dates effect on yield and yield components of different wheat varieties. Journal of Agricultural Research. 2008;**46**(2):135-140

[79] Jadhav AG, Karanjikar PN. Response of new wheat genotypes to different dates of sowing under irrigated conditions. Annals of Agricultural Research. 2001;**22**:295-296

[80] Anureet K, Pannu RK, Buttar GS. Splitting of nitrogen dose affects yield and net returns in wheat sown on different dates. Indian Journal of Ecology. 2010;**37**(1):18-22

[81] Tahir M, Ali A, Nadeem MA, Hussain A, Khalid F. Effect of different sowing dates on growth and yield of wheat (*Triticum aestivum* L.) varieties in district Jhang, Pakistan. Pakistan Journal of Life and Social Sciences. 2009;**7**(1):66-69

[82] Farooq M, Hussain M, Usman M, Farooq S, Alghamdi SS, Siddique KH. Impact of abiotic stresses on grain composition and quality in food legumes. Journal of Agricultural and Food Chemistry. 2018;**66**(34):8887-8897

[83] Nahar K, Ullah SM. Effect of water stress on moisture content distribution in soil and morphological characters of two tomato (*Lycopersicon esculentum* Mill) cultivars. Journal of Scientific Research. 2011;**3**(3):677-682. DOI: 10.3329/jsr.v3i3.7000

[84] Kabir MR, Hossain A, Rahman MM, Hakim MA, Sarker MA. Stability analysis of wheat for grain yield affected by different environment. Bangladesh Research Publication Journal. 2009;**3**:833-840

[85] Slafer GA, Andrade FH. Changes in physiological attributes of the dry matter economy of bread wheat (*Triticum aestivum* L.) through genetic improvement of grain yield potential at different regions of the world. Euphytica. 1991;**58**:37-49

[86] Dixit AK, Gupta AK. Influence of methods of sowing and seed rates on growth and yield of wheat (*Triticum aestivum* L.) under different dates of sowing. Environment and Ecology. 2004;**22**(3):407-410

[87] Knapowski T, Ropińska P, Kazek M, Wenda-Piesik A. Flour and bread quality of spring wheat cultivars (*Triticum aestivum* L.) sown at facultative and spring sowing dates. Acta Scientiarum Polonorum Seria Agricultura. 2018;**17**(3):133-142

[88] Motzo R, Fois S, Giunta F. Protein content and gluten quality of durum wheat (*Triticum turgidum* subsp. durum) as affected by sowing date. Journal of the Science of Food and Agriculture. 2007;**87**(8):1480-1488. DOI: 10.1002/jsfa.2855

[89] Stępniewska S. Technological value of grain of selected wheat varieties. Acta Agrophysica. 2015;**22**(1):103-114

[90] Knapowski T, Kozera W, Murawska B, Wszelaczyńska E, Pobereżny J, Mozolewski W, et al. Evaluation of technological parameters of selected winter wheat cultivars in baking terms. Chemical Engineering and Equipment. 2015;**54**(5):255-256

[91] Knapowski T, Szczepanek M, Wilczewski E, Pobereżny J. Response of wheat to seed dressing with humus and foliar potassium fertilization. Journal of Agricultural Science and Technology. 2015;**17**(6):1559-1569

[92] Sułek A. Impact of sowing and harvesting dates on yielding and protein content in spring wheat grain of the Nawra variety. Fragmenta Agronomica. 2009;**26**(2):138-144

[93] Sułek A, Nieróbca A, Cacak-Pietrzak G. Impact of the autumn sowing date on the yield and quality of spring wheat grain. Polish Journal of Agronomy. 2017;**29**:43-50

[94] Wenda-Piesik A, Knapowski T, Ropińska P, Kazek M. Grain quality of common wheat varieties (*Triticum aestivum* L. Emend. Fiori et Paol.) sown in late autumn and spring. Acta Agrophysica. 2017;**24**(4):601-612

[95] Balla K, Veisz O. Changes in the quality of cereals in response to heat and drought stress. Acta Agronomica Óvariensis. 2007;**49**(2):451-455

[96] Labuschagne MT, Elago O, Koen E. The influence of temperature extremes on some quality and starch characteristics in bread, biscuit and durum wheat. Journal of Cereal Science. 2009;**49**(2):184-189. DOI: 10.1016/j.jcs.2008.09.001

[97] Hrušková M, Švec I. Wheat hardness in relation to other quality factors. Czech Journal of Food Sciences. 2009;**27**(4):240-248

[98] Pacuta V, Bajci P. State and changes in Slovak sugar beet production in recent years. Listy Cukrovarnicke a Reparske. 1998;**114**(2):46-49

[99] Wardlaw IF, Moncur LJFPB. The response of wheat to high temperature following anthesis. I. The rate and duration of kernel filling. Functional Plant Biology. 1995;**22**(3):391-397. DOI: 10.1071/PP9950391

[100] DuPont FM, Altenbach SB. Molecular and biochemical impacts of environmental factors on wheat grain development and protein synthesis. Journal of Cereal Science. 2003;**38**(2):133-146. DOI: 10.1016/S0733-5210(03)00030-4

[101] Sial MA, Arain MA, Khanzada SH, Naqvi MH, Dahot MU, Nizamani NA. Yield and quality parameters of wheat genotypes as affected by sowing dates and high temperature stress. Pakistan Journal of Botany. 2005;**37**(3):575-584

[102] Singh MP, Singh NB. Thermal requirement of Indian mustard (*Brassica juncea*) at different phonological stages under late sown condition. Indian Journal of Plant Physiology. 2014;**19**(3):238-243

[103] Islam MR, Barutcular C, Çig F, El Sabagh A, Alam MA, Kamal MM, et al. Assessing impact of thermal units on growth and development of mustard varieties grown under optimum sown conditions. Journal of Agrometeorology. 2019;**21**(3):270-281

[104] Akhter S, Singh L, Saxena A, Rasool R, Ramzan S, Showqi I. Agro meteorological indices for brown sarson (*Brassica rapa* L.) sown under different dates of sowing in temperate region of Kashmir. The Bioscan. 2016;**11**(1):279-283

[105] Li QY, Jun YI, Liu WD, Zhou SM, Lei LI, Niu JS, et al. Determination of optimum growing degree-days (GDD) range before winter for wheat cultivars with different growth characteristics in North China plain. Journal of Integrative Agriculture. 2012;**11**(3):405-415

[106] Srivastava AK, Chakravarty NV, Sharma PK, Bhagavati G, Prasad RN, Gupta VK, et al. Relation of growing degree-days with plant growth and yield in mustard varieties grown under a semi-arid environment. Journal of Agricultural Physics. 2005;**5**(1):23-28

[107] Warthinhton CM, Hatchinson CM. Accumulated degree days as a model to determine key development stages and evacuate yield and quality of potato in Northeast Florida. Proceedings of the Florida State Horticultural Society. 2005;**118**:98-101

[108] Ram H, Singh G, Mavi GS, Sohu VS. Accumulated heat unit requirement and yield of irrigated wheat (*Triticum aestivum* L.) varieties under different crop growing environment in Central Punjab. Journal of Agrometeorology. 2012;**14**(2):147-153

[109] Al-Karaki GN. Phenological Development-Yield Relationships in Durum Wheat Cultivars under Late-Season High-Temperature Stress in a Semiarid Environment. International Scholarly Research Network ISRN Agronomy; 2012. 7p. Article ID: 456856. doi: 10.5402/2012/456856

[110] Amrawat T, Solanki NS, Sharma SK, Jajoria DK, Dotaniya ML. Phenology growth and yield of wheat in relation to agrometeorological indices under different sowing dates. African Journal of Agricultural Research. 2013;**8**(49):6366-6374

[111] Jahan MA, Hossain A, Timsina J, da Silva JA. Evaluation of tolerance of six irrigated spring wheat (*Triticum aestivum* L.) genotypes to heat stress using stress tolerance indices and correlation analysis. International Journal of Agricultural Research. 2018;**13**:39-52

[112] Khan AA, Kabir MR. Evaluating wheat genotypes using different stress tolerance indices. Cercetări Agronomice în Moldova. 2014;**XLVII**(4):160. DOI: 10.1515/cerce-2015-0004

[113] Hussain S, Jamil M, Napar AA, Rahman R, Bano A, Afzal F, et al. In: Azooz MM, Ahmad P, editors. Plant-Environment Interaction: Responses and Approaches to Mitigate Stress. Chichester, UK: John Wiley and Sons, Ltd. DOI: 10.1002/9781119081005.ch9

[114] Rahman MM, Hasan MA, Chowdhury MF, Islam MR, Rana MS. Performance of wheat varieties under late planting-induced heat stress condition. Bangladesh Agronomy Journal. 2018;**21**(1):9-24

[115] Fischer RA, Maurer R. Drought resistance in spring wheat cultivars. I. Grain yield responses. Australian Journal of Agricultural Research. 1978;**29**(5):897-912

[116] Ramirez-Vallejo P, Kelly JD. Traits related to drought resistance in common bean. Euphytica. 1998;**99**(2):127-136

[117] Clarke JM, McCaig TN. Evaluation of techniques for screening for drought resistance in wheat. Crop Science. 1982;**22**(3):503-506

[118] Rosielle AA, Hamblin J. Theoretical aspect of selection for yield in stress and

non-stress environment. Crop Science. 1981;**21**:943-946

[119] Fernandez GCJ. Effective selection criteria for assessing stress tolerance. In: Kuo CG, editor. Proceedings of the International Symposium on Adaptation of Vegetables and Other Food Crops in Temperature and Water Stress, Publication; Tainan, Taiwan; 1992

[120] Abdi N, Darvishzadeh R, Maleki HH. Effective selection criteria for screening drought tolerant recombinant inbred lines of sunflower. Genetika. 2013;**45**(1):153-166. DOI: 10.2298/gensr1301153a

[121] Schneider KA, Rosales-Serna R, Ibarra-Perez F, Cazares-Enriquez B, Acosta-Gallegos JA, Ramirez-Vallejo P, et al. Improving common bean performance under drought stress. Crop Science. 1997;**37**(1):43-50

[122] Gavuzzi P, Rizza F, Palumbo M, Campanile RG, Ricciardi GL, Borghi B. Evaluation of field and laboratory predictors of drought and heat tolerance in winter cereals. Canadian Journal of Plant Science. 1997;**77**(4):523-531

[123] Lin CS, Binns M, Lefkovitch LP. Stability analysis where do we stand? Crop Science. 1986;**26**:894-900

[124] Bouslama M, Schapaugh WT. Stress tolerance in soybean. Part 1: Evaluation of three screening techniques for heat and drought tolerance. Cvrop Science. 1984;**2**:933-937

[125] Anwar J, Subhani GM, Hussain M, Ahmad J, Hussain M, Munir M. Drought tolerance indices and their correlation with yield in exotic wheat genotypes. Pakistan Journal of Botany. 2011;**43**(3):1527-1530

[126] Nouri A, Etminan A, Teixeira da Silva JA, Mohammadi R. Assessment of yield, yield-related traits and drought tolerance of durum wheat genotypes (*Triticum turjidum* var. durum Desf.). Australian Journal of Crop Science. 2011;**5**(1):8-16

[127] Ali MB, El-Sadek AN. Evaluation of drought tolerance indices for wheat (*Triticum aestivum* L.) under irrigated and rainfed conditions. Communications in Biometry and Crop Science. 2016;**11**:77-89

[128] Sharma A, Rawat R, Verma J, Jaiswal J. Correlation and heat susceptibility index analysis for terminal heat tolerance in bread wheat. Journal of Central European Agriculture. 2013;**14**(2):57-66

[129] Singh S, Sengar RS, Kulshreshtha N, Datta D, Tomar RS, Rao VP, et al. Assessment of multiple tolerance indices for salinity stress in bread wheat (*Triticum aestivum* L.). Journal of Agricultural Science. 2015;**7**(3):49-57. DOI: 10.5539/jas.v7n3p49

[130] Mitra J. Genetics and genetic improvement of drought resistance in crop plants. Current Science. 2001;**25**:758-763

[131] Toorchi M, Naderi R, Kanbar A, Shakiba MR. Response of spring canola cultivars to sodium chloride stress. Annals of Biological Research. 2012;**2**(5):312-322

[132] Malekshahi F, Dehghani HA, Alizadeh BA. A study of drought tolerance indices in canola (*Brassica napus* L.) genotypes. Journal of Science and Technology of Agriculture and Natural Resources. 2009;**13**(48(B)):77-90

[133] Khalili M, Naghavi MR, Aboughadareh AP, Talebzadeh SJ. Evaluating of drought stress tolerance based on selection indices in spring canola cultivars (*Brassica napus* L.). Journal of Agricultural Science. 2012;**4**(11):78-85

Chapter 9

Maize Adaptability to Heat Stress under Changing Climate

Ayman EL Sabagh, Akbar Hossain, Muhammad Aamir Iqbal, Celaleddin Barutçular, Mohammad Sohidul Islam, Fatih Çiğ, Murat Erman, Oksana Sytar, Marian Brestic, Allah Wasaya, Tasmiya Jabeen, Maham Asif Bukhari, Muhammad Mubeen, Habib-ur-Rehman Athar, Faraz Azeem, Hakki Akdeniz, Ömer Konuşkan, Ferhat Kizilgeci, Muhammad Ikram, Sobhy Sorour, Wajid Nasim, Mabrouk Elsabagh, Muhammad Rizwan, Ram Swaroop Meena, Shah Fahad, Akihiro Ueda, Liyun Liu and Hirofumi Saneoka

Abstract

The rapidly increasing human population is an alarming issue and would need more food production under changing climate. Abiotic stresses like heat stress and temperature fluctuation are becoming key issues to be addressed for boosting crop production. Maize growth and productivity are sensitive to temperature fluctuations. Grain yield losses in maize from heat stress are expected to increase owing to higher temperatures during the growing season. This situation demands the development of maize hybrids tolerant to heat and drought stresses without compromising grain yield under stress conditions. The chapter aimed to assess the updates on the influence of high-temperature stress (HTS) on the physio-biochemical processes in plants and to draw an association between yield components and heat stress on maize. Moreover, exogenous applications of protectants, antioxidants, and signaling molecules induce HTS tolerance in maize plants and could help the plants cope with HTS by scavenging reactive oxygen species, upregulation of antioxidant enzymes, and protection of cellular membranes by the accrual of compatible osmolytes. It is expected that a better thought of the physiological basis of HTS tolerance in maize plants will help to develop HTS maize cultivars. Developing HTS-tolerant maize varieties may ensure crops production sustainability along with promoting food and feed security under changing climate.

Keywords: corn, exogenous applications, heat tolerance, grain quality, grain yield, changing climate

1. Introduction

Since the turn of the twentieth century, the air temperature has risen, expected to proceed to rise as a result of climatic variability. These rises in temperatures may trigger high-temperature stress (HTS): serious damage to plants [1, 2]. As a result, food and feed security have become a crucial challenge under current prevailing agro-climatic conditions [3–5]. Climate modeling has indicated that high temperature during the day and night is threatening global agriculture production system [6]. The result is that maize crop yield is reduced globally [7, 8]. Maize is one of the important crops being cultivated globally with a wide range of uses, and it is an important food crop in the world [9–11], it has been primarily aimed for increasing yield, quality, and stability under different environments [12–15]. Maize is an important component of human food, animal feed, and biofuel industries [5]. It ranks top among cereal crops globally and becomes raw material of numerous food and feed industries. Among growth limiting factors, heat stress has a major effect on maize growth and nutrient composition at different developmental stages. Since several abiotic stresses occur simultaneously, such as drought stress and heat stress, the development of improved breeding procedures is essential for increasing the maize productivity and quality [16]. There is a crucial need for further research to develop maize genotypes tolerant to high temperature and drought stress.

Various physiological and biochemical processes govern plant growth and yield. Stomatal conductance, for example, regulates water loss as transpiration as well as an influx of CO_2 for its fixation in the Calvin cycle. Several researchers had suggested that the stomatal conductance is an important indirect heat-tolerant selection criterion in crops [17]. Similarly, osmoprotectants and chaperone proteins got an important part in the adaptive reaction of maize to heat stress and combined stresses. Moreover, leaf senescence-related proteins enhance maize tolerance to combined heat and drought stress [18]. Introgression of these traits in locally acclimated maize hybrids through potential donor hybrids helps in developing maize hybrids tolerant to heat and drought stress. Moreover, identification of donor genotypes possessing favorable traits is important in heat stress breeding programs [19]. Therefore, the present review aimed to evaluate the updates on the effect of heat stress on different plant developmental stages, some physiological and biochemical traits, yield and yield traits of maize. Moreover, this review included updates on various strategies used to improve crop tolerance against heat stress including, conventional breeding strategies, management practices, shotgun approaches, and molecular biology-based strategies. Given the critical analysis of success and limitations for improving maize crop productivity under heat stress, future directions for research are also suggested.

2. Adverse effects of heat stress on growth, development, and yield performance of maize

2.1 Morpho-physiological responses

Temperature above 35^0C for a prolonged period is considered unfavorable for crop growth and development and, particularly 40^0C during flowering and grain filling have severe negative impacts on grain yield [5]. Plants under heat stress exhibited significantly reduced stomatal conductance resulting in a reduced rate of photosynthesis. Excessive heat also causes a reduction in net photosynthesis, leaf area, reduced biomass accumulation and seed weight [20]. However, heat-tolerant maize varieties that produced the highest metabolites are not usually high yielding varieties. The heat-tolerant maize varieties are usually characterized by the reduced plant height, leaves plant^{-1}, and leaf area index ultimately reduced the yield. Therefore, several factors should be put into consideration when selecting for heat tolerance in maize. At the cellular level, HTS triggers the appearance of certain genes and increases the accumulation of certain metabolites that may enhance the heat enduring ability of plants [21]. Generally, remarkable genotypic variations in the stomatal conductance were observed [22, 23]. Stomatal conductance, which is a key trait of the photosynthetic leaf, was significantly influenced by abiotic stresses [24]. Delay canopy senescence due to various light interceptions by green leaf area has been reported to be necessary for high productivity of hybrid maize under normal watering and drought stress [16]. The impinging of high-intensity light to plants can lead to permanent damage to membrane structure [20]. The cell membrane is considered the first physiologically sensitive structure to the high temperature and becomes functionally inactive at heat stress [25]. Membrane function and cell wall stretch have inverse relation [26, 27]. Continuous damage in the biological membrane may downregulate the mobility of water, ions, and soluble organic solid molecules within plant cell membranes; hence carbon of production, transport, and accumulation may be affected by these factors. Membrane stability could be used as an assessment of high-temperature tolerance of plants. It is the most appropriate and convenient test; leakages of electrolytes at a high temperature can be measured by this test [28].

Soil plant analyses development (SPAD) value and grain yield have a significant relationship after anthesis, but no positive association has been noticed during the middle and later grain-filling stages [29, 30]. During HTS, the chlorophyll biosynthesis gene gets downregulated [31]. Experimental observation has suggested that the differences among net photosynthetic ratio after exposure to high temperatures were related to the conversion of the chlorophyll "a" into chlorophyll b ratio; due to low chlorophyll "a" and rapid leaf senescence, the photosynthetic rate is negatively affected [32]. HTS induces several metabolic events at the cellular and subcellular levels. The heat stress influences the production of ROS and oxidative stress as well [33–35]. The antioxidative defense system includes both enzymatic and nonenzymatic antioxidants that are shown to participate in response to the development of oxidative stress influenced by heat stress [21].

Scientists showed that rather extreme heat intensity could cause serious tissue damage as well as mortality may arise in a matter of minutes and could ultimately be due to a massive collapse of cell organization [36]. Damages can occur just after deep-term exposures at moderate to maximum heat stress. Informal and gradual damages caused by high temperatures include chlorophyll and mitochondrial destruction of enzymatic activity, protein catabolism impairment, protein deterioration, and cell turgidity looseness [37]. As can be seen in studies, with either the introduction of heat-shocked proteins, plants and animals react to high-temperature pressure [38, 39]. These are intended to avoid species from the harmful impacts of heat stress as well as other sources of pressure [40]. A simple reaction to high-temperature stress is a reduction in regular cellular metabolism. This drop is especially marked at 45°C. The fall in the natural production of protein also goes hand in hand with increased expression and transcription of a fresh set of molecules identified as heat-shock proteins (HSPs) [41]. Previous studies demonstrated that in *Zea mays,* high-temperature stress reduced the protein production and changes the chemical structure of these proteins [42]. Heat stress at the reproduction phase negatively affects the physiology of plants like flower initiation, source-sink relationship, and falling of pods, which ultimately decreases the number of seeds [43]. High-temperature stress is most crucial for the physiological traits of crop plants. High temperature reduced the number of ears, number of kernels, chlorophyll efficiency, firing of leaf, and blasting of the tassel [44]. Climatic stress like high-temperature stress severely reduces the growth and yield of several crops belongs to Leguminosae (Fabaceae). Heat stress severely reduced the physiological growth development and production of *Vigna radiata*. Heat stress reduced dry matter production and other yield attributes [45].

2.2 Effect on seed germination and seedling development

HTS hampers the plant growth; particularly germination and seedling emergence are more sensitive [46]. Stressful environment severely reduces the germination and early seedling growth in several crop plants [47, 48]. However, seeds of sensitive crops exposed to 24 and 48 h moderate heat stress exhibited a higher germination rate. Such an increase in seed germination rate due to short-term exposure to moderate heat stress was attributed to the altered expression of gibberellin and abscisic acid biosynthesis genes [49]. The seedling stage is generally considered as the most sensitive stage to stress in maize development [50]. However, the detrimental impact of water deficit stress on the initial phase of growth and seedling establishment of maize plants cannot be underestimated [51–53].

The appropriate sowing date is important for seed germination and seedling establishment to physiological maturity. The heat-tolerant maize varieties germinated earlier than the non-drought tolerant maize varieties under the critical level of watering. During germination, HTS is associated with an impaired emergency, and a reduced plant stand and plant density [54]. Biochemical components such as soluble sugar and proline increased with increased stress, while starch content and relative water content reduced with increased water deficit [55]. Fluctuations in mean daily temperature (either it is maximum or minimum) disturb seed germination ability [56]. High-temperature stress is the main cause of the reduction in plant yield due to poor germination. [57, 58] studied the impact of high temperature on various developmental phases, especially at seedling emergence in various crop genotypes. Critical periods of stress in maize include seedling establishment stages, rapid growth period, pollination and grain-filling stage. It is proven that in the maize plant with the implementation of stress, not only the leaf area is reduced, but also its growth rate is affected and the appearance of each leaf is delayed [59].

2.3 Grain-filling stage

HTS at the grain-filling stage in spring maize is the main obstacle [60]. Temperature beyond 40°C, mainly during flowering and grain filling has a severe impact on plant grain productivity [5]. Grain filling is highly sensitive to drought and heat, due to the involvement of the array of diverse enzymes and transporters, located in the leaves and seeds [45]. During HTS, the stability of the thylakoid membrane structure is reduced, resulting in degrading chlorophyll, which reduces light energy absorption, transfer, and photosynthetic carbon assimilation, and ultimately photosynthesis is reduced. Inhibited photosynthesis decreases the supply of photosynthates to the grain, leading to a serious reduction of kernel weight and grain yield [60–62]. Delay in the development of reproductive organs might be the result of the reduced cell division and cell elongation processes due to reduced supply of photosynthates and carbohydrate metabolism during the active vegetative growth stages [63].

2.4 Yield components and grain yield

A projection based on the increased daily maximum temperatures concluded that to increase the maize yields by 12% for the period 2016–2035, improved technologies would be needed [64]. Maize plant can face moderate to high temperature, but temperature above 35°C for a long duration is considered unfavorable for crop growth and development, and temperature beyond 40°C, mainly during flowering and grain filling will have a severe impact on plant grain productivity [5]. Meanwhile, early season temperature increases have induced the maize reproductive period to start earlier, developing the risk of water and heat stress. Declines in time to maturation of maize shown of independence of effects to availability of water, the potential of yield which becoming increasingly limited by warming itself [65]. Irrigation regimes were the major determinant of grain yield during the grain-filling stage in maize while significant differences in the number of kernels per row were obtained among irrigation regimes [66]. A large difference in grain yield is caused due to HTS, which is shown in **Figure 1**. Tissue injuries inversely influence the photosynthetic rate during heat stress, which can cause leaf damaging and increase the rate of leaf senescence that largely results in decreasing photosynthetic efficiency [44]. Reduced chlorophyll content, including grain yields and oxidative damages, possibly had a direct correlation under heat stress [5, 67]. Previous research studies indicate that high temperature has a severe effect on the cob growth rate as well as biomass partitioning [68]. Many factors including duration of pollen viability, increased kernel abortion rate, lower the rate of cell division in storage tissue (endosperm), decrease in starch synthesis, downregulate the sink capacity of developing kernel, increased rate of sugar accumulation, kernel development, and less/higher enzyme activities could be responsible for the reduction in kernel per row under heat stress [44, 67]. Stress environment leads to a severe reduction in yield of crop plants probably by disrupting leaf gas exchange properties, which not only limit the size of the source and sink tissues, but the phloem loading, assimilate translocation, and dry matter partitioning are also impaired [46]. Unsuccessful fertilization reduces the seed size and increases flower abortion rate owing to high temperature and it has negative effects on plant reproductive phase [69, 70]. Temperature range 0–35°C, is considered suitable for leaf growth, the temperature range 35–40°C has an inverse relation with leaf growth. Temperature beyond 35-40°C is responsible for lower net photosynthetic rate, which further leads to protein aggregation, enzyme inactivation, inhibition of protein synthesis leading to the degradation of protein synthesis [69, 71]. Eventually, an increase in temperature

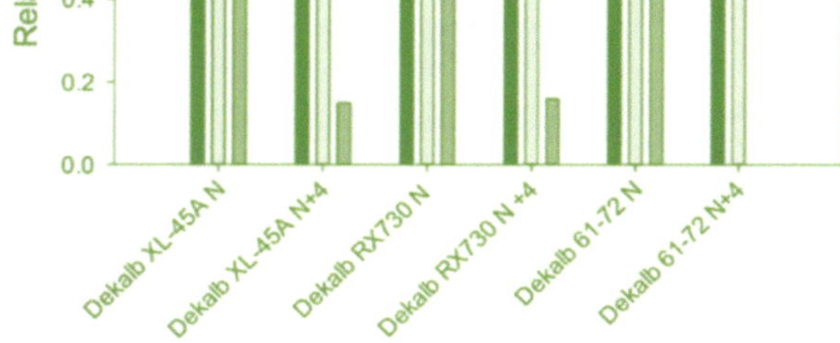

Figure 1.
Differences in total leaf collars, cumulative leaf area, and grain yield of three corn hybrids grown under normal Ames, Iowa temperatures and normal +4°C temperatures.

beyond its critical value leads to generating a heat stress that harms the morphological growth, grain yield, and yield-related attributes of two maize cultivars "Xida 319" and "Xida 889" [72].

2.5 Quality traits of maize

Temperatures higher than 35°C negatively affect maize grain quality. Grain quality, which is governed by factors including the duration and rate of grain filling and the availability of assimilates, is negatively influenced under water deficit conditions. Similar negative effects of stress were reported on the grain weight of wheat [15, 73–75]. Variations in flour quality in a hard-grained crop could be related to changes in protein composition due to heat stress during the grain-filling stage [76]. As per the findings of Mousavi et al. [77], heat stress at the flowering stage greatly reduced the starch content due to the reduction in the photosynthetic activities leading to an increase in the grain protein ratio. Usually, maize quality properties are affected by genotypes, environmental factors, and their interactions (**Figure 2**). Therefore, growth and development of maize are dramatically affected by heat stress leading to reduced grain weight with low starch, crude oil, and protein contents [30]. Grain filling is the most environmentally sensitive phase in maize, which strongly affects grain development quantitatively and qualitatively [7, 15]. Oury and Godin [78] reported a negative correlation between protein contents and grain weight in maize under stress conditions. Association analysis revealed that cob length, thousand-grain weight, and protein contents had a significant relationship with grain yield of maize [79].

In the previous study, the starch content in waxy maize grain was decreased, whereas protein content was increased, resulting in the change of grain quality [80]. However, the activities of enzymes involved in the synthesis of starch and protein are still lacking [81]. The qualitative and quantitative characteristics of grain productivity are mainly influenced by the environmental fluctuation and these changes inversely influence the development and maturing of seed that affect the seed-filling process and deposition of reserves [80]. Generally, high impinging of light affects negatively in plant productivity by causing premature senescence, decreased seed-filling duration, and enhancing remobilization of photosynthates from source to sink [82]. These factors combined, mainly lowers plant biomass and productivity, and finally lowers the assimilate production and mobilization of the reserve to different developing crops [83]. Generally, it is predicted that gene controlling cell division gets downregulated due to water stress, which could be responsible for the decreased cell number in cotyledons along with endosperm. However, further research is required to find out the actual

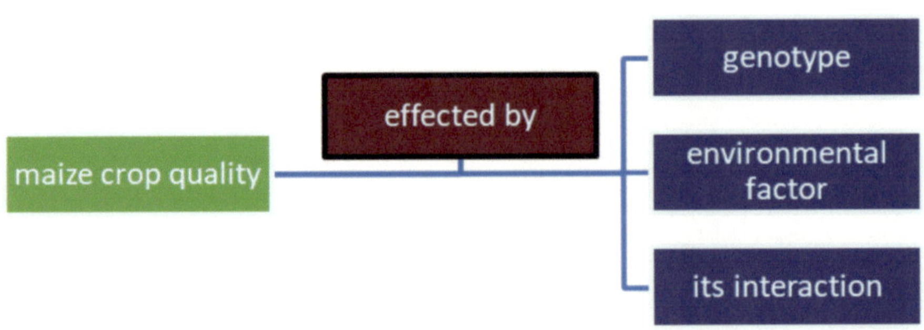

Figure 2.
Quality of maize is influenced by genotype, environment, and their interaction.

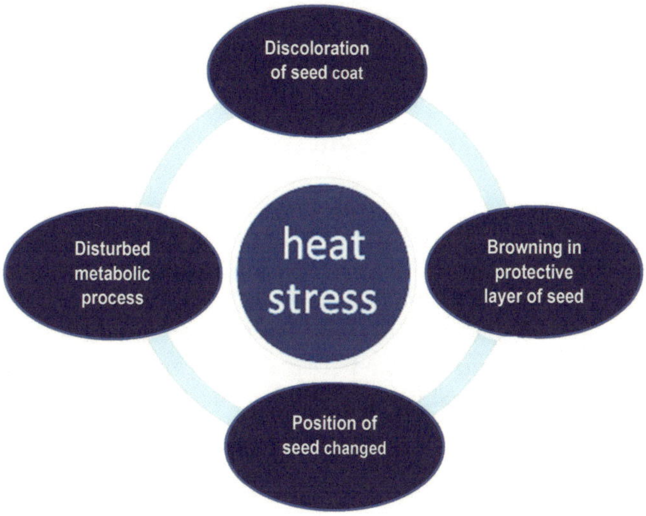

Figure 3.
Quality of maize is deteriorated due to heat stress.

mechanisms controlling these events. Probably due to low enzyme efficiency or high km carbohydrate gene gets downregulated in developing seedling, resulting in limited availability of sucrose, finally producing reduced seed size [45]. The time of seed filling reduced in pea, soybean, and white lupin, resulting in smaller grains [84]. Heat stress during grain filling markedly decreased starch accumulation in wheat [85] and rice [86].

High-temperature stress decreases the protein concentration in the wheat seeds during seed formation stage [76]. Carbon and nitrogen transmission in the seed is improved with the maximum temperature but C transfer is reduced by the daily temperature fluctuations [87]. Temperature variability effects are more visible on the size of seed than seed N contents [87]. Size of seed and protein concentration in the seed are inversely proportional to each other [88]. High-temperature stress reduces seed production, which ultimately declines the seed protein contents [89]. Protein accumulation in the seeds depends upon high-temperature stress [89]. When high-temperature stress occurs at the seed-filling stage it declines the seed protein contents [89]. When wheat crops are exposed to the high-temperature, glutenin protein production is decreased while gliadins protein production remains stable [90]. Seed protein contents of various crops are decreased after imposing the high-temperature stress, but various amino acid concentrations become low [91]. Heat stress damaged the protective layer of seed and food storage tissues of seed, which is why the quality of seed was deteriorated (**Figure 3**).

3. Adaptation and management strategies of maize under heat stress

Enhancement of the antioxidant defense system is an important strategy to scavenge ROS by antioxidant enzymes [92]. Similar to antioxidant defense, phytohormones such as auxin (indole acetic acid, IAA), cytokinins (CKs), abscisic acid (ABA), ethylene (ET), gibberellins (GAs), salicylic acid (SA), brassinosteroids (BRs), and jasmonates (JAs) have key roles in coordinating various signal transduction pathways during the abiotic-stress response [93]. Many studies have shown that altering cultural practices, such as planting rate [94], planting date [95, 96], the phenological variation of crop cultivars [60, 95] soil management [97], nutrient management [60], and irrigation [60] can positively or negatively modify maize yield response to climate change.

3.1 Avoid high-temperature stress by adjusting the sowing date

Advancing or delaying the sowing date may be a potent, farmer-friendly and biologically viable strategy to avoid HTS. Earlier findings reported that earlier sowing dates and longer season varieties have overcome the negative effects of climate warming on spring maize yield [95]. Similarly, other findings reported by [98] showed that by changing sowing date from late April to late May, the mean daily temperature decreased 1.7 and 4.3°C whereas the diurnal temperature increased 4.3 and 3.1°C during grain-filling middle stage (16-45 days after silking) and grain-filling late stage (45 days after silking to maturity), respectively.

3.2 Optimizing irrigation

High air temperatures during the crop growing season can reduce harvestable yields. However, crop varieties with improved heat tolerance traits as well as crop management strategies at the farm scale are thus needed for climate change mitigation. Therefore, to mitigate the negative impact of increased growing season temperatures on crop growth and yield, especially in low latitude regions, heat-tolerant crop varieties, as well as modified farm management practices are needed, especially in the areas when irrigation is needed for crop production and irrigation water depends on the underground aquifers [99]. They also observed that applied irrigation at nighttime through subsurface drip reduced the root-zone soil temperature, which helped plant for improving plant growth and yield of corn. Optimizing irrigation has the potential to improve the water use efficiency of maize leading to enhanced heat tolerance [60]. Soil drought stress and atmospheric high temperature in the vegetative growth period could delay the process of growth of spring maize and shorten the reproductive stage, but those get improved when the soil moisture content in the maize field is maintained 65% field capacity by drip irrigation [100].

3.3 Accumulation of heat-stress defensive phytohormones in plant tissues

Plant growth hormones and exogenous chemicals (e.g., ABA and $CaCl_2$) play important roles in strengthening heat tolerance in maize under HTS [60]. Exogenous ABA induces maize to produce HSPs, strengthening PSII heat tolerance [101]. An exogenous $CaCl_2$ increases the maize cell membrane antioxidant capacity to improve heat tolerance [102]. Phytohormones such as auxin (IAA), cytokinins (CKs), abscisic acid (ABA), ethylene (ET), gibberellins (GAs), salicylic acid (SA), brassinosteroids (BRs), and jasmonates (JAs) have key roles in coordinating various signal transduction pathways during the abiotic-stress response [93].

Auxin or indole-3-acetic acid (Aux/IAA) acts as a chemical messenger to communicate cell activities when crops face different environmental stresses, including salinity, drought, waterlogging, extreme temperatures (heat, chilling, and freezing), heavy metals, light (intense and weak), and radiation (UV-A/B) [92, 103, 104]. Cytokinin (CK) is one of them, which functions solely and or with other hormones to mediate different mechanisms within plants in response to environmental fluctuations. During heat stress, protein denaturation and metabolic imbalance are occurred due to the excessive production of ROS. While to survive against heat stress, plants stimulate heat-shock proteins as a protective measure to prevent protein denaturation [105]. For example, the upregulation of heat-shock proteins in tobacco and bentgrass was recorded due to the enhancement of the antioxidant activity as a result of higher CK in plant cells [106]. Besides this, external application of CK inhibits the damage in photosynthesis under heat stress in maize, rice, and passion fruit [107, 108]. Salicylic acid (SA) is a naturally occurring phenolic compound [109] which plays a crucial part in the regulation of growth and development of the plants, and also a defensive mechanism to survive against abiotic stresses [110]. Similar to SA, abscisic acid (ABA) plays a vital role in plants' physiological adjustments such as against abiotic stresses [111, 112] along with increasing seedling growth, endogenous levels of ABA, and reduced oxidative damage to plants due to heat stress. Similarly, Hasanuzzaman et al. [21] observed that ABA is a signaling molecule and also enhance the number of other signaling molecules such as nitric oxide for thermos-tolerance. Similar to other phytohormones, gibberellic acid (GAs) also interacts with other phytohormones in numerous developmental and stimulus-response processes in plants. GAs have been reported to alleviate the adverse effects of abiotic stress in plants, including rice as reported by Yamaguchi [113]. Brassinosteroids (BRs) is a new group of phytohormones, present in almost every part of the plants [114]. Similar to other phytohormones, BRs have shown tremendous potential against the abiotic stress-induced oxidative stress [103] including high temperature [115].

3.4 Nutrient management

Inadequate and imbalanced nutrients and impaired soil fertility are associated with mineral-nutrient deficiencies and toxicities [116–118]. Adequate nutrition is essential for the integrity of plant structure and key physiological processes. For example, nitrogen (N) and magnesium are a structural part of chlorophyll and these are needed for photosynthesis. Nitrogen plays a very crucial role in temperature stress tolerance. At higher temperatures, the intensity of light is also very high. So, high light intensity, as a function of high temperature, which affects the uptake of mineral nutrients, ultimately influences the plant growth negatively. Since N plays a major role in the utilization of absorbed light energy and photosynthetic carbon metabolism [119, 120]. Whereas phosphorus is needed for energy production and storage; it is a structural part of nucleic acids and potassium is needed for osmotic regulation and activation of enzymes [117, 118]. Maize physiological function decreases under abiotic stress but can be compensated by nutritional management, for example, adequate potassium fertilizer improves cell membrane stability, turgor pressure, water potential in maize under water-deficit conditions [60]. Thus, a strategy to improve heat tolerance in maize at the grain-filling stage is to regulate nutrition.

3.5 Selection of high-temperature stress-resistant varieties

Selection criteria have been proposed in traditional breeding to facilitate the detection of heat-tolerant maize variety. As different varieties respond differently to HTS, breeding heat-tolerant varieties is an effective strategy to improve heat tolerance at the spring maize grain-filling stage [60]. Screening of various cultivars was done to screen the warmness of the plant canopy, stomata behavior of upper most leaf (flag leaf), and photosynthesizing efficiency that are closely related to each other for the production maximum grain production under high-temperature stress conditions [121–123].

3.6 Morpho-physiological mechanisms

Under HT conditions, plants exhibit various mechanisms for surviving, which include long-term evolutionary phenological and morphological adaptations and short-term avoidance or acclimation mechanisms such as changing the leaf orientation, transpirational cooling, or alteration of membrane lipid compositions [92]. Also, high-temperature stress can be avoided by crop management practices such as selecting proper sowing methods, choice of sowing date, cultivars, irrigation methods, etc. It was discussed that combined hotter and drier climate change scenarios cause a greater maize yield reduction than hotter only scenarios. The incorporating drought and heat tolerance into maize germplasm has the potential to offset predicted yield losses and sustain maize productivity under climate change [19].

Tao and Zhao [60] reported that superoxide dismutase (SOD) increased and malonic dialdehyde (MDA) decreased in maize ear leaf for enhancing the stability of cell membrane, which helps to improve photosynthesis for good grain-filling characteristics (long quickly increase period and high mean rate of grain filling). It also produced high kernel weight under HTS [124, 125] leading to reporting of new origins of genetic engineering which exhibited leakage of electrolytes and MSI are the two basic parameters to screen the temperature stress-tolerant cultivars of various crops [126]. Electrical ions were gathered from the affected plants and were washed out with pure water to measure the membrane stability index MSI [127]. Seed production ability and stability index of the membrane were closely related to each other [3]. Mitochondrial tetrazolium is a very useful indicator of HTS sensitivity. Leaves' tissues were dipped in triphenyl tetrazolium chloride chemical mixture during HTS. The spectrographic technique was used to quantify the related rates of triphenyl tetrazolium chloride reduction to formazan and tissues viability [128]. Heat tolerance (HT) of the crop is generally defined as the ability of the plant to grow and produce an economic yield under HS. This is a highly specific trait, and closely related to the species, even different organs and tissues of the same plant, may vary significantly in this respect. Plants have evolved various mechanisms for thriving under higher prevailing temperatures. They include short-term avoidance/acclimation mechanism or long-term evolutionary adaptations [92]. Many alternative traits related to heat resistance in *Zea mays* have been identified, including leaf kinetics, net photosynthesis rate (Pn), leaf anatomy at seedling stage [129] anther emergence [130], pollen grain viability [131], etc. However, the utility of those traits in stress breeding is not well established to date. Furthermore, most of the research focused on the heat stress on temperate maize, whereas only limited information is available on tropical maize [42].

One of the ways to deal with the adverse effects of heat stress may involve exploring some molecules that have the potential to protect the plants from the harmful effects of HT. In recent decades, exogenous application of protectants such as osmoprotectants, phytohormones, signaling molecules, trace elements, etc., have shown a beneficial effect on plants grown under HTS and these protectants have growth-promoting and antioxidant capacity [21, 92]. Exogenous applications of several phytohormones were found to be effective in mitigating heat stress in plants. Accumulation of osmolytes such as proline (Pro), glycine betaine (GB), and trehalose (Tre) is a well-known adaptive mechanism in plants against abiotic stress conditions including HT [92]. Supplementation with Pro and GB considerably reduced the H_2O_2 production, improved the accumulation of soluble sugars, and protected the developing tissues from heat stress effects. At the field level, managing or manipulating cultural practices, such as the timing and methods for sowing, irrigation management, and selection of cultivars and species, can also considerably decrease the adverse effects of HT stress. In recent decades, exogenous applications of protectants such as osmoprotectants, phytohormones, signaling molecules, trace elements, etc., have shown beneficial effects on plants growing under HT, due to the growth-promoting and antioxidant activities of these compounds [21, 92].

3.7 Molecular markers utilization

The genetic analytical study depends upon the genetic markers. Information about genetic reproduction aids to identify potential gene markers [132]. To mitigate the harmful effects of high-temperature several gene markers like a random polymorphic amplifier, AFLP (amplifier fragmentation length polymorphism), as well as sequenced simple repeats SSR, were used to increase the crop production under heat-stress [133, 134]. During genetic breeding, the SNP marker was used because of its genetic sequence in legumes to identify resistant genotypes against heat stress [135]. QTL chromosome numbers and their origin were very useful to mitigate the effects of heat stress [132]. Different molecular markers are studied in population genomics across the environment in many individuals to find out novel variation patterns and help to find if the genes have functions in significant ecological traits. Genome-wide association study (GWAS) is a powerful tool for understanding the complete set of genetic variants in different crop cultivars to recognize allelic variant linked with any specific [136]. GWASs generally highlight linkage among SNPs single nucleotide polymorphism marker and traits and based on GWAS design, genotyping tools, statistical models for examination, and results in interpretation [137].

3.8 Accumulation of antioxidants and heat-shock proteins

Heat stress disturbed the crop metabolic activities by changing tissue balance. Heat stress directly produced toxic substances in plant tissues call ROS due to which plant suffers from oxidative stress. Moreover, to reduce oxidative damage resulting from heat-induced oxidative stress, plants have developed different adaptive mechanisms, via the biosynthesis of enzymatic and non-enzymatic antioxidants and the sequestering of other materials in crop tissues. Enhancement of antioxidant defense system is an important strategy to scavenge ROS by antioxidant enzymes such as ascorbate peroxidase (APX), ascorbate reductase (AR), catalase (CAT), glutathione reductase (GR), glutathione peroxidase (GPX), and superoxide dismutase (SOD) and with non-enzymatic antioxidants such as ascorbate (AsA), glutathione (GSH), carotenoids, flavanones, and anthocyanins [92]. Furthermore, adaptation to temperature changes, at the molecular level, was accompanied by the degradation of the normal proteins and the synthesis of HSPs involved in the mechanism of defense in plants. Seed germination is the most critical growth stage of the whole plant life cycle because it is the first step to carry out whole-plant growth and development, but heat stress is the main reducing factor of seedling emergence in semiarid areas [138, 139].

4. Conclusion

Heat stress and unprecedented climate changes have become a major challenge for sustainable crop production globally. Plant growth, development, and productivity get compromised due to heat stress. Elucidating maize hybrid for temperature tolerance could be an indispensable step toward a balanced yield. Tolerance and avoidance of stress could be an easy way to boost crop production under a changing climate; for example photosynthetic rate can be improved by targeting candidate traits and candidate genes involved in photosynthesis at a molecular level. It could lead to high assimilates production, more transportation of sugar to grain; finally, it decreases grain-filling rate, improves kernel size, and could be very useful to improve plant productivity. Heat-insensitive maize hybrids can be developed by gene editing *CRISPER-CAS9* system through targeting a gene that is responsible for heat sensitivity. The base of further research should be focused on spring maize crops. Field experiments regarding the sowing date are essential by analyzing the impact of meteorological factors on maize growth and grain yield. Application of osmoprotectants, nanotechnology, and the use of sustainable agriculture agents have become necessary for further research. Further, interdisciplinary studies that include agronomy, animal sciences, and climate modeling are warranted to assess the impact of the feeding of both the HTS-tolerant maize varieties and those grown under heat stress on animal health and production. This review could encourage such interdisciplinary approaches to develop maize hybrids with high nutritional values and are not prone to drastic yield reductions owing to fluctuations in agro-climatic factors (especially temperature) and the outcome may lead to sustainable maize production in the tropics under changing climate.

Conflicts of interest

The authors declare no conflicts of interest.

Disclaimer

We hereby declare that the book chapter does not have any material which has been accepted to publish any journal or publisher, and also has no copy of any material in previously published, except where due permission and reference is made in the text.

Author details

Ayman EL Sabagh[1,6]*, Akbar Hossain[2]*, Muhammad Aamir Iqbal[3], Celaleddin Barutçular[4], Mohammad Sohidul Islam[5], Fatih Çiğ[6], Murat Erman[6], Oksana Sytar[7,8], Marian Brestic[8], Allah Wasaya[9], Tasmiya Jabeen[10], Maham Asif Bukhari[10], Muhammad Mubeen[10], Habib-ur-Rehman Athar[11], Faraz Azeem[12], Hakki Akdeniz[13], Ömer Konuşkan[14], Ferhat Kizilgeci[15], Muhammad Ikram[16], Sobhy Sorour[1], Wajid Nasim[17], Mabrouk Elsabagh[18,19], Muhammad Rizwan[20], Ram Swaroop Meena[21], Shah Fahad[22], Akihiro Ueda[23], Liyun Liu[23] and Hirofumi Saneoka[23]

1 Faculty of Agriculture, Department of Agronomy, Kafrelsheikh University, Kafr El-Sheikh, Egypt

2 Bangladesh Wheat and Maize Research Institute, Bangladesh

3 Faculty of Agriculture, Department of Agronomy, University of Poonch Rawalakot (AJK), Pakistan

4 Faculty of Agriculture, Department of Field Crops, Cukurova University, Adana, Turkey

5 Department of Agronomy, Hajee Mohammad Danesh Science and Technology University, Bangladesh

6 Faculty of Agriculture, Department of Field Crops, Siirt University, Turkey

7 Department of Plant Biology, Taras Shevchenko National University of Kyiv, Institute of Biology, Kyiv, Ukraine

8 Department of Plant Physiology, Slovak University of Agriculture, Nitra, Slovak Republic

9 College of Agriculture, BZU, Bahadur Sub-Campus Layyah, Pakistan

10 Department of Environmental Sciences, COMSATS University Islamabad, Vehari Campus, Pakistan

11 Botany Department, Institute of Pure and Applied Biology, Bahauddin Zakariya University, Multan, Pakistan

12 National Institute of Plant Genome Research, New Delhi, India

13 Igdir Universitesi, Ziraat Fakultesi, Tarla Bitkileri Bölümü, Igdir, Turkey

14 Faculty of Agriculture, Department of Field Crops, Mustafa Kemal University, Hatay, Turkey

15 Kiziltepe Vocational School, Mardin Artuklu University, Mardin, Turkey

16 Faculty of Agricultural Sciences and Technology, Department of Agronomy, Bahauddin Zakariya University, Multan, Pakistan

17 Department of Agronomy, University College of Agriculture and Environmental Sciences, The Islamia University of Bahawalpur (IUB), Bahawalpur, Pakistan

18 Faculty of Agricultural Sciences and Technologies, Department of Animal Production and Technology, Niğde Ömer Halisdemir University, Niğde, Turkey

19 Faculty of Veterinary Medicine, Department of Nutrition and Clinical Nutrition, Kafrelsheikh University, Kafr El-Sheikh, Egypt

20 Department of Agronomy, PMAS-Arid Agriculture University, Rawalpindi, Pakistan

21 Department of Agronomy, Institute of Agricultural Sciences, Banaras Hindu University, Varanasi, UP, India

22 Agriculture Department, The University of Swabi, Pakistan

23 Laboratory of Plant Nutritional Physiology, Graduate School of Biosphere Science, Hiroshima University, Japan

*Address all correspondence to: aymanelsabagh@gmail.com
and akbarhossainwrc@gmail.com

References

[1] Farooq M, Bramley H, Palta JA, Siddique KHM. Heat stress in wheat during reproductive and grain-filling phases. Critical Reviews in Plant Sciences. 2011;**30**(6):491-507

[2] Molla MSH, Nakasathien S, Ali MA, Khan ASMMR, Alam MR, Hossain A, et al. Influence of nitrogen application on dry biomass allocation and translocation in two maize varieties under short pre-anthesis and prolonged bracketing flowering periods of drought. Archives of Agronomy and Soil Science. 2019;**65**(7):928-944

[3] Rojas-DowningMM,NejadhashemiAP, Harrigan T, Woznicki SA. Climate change and livestock: Impacts, adaptation, and mitigation. Climate Risk Management. 2017;**16**:145-163

[4] Vogel E, Donat MG, Alexander LV, Meinshausen M, Ray DK, Karoly DG, et al. The effects of climate extremes on global agricultural yields. Environmental Research Letters. 2019;**14**(5). DOI: 10.1088/1748-9326/ab154b

[5] Shiferaw B, Prasanna BM, Hellin J, Bänziger M. Crops that feed the world: Past successes and future challenges to the role played by maize in global food security. Food Security. 2011;**3**:307-311

[6] Khaliq A, Iqbal MA, Zafar M, Gulzar A. Appraising economic dimension of maize production under coherent fertilization in Azad Kashmir, Pakistan. Custos e Agronegocio. **2019;15**(2):243-253

[7] El Sabagh A, Barutcular C, Islam MS. Relationships between stomatal conductance and yield under deficit irrigation in maize (*Zea mays* L.). Journal of Experimental Biology and Agricultural Sciences. 2017;**5**:15-21. DOI: 10.18006/2017.5 (1).014.021

[8] Abdelaal KA, Hafez YM, El Sabagh A, Saneoka H. Ameliorative effects of abscisic acid and yeast on morpho-physiological and yield characteristics of maize plant (*Zea mays* L.) under water deficit conditions. Fresenius Environmental Bulletin. 2017;**26**(12):7372-7383

[9] Majid MA, Islam MS, EL Sabagh A, Hasan MK, Saddam MO, Barutcular C, et al. Influence of varying nitrogen levels on growth, yield and nitrogen use efficiency of hybrid maize (*Zea mays* L.). Journal of Experimental Biology and Agricultural Sciences. 2017;**5**(2):134-142

[10] El Sabagh A, Hossain A, Barutçular C, Abdelaal K, Fahad S, Anjorin F, et al. Sustainable maize (*Zea mays* L.) production under drought stress by understanding its adverse effect, survival mechanism and drought tolerance indices. Journal of Experimental Biology and Agricultural Sciences. 2018;**6**(2):282-295

[11] El Sabagh A, Barutçular C, Hossain A, Islam MS. Response of maize hybrids to drought tolerance indices in relation to grain weight. Fresenius Environmental Bulletin. 2018;**27**(4):2476-2482

[12] Ignjatovic-Micic D, Kostadinovic M, Bozinovic S, Andjelkovic V, Vancetovic J. High grain quality accessions within a maize drought-tolerant core collection. Scientia Agricola. 2014;**71**:345-355

[13] Ranum P, Pena-Rosas JP, Garcia-Casal MN. Global maize production, utilization, and consumption. Annals of the New York Academy of Sciences. 2014;**1312**(1):105-112

[14] Barutçular C, El Sabagh A, Konuskan O, Saneoka H,

Yoldash KM. Evaluation of maize hybrids to terminal drought stress tolerance by defining drought indices. Journal of Experimental Biology and Agricultural Sciences. 2016;**4**(6):610-616

[15] Barutçular C, Yıldırım M, Koç M, Akıncı C, Toptaş I, Albayrak O, et al. Evaluation of SPAD chlorophyll in spring wheat genotypes under different environments. Fresenius Environmental Bulletin. 2016;**25**(4):1258-1266

[16] Cairns JE, Sonder K, Zaidi PH, Verhulst N, Mahuku G, Babu R. Maize production in a changing climate. Advances in Agronomy. 2012;**144**:1-58

[17] Iqbal A, Iqbal MA, Iqbal A, Zubair A, Muhammad M, Zahoor A, et al. Boosting forage yield and quality of maize (*Zea mays* L.) with multi-species bacterial inoculation in Pakistan. Phyton International Journal of Experimental Botany. 2017;**86**:84-88

[18] Zhao F, Zhang D, Zhao Y, Wang W, Yang H, Tai F, et al. The difference of physiological and proteomic changes in maize leaves adaptation to drought, heat, and combined both stresses. Frontiers in Plant Science. 2016;7:1471. DOI: 10.3389/fpls.2016.01471

[19] Cairns JE, Hellin J, Sonder AJL, MacRobert JF, Thierfelder C, Prasanna BM. Adapting maize production to climate change in sub-Saharan Africa. Food Security. 2013;**5**:345. DOI: 10.1007/s12571-013-0256-x

[20] Meena H, Meena RS, Rajput BS, Kumar S. Response of bio-regulators to morphology and yield of clusterbean [*Cyamopsis tetragonoloba* (L.) Taub.] under different sowing environments. Journal of Applied and Natural Science. 2016;**8**:715-718

[21] Hasanuzzaman M, Gill SS, Fujita M. Physiological role of nitric oxide in plants grown under adverse environmental conditions. In: Gill SS, Tuteja N, editors. Plant Acclimation to Environmental Stress. New York: Springer; 2013. pp. 269-322

[22] Bahar B, Yildirim M, Barutçular C. Relationships between stomatal conductance and yield components in spring durum wheat under Mediterranean conditions. Notulae Botanicae Horti Agrobotanici Cluj-Napoca. 2009;**37**:45-48. DOI: 10.15835/nbha3723084

[23] Barutçular C, Yildirim M, Koç M, Akinci C, Tanrikulu A, El-Sabagh A, et al. Quality traits performance of bread wheat genotypes under drought and heat stress conditions. Fresenius Environmental Bulletin. 2016;**25**(10):1-7

[24] Jiang Q, Roche D, Monaco TA, Hole D. Stomatal conductance is a key parameter to assess limitations to photosynthesis and growth potential in barley genotypes. Plant Biology. 2006;**8**:515-521. DOI: 10.1055/s-2006-923964

[25] Blum A. Crop responses to drought and the interpretation to adaptation. Plant Growth Regulation. 1996;**20**:135-148

[26] Khalili M, Moghaddam M, Kazemi Arbat H, Shakiba MR, Kanooni H, Choukan R. Effect of drought stress on different corn genotypes. Journal of Agricultural Science. 2010;**2**(20):67-84

[27] Yassin M, Mekawy AM, El Sabagh A, Islam MS, Hossain A, Barutcular C, et al. Physiological and biochemical responses of two bread wheat (*Triticum aestivum* L.) genotypes grown under salinity stress. Applied Ecology and Environmental Research. 2019;**17**(2):5029-5041

[28] Kumar S, Meena RS, Bohra JS. Interactive effect of sowing dates and nutrient sources on dry matter

accumulation of Indian mustard (*Brassica juncea* L.). Journal of Oilseed Brassica. 2018;**9**(1):72-76

[29] Monneveux P, Sanchez C, Tiessen A. Future progress in drought tolerance in maize needs new secondary traits and cross combinations. Journal of Agricultural Science. 2008;**146**:287-300. DOI: 10.1017/S0021859608007818 P

[30] Barutcular C, El Sabagh A, Koc M, Ratnasekera D. Relationships between grain yield and physiological traits of durum wheat varieties under drought and high temperature stress in Mediterranean environments. Fresenius Environmental Bulletin. 2017;**26**:4282-4291

[31] Havaux M. Carotenoids as membrane stabilizers in chloroplasts. Trends in Plant Science. 1998;**3**(4):147-151. DOI: 10.1016/S1360-1385(98)01200-X

[32] Yildiz M, Terzi H. Determination of plants' tolerance to high temperature stress by cell viability and photosynthetic pigmentation tests. Erciyes UFBE Der. 2007;**23**(1-2):47-60

[33] Zandalinas SI, Balfagón D, Arbona V, Gómez-Cadenas A. Modulation of antioxidant defense system is associated with combined drought and heat stress tolerance in citrus. Frontiers in Plant Science. 2017;**8**:953. DOI: 10.3389/fpls.2017.00953

[34] Soengas P, Rodríguez VM, Velasco P, Cartea ME. Effect of temperature stress on antioxidant defenses in *Brassica oleracea*. ACS Omega. 2018;**3**(5):5237-5243. DOI: 10.1021/acsomega.8b00242

[35] Thirunavukkarasu N, Sharma R, Singh N. Genome wide expression and functional interactions of genes under drought stress in maize. International Journal of Genomics. 2017;**2568706**. DOI: 10.1155/2017/2568706

[36] Schoffl F, Prandl R, Reindl A. Molecular responses to heat stress. In: Shinozaki K, Yamaguchi-Shinozaki K, editors. Molecular Responses to Cold, Drought, Heat and Salt Stress in Higher Plants. Austin, Texas: R.G. Landes Co.; 1999. pp. 81-98

[37] Howarth CJ. Genetic improvements of tolerance to high temperature. In: Ashraf M, Harris PJC, editors. Abiotic Stresses Plant Resistance Through Breeding and Molecular Approaches. New York: Howarth Press Inc.; 2005

[38] Lindquist S, Craig EA. The heat-shock proteins. Annual Review of Genetics. 1988;**22**:631-677

[39] Vierling E. The roles of heat shock proteins in plants. Annual Review of Plant Physiology and Plant Molecular Biology. 1991;**42**:579-620

[40] Mariamma M, Muthukumar B, Veluthambi K, Gnanam A. Effects of high temperature stress on the expression of low molecular weight heat shock proteins in rice leaves. Journal of Plant Physiology. 1997;**151**:763-765

[41] Perras M, Sarhan F. Synthesis of freezing tolerance proteins in leaves, crown and roots during cold acclimation of wheat. Plant Physiology. 1989;**89**:577-585

[42] Monjardino P, Smith AG, Jones RJ. Heat stress effects on protein accumulation of maize endosperm. Crop Science. 2005;**45**:1203-1210

[43] Duthion C, Pigesire A. Seed length corresponding to the final stage seed abortion of three grain legumes. Crop Science. 1991;**31**:1579-1583

[44] Noor JJ, Vinayan MT, Umar S, Devi P, Iqbal M, Seetharam K, et al. Morpho-physiological traits associated with heat stress tolerance in tropical maize (*Zea mays* L.) at reproductive

stage. Australian Journal of Crop Science. 2019;**13**(04):536-545

[45] Sehgal A, Sita K, Siddique KHM, Kumar R, Bhogireddy S, Varshney RK, et al. Drought or/and heat-stress effects on seed filling in food crops: Impacts on functional biochemistry, seed yields, and nutritional quality. Frontiers in Plant Science. 2018;**9**:1705. DOI: 10.3389/fpls.2018.01705

[46] Farooq M, Wahid A, Kobayashi N, Fujita D, Basra SMA. Plant drought stress: Effects, mechanisms and management. Agronomy for Sustainable Development. 2009;**29**:185-212

[47] Wahid A, Gelani S, Ashraf M, Foolad M. Heat tolerance in plants: An overview. Environmental and Experimental Botany. 2007;**61**:199-223. DOI: 10.1016/j.envexpbot.2007.05.011

[48] Borlu HO, Celiktas V, Duzenli S, Hossain A, El Sabagh A. Germination and early seedling growth of five durum wheat cultivars (*Triticum durum* desf.) is affected by different levels of salinity. Fresenius Environmental Bulletin. 2018;**27**(11):7746-7757

[49] Begcy K, Sandhu J, Walia H. Transient heat stress during early seed development primes germination and seedling establishment in rice. Frontiers in Plant Science. 2018;**9**:17-68. DOI: 10.3389/fpls.2018.01768

[50] Li R, Zeng Y, Xu J, Wang Q, Wu F, Cao M, et al. Genetic variation for maize root architecture in response to drought stress at the seedling stage. Breeding Science. 2015;**65**(4):298-307. DOI: 10.1270/jsbbs.65.298

[51] Anjum SA, Wang LC, Farooq M, Hussain M, Xue LL, Zou CM. Brassinolide application improves the drought tolerance in maize through modulation of enzymatic antioxidants and leaf gas exchange. Journal of Agronomy and Crop Science. 2011;**197**(3):177-185. DOI: 10.1111/j.1439037X.2010.00459.x

[52] Shao HB, Chu LY, Shao MA, Jaleel CA, Hong-Mei M. Higher plant antioxidants and redox signaling under environmental stresses. Comptes Rendus Biologies. 2008;**331**(6):433-441

[53] Zivcak M, Brestic M, Sytar O. Osmotic adjustment and plant adaptation to drought stress. In: Hossain M, Wani S, Bhattacharjee S, Burritt D, Tran LS, editors. Drought Stress Tolerance in Plants. Vol. 1. Cham: Springer; 2016

[54] Buriro M, Oad FC, Keerio MI, Tunio S, Gandahi AW, Ul Hassan SW, et al. Wheat seed germination under the influence of temperature regimes. Sarhad Journal of Agriculture. 2011;**27**(4):539-543

[55] Anjorin FB, Adejumo SA, Agboola L, Samuel YD. Proline, soluble sugar, leaf starch and relative water contents of four maize varıetıes ın response to dıfferent watering regimes. Cercetri Agronomice în Moldova. 2016;**49**(3):51-62

[56] Bewley JD, Black M. Seeds: Physiology of Development and Germination. Vol. 2. Berlin: Springer-Verlag; 1986. pp. 297-304

[57] Joshi AK, Mishra B, Chatrath R, Ferrara GO, Singh RP. Wheat improvement in India: Present status, emerging challenges and future prospects. Euphytica. 2007;**157**(3):431-446

[58] Essemine J, Ammar S, Bouzid S. Impact of heat stress on germination and growth in higher plants: Physiological, biochemical and molecular repercussions and mechanisms of defence. Journal of Biological Sciences. 2010;**10**(6):565-572

[59] Nejad SMH, Alizadeh O, Amiri B, Barzegari M, Bayat ME. The effects of drought and heat stress on some physiological and agronomic characteristics of new hybrids of corn in the north of Khuzestan Province (Iran). EurAsian Journal of BioSciences. 2017;**11**:32-36

[60] Tao F, Zhao Z. Adaptation of maize production to climate change in North China Plain: Quantify the relative contributions of adaptation options. European Journal of Agronomy. 2010;**33**:103-116

[61] Muchow RC. Effect of high temperature on grain-growth in field-grown maize. Field Crops Research. 1990;**23**:145-158

[62] Singletary GW, Banisadr R, Keeling PL. Heat stress during grain filling in maize, effects on carbohydrate storage and metabolism. Australian Journal of Plant Physiology. 1994;**21**:829-841

[63] Suwa R, Hakata H, Hara H, El-Shemy HA, Adu-Gyamfi JJ, Nguyen NT, et al. High temperature effects on photosynthate partitioning and sugar metabolism during ear expansion in maize (*Zea mays* L.) genotypes. Plant Physiology and Biochemistry. 2010;**48**:124-130. DOI: 10.1071/PP9910259

[64] Hawkins E, Fricker TE, Challinor AJ, Ferro CA, Ho CK, Osborne TM. Increasing influence of heat stress on French maize yields from the 1960s to the 2030s. Global Change Biology. 2013;**19**(3):937-947. DOI: 10.1111/gcb.12069

[65] Harrison L, Michaelsen J, Funk C, Husak G. Effects of temperature changes on maize production in Mozambique. Climate Research. 2011;**46**(3):211-222

[66] Hatfield JL, Dold C. Climate change impacts on corn phenology and productivity. In: Amanullah K, Fahad S, editors. Corn: Production and Human Health in Changing Climate. 2018. p. 95. DOI: 10.5772/intechopen.76933

[67] Alam A, Seetharam K, Zaidi PH, Dinesh A, Vinayan MT, Kumar NU. Dissecting heat stress tolerance in tropical maize (*Zea mays* L.). Field Crops Research. 2017;**204**:110-119

[68] Edreira JI, Mayer LI, Otegui ME. Heat stress in temperate and tropical maize hybrids: Kernel growth, water relations and assimilate availability for grain filling. Field Crops Research. 2014;**166**:162-172

[69] Rahman HU. Genetic analysis of stomatal conductance in upland cotton *(Gossypium hirsutum* L.*)* under contrasting temperature regimes. Journal of Agriculture Science. 2005;**143**:161-168

[70] Talwar HS, Takeda H, Yashima S, Senboku T. Growth and photosynthetic responses of groundnut genotypes to high temperature. Crop Science. 1999;**39**:60-466. DOI: 10.2135/cropsci1999.0011183X0039000200027x

[71] Ristic Z, Momčilović I, Bukovnik U, Prasad PV, Fu J, DeRidder BP, et al. Rubisco activase and wheat productivity under heat-stress conditions. Journal of Experimental Botany. 2009;**60**(14):4003-4014

[72] Hussain HA, Men S, Hussain S, Chen Y, Ali S, Zhang S, et al. Interactive effects of drought and heat stresses on morpho-physiological attributes, yield, nutrient uptake and oxidative status in maize hybrids. Scientific Reports. 2019;**9**:3890. DOI: 10.1038/s41598-019-40362-7

[73] Brdar MD, Kraljevíc-Balalíc MM, Kobiljski BD. The parameters of grain filling and yield components in common wheat (*Triticum aestivum* L.) and

durum wheat (*Triticum turgidum* L.*var*. durum). Central European Journal of Biology. 2008;**3**:75-82. DOI: 10.2478/s11535-007-0050-x

[74] Pierre CS, Peterson CJ, Ross AS, Ohm J, Verhoeven MC, Larson M, et al. Wheat grain quality changes with genotype, nitrogen fertilization, and water stress. Agronomy Journal. 2008;**100**:414-420. DOI: 10.2134/agrojnl2007.0166

[75] El Sabagh A, Barutçular C, Saneoka H. Assessment of drought tolerance maize hybrids at grain growth stage in Mediterranean area. International Journal of Biological, Biomolecular, Agricultural, Food and Biotechnological Engineering. 2015;**9**:989-992

[76] Gooding MJ, Ellis RH, Shewry PR. Effects of restricted water availability and increased temperature on the grain filling, drying and quality of winter wheat. Journal of Cereal Science. 2003;**37**:295-309. DOI: 10.1006/jcrs.2002.0501

[77] Mousavi H, Lack S, Alavi FM. Analysis of correlation and stepwise regression between grain protein yield and related traits of maize in conditions of drought stress and zinc sulfate spraying. International Journal of Agriculture and Crop Sciences. 2013;**5**:2783-2788

[78] Oury F, Godin C. Yield and grain protein concentration in bread wheat: How to use the negative relationship between the two characters to identify favorable genotypes? Euphytica. 2007;**157**:45-57. DOI: 10.1007/s10681-0079395-5

[79] Yousaf MI, Hussain K, Hussain S, Ghani A, Arshad M, Mumtaz A, et al. Characterization of indigenous and exotic maize hybrids for grain yield and quality traits under heat stress. International Journal of Agriculture and Biology. 2018;**20**(2):333-337. DOI: 10.17957/IJAB/15.0493

[80] Lu D, Cai X, Yan F, Sun X, Wang X, Lu W. Effects of high temperature after pollination on physicochemical properties of waxy maize flour during grain development. Journal of the Science of Food and Agriculture. 2014;**94**:1416-1421. DOI: 10.1002/jsfa.6433

[81] Yang JC, Zhang JH. Grain filling of cereals under soil drying. New Phytologist. 2006;**169**:223-236. DOI: 10.1111/j.1469-8137.2005.01597.x

[82] Plaut Z, Butow BJ, Blumenthal CS, Wrigley CW. Transport of dry matter into developing wheat kernels and its contribution to grain yield under post-anthesis water deficit and elevated temperature. Field Crops Research. 2004;**86**(2-3):85-198

[83] Zare M, Ghahremaninejad M, Bazrafshan F. Influence of drought stress on some traits in five mung bean (*Vigna radiata* (L.) R. Wilczek) genotypes. International Journal of Agronomy and Plant Production. 2012;**3**:234-240

[84] Duthion C, Pigeaire A. Seed lengths corresponding to the final stage in seed abortion of three grain legumes. Crop Science. 1991;**31**:1579-1583. DOI: 10.2135/cropsci1991.0011183X003100060040x

[85] Hurkman WJ, McCue KF, Altenbach SB, Korn A, Tanaka CK, Kothari KM. Effect of temperature on expression of genes encoding enzymes for starch biosynthesis in developing wheat endosperm. Plant Science. 2003;**164**:873-881. DOI: 10.1016/S0168-9452(03)00076-1

[86] Yamakawa H, Hakata M. Atlas of rice grain filling-related metabolism under high temperature: Joint analysis of metabolome and transcriptome

demonstrated inhibition of starch accumulation and induction of amino-acid accumulation. Plant Cell Physiology. 2010;**51**:795-809

[87] Daniel C, Triboi E. Effects of temperature and nitrogen nutrition on the grain composition of winter wheat: Effects on gliadin content and composition. Journal of Cereal Science. 2000;**32**:45-56

[88] Erekul O, Köhn W. Effect of weather and soil conditions on yield components and bread making quality of winter wheat (*Triticum aestivum* L.) and winter triticale (*Triticosecale Wittm.*) varieties in North-East Germany. Journal of Agronomy and Crop Science. 2006;**192**(6):452-464

[89] Castro M, Peterson CJ, Dalla Rizza M, Dellavalle PD, Vázquez D, Ibanez V, et al. Influence of heat stress on wheat grain characteristics and protein molecular weight distribution. In: Buck HT, Nisi JE, Salomon N, editors. Wheat Production in Stressed Environment. The Netherlands: Springer; 2007. pp. 365-371

[90] Majoul T, Bancel E, Triboï E, Ben Hamida J, Branlard G. Proteomic analysis of the effect of heat stress on hexaploid wheat grain: Characterization of heat-responsive proteins from total endosperm. Proteomics. 2003;**3**:175-183

[91] Dias AS, Bagulho AS, Lidon FC. Ultra-structure and biochemical traits of bread and durum wheat grains under heat stress. Brazzilian Journal of Plant Physiology. 2008;**20**:323-333

[92] Hasanuzzaman M, Nahar K, Alam MM, Roychowdhury R, Fujita M. Physiological, biochemical, and molecular mechanisms of heat stress tolerance in plants. International Journal of Molecular Sciences. 2013;**14**(5):9643-9684. DOI: 10.3390/ijms14059643

[93] Wani SH, Kumar V, Shriram V, Sah SK. Phytohormones and their metabolic engineering for abiotic stress tolerance in crop plants. Crop Journal. 2016;**4**(3):162-176

[94] Finger R, Hediger W, Schmid S. Irrigation as adaptation strategy to climate change—A biophysical and economic appraisal for Swiss maize production. Climatic Change. 2011;**105**(3-4):509-528

[95] Liu Z, Hubbard KG, Lin X, Yang X. Negative effects of climate warming on maize yield are reversed by the changing of sowing date and cultivar selection in Northeast China. Global Change Biology. 2013;**19**(11):3481-3492

[96] Challinor AJ, Watson J, Lobell DB, Howden SM, Smith DR, Chhetri N. A meta-analysis of crop yield under climate change and adaptation. Nature Climate Change. 2014;**4**(4):287. DOI: 10.1038/nclimate2153

[97] Aguilera E, Lassaletta L, Gattinger A, Gimeno BS, et al. Managing soil carbon for climate change mitigation and adaptation in Mediterranean cropping systems: A meta-analysis. Agriculture, Ecosystems & Environment. 2013;**15**(168):25-36

[98] Dai MH, Shan CG, Wang P. Effect of temperature and solar ecological factors on spring maize production. Journal of China Agricultural University. 2009;**14**:35-41

[99] Dong X, Xu W, Zhang Y, Leskovar D. Effect of irrigation timing on root zone soil temperature, root growth and grain yield and chemical composition in corn. Agronomy. 2016;**6**(2):34. DOI: 10.3390/agronomy6020034

[100] Yuan BZ, Sun J, Nishiyama S. Effect of drip irrigation on strawberry growth and yield inside a plastic

greenhouse. Biosystems Engineering. 2004;**87**(2):237-245

[101] Maestri E, Klueva N, Perrotta C. Molecular genetics of heat tolerance and heat shock proteins in cereals. Plant Molecular Biology. 2002;**48**:667-681. DOI: 10.1023/A:1014826730024

[102] Gong M, Chen SN, Song YQ, Li ZG. Effect of calcium and calmodulin on intrinsic heat tolerance in relation to antioxidant systems in maize seedlings. Australian Journal of Plant Physiology. 1997;**24**:371-379

[103] Vardhini BV, Anjum NA. Brassinosteroids make plant life easier under abiotic stresses mainly by modulating major components of antioxidant defense system. Frontiers in Environmental Science. 2015;**2**:1-16

[104] Sharma N, Hundal GS, Sharma I, Bhardwaj R. 28-Homobrassinolide alters protein content and activities of glutathione-s-transferase and polyphenol oxidase in *Raphanus sativus* L. plants under heavy metal stress. Toxicology International. 2014;**21**(1):44-50. DOI: 10.4103/0971-6580.128792

[105] Mittler R, Finka A, Goloubinoff P. How do plants feel the heat? Trends in Biochemical Sciences. 2011;**37**:118-125

[106] Xu Y, Gianfagna T, Huang B. Proteomic changes associated with expression of a gene (ipt) controlling cytokinin synthesis for improving heat tolerance in a perennial grass species. Journal of Experimental Botany. 2010;**61**:3273-3289

[107] Dhakal Y, Meena RS, De N, Verma SK, Singh A. Growth, yield and nutrient content of mungbean (*Vigna radiata* L.) in response to INM in eastern Uttar Pradesh, India. Bangladesh Journal of Botany. 2015;**44**(3):479-482

[108] Dhakal Y, Meena RS, Kumar S. Effect of INM on nodulation, yield, quality and available nutrient status in soil after harvest of green gram. Legume Research. 2016;**39**(4):590-594

[109] Miura K, Tada Y. Regulation of water, salinity, and cold stress responses by salicylic acid. Frontiers in Plant Science. 2014;**5**:4. DOI: 10.3389/fpls.2014.00004

[110] Hara M, Furukawa J, Sato A, Mizoguchi T, Miura K. Abiotic stress and role of salicylic acid in plants. In: Ahmad P, Prasad M, editors. Abiotic Stress Responses in Plants. New York, NY: Springer; 2012. pp. 235-251

[111] Meena RS, Yadav RS. Yield and profitability of groundnut *(Arachis hypogaea* L) as influenced by sowing dates and nutrient levels with different varieties. Legume Research. 2015;**38**(6):791-797

[112] Meena RS, Dhakal Y, Bohra JS, Singh SP, Singh MK, Sanodiya P. Influence of bioinorganic combinations on yield, quality and economics of mungbean. American Journal of Experimental Agriculture. 2015;**8**(3):159-166

[113] Yamaguchi S. Gibberellin metabolism and its regulation. Annual Review of Plant Biology. 2008;**59**:225-251

[114] Choudhary SP, Yu JQ, Yamaguchi-Shinozaki K, Shinozaki K, Tran LS. Benefits of brassinosteroid crosstalk. Trends in Plant Science. 2012;**17**(10):594-605

[115] Singh A, Meena RS. Response of bioregulators and irrigation on plant height of Indian mustard (*Brassica juncea* L.). Journal of Oilseed Brassica. 2020;**11**(**1**):9-14

[116] Cakmak I. Plant nutrition research: Priorities to meet human needs for food in sustainable ways. Plant and Soil. 2002;**247**:3-24

[117] Waraich EA, Ahmad R, Ashraf MY, Saifullah AM. Improving agricultural water use efficiency by nutrient management in crop plants. Acta Agriculturae Scandinavica, Section B: Plant Soil Science. 2011;**61**(4):291-304

[118] Waraich EA, Ahmad R, Halim A, Aziz T. Alleviation of temperature stress by nutrient management in crop plants: A review. Journal of Soil Science and Plant Nutrition. 2012;**12**(2):221-244

[119] Meena RS, Lal R, Yadav GS. Long term impacts of topsoil depth and amendments on soil physical and hydrological properties of an Alfisol in central Ohio, USA. Geoderma. 2020;**363**:1141164

[120] Meena RS, Kumar S, Datta R, Lal R, Vijayakumar V, Britnicky M, et al. Impact of agrochemicals on soil microbiota and management: A review. Land. 2020;**9**:34. DOI: 10.1016/j.geoderma.2019.114164

[121] Reynolds MP, Singh RP, Ibrahim A, Ageeb OAA, Larque-Saavedra A, Quick JS. Evaluating physiological traits to complement empirical selection for wheat in warm environments. Euphytica. 1998;**100**:85-94

[122] Reynolds MP, Rebetzke G, Pellegrinesci A, Trethowan R. Dought adaptation in wheat. In: Ribaut JM, editor. Drought Tolerance in Cereals. New York: Haworth Food & Agricultural Products Press; 2006. pp. 402-436

[123] Amani I, Fischer RA, Reynolds MP. Canopy temperature depression associated with yield of irrigated spring wheat cultivars in a hot climate. Journal of Agronomy and Crop Science. 1996;**176**:119-129

[124] Freeha A, Abdul W, Farrukh J, Muhammad A. Influence of foliar applied thiurea on flag leaf gas exchange and yield parameters of bread wheat (*Triticum aestivum* L.) cultivars under salinity and heat stress. International Journal of Agriculture & Biology. 2008;**10**(6):619-626

[125] Zhao H, Dai T, Jiang D, Cao W. Effects of high temperature on key enzymes involved in starch and protein formation in grains of two wheat cultivars. Journal of Agronomy and Crop Science. 2008;**194**:47-54

[126] Shanahan JF, Edwards IB, Quick JS, Fenwick JR. Membrane thermostability and heat tolerance of spring wheat. Crop Science. 1990;**30**:247-251

[127] Ibrahim AMH, Quick JS. Genetic control of high temperature tolerance in wheat as measured by membrane thermal stability. Crop Science. 2001;**41**:1405-1407

[128] Towill LE, Mazur P. Studies on the reduction of 2,3,5-triphenyl tetrazolium chloride as a viability assay for plant tissue culture. Canadian Journal of Botany. 1974;**53**:1097-1102

[129] Yu Q, Hlavacka A, Matoh T, Volkmann D, Menzel D, Goldbach HE, et al. Short-term boron deprivation inhibits endocytosis of cell wall pectins in meristematic cells of maize and wheat root apices. Plant physiology. 2002;**30**(1):415-421. DOI: 10.1104/pp.006163

[130] Schoper JB, Lambert RJ, Vasilas BL, Westgate ME. Plant factors controlling seed set in maize—The influence of silk, pollen, and ear-leaf water status and tassel heat-temperature at pollination. Plant Physiology. 1987;**83**:121-125

[131] Frova C, Portaluppi P, Villa M, Sari GM. Sporophytic and gametophytic components of thermotolerance affected by pollen selection. Journal of Heredity. 1995;**86**(1):50-54

[132] Meena RS, Kumar V, Yadav GS, Mitran T. Response and interaction

of *Bradyrhizobium japonicum* and *Arbuscular mycorrhizal* fungi in the soybean rhizosphere: A review. Plant Growth Regulators. 2018;**84**:207-223

[133] Pottorff M, Roberts PA, Close TJ, Lonardi S, Wanamaker S, Ehlers JD. Identification of candidate genes and molecular markers for heatinduced brown discoloration of seed coats in cowpea [*Vigna unguiculata* (L.) Walp]. BMC Genomics. 2014;**15**:328. DOI: 10.1186/1471-2164-15-328

[134] Lavania D, Siddiqui MH, Al-Whaibi MH, Singh AK, Kumar R, Grover A. Genetic approaches for breeding heat stress tolerance in faba bean (*Vicia faba* L.). Acta Physiologiae Plantarum. 2015;**37**:1737. DOI: 10.1007/s11738-014-1737-z

[135] Bett K, Ramsay L, Sharpe A. Lentil genome sequencing: Establishing a comprehensive platform for molecular breeding. In: 6th International Food Legume Res. In Conf. and 7th International Conf. on Legume Genetics. Saskatoon, Saskatchewan, Canada. Crop Development Center: Saskatoon, SK. 2014. p. 19

[136] Manolio TA. Genome wide association studies and assessment of the risk of disease. New England Journal of Medicine. 2010;**363**:66-176

[137] Bush WS, Moore JH. Genome-wide association studies. PLoS Computational Biology. 2012;**8**:e1002822

[138] Huang Z, Zhang X, Zheng G, Gutterman Y. Influence of light, temperature, salinity and storage on seed germination of Haloxylon ammodendron. Journal of Arid Environments. 2003;**55**(3):453-464

[139] Iqbal MA, Mian MA. Boosting spring planted irrigated maize (*Zea mays* L.) grain yield with planting patterns adjustment. American-Eurasian Journal of Agricultural & Environmental Sciences. 2015;**15**(3):315-319

Section 3

Molecular Mechanisms against Abiotic Stresses

Chapter 10

Cys2His2 Zinc Finger Proteins Boost Survival Ability of Plants against Stress Conditions

Kemal Yuce and Ahmet Ismail Ozkan

Abstract

Zinc finger (ZnF) proteins are the largest transcription factors family. They constitute of nine sub-groups including Cys2His2, Cys3His, Cys3HisCys4, Cys2HisCys5, Cys4HisCys3, Cys2HisCys, Cys4, Cys6 and Cys8. ZnFs perform tasks of recognizing DNA, packaging RNA, transcriptional activity, regulating apoptosis, folding and collecting proteins, and binding lipids. One of the largest sub-groups of these proteins is ZF-Cys2His2, containing SIZ1, ZAT, ZAT7, ZFP1, ZFP252, DST, ZFP1, SIZF3, ZFP179, ZjZFN1, SICZFP1, and ZF-Cys2His2 proteins are found in plants tissues and fulfill important tasks in their defense to struggle with biotic and abiotic stresses (i.e., salt, drought, cold, oxidative). The aim of this chapter is to reveal importance of ZF-Cys2His2 proteins against various stress conditions.

Keywords: Cys2His2 Zinc finger proteins, plant stress physiology, salinity stress, cold stress, oxidative stress

1. Introduction

Cys2His2 zinc finger (ZF-Cys2His2) proteins have been found in a number of plants including *Arabidopsis*, cotton, rice and wheat. The ZF-Cys2His2 is built by two Cys and two His residues. This structure provides a conservative motif together with Zn^{2+}. The core has an α-helix and an anti-parallel double-stranded β-sheet so that the ZF-Cys2His2 proteins have stable and relatively independent protein domains. ZF-Cys2His2 proteins form a relatively wide family of transcriptional factors in plants. Current studies have revealed that ZF-Cys2His2 proteins operate as important transcriptional regulators in plant responses to a large broad of stress conditions (like drought, excessive light, extreme temperatures, oxidative stress, salinity, and silique shattering) [1]. Over-expression of some ZF-Cys2His2 genes has led to an increased tolerance to various stresses and activation of some stress-related genes [2]. Abscisic acid (ABA) is an important phytohormone involved in regulating stress responses and plant growth and development. In addition, ABA is involved in many important processes of plants such as stomatal closure, leaf senescence, cuticle wax accumulation, bud dormancy, osmotic regulation, seed germination and growth inhibition. Since the ABA regulates downstream responses to biotic and abiotic environmental changes through both transcriptional and post-transcription mechanisms, the responses of ABA and ZF-Cys2His2 proteins to various stresses are also mentioned here [3].

The world population has exceeded 7.5 billion. To supply the nutritional needs of this population, it is important to know the proteins and genes related to the response of plants to stress conditions. In this context, to increase the durability and yield of plants, transgenic plant researches are carried out. And the most commonly focused proteins in these researches are ZF-Cys2His2 proteins. For this reason, in this chapter, the changes occurring in ZF-Cys2His2 proteins with transgenic methods and various stress conditions and what these changes bring to the plant have been discussed (**Table 1**).

The kind of C2H2 zinc finger proteins	The kind of stress	References
AtSIZ1	Salt stress	[4]
GmZAT4	Salt stress	[5]
ZAT7	Salt stress	[6]
AhZFP1	Salt stress	[7]
ZFP252	Salt stress	[8]
ZmZF1	Salt stress	[9]
DST	Salt stress	[10]
GhZFP1	Salt stress	[11]
SIZF3	Salt stress	[12, 13]
ZFP179	Salt stress	[14]
ZjZFN1	Salt and cold stress	[15]
TaZNF	Salt stress	[16]
SICZFP1	Salt and cold stress	[17]
TaDi19A	Salt osmotic and cold stress	[18]
AtDi19-3	Salt and drought stress	[19]
AZF2 STZ	Salt, cold and drought stress	[20]
GsZFP1	Cold and drought stress	[21]
ZFP245	Cold and drought stress	[22]
GbZF1	Cold stress	[23]
PeSTZ1	Cold and oxidative stress	[24]
SCOF-1	Cold stress	[25]
OsCTZFP8	Cold stress	[26]
ZAT12 ZAT7	Oxidative stress	[27] [28]
ZFP36	Oxidative stress	[29]
GsZFP1	Cold stress	[21]
ZAT18	Oxidative and drought stress	[30]
OsMSR15	Drought stress	[31]
VTA2	Oxidative and fungal stress	[32]
MtSTOP	pH and aluminum stress	[33]
ART1	Aluminum stress	[34]

Table 1.
C2H2 zinc finger proteins related with plant stress.

2. Salinity stress and Cys2His2 zinc finger

One of the stress factors is salt. Plants are affected with development and yield from salt stress. Transgenic plant studies in combating salt stress have become one of the most important issues of our century and ZnF proteins attract a lot of attention in this context. Salt application has induced expression of AtSIZ1. The germination energy, index and rate, cotyledon growth rate and root length were found to be importantly higher than wild-type in lines where AtSIZ1 was over-expressed under various stress applications at the germination stage. However, these indicators decreased significantly in AtSIZ1 mutants. Higher proline, potassium and soluble sugar, lower sodium, malondialdehyde, sodium/potassium ratios were observed in the lines where over-expression occurred in the mature seedling stage compared to the wild-type. Stress-related marker genes such as AtGSTU5, AtP5CS1, COR15A, RD29A, RD29B and SOS1 have been found to be over-expressed in lines with an excessive expression than those of wild-type and mutant under salt application. Due to the results AtSIZ1 functions in maintaining both ionic homeostasis and osmotic balance to improve salt tolerance in *Arabidopsis* [4]. It has also been shown that GmZAT4 plays a significant role in both polyethylene glycol and sodium chloride stresses tolerance and ABA responses in both *A. thaliana* and soybean. Over-expression of GmZAT4 increased the tolerance of *A. thaliana* to 20% polyethylene glycol and 150 mM sodium chloride and increased the germination rate after 1 or 2 μM ABA administration [5]. The EAR motif for ZAT7 has a significant role of defense response to salt stress in *Arabidopsis*. Suppressing growth and being more tolerant to salt stress have observed in transgenic *Arabidopsis* plants expressing ZF-Cys2His2 protein ZAT7. Mutation or deletion of ZAT7's EAR motif did not affect growth suppression, but eliminated salt tolerance. These results showed that ZAT7's EAR motif is important in increasing salt stress tolerance of the transgenic plants. On the other hand, the EAR motif appears not to be involved in suppressing transgenic plants growth. Other analyzes of ZAT7 using RNAi lines suggested that ZAT7 functions as a suppressor of the defense response's suppressor [6]. Semi-quantitative RT-PCR studies have revealed that AhZFP1 expression is stimulated by salt stress in the peanut root, stem and leaves [7]. Excessive expression of ZFP252 in rice increased the amount of free proline, soluble sugar, stress defense genes expression and improved the plant's response to both salt and drought stress. This result showed that ZFP252 plays a significant role in both salt and drought stresses of rice plant. And it is also useful for genetically modified plants to increase both salt and drought stresses tolerance [8]. ZmZF1 transcripts are strongly stimulated by salt stress. It is also stimulated by drought stress and ABA. Over-expression of ZmZF1 under cauliflower mosaic virus 35S promoter control increased both salt and drought tolerance in young seedling phase of transgenic *Arabidopsis* [9]. Stomata control the absorption of carbon-dioxide and improve water use efficiency. Thus, it plays an important role in abiotic stress tolerance. H_2O_2, stimulates stomatal closure, is an important signaling molecule. Another ZFP, DST is directly organizing genes related to H_2O_2 homeostasis. Therefore, the stomatal closure is regulated negatively. The loss of DST function decreases stomatal density and increases stomatal closure. As a result, increased salt and drought tolerance appears in the rice plant. These findings provide a different perspective on the abiotic stress tolerance mechanism associated with stomata and also a significant genetic engineering approach in developing abiotic stress tolerance [10]. GhZFP1 over-expression in transgenic tobacco plant increased salt stress tolerance and *Rhizoctonia solani* resistance. This indicates that the plant can act as a significant regulator in responding to biotic and abiotic stress [11]. Ascorbic acid supports stress tolerance by breaking down reactive oxygen species (ROS). ROS degradation capacity of ascorbic acid is increased in plants that over-expressed SIZF3. This has increased the salt tolerance of plants. Findings in the studies showed that SIZF3 simultaneously supports both the collection of ascorbic acid and improves the plant salt tolerance response [12]. ZFP3 expression level was found to be highly suppressed by mannitol, sodium chloride and sucrose. The mutant *Arabidopsis* exhibited a reduced tolerance condition in terms of ZFP3. Excessive expression of ZFP3 reduced stress-related genes (such as AtP5CS1, KIN1, RD22 and RD29B) expression. These results showed that ZFP3 is important in osmotic and salt stress response [13]. Over-expression of ZFP179 increased salt tolerance in rice plant. And the transgenic seedlings showed hypersensitive activity to exogenic ABA. Increased amount of soluble sugars and free proline under salt stress have been observed in transgenic plants. In ZFP179 transgenic rice plant the oxidative stress tolerance, ability to break down ROS and expression level of stress-related genes such as OsDREB2A, OsLea3, OsP5CS and OsProT increased under salt stress. These findings showed that ZFP179 plays an important role in salt stress in plants and is useful for developing transgenic plants that are highly tolerant to salt stress [14]. Expression of ZjZFN1 was found to be increased by ABA, cold and salt. ZjZFN1 expression improved seed germination, plant adaptation to salt stress by growth under salt stress and increased green cotyledons in *Arabidopsis thaliana*. Physiological and transcriptional analyzes suggest that ZjZFN1 can affect the collection of ROS and regulate the transcription of genes associated with salt response. ZjZFN1 over-expressing plants RNA sequence analysis revealed that ZjZFN1 could function as a transcriptional activator in the regulation of stress response pathways such as alpha-linoleic acid metabolism, phenyl alanine metabolism and phenylpropanoid biosynthesis pathways. The results showed that ZjZFN1 plays an important role in the formation of salt stress response in plants [15]. In another study, it has been proven that TaZnF, a ZF-Cys2His2 protein, significantly improves salt tolerance of transgenic *Arabidopsis*. Various physiological indexes of transgenic plants showed improvement under salt stress compared to the control group. The noninvasive micro-test (NMT) technique showed the excretion of Na is significantly accomplished by the root end of the transgenic *Arabidopsis*. TaZnF is mainly found in the nucleus and has demonstrated transcriptional activity [16].

3. Cold stress and C2H2 zinc finger proteins

GsZFP1 was found to be stimulated by ABA (100 μM), cold (4°C) and salt (200 mM sodium chloride) in leaves and by ABA (100 μM), cold (4°C) and drought (30% PEG 6000) in root. It was found that over-expression of GsZFP1 in transgenic *Arabidopsis* resulted in higher tolerance in cold and drought stress and a reduced rate of water loss. Over-expression of GsZFP1 was found to increase the expression of CBF1, CBF2, CBF3, COR47, NCED3 and RD29A stress response genes in response to cold stress, and increased expression of NCED3, P5CS, RAB18, RD22 and RD29A in response to drought stress [17]. It was revealed by the semi-quantitative RT-PCR experiment that ZFP245 was strongly stimulated after 6 hours of cold and drought stress and then decreased to normal level. ZFP245 did not occur by the application of high salt or ABA [18]. High degree of aggregation of GmZF1 mRNAs stimulated by exogenic ABA suggested that GmZF1 was involved in the ABA-dependent signal transduction pathway. GmZF1 over-expression increased expression of the cold-induced cor6.6 gene by recognizing the likely protein DNA binding site. This suggests that GbZF1 of soybean can improve *Arabidopsis* cold stress tolerance by regulating cold regulatory genes expression in transgenic *Arabidopsis* [19]. PeSTZ1 increases freezing tolerance through organizing the breakdown of ROS by directly regulating PeAPX. PeSTZ1 is preferably expressed in young roots. This ZFP has been upregulated in cold applications. PeSTZ1 functions as a transcriptional activator to increase cold tolerance. PeSTZ1 upregulation decreases malondialdehyde and ROS aggregation by activating antioxidant systems. This is thought to be achieved through direct regulation of PeAPX2 expression [20]. SCOF-1 transcription is particularly stimulated by ABA and low temperature, but not neither with dehydration nor high salt. SCOF-1 over-expression stimulated the expression of genes regulated by cold and improved the cold tolerance of tobacco plants and non-acclimated transgenic *Arabidopsis.* SCOF-1 can function as a positive regulator of COR gene expression regulated by ABRE through protein-protein interaction and thus can improve the cold tolerance of plants [21]. SICZFP1 is strongly stimulated by cold stress, dehydration and salt stress, but not by ABA. SICZFP1 over-expression in transgenic rice and *Arabidopsis* plants increased expression of cold-response-related genes. This suggests that SICZFP1 plays an important role in the response of plants to cold stress [22]. TaDi19A has been found to be expressed in both leaves and roots of wheat seedlings grown under stress-free conditions, but is significantly upregulated in salt, osmotic and cold stress conditions, or in hormone applications such as stress-related ethylene and ABA. Heterologous over-expression of TaDi19A in *Arabidopsis thaliana* increased salinity, mannitol and ABA stress sensitivity during the germination of plants. Root elongation in these transgenic lines showed less salinity stress tolerance and less ethephon sensitivity. The expression of ABA signal path genes such as ABA1, ABF3, ERD15, RAB18 and SOS2 (SOS pathway) have changed in transgenic plants [23]. Transgenic rice expressing excessive OsCTZFP8 exhibited cold-tolerant phenotypes with significantly higher pollen fertility and seed placement rates than non-transgenic control plants. Yield of lines expressing OsCTZFP8 per plant was significantly higher ($p < 0.01$) than non-transgenic control plants under cold application. This result shows that OsCTZFP8 is a C2H2 zinc finger transcription factor that plays an important role in cold tolerance in rice [24].

4. Oxidative stress and C2H2 zinc finger proteins

Cytosolic Apx1, ascorbate peroxidase 1, is an important H_2O_2-removing enzyme in plants. Both WRKY transcription factor (WRKY25) and two ZnF proteins (ZAT12 and ZAT7) expressions have increased in Apx1 gene suppressed plants that grown under the controlled conditions. When cells were exposed to oxidative stress, heat shock, and injury, the WRKY25, ZAT7 and ZAT12 expressions increased together. However, light and osmotic stresses did not increase them. Transgenic plants expressing ZAT7 and ZAT12 were able to tolerate oxidative stress. WRKY25, ZAT7 or ZAT12 expression in transgenic plants did not increase Apx1 expression under controlled conditions. Plants without ZAT12 could not increase Apx1, WRKY25 and ZAT7 expressions in response to H_2O_2, so that plants without ZAT12 have become more sensitive to H_2O_2 applications than normal plants. It has been revealed that ZAT12 is an important component of oxidative stress signal transmission in *Arabidopsis* and needs Apx1, WRKY25 and ZAT7 during oxidative stress [25]. Transcription profiles of plants that are exposed to H_2O_2 stress, expressing excessive ZAT12 and normal have revealed that the basic expression of ZAT12 in *Arabidopsis* results in increased expression in oxidative and mild stress transcripts. Thus, ZAT12 has been shown to play a key role in reactive oxygen and abiotic stress signal in *Arabidopsis* [26]. ABA application stimulated the increase of both ascorbate peroxidase and superoxide dismutase activities and OsMPK1, OsMPK5 and ZFP182 expressions in rice plant leaves. It has been noted that ABA-induced antioxidant defense needs ZFP182. And the ZFP182 expression is regulated by rice MAPKs in ABA signaling [27]. ABA-activated mitogen active protein kinases

(MAPKs) and ABA-induced H_2O_2 production have been shown to regulate ZFP36 expression in ABA signaling, on the other hand, NADPH oxidase genes expression regulated by ZFP36, H_2O_2 production and the expression of OsMPK genes in ABA signaling. It has been noted that rice plants requires to ZFP36 for tolerance of both water stress and oxidative stresses, for ABA-induced antioxidant defense and for cross-interaction between NADPH oxidase, H_2O_2 and MAPK in the ABA signal [28].

5. Drought stress and C2H2 zinc finger proteins

Over-expression of ZAT18 has increased drought tolerance in *Arabidopsis* while ZAT18 mutation reduced plant tolerance to drought stress. It was found that ZAT18 is expressed in stems and herbal rosette leaves. Sub-cellularly, it has been noted that ZAT18 protein is mainly localized in the nucleus. Overexpressed ZAT18 plants exhibited higher leaf water content, lower ROS content, higher antioxidant enzyme activity compared to wild-type after drought application. RNA sequencing analysis showed that the 423 and 561 genes were transcriptionally regulated by the ZAT18 transgene before and after drought application, respectively. Path enrichment analysis showed that hormone metabolism, stress and signal were over represented in ZAT18 overexpression lines. Various stress-sensitive genes (including COR47, ERD7, LEA6 and RAS1) and hormone signal transduction-related genes (including JAZ7 and PYL5) have been identified as target genes of ZAT18. Taken together, ZAT18 acts as a positive regulator and plays an important role in the plant's response to drought stress [29]. Di19 family, protein 19 induced by drought, is a ZF-Cys2His2 protein. AtDi19-3 is a transcription activator. In plants, AtDi19-3 is significantly stimulated by ABA, mannitol and sodium chloride. Mutation of AtDi19-3 increased tolerance of plants against ABA, drought and high salinity. Excessive expression of AtDi19-3 caused drought, salt and ABA sensitivity. Seed germination and cotyledon greening rates were higher in the Atdi19-3 mutant under the condition of applying sodium chloride, mannitol or ABA. But this data in transgenic plants, in which AtDi19-3 was over-expressed, was lower than in normal plants. Although the roots of AtDi19-3 mutant seedings were longer, the roots of AtDi19-3 over-expressing transgenic seedlings were shorter than in normal plants. Although the chlorophyll and proline contents in the AtDi19-3 mutant were higher, these data of AtDi19-3 over-expressed seedlings were lower than that of normal plants. While the AtDi19-3 mutant tolerates more droughts, transgenic plants that over-express AtDi19-3 showed more susceptibility to drought than normal plants. In addition, ABA related genes expression signaling pathway varied in the Atdi19-3 mutant and in over-expressing plants of AtDi19-3 [30]. AZF2 and STZ expression has been shown to be strongly stimulated by ABA, cold and high salt stresses, and dehydration. Transgenic *Arabidopsis*, which over-expresses STZ, has become tolerant to growth delay and drought stress. In that study, it was suggested that STZ and AZF2 act as transcriptional suppressors on increased stress tolerance during the growth delay [31]. The expression of OsMSR15 in *Arabidopsis* provides it with drought resistance. Hypersensitive activity to exogenic ABA in this transgenic plant growth during sprouting and seed germination. Transgenic plants also showed less electrolyte leakage, high levels of free proline and increased expressions of a number of stress sensitive genes (such as LEA3, DREB1A, P5CS1 and RD29A) under drought stress. The results showed that OsMSR15 is a significant regulator that plays a role against plant drought stress [32]. In another study, over-expression of TaZFP34 has been shown to have a negative effect on yield performance and mimic the negative effect of drought stress on wheat productivity [33]. STF-2 is also a ZF-Cys2His2 protein.

And its transgenically over-expression increased significantly drought tolerance in transgenic tobacco. The results showed that STF-2 plays a significant role in the response of soybean's to drought stress [34].

6. Other stress and C2H2 zinc finger proteins

VTA2 contains the ZF-Cys2His2 proteins. It controls H_2O_2 detoxification and host plant root infection. And it is an important regulator of fungal pathogenesis [35]. StZFP2 is a Q-type C2H2 zinc finger transcription factor induced by injury and invasion. Previous studies show that Q-type C2H2 TFs are involved in responding to stress and may be protective against drought, salinity, or pathogenic infections when overexpressed. The increase in StPIN2, a classic marker for insect defense in potatoes, was consistent with decreases in larval weight gain [36]. MtSTOP is also a C2H2 zinc finger protein and regulates Medicago's response to H^+ and Al^{3+} toxicity. MtSTOP is expressed in root, stem, nodule and other tissues. MtSTOP is upregulated with acidic pH and Al^{3+} stress or a combination of both. Growth or morphology in mtSTOP mutants did not change under normal conditions; however, mutant seedlings are characterized by significantly reduced root elongation and are sensitive to low pH (pH 4.3) and Al^{3+} stress. Compared to its control, more Al accumulated in the mutant roots and citric acid secreted from the mutant roots was significantly lower in both normal and Al stress conditions. This indicates that MtSTOP hair roots synthesize more citric and malic acids [37]. Regarding aluminum, Rice (*Oryza sativa*) is one of the aluminum tolerant species among small grain cereals. This type of aluminum tolerance occurs with many genes involved in the detoxification of Al at the cellular level. These findings emerged with ART1. ART1 is a C2H2 type zinc finger transcription factor and regulates the expression of 31 genes as down-stream [38].

Conflict of interest

The authors declare no conflict of interest.

Financial support

This work was supported by the Selcuk University and Dicle University.

Author details

Kemal Yuce[1*] and Ahmet Ismail Ozkan[2]

1 Faculty of Medicine, Department of Physiology, Selcuk University, Konya, Turkey

2 Graduate School of Natural and Applied Sciences, Department of Biology, Dicle University, Diyarbakir, Turkey

*Address all correspondence to: yuce@selcuk.edu.tr

References

[1] Wang K, Ding Y, Cai C, Chen Z, Zhu C. The role of C2H2 zinc finger proteins in plant responses to abiotic stresses. Physiologia Plantarum. 2018;**165**(4):690-700

[2] Kiełbowicz-Matuk A. Involvement of plant C2H2-type zinc finger transcription factors in stress responses. Plant Science. 2012;**185-186**:78-85

[3] Chen K, Li GJ, Bressan RA, Song CP, Zhu JK, Zhao Y. Abscisic acid dynamics, signaling, and functions in plants. Journal of Integrative Plant Biology. 2020;**62**(1):25-54

[4] Han G, Yuan F, Guo J, Zhang Y, Sui N, Wang B. AtSIZ1 improves salt tolerance by maintaining ionic homeostasis and osmotic balance in *Arabidopsis*. Plant Science. 2019;**285**:55-67

[5] Sun Z, Liu R, Guo B, Huang K, Wang L, Han Y, et al. Ectopic expression of GmZAT4, a putative C2H2-type zinc finger protein, enhances PEG and NaCl stress tolerances in *Arabidopsis thaliana*. 3 Biotech. 2019;**9**(5)

[6] Ciftci-Yilmaz S, Morsy MR, Song L, Coutu A, Krizek BA, Lewis MW, et al. The EAR-motif of the Cys2/His2-type zinc finger protein Zat7 plays a key role in the defense response of *Arabidopsis* to salinity stress. Journal of Biological Chemistry. 2007;**282**(12):9260-9268

[7] Pan L, Yang Q, Chi X, Chen M, He Y, Yu S. AhZEP1, a cDNA encoding C2H2-type zinc finger protein, induced by salt stress in peanut (*Arachis hypogaea* L.). In: 2010 4th International Conference on Bioinformatics and Biomedical Engineering. 2010. pp. 1-7

[8] Xu D-Q, Huang J, Guo S-Q, Yang X, Bao Y-M, Tang H-J, et al. Overexpression of a TFIIIA-type zinc finger protein geneZFP252enhances drought and salt tolerance in rice (*Oryza sativa* L.). FEBS Letters. 2008;**582**(7):1037-1043

[9] Huai J, Zheng J, Wang G. Overexpression of a new Cys2/His2 zinc finger protein ZmZF1 from maize confers salt and drought tolerance in transgenic *Arabidopsis*. Plant Cell, Tissue and Organ Culture (PCTOC). 2009;**99**(2):117-124

[10] Huang XY, Chao DY, Gao JP, Zhu MZ, Shi M, Lin HX. A previously unknown zinc finger protein, DST, regulates drought and salt tolerance in rice via stomatal aperture control. Genes & Development. 2009;**23**(15):1805-1817

[11] Guo Y-H, Yu Y-P, Wang D, Wu C-A, Yang G-D, Huang J-G, et al. GhZFP1, a novel CCCH-type zinc finger protein from cotton, enhances salt stress tolerance and fungal disease resistance in transgenic tobacco by interacting with GZIRD21A and GZIPR5. New Phytologist. 2009;**183**(1):62-75

[12] Li Y, Chu Z, Luo J, Zhou Y, Cai Y, Lu Y, et al. The C2H2 zinc-finger protein SlZF3 regulates AsA synthesis and salt tolerance by interacting with CSN5B. Plant Biotechnology Journal. 2018;**16**(6):1201-1213

[13] Wu K, Zhang A, Liu D, Hua C, Yan A, Liu B, et al. The *Arabidopsis* gene zinc finger protein 3(ZFP3) is involved in salt stress and osmotic stress response. PLOS One. 2016;**11**(12):e0168367

[14] Sun SJ, Guo SQ, Yang X, Bao YM, Tang HJ, Sun H, et al. Functional analysis of a novel Cys2/His2-type zinc finger protein involved in salt tolerance in rice. Journal of Experimental Botany. 2010;**61**(10):2807-2818

[15] Teng K, Tan P, Guo W, Yue Y, Fan X, Wu J. Heterologous expression of a novel *Zoysia japonica* C2H2 zinc finger

gene, ZjZFN1, improved salt tolerance in *Arabidopsis*. Frontiers in Plant Science. 2018;**9**

[16] Ma X, Liang W, Gu P, Huang Z. Salt tolerance function of the novel C2H2-type zinc finger protein TaZNF in wheat. Plant Physiology and Biochemistry. 2016;**106**:129-140

[17] Luo X, Bai X, Zhu D, Li Y, Ji W, Cai H, et al. GsZFP1, a new Cys2/His2-type zinc-finger protein, is a positive regulator of plant tolerance to cold and drought stress. Planta. 2011;**235**(6):1141-1155

[18] Huang J, Wang JF, Wang QH, Zhang HS. Identification of a rice zinc finger protein whose expression is transiently induced by drought, cold but not by salinity and abscisic acid. DNA Sequence. 2009;**16**(2):130-136

[19] Chen F, Yu G-H, Jiang L-L, Ma X-F, Xu Z-S, Liu M-M, et al. A soybean C2H2-type zinc finger gene GmZF1 enhanced cold tolerance in transgenic *Arabidopsis*. PLOS One. 2014;**9**(10):e109399

[20] He F, Li HG, Wang JJ, Su Y, Wang HL, Feng CH, et al. Pe STZ 1, a C2H2-type zinc finger transcription factor from *Populus euphratica*, enhances freezing tolerance through modulation of ROS scavenging by directly regulating Pe APX 2. Plant Biotechnology Journal. 2019;**17**(11):2169-2183

[21] Kim JC, Lee SH, Cheong YH, Yoo C-M, Lee SI, Chun HJ, et al. A novel cold-inducible zinc finger protein from soybean, SCOF-1, enhances cold tolerance in transgenic plants. The Plant Journal. 2001;**25**(3):247-259

[22] Zhang X, Guo X, Lei C, Cheng Z, Lin Q, Wang J, et al. Overexpression of SlCZFP1, a novel TFIIIA-type zinc finger protein from tomato, confers enhanced cold tolerance in transgenic *Arabidopsis* and rice. Plant Molecular Biology Reporter. 2010;**29**(1):185-196

[23] Li S, Xu C, Yang Y, Xia G. Functional analysis ofTaDi19A, a salt-responsive gene in wheat. Plant, Cell & Environment. 2010

[24] Jin Y-M, Piao R, Yan Y-F, Chen M, Wang L, He H, et al. Overexpression of a new zinc finger protein transcription factors CTZFP8 improves cold tolerance in rice. International Journal of Genomics. 2018;**2018**:1-13

[25] Rizhsky L, Davletova S, Liang H, Mittler R. The zinc finger protein Zat12 is required for cytosolic ascorbate peroxidase 1 expression during oxidative stress in *Arabidopsis*. Journal of Biological Chemistry. 2004;**279**(12):11736-11743

[26] Davletova S, Schlauch K, Coutu J, Mittler R. The zinc-finger protein Zat12 plays a central role in reactive oxygen and abiotic stress signaling in *Arabidopsis*. Plant Physiology. 2005;**139**(2):847-856

[27] Zhang H, Ni L, Liu Y, Wang Y, Zhang A, Tan M, et al. The C2H2-type zinc finger protein ZFP182 is involved in abscisic acid-induced antioxidant defense in rice. Journal of Integrative Plant Biology. 2012;**54**(7):500-510

[28] Zhang H, Liu Y, Wen F, Yao D, Wang L, Guo J, et al. A novel rice C2H2-type zinc finger protein, ZFP36, is a key player involved in abscisic acid-induced antioxidant defence and oxidative stress tolerance in rice. Journal of Experimental Botany. 2014;**65**(20):5795-5809

[29] Yin M, Wang Y, Zhang L, Li J, Quan W, Yang L, et al. The *Arabidopsis* Cys2/His2 zinc finger transcription factor ZAT18 is a positive regulator of plant tolerance to drought stress. Journal of Experimental Botany. 2017;**68**(11):2991-3005

[30] Qin LX, Li Y, Li DD, Xu WL, Zheng Y, Li XB. *Arabidopsis* drought-induced protein Di19-3 participates in plant response to drought and high salinity stresses. Plant Molecular Biology. 2014;**86**(6):609-625

[31] Sakamoto H, Maruyama K, Sakuma Y, Meshi T, Iwabuchi M, Shinozaki K, et al. *Arabidopsis* Cys2/His2-type zinc-finger proteins function as transcription repressors under drought, cold, and high-salinity stress conditions. Plant Physiology. 2004;**136**(1):2734-2746

[32] Zhang X, Zhang B, Li MJ, Yin XM, Huang LF, Cui YC, et al. OsMSR15 encoding a rice C2H2-type zinc finger protein confers enhanced drought tolerance in transgenic *Arabidopsis*. Journal of Plant Biology. 2016;**59**(3):271-281

[33] Chang H, Chen D, Kam J, Richardson T, Drenth J, Guo X, et al. Abiotic stress upregulated TaZFP34 represses the expression of type-B response regulator and SHY2 genes and enhances root to shoot ratio in wheat. Plant Science. 2016;**252**:88-102

[34] Song B, Zhang Y, Li Y, Fu Y, Wang P. Novel drought-inducible Cys2/His2-type zinc finger protein STF-2 from soybean (*Glycine max*) enhances drought tolerance in transgenic plants. Pakistan Journal of Botany. 2019;**51**(3)

[35] Tran V-T, Braus-Stromeyer SA, Kusch H, Reusche M, Kaever A, Kühn A, et al. Verticillium transcription activator of adhesion Vta2 suppresses microsclerotia formation and is required for systemic infection of plant roots. New Phytologist. 2014;**202**(2):565-581

[36] Lawrence SD, Novak NG. Over-expression of StZFP2 in *Solanum tuberosum* L. var. Kennebec (potato) inhibits growth of Tobacco Hornworm larvae (THW, *Manduca sexta* L.). Plant Signaling & Behavior. 2018;**13**(7):e1489668

[37] Wang J, Hou Q, Li P, Yang L, Sun X, Benedito VA, et al. Diverse functions of multidrug and toxin extrusion (MATE) transporters in citric acid efflux and metal homeostasis in *Medicago truncatula*. The Plant Journal. 2017;**90**(1):79-95

[38] Tsutsui T, Yamaji N, Feng MJ. Identification of a cis-acting element of ART1, a C2H2-type zinc-finger transcription factor for aluminum tolerance in rice. Plant Physiology. 2011;**156**(2):925-931

Chapter 11

Genes for Different Abiotic Stresses Tolerance in Wheat

Sudhir Kumar, Shampa Purkyastha, Chandan Roy, Tushar Ranjan and Rakesh Deo Ranjan

Abstract

In the recent past years, global warming and climate change have drastically affected the agricultural crop productivity grown in tropical and subtropical areas globally by appearing to several new biotic and abiotic stresses. Among the abiotic stresses, heat, drought, moisture, and salt stresses are most prevalent. Wheat is the most common and widely used crops due to its economic and social values. Many parts of the world depend on this crop for food and feed, and its productivity is highly vulnerable to these abiotic stresses. Improving tolerance to these abiotic stresses is a very challenging assignment for wheat researchers, and more research is needed to better understand these stresses. The progress made in understanding these abiotic stress tolerances is due to advances in three main research areas: physiology, genetic, and breeding research. The physiology research focused on the alternative physiological and biochemical metabolic pathways that plants use when exposed to abiotic stresses. Identifying genes contributing to particular stress tolerance is very important. New wheat genotypes having a high degree of abiotic stress tolerance are produced through marker-assisted breeding by making crosses from promising concerned stress-tolerant genotypes and selecting among their progeny using gene-specific markers.

Keywords: climate change, abiotic stress, wheat, physiology, genetic, marker assisted breeding, phytohormones

1. Introduction

Wheat is the second most important cereal crops of the world occupying about 220 million hectares area (mha) with a production of 716 million tons of food grain with a productivity of 3.2 tons per hectare [1]. It is extensively grown in Asia particularly in China and India. In India its production is enhances after the green revolution of late 1960s followed by another green evolution during 1980s. During these two green revolutions, the rate of annual growth in wheat production globally was ~3%, but in recent years it is declined to <0.9% due to appearance of new biotic and abiotic stresses. Although currently, the global wheat production has been able to meet the current demand and consumption, but we will have to enhance production and achieve the targets of at least ~858 Mt to meet the demand in 2050, as against current global production of 763 Mt. It comprises amounts to at least ~15% desired increase in global wheat production (1.5% annual increase)

IntechOpen

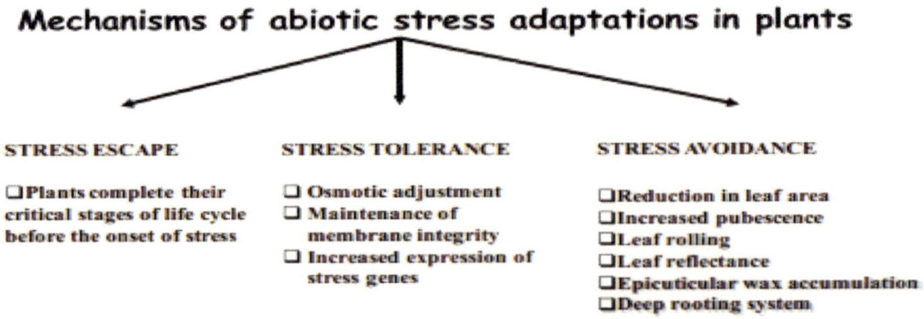

Figure 1.
Abiotic stress adaptive mechanism and their associated traits in plants heat stress tolerance.

during the next three decades to feed the global human population, which is estimated to reach ~9.7 billion in 2050 (https://population.un.org/wpp/). It is quite challenging to achieve this target production despite of shrinkage in arable land due to urbanization, and the probable negative impact of climate change. Due to its significant contribution to global food security, it is very much essential to improve its production and productivity to feed the ever increasing population on limited cultivated land. However, the most remarkable environmental concern in agriculture is the increase of global temperature. With regard to global climate models, the mean ambient temperature is predicted to increase by 1–6°C by the end of twenty-first century [2]. Such increase of global temperature may have a significant influence on agricultural productivity in accordance with the severity of the high temperature, drought, salinity, water logging, and mineral toxicity stresses (**Figure 1**).

2. Heat stress tolerance

High temperature-induced heat stress is expressed as the rise in air temperature beyond a threshold level for a particular period which is sufficient to cause injury or irremediable damage of crop plants in general [3]. The heat stress situation is become more intense when soil temperature increases due to increase in air temperature associated with decline in soil moisture. It negatively affects the yield attributing traits and ultimately results in reduction in wheat productivity. Some indicators of heat stress effects in wheat are illustrated in **Figure 2**. Wheat is very sensitive to heat stress particularly in some physiological growth stages. It has been estimated that reduction in global wheat yield falls by 6% for each 1°C of further temperature rise [4]. The low latitudes showed a distinct increase in simulated yield variability with higher temperature than that observed at high latitudes. This greater relative yield decline was due to the higher reference temperature [5]. The effects of heat stress on plants are very complex resulting in alteration of growth and development, changes in physiological functions, and reduced grain formation and yield.

Heat stress leads to changes in plant water relations, reduction of photosynthetic capacity, decreases of metabolic activities and changes of hormones, production of oxidative reactive species, promotion of ethylene production, reduction of pollen tube development, and increases of pollen mortality [1] in wheat. During the period from 1880 to 2012, the Earth's system warmed by 0.85°C [6]. This warming period will be continue and is predicted to rise between the range of 1.5–4.0°C in the future [7]. The changes in climatic factors like temperature, precipitation, CO_2, weather

Figure 2.
Major effects of heat stress on plants growth and development. Pn, Rs, and Ci indicate photosynthesis, stomatal conductance, and intercellular CO_2 concentration respectively.

variability, and soil moisture deficit would have positive or negative effects on crop system which will appears in its production level. The deleterious impacts of climate change on crop production are challenging the food security of the world and it is predicted that sustaining wheat production will be impacted more by increasing temperature. High temperature affects crops in different ways including poor germination and plant establishment, reduced photosynthesis, leaf senescence, decreased pollen viability, and consequently production of less grain with smaller grain size. Degree of such effect varies depending on the crops, cultivars, phenological stages, sowing dates and management practices. Some other adaptation measures are related to surface cooling by irrigation, antioxidants defense [8] and osmoprotectants [3, 5] minimizes the effects of heat stress. However, development of heat-tolerant wheat varieties and generation of improved pre-breeding materials for any breeding program in future is crucial in meeting the food security [9]. Proteomic and transcriptomic data are important to identifying genes and proteins that respond to environment, and affects yield and quality of wheat.

2.1 Genetic management

Breeding is a strategy for genetic manipulation of crop and its adaptation response under changing environment. Therefore, it requires the evaluation of genetic diversity of existing germplasm for the selection and induction of stress inducible genes/QTLs of genetic resources for developing new varieties in the production systems.

Sl. no.	Traits/QTL	Phenotypic variance (%)	Linked marker (position in cM)	Physical position (Mbp)[d]	References
I. Agronomic traits					
	1. Grain yield				
	a. Q.Yld.aww-3B-2	22	XWPT8021-Xgwm0114B (190.7)	802.3	—
	2. Thousand grain weight				
	a. Qtgw.iiwbr-2A	23.7	Xgwm12280.8	(174.41)	—
	b. QHthsitgw.bhu-7B	20.3	Xgwm1025–Xgwm745 (144.1)	ND	[12]
	c. 2A (36.1)[c]		224,948|F|0-9:T > A-9:T > A-kukri_c22235_1549 (21–24)	ND	[13, 14]
	3. Grain weight per spike				
	a. Qtgws.iiwbr-2A	28.9	Xgwm497.1 (41.61)	684	—
	b. Qgws.iiwbr-2A	19.9	Xgwm122 (171.41)	80.8	—
	4. Grain number per spike				
	a. Qlgns.iiwbr-2A	23.16	Xgwm372 (149.01)	203.3	—
	b. Qgns.iiwbr-2A	20.04	Xgwm448 (166.51)	154.4	—
	5. Kernel number per spike				
	a. QHknm.tam-2B	21.6	Xgwm111.2 (36.9)	786.6	[15]
	6. Kernel weight per main spike				
	a. QHkwm.tam-3B	19	Xwmc527 (89.8) 540.2		[15]
	b. QHkwm.tam-3B	21.2	Xwmc326 (123.6)	778.7	[15]
	7. Single kernel weight of main spike				
	a. QHskm.tam-1A	22.6	Xcfa2129 (43.2)	513.7	[15]
	b. QHskm.tam-2A	21	Xgwm356 (129.5)	670.6	[15]
II. Physiological traits					
	1. Grain filling duration				
	a. QHgfd.iiwbr-5A	22	X1079678|F|0 (107.5)	ND	[16]
	b. QHthsigfd.bhu-2B	20.2	Xgwm935–Xgwm1273 (385.3)	ND	[12]
	2. Ear emergence time				
	a. Q.Eet.aww-7A-2	39	XPPDD1-XWPT0330 (35)	63.5	—
	3. Canopy temperature: grain filling				
	a. Q.Ctgf.aww-3B	21	XWPT-8021–Xgwm0114B (192.7)	802.3	—
	4. Canopy temperature depression				
	a. QHtctd.bhu-7B	19.8	Xgwm1025–Xgwm745 (144.1)	ND	[12]

Table 1.
List of major and stable QTL for heat tolerance-related traits in wheat.

2.1.1 Stable QTLs for heat stress-related traits

Recent advances in molecular science play an important role to understand the complexity of stress response mechanisms under heat stress conditions and emphasized on the knowledge of molecular pathways and protective mechanisms to breed heat stress tolerant plants. Heat tolerance is obviously a polygenic trait, and the molecular techniques also help in analyzing the genetic basis of plant thermo tolerance. QTL mapping and subsequent marker-assisted selection made it possible to better understanding the heat tolerance in plants [10]. Recently several QTLs for different yield component traits have been identified which can be used for developing heat tolerance in wheat. For example, QTLs for heat tolerance has been identified for grain weight and grain-filling duration, senescence-related traits and canopy temperature. Besides others recognized QTLs present on chromosomes 2B, 5B and 4A in wheat under heat stress conditions [11]. The electrolyte leakage is an indication of reduced cell membrane thermo stability (CMT) which reflects the performance of wheat genotypes under heat shock. Genotypes generating heat shock proteins (HSPs) can withstand heat stress as they protect proteins from heat-induced damage. It has been also suggested that the abundance of small heat shock protein and superoxide dismutase during milky-dough stage plays a vital role in the biosynthesis of starch granule, and this will help to develop heat-tolerant wheat cultivars containing high grain quality. A large number major and stable QTLs were reported (**Table 1**), which included for agronomic traits and for physiological traits showing ≥20% phenotypic variances. These QTLs may prove useful for improvement of such traits using marker assisted selection (MAS).

2.1.2 Biotechnological approach for improving heat tolerance

Genetic engineering and transgenic approaches can diminish the adverse effects of heat stress by improving heat tolerance mechanisms [17]. It involves the incorporation of genes for heat tolerance into the desired plants [18]. However, the complexity of the genomic pattern makes it difficult to research for genetic modification in wheat. Prolong exposes to heat stress leads to increases in production of protein synthesis elongation factor (EF-Tu) in chloroplast which is associated with heat tolerance in wheat. The constitutive expression of EF-Tu in transgenic wheat protected leaf proteins against thermal aggregation, reduced thylakoid membranes disruption, enhanced photosynthetic capability, and resisted pathogenic microbes infection [19], hence the wheat genotypes having more EF-Tu showed better tolerance to heat stress as compared to genotypes with less EF-Tu [20]. Recently, it have been found that many transcription factors (TFs) involved in various abiotic stresses and engineered to improve stress tolerance in crops [21].

3. Drought stress tolerance

Drought stress can be simply defined as a scarcity of water which leads to dramatic changes in morphological, biochemical, physiological, and molecular features [22]. All of these changes hamper plant growth and crop production. Negative impact of drought stress appears at any growth stage and level of adverse effects depends on stage specific stresses and local environment. Therefore, genotypes may be tested for their drought tolerance at different particular growth stages. Severity of drought induced damage on plants depending on plant genotype and growth stage. Some genotypes may show tolerance to drought at germination or seedling stage, but these may be very sensitive to drought at the flowering stage or vice versa. Globally, more than 50% of the wheat cultivated land is exposed to periodic drought which causes losses up to 9–10% in production. Furthermore, decrease in precipitation and increasing evaporation as a consequence of global warming may expected to increase in frequency of drought and its severity in the future. Therefore, understanding the drought induced damages in wheat plants and approaches to improve drought tolerance is crucial to increase wheat productivity. Drought stress imposes damaging effects on several plants physiological processes occur in its different growth stages such as germination, vegetative growth, reproductive, and maturity. Under such stress conditions plant restricts the photosynthesis, respiration, transpiration, uptake and transportation of water and nutrient and translocation of assimilates. Drought stress damages the cell membrane structure, disorganization of ultra-structural cellular components and disruption of its properties, enzyme activities and anion and cationic imbalance are some of the major reasons for disturbing plant physiological processes. Drought stress usually leads to the production of reactive oxygen species (ROS). Hydrogen peroxide (H_2O_2), superoxide (O_2^-), singlet oxygen ($^1O_2$), and hydroxyl radicals (OH^-) are the most common species which are generated due to iron-catalyzed Fenton reaction due to the activities of lipoxygenases, peroxidases (POX), NADPH oxidase, and xanthine oxidase. The ROS in any form causes substantial damage to cell components and can cause cell death [23]. Plants have a very much evolved antioxidant defense system to rummage and keep up a reasonable degree of ROS to keep cells from oxidative harm. Under cell antioxidant defense system, it have some nonenzymatic antioxidants (ascorbic acid, AsA; glutathione, GSH; phenolic compounds; alkaloids; non protein amino acids; and α-tocopherols) and some antioxidant enzymes (super oxide dismutase, SOD; catalase, CAT; ascorbate peroxidase, APX; glutathione reductase, GR; monodehydroascorbate reductase, MDHAR; dehydroascorbate reductase, DHAR; glutathione peroxidase, GPX; and glutathione-S-transferase, GST) which work coordinately to eliminate ROS in an efficient way. Biotechnological approaches also helpful in enhancing the antioxidant system to confer oxidative as well as abiotic stress tolerance. Performances of drought-affected plants are remarkably improved by exogenous application of osmolytes, hormones, antioxidants and signaling molecules.

3.1 Genetics of drought tolerance

Drought stress tolerance is a complex trait influenced by genetic with many quantitative trait loci (QTLs) and environmental factors. Genetic analyses of drought tolerance have been studied through the development of molecular markers and genome sequencing in wheat. Such analyses include several approaches, e.g., QTL-mapping, association-mapping, genome-wide analyses and expression analysis aim to identify QTL or gene-related traits for drought stress tolerance. Revealing the genetic basis underlying the drought tolerance in wheat requires a phenotypic and genetic variation of relevant traits in large populations with dense genetic maps. The genetic basis of drought tolerance is due to polygenic inheritance, where each gene has small effect with high GXE interaction, hence low-heritability. Furthermore, the genetic independence of drought tolerance at different developmental stages makes the detected QTL less useful in crop improvement. Therefore, several QTLs have been discovered for drought tolerance-related traits, but a limited number of QTLs are genetically characterized or cloned and incorporated in breeding programs. Identifying stable QTL with large-effect that controls many drought tolerances-related traits at different developmental stages would be a great effort for crop improvement, but has not been found.

3.1.1 Quantitative trait locus (QTL) of drought tolerance

Quantitative trait loci (QTL) are location from where some genes influence a phenotype of quantitatively inherited trait. Genetic variations of a crop can be clarified through QTL mapping (polygenes). Mapping of QTL allows the estimation of the places, quantity, level of effects for the phenotype, gene activity pattern and important genomic regions. Multi-environmental field conditions are commonly used to evaluate the genotype performance [24, 25] using a different type of bi-parental population, e.g., recombinant inbred line (RIL) population, doubled haploid (DH) population [26, 27] or advanced backcross [28]. Different DNA molecular markers have been used to genotype the populations and identify QTL [26, 29]. Recently, a high-density genetic SNP map [28] (SNP array or genotyping by sequencing (GBS)) have been used to genotype the population [27]. Numerous QTLs have been identified for grain yield on chromosomes one, three and six, grain number per spike on chromosome two, three and six and spikelet number for each spike on two, five and six. Such major QTL controlling grain yield can be utilized in marker-helped determination rearing for yield improvement under dry spell pressure. QTL studies using a biparental mapping population have also discovered the genetic factors of other physiological and adaptive traits (**Table 2**), e.g., leaf chlorophyll content, leaf waxiness and leaf rolling in wheat, transpiration efficiency, water-use efficiency, biomass, leaf area, and growth rate-related traits in wheat. Meta-QTL (MQTL) analysis on drought tolerance in wheat has revealed QTLs for, photosynthesis, soluble carbohydrates, water status, carbon isotope discrimination, canopy temperature, coleoptiles vigor and stay-green.

QTL investigation is so basic to target characteristics and for doing this a couple of stages are required. Initially, phenotypic evaluation of reasonably huge population for markers which are polymorphic is required. Besides, genotyping of the population is noteworthy. Thirdly, there is a prerequisite for quantifiable examination to distinguish the loci that are influencing the target trait. Several studied has been done and recognized >1200 QTLs for various characteristics conveyed over every one of the 21 chromosomes engaged with dry season resilience. Most extreme number of QTLs has been accounted for agronomic attributes, trailed by physiological qualities and root characteristics. Among agronomic qualities, most extreme QTL are known for thousand grain weight (TGW) trailed by grain yield and different attributes recorded under dry season conditions just as should be expected conditions. Among physiological qualities, most extreme number of QTLs are accessible for SPAD/chlorophyll content (82 QTL) trailed by water-dissolvable starches (76 QTL), coleoptile length (68 QTL). Among the root characteristics, greatest number of QTL is known for root length. Just 70 of these detailed QTL are major (clarifying ~>20% PVE), and just 19 QTL (counting 14 QTL for agronomic qualities, 5 for physiological attributes) are steady QTL utilized for QTL examination. The root attributes display high QTL × environment interaction, which recommends non accessibility of stable QTL for these characteristics. Fourteen stable major QTL were accounted for five agronomic attributes, with phenotypic fluctuation for individual QTL extending from 19.60% (grain yield QTL qGYWD.3B.2) to 45.20% (1000-grain weight QTL on 3B). These QTL can be utilized for development of dry spell resistance utilizing marker assisted selection (MAS). Two of the five QTL for grain yield that respond to dry season/heat stress cover a specific Mega QTL; these two QTL are found one each on chromosomes 4A and 7A [39] in areas, which likewise harbor QTL for the accompanying 14 qualities, which add to seedling rise, grain yield and reception to dry spell conditions: (1) days to heading, (2) days to development, (3) remain green propensity, (4) biomass, (5) shelter temperature; (6) carbon isotope separation, (7) coleoptile energy, (8) grain filling, (9) plant stature, (10) portion number, (11) spike thickness, (12) 1000-bit weight, (13) water-solvent sugars and (14) grain yield. Two other QTL for kernel width/thickness proportion on chromosome 5A cover a MQTL on 5A which represent to QTL for plant stature, spike weight and TGW [39]. The four stable major QTL for dry spell resilience incorporate two QTL for grain yield and two

Traits	Chromosome	Reference
Grain yield	1B, 1D, 3B, 4A, 6D, 7D	[30]
Grain weight per spike	1B, 1D	[31]
Thousand grain weight	1B, 1D, 2A, 2B, 3A, 3B,4A, 4D, 6A, 6D, 7B, 7D	[32]
Grain number (m^{-2})	1B, 5A, 5B, 7D	[33]
Grain number per spike	1A, 2A, 2B, 3A, 6B	[33, 34]
Harvest index	1B, 2D, 4BS, 5A	[32]
Spike number per plant	1A, 2A, 2B, 2D, 4B, 5A,7B	[32]
Spikelet compactness	1A, 1B, 2B, 5A, 5B, 6A,6B, 7A	[32]
Spikelet number per spike	1B, 1D, 2B, 3B, 4B, 5A, 6B, 7D	[32]
Sterile spikelet number per spike	7A	[32]
Fertile spikelet spike per spike	2A	[32]
Spike length	2B, 7A, 7B	[32]
Biomass	1B	[32]
Shoot biomass	4B	[35]
Spike length	2B, 7A, 7B	[32]
Physiological traits		
Leaf area, growth rate, transpiration efficiency, water-use efficiency	2A, 2D, 3A, 4B, 6A	[36]
Stomatal density, index, aperture area, length; guard cell area and length	2B, 4AS, 5AS, 7AL, 7BL;1BL, 4BS, 5BS, 7AS	—
Stomatal conductance, net photosynthetic rate	5A, 6B	[33]
Root length	2D, 4B, 5D, 6B	[35]
Root biomass	2D, 4BS	[35]
Metabolite traits		
Abscisic acid (ABA)	1B, 2A, 3A, 4D, 5A, 6D,7B	[37]
Jasmonic acid (JA), salicylic acid (SA), ethylene	6A	[38]

Table 2.
The detected quantitative trait loci (QTLs) for agronomic, physiological and metabolite traits in wheat using bi-parental mapping populations.

QTL for kernel width/thickness proportion. In an ongoing report, after broad field tests directed under pressure conditions in India, Australia and Mexico, a fundamental impact yield QTL (QYld.aww-1B.2) was fine-mapped to 2.9-cM locale relating to 2.2-Mbp genomic area containing 39 predicted genes (Tura et al., 2020). This QTL could be exploited in wheat breeding. The QTL for TGW, which is a significant segment of grain yield and have high heritability as well as stability, can be exploited for development of grain yield under water stress. Four QTL for days to heading and days to maturity may likewise be exploited utilizing MAS. Five significant and stable QTL for three physiological characteristics (SPAD/chlorophyll content, stem save assembly and water-solvent starches) each clarified PV running from ~20 to ~60% (**Table 3**). These attributes add to grain filling/advancement and thus to grain yield. The markers related with QTL for these characteristics are additionally acceptable possibility for marker assisted selection (MAS).

Sl. no.	QTL/trait	PVE %	Linked marker (position in cM)	Physical position (Mbp)	References
I. Agronomic traits					
1. Grain yield					
	a. qGYWD.3B.2	19.6	Xgpw7774 (97.6)	16.2	—
	b. 4A	20	Xwmc420 (90.4)	538.2	—
	c. 4A-a	23.9	Xgwm397 (6)	708.6	[11]
	d. Qyld.csdh.7AL	20.0	Xgwm332 (155.9)	681.6	[40]
	e. 6D	26.6	2,265,648|F|0-60:A>G-60:A>G-RAC875_c57371_238 (73)	ND	[14]
2. 1000 grain weight					
	a. 2A	36.1	2,264,948|F|0-9:T > A-9:T > A-Kukri_c22235_1547 (21.0-24.0)	ND	[14]
	b. 3B	45.2	Xbarc101 (86.1)	34.3	[41]
	c. QTgw-7D-b	21.9	XC29-P13 (12.5)	ND	[42]
3. Days to heading					
	a. QDh-7D.b	22.7	XC29-P13 (12.5)	ND	[42]
	b. QHd.idw-2A.2	32.2	Xwmc177 (46.1)	33.7	[29]
	c. 5D	21.4	1,126,619|F|0-21:A > T-21:A > T-wsnp_Ex_c1278_2449191 (162)	ND	[43]
4. Kernel width/thickness ratio					
	a. qWTR-5A-1	33.09	Xwmc74-Xgwm291 (61)	702.5–698.1	[44]
	b. qWTR-5A-2	23.59	Xgwm291-Xgwm410 (71)	698.1	—
5. Days to maturity					
	a. QDm-7D.b	22.7	X7D-acc/cat-10 (2.7)	ND	[29]
II. Physiological traits					
1. Stem reserve mobilization					
	a. QSrm.ipk-2D	42.2	Xgwm249a (142)	141.1	[45]
	b. QSrm.ipk-5D	37.5	Xfbb238b (19)	ND	[45]
	c. QSrm.ipk-7D	21	Xfbb189b (338)	ND	[45]
2. Water-soluble carbohydrates					
	a. QWsc-c.aww-3A	19	Xwmc0388A (64.9)	208	—
3. SPAD/chlorophyll content					
	a. Qchl.ksu-3B	59.1	Xbarc68 (67.2)	76.1	[46]

PVE shows phenotypic variation explained; c means position of linked flanking marker was given if either the second marker or its sequence was not available; ND explain the physical position of QTL could not be determined due to lack of linked marker sequence information.

Table 3.
A list of major and stable QTL (PVE ranging from 19 to 59%) for agronomic and physiological traits identified under drought/water stress.

3.1.2 Genomics analyses of drought tolerance

As of late, genome-wide investigations fuse genome-wide association study (GWAS) and genomic selection (GS) has been used to grasp the inherited multifaceted nature and breed for drought tolerance. GWAS approaches can be utilized with huge quantities of SNPs that produce a high-thick guide in an enormous and various assortments that give an elective way to deal with distinguish explicit qualities while the GS can be utilized in both bi-parental and different populaces. A predetermined number of studies have concentrated on physiological attributes, e.g., leaf green region, leaf water substance and water-soluble carbohydrates with around 12 MTAs have been distinguished. Chromosome 1A was found to contain a significant genomic region for physiological attributes, for example, water-dissolvable starches. Recently, utilized the most recent wheat genome sequences to physically map the most consistent and significant genomic regions that related with numerous agronomic and physiological attributes under drought stress in wheat. For example, the physical region of 1A was as a highly significant region for grain weight, flag leaf area and flag leaf width.

4. Salt stress tolerance

Globally, over 20% of the cultivable land is influenced by salinity. Because of environmental change and anthropogenic exercises, the salt influenced region is tended to increase day by day. A saline soil is commonly characterized as one in which the electrical conductivity (EC) of the saturation extract in the root zone surpasses 4 dS m^{-1} (roughly 40 mM NaCl) at 25°C and has a exchangeable sodium of 15%. It has been assessed that overall 20% of all out developed and 33% of irrigated agricultural lands are influenced by high salinity. Salt affected soils currently constitute 6.74 million ha in various agro ecological regions, the zone is probably going to increment to 16.2 million ha by 2050. Abiotic stresses (including salinity) are responsible for more than 50% yield reduction [47]. In opposite, because of fast increment of worldwide population, food production ought to be expanded by over 70% by 2050 [48]. Wheat (*Triticum* spp.) positions first on the world's grain production. Wheat is expended as staple food by over 36% of world population. Wheat gives almost 55% of the carbohydrates and 20% of the food calories consumed globally. The productivity of wheat is frequently unfavorably influenced by salt stress. The yield of wheat begins to decay at 6–8 dS m^{-1} [49].

4.1 Effect of salinity on growth: two phases of the growth response

In saline soil plant development is restrained by two reasons. To begin with, it decreases the plant's capacity to take up water, and this prompts more slow development. This is the osmotic stress or water-deficiency impact of salinity. Second, it might enter the transpiration stream and in the end harm cells of leaves includes in the transpiration prompts further reducing development. This is the salt-specific or ion-excess effect of salinity. The two impacts give rise to a two-stage development response to salinity (**Figure 3**). The outline shows the development reaction to salt that is included step by step.

Phase 1: The primary period of the development reaction results from the impact of salt present in the soil solution lessens leaf development and less significantly, root development [50]. The cell and metabolic processes included are in common to dry season influenced plants. Neither Na^+ nor Cl^- develops in developing tissues at concentrations that hinder development: meristematic tissues are

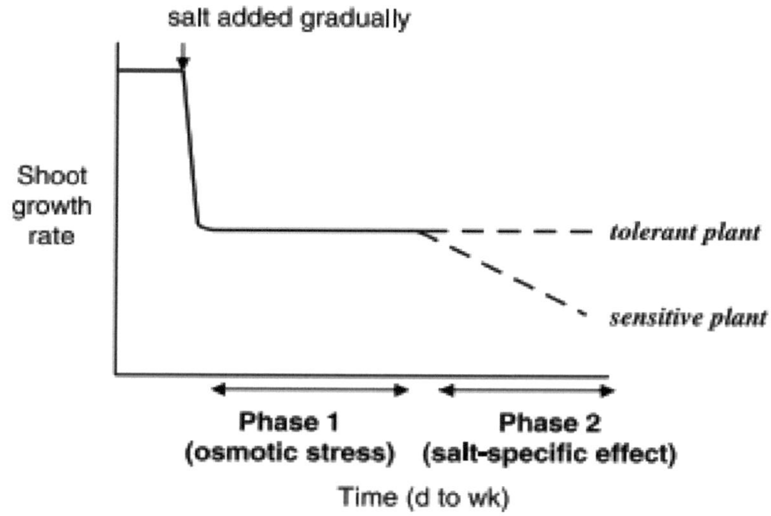

Figure 3.
Schematic outline of the two-stage development reaction to salinity for genotypes that differ in the rate at which salt arrives at harmful levels in leaves. For annual species, the time scale is d or wk., depending upon species and salinity level. For perennial species, the timescale is months or yr. During stage 1, development of the two genotypes is decreased in light of the osmotic stress of the saline solution outside the roots. During stage 2, leaves in the more sensitive genotype die and decrease the photosynthetic limit of the plant. This applies an extra impact on development [50]. In the event that salt is included one stage, the development rate dives to zero or below and takes 1–24 h to recover to the new consistent rate, contingent upon the level of the osmotic shock [51].

taken care of to a great extent in the phloem, from which salt is viably avoided and quickly elongating cells can accommodate the salt that shows up in the xylem inside their extending vacuoles.

Phase 2: The second phase of the development reaction results from the toxic effect of salt inside the plant. The salt taken up by the plant moves in old leaves: proceeded with transport into transpiring leaves brings about extremely high Na^+ and Cl^- concentrations, and the leaves become die. The reason for injury is presumably the salt burden surpassing the capacity of cells to compartmentalize salts in the vacuole. Salts would then develop quickly in the cytoplasm and inhibit enzyme activity. On the other hand, they may develop in the cell walls and get dried out the cell. The rate of leaf death is crucial for survival of the plant. In the event that new leaves are ceaselessly created at a rate more prominent than that at which old leaves die, there will be sufficient photosynthesizing leaves for the plant to produce flowers and seeds, in spite of the fact that in decreased numbers. In any case, if old leaves die more rapidly than new ones create, the plant may not get by to produce seed. For an annual plant there is a competition against time to initiate flowers and form seeds, while the leaf region is as yet sufficient to supply the important photosynthates. For perennial species, there is a chance to enter a state like to dormancy and survive under the stress. Salt stress not just prompts the decrease of harvest yield yet it additionally influences the metabolic processes in plants through disability of water potential of cells, ion toxicity, take-up of fundamental mineral supplements, membrane integrity and function. NaCl is the most dissolvable and across the board salt and collection of sodium particle (Na^+) in plant tissues is one of the most hindering impacts of saltiness. The take-up of fundamental micronutrients, for example, potassium (K^+) and calcium (Ca^+) from soil is restrained by higher centralization of Na^+ [52]. K^+ is required for development or improvement of plants and for keeping up high K^+/Na^+ ratio in shoot which is the significant technique received by plants to adapt up to salt stress. K^+ and Na^+ however having comparative compound properties, both have distinctive physiological effect on plant development. Under salt pressure, hyperosmotic and hyperionic (particle harmfulness) stresses happen because of low water potential of soil and abundance sodium particle amassing inside the plant. Ionic stress is additionally connected with nourishing irregularity. Salt stress additionally causes diminished germination rate, decreased development, altered reproductive behavior and diminished yield. Modified enzymatic movement, disturbed photosynthesis, oxidative pressure, disrupted biomembrane structure and function, harm of ultrastructural cell components, and hormonal imbalance are a few explanations behind diminishing generally speaking development and improvement of plants under salt pressure.

4.2 Mechanisms of salt tolerance

Salt tolerant is a polygenic trait directed by multiple factors/genes. There are various systems for salt resilience helps in decreasing Na^+ gathering in the cytoplasm by restricting Na^+ section into the cell, effectively moving Na^+ out of the cell, and compartmentalizing Na^+ into the vacuole. High-affinity potassium transporters (HKTs) are most active at level of plasma membrane and act as Na^+/K^+ symporters as well as Na^+ particular uniporter. Significant two subfamilies of HKTs: HKT1 and HKT2 are being investigated phylogentically [53]. HKT1 are only permeable to Na^+ but HKT2 are penetrable to both Na^+ and K^+. The group of HKTs having a place HKT/Trk/Ktr-type K^+ transporter superfamily are found generally in microorganisms and plants. In numerous plants, Na^+ and Cl^- are avoided by roots and water is taken up from the soil. This avoidance at higher salinities is kept up by halophytes. For example, sea grain grass, *Hordeum marinum*, avoids both Na^+ and Cl^- until at least 450 mM NaCl. Receptive oxygen species (ROS), made during the stress causes chlorophyll degradation and membrane lipid peroxidation. Malondialdehyde (MDA) is one of the final products of peroxidation of polyunsaturated fatty acids in the cell layers. The increase in free-radicals causes the overproduction of MDA which is the most notable marker of the oxidative stress. Plants accumulate different kind of metabolites on introduction to stressful conditions. The enormous changes under abiotic stress are showed up by soluble sugar, proline, phenolic compounds, chlorophyll substance, K^+/Na^+, shoot-root biomass proportion, etc. Total soluble sugar is an essential part of carbohydrate metabolism. It shows a close connection among photosynthesis and plant productivity and reflects the ability of grains to use assimilates. Proline is the fundamental amino acid act as excellent osmolyte and besides fill in as metal chelator anti-oxidative defense molecule and signaling molecule.

Thereby it maintains concentration of ROS in normal range and prevent oxidative burst in plants. Phenolic compounds also show important role in neutralizing the free radicals, quenching singlet oxygen and decomposing peroxides. Different approaches have been adopted to improve plant performance under salt stress; introduction of genes, screening of better performing genotypes, and crop improvement through conventional breeding methods which are often not so successful and not suitable due to time consuming or reduction of plant vigor with the succession of time. Uses of exogenous phytoprotectants, seed priming, nutrient management, and application of plant hormones are convenient for improving plant performances. These approaches are being also popular for stress management practices including the salt stress.

In this manner it keeps up concentration of ROS in ordinary range and prevent oxidative burst in plants. Phenolic compounds additionally show significant job in neutralizing the free radicals, extinguishing singlet oxygen and breaking down peroxides. Various methodologies have been adopted to improve plant performance under salt stress; introduction of genes, screening of better performing genotypes and crop improvement through traditional breeding techniques which are frequently not all that fruitful and not reasonable because of tedious or decrease of plant vigor with the progression of time. Uses of exogenous phytoprotectants, seed priming, supplement management, and utilization of plant hormones are advantageous for improving plant exhibitions. These methodologies are being also popular for stress management practices including the salt stress.

4.3 HKT-type transporters and genes response to salinity

Class 1HKT genes are involved in regulating transport of Na^+ in higher plants. Several HKT1 genes including HKT1; 1/2-like, HKT1; 3-like, HKT1; 4-like, and HKT1; 5-like, have been identified and mapped to wheat homologous chromosome groups 2, 6, 2 and 4 respectively. Among these, Nax1 in chromosome 2AL co-segregated with sodium transporter gene HKT1; 4-A2, which was shown to control Na^+ unloading from xylem in roots and sheaths. Nax2 was mapped to the distal region of chromosome 5AL that is homologous to a region on chromosome 4DL containing Kna1 [54]. Based on synteny and phylogeny analysis with Nax2, TmHKT1; 5-A significantly reduced leaf sodium content and increased durum wheat grain yield by 25% compared to lines without the Nax2 locus. Furthermore, decreased expression of TaHKT1; 5-D, which is homoeologous to TmHKT1; 5-A and underlies Kna1 locus in bread wheat, caused by target-specific RNA interference-induced silencing (RNAi) led to an accumulation of Na^+ in leaves, strongly suggesting that TaHKT1; 5-D should be the candidate gene of Kna1.

Class 1HKT genes are engaged with managing transport of Na^+ in higher plants. A few HKT1 genes including HKT1; 1/2-like, HKT1; 3-like, HKT1; 4-like, and HKT1; 5-like, have been recognized and mapped to wheat homologous chromosome groups 2, 6, 2 and 4 respectively. Among these, Nax1 in chromosome 2AL co-segregated with sodium transporter gene HKT1; 4-A2, which was appeared to control Na^+ emptying from xylem in roots and sheaths. Nax2 was mapped to the distal region of chromosome 5AL that is homologous to an region on chromosome 4DL containing Kna1 [54]. In view of synteny and phylogeny investigation with Nax2, TmHKT1; 5-An altogether decreased leaf sodium content and expanded durum wheat grain yield by 25% contrasted with lines without the Nax2 locus. Besides, diminished articulation of TaHKT1; 5-D, which is homoeologous to TmHKT1; 5-An and underlies Kna1 locus in bread wheat, brought about by target-explicit RNA obstruction actuated hushing (RNAi) prompted a collection of Na^+ in leaves, firmly proposing that TaHKT1; 5-D ought to be the applicant quality of Kna1. A major mechanism in salinity tolerance of wheat is Na^+ exclusion mediated by HKT genes. AtHKT1 is regulated by small RNA and DNA methylation. Moreover, DNA methylation also participates in the response of TaHKT1; Transcription factors such as AtAB14 and OsMYBc were shown to regulate HKT genes in plants, offering more candidate targets for enhancing salinity tolerance.

4.4 Genes involved between salinity response and other environmental and developmental signals in wheat

When there is high concentration of salt in plant system, the activation of complex physiological responses such as phytohormone signaling pathways and developmental signals starts to adapt the stress; therefore it is essential to identify the environmental and developmental signals. First of all an attempt was performed by looking at phytohormones, as most phytohormones are regulatory factors of both developmental process and stress response. For example, the wheat gene

TaAOC1, encoding cyclase involved in jasmonic acid synthesis, was induced by high salinity. Constitutive expression of TaAOC1 in both wheat and Arabidopsis restricted root growth, but enhanced salt tolerance and Jasmonic acid content. It indicates the different branches of metabolic pathway participate in a single process but controlled by different mechanisms. Light is an essential factor that positively affects the development and growth of plants. TaGBF1, a blue light specific responsive G-box binding factor, was prompted after exposure to salt. TaGBF1 caused salt affectability and advanced light blue interceded photomorphogenesis, indicating that it was a typical segment of the blue light and salt stress responsive signaling pathways. Curiously hereditary examination recommended that the job of TaGBF1 because of salt depended on AB15, a key part of ABA signaling pathway. The extensive studied has been done for the identification of salt tolerant QTLs. The available studies led to identification of ~500 QTL (excluding those involved in digenic

Sl. no.	Traits	QTL/locus	PVE %	Linked marker	Physical position (Mbp)[a]	References
	Na^+ exclusion	Kna1	—	Xwg199, Xabc305, Xbcd.402, Xpsr567, Xpsr375	390.2	[55]
	Na^+ exclusion	Nax1	38	Xgwm312, Xwmc170	709.0–711.5	[56]
	Dry weight of plumule at germination	Qpdwg-4D.1	19.8	Xfbb226–Xfba177	ND	[57]
	Na^+ exclusion	QNax.aww-7AS	41	Xwmc083–Xcdo595	89.9	[58]
	Booting	QB.uabcs-2D	23.6	Xcdo1379		[59]
	Ear emergence time	QEet.uabcs-2D	27.1	Xcdo1379	ND	
	Flowering	QFl.uabc-2D	26.7	Xbcd102a	ND	
	Maturity	QM.uabc-2D	28.9	Xcdo137	ND	
	Ear length	QEl.uabc-2D	21.5	Xbcd102a	ND	
	Seedling shoot fresh weight	3B-1	19.2	wPt-798,970-wPt-8303	ND	
	Na^+ exclusion value	qSNAX.7 A.3	18.79	AX-95248570–AX-95002995	700.6	[60]
	3rd leaf Na^+ and K^+ concentration and K^+/Na^+ ratio	4B	18, 20, 27	Xm564	657.1	[61]
	3rd leaf Na^+ concentration	3B	18	Xm551	701.9	
	K^+ μmol/g dry weight	QK.asl-5A	28.2	Vrn-A1	587.4	[62]

PVE: phenotypic variation explained; "–"explain PVE% not available; ND shows physical position of QTL could not be determined due to lack of linked marker sequence information.[a]*Position of one flanking marker was given if either the second marker or its sequence was not available.*

Table 4.
A list of major QTL/loci (PVE of ~>20%) for plant traits under salt stress condition in bread and durum wheat.

epistatic interactions and QTL × treatment interactions); these QTL are spread over all the 21 wheat chromosomes and could prove useful resource for MAS intended at improving salt tolerance in wheat. The phenotypic variance (PV) explained by individual QTL ranged from 8.4% to 38.0%, and only a dozen major QTL have been reported (**Table 4**). The traits used for QTL analysis included Na^+ exclusion/content, K^+ content and K^+/Na^+ ratio, etc., both at the seedling and adult plant stages. Since several studies in different plant systems including wheat have demonstrated that Na^+ concentration is not necessarily associated with salinity tolerance, other additional mechanisms (tissue tolerance and osmotic adjustment) may also be examined in future in order to breed for salinity tolerance in bread wheat. It has been studied that bread wheat exhibit low rates of Na^+ transport, which leads to high K^+/Na^+ ratio in leaves. A high K^+/Na^+ discrimination provides tolerance to salinity stress. The extensive studied has been accomplished for the ID of salt open minded QTLs. The accessible examinations prompted identification of ~500 QTL (barring those associated with digenic epistatic collaborations and QTL × treatment communications); these QTL are spread over all the 21 wheat chromosomes and could demonstrate valuable asset for MAS expected at improving salt resilience in wheat. The phenotypic difference (PV) clarified by individual QTL extended from 8.4% to 38.0%, and just 12 significant QTL have been accounted (**Table 4**). The qualities utilized for QTL investigation included Na^+ rejection/content, K^+ substance and K^+/Na^+ proportion, and so forth., both at the seedling and grown-up plant stages. Since a few investigations in various plant frameworks including wheat have exhibited that Na^+ fixation is not really connected with saltiness resilience, other extra components (tissue resistance and osmotic alteration) may likewise be analyzed in future so as to raise for saltiness resistance in bread wheat. It has been contemplated that bread wheat show low paces of Na^+ transport, which prompts high K^+/Na^+ proportion in leaves. A high K^+/Na^+ segregation gives resilience to saltiness stress.

5. Pre-harvest sprouting (PHS) tolerance

Germination of wheat inside the grain ear head before reap is called pre-gather sprouting (PHS). Exposure of prolonged precipitation and high humidity after the grain has matured and before it very well may be collected can prompts pre-harvest sprouting (PHS), which can be thought of as an premature germination. Germination can start as a wheat seed retains moisture and swells. A noticeable sign of PHS incorporates kernel swelling, germ discoloration, seed-coat parting, and the root and shoot emergence.

5.1 Effect of preharvest sprouting in wheat quality

Pre-collect growing in bread wheat (*Triticum aestivum* L.) is a setback that happens everywhere throughout the world to varying degrees. The issue happens when high humidity goes with precipitation on standing full grown wheat crops before harvest, and seeds in the spike sprout. As the outcome of this, wheat qualities as well as quantity are affected, diminishing healthy benefit and yield. Changes in sugar content, total protein and composition of amino acids joined by enzymatic activities are the explanations behind the degradation in quality and yield. Many early wheat scientists reported that pre-harvest sprouting is negatively correlated with yield, seed viability, seedling vigor, flour yield and baking quality. Pre-harvest sprouting results in lower yields due to decreased test weights, and it limits end-use applications for wheat due to decreased grain quality. Reduced grain quality, coupled with decreased yields, can result in substantial financial losses to farmers and food processors. Products made from germinated seeds can be spongy, soggy, off-color and of inferior quality [63]. Sprouted seed baked to Compact interior and smaller volume breads due to higher α-amylase activity results in starch degradation, hence producing lower quality of bread that is below the accepted standards of consumers. Numerous early wheat researchers revealed that pre-harvest sprouting is negatively correlated with yield, seed suitability, seedling force flour yield and preparing quality. Pre-harvest sprouting outcomes in lower yields because of diminished test weight and it limits end-use applications for wheat because of diminished grain quality. Diminished grain quality, combined with diminished yields, can bring about significant financial losses to farmers and food processors. Items produced using germinated seeds can be spongy, soggy, off-color and of inferior quality [63]. Germinated seed baked to Compact inside and smaller volume breads because of higher α-amylase activity brings about starch degradation, thus creating lower quality of bread that is underneath the acknowledged norms of customers.

5.2 Mechanism of preharvest sprouting resistance

Pre-harvest sprouting is controlled by genetic factors, environmental conditions and their interactions. The protection from germination is fundamentally connected with an adequate level of kernel dormancy. Pre-harvest sprouting depends significantly on (1) hereditary attributes like kernel coat, protecting structures of spike and straightness of spike, (2) natural conditions like temperature and precipitation, and (3) agronomic perspective like fertilization. The main considerations next to conditions influencing the resilience to PHS are seed dormancy, seed coat penetrability and color, α-amylase activities, endogenous hormones levels, genes and QTLs. Dormancy was seen as the fundamental internal factor which lead to the wheat resistance from PHS [64–66]. The seed coat permeability is the essential guaranteeing divider which could increase the wheat PHS resilience. The seed coat color additionally assumes a critical activity in PHS. All around, white wheat varieties have higher germination rates than the red ones [67]. Cultivars having red kernels are more impervious to growing than white ones. Accordingly, red kernel shading is consistently used as an indicator of sprouting resistance in wheat. The α-amylase viewed as one of the significant elements that influence wheat germination rate, cold versatility and production. Some extraordinary endogenous factors like gibberellic acid (GA), abscisic acid (ABA) and indole acidic acid (IAA) could in like manner impact PHS through a wide scope of ways. PHS is a quantitative characteristic compelled by various genes. Viviparous-1 (Vp-1) has been recognized as the main gene that coordinated seed germination and dormancy. Some different genes were also regarded to participate in embryos maturing, seed dormancy and germination through system guideline with Vp-1 to control PHS. QTLs for dormancy and PHS were found in different materials through molecular markers. During kernel development, the Vp-1 gene expressed in cytoplasm subsequent to flowering controlled seed dormancy at the transcriptional level, advanced the seed development and checked the outflow of germination-related genes [68]. There were numerous allelic variety of Vp-1 gene in various grain crops, however the anticipated protein of Vp-1 was monitored with four DNA binding regions A1, B1, B2, and B3. Three alleles Vp-1A, Vp-1B, Vp-1D of Vp-1, situated on 3A, 3B and 3D homologous chromosomes in wheat, separately, have been identified [66, 69]. Numerous investigations additionally centered on the allele's variety of Vp-1 to clarify how Vp-1 managed the resistance to PHS. Six alleles of Vp-1A, namely Vp-1Aa, Vp-1Ab, Vp-1Ac, Vp-1Ad, Vp-1Ae and Vp-1Af, were found in 81 wheat cultivars and advanced lines [69]. Six alleles of Vp-1B named Vp-1Ba, Vp-1Bb, Vp-1Bc, Vp-1Bd, Vp-1Be and Vp-1Bf were likewise found in wheat [69, 70]. However, no alleles of Vp-1D were found in wheat. The wheat varieties with alleles of Vp-1Ab and Vp-1Ad were regarded to have low germination index (GI) and strong PHS tolerance [69]. However, the wheat varieties with the allele Vp-1Ba have higher germination index and more sensitive to PHS than the other five ones, which even positively influenced on the decrease of germination rate [69, 70]. More than 47 investigations on QTL interval mapping for PHS resistance and related characteristics including ~40 distinct population derived from bread wheat (including synthetic wheat), durum wheat and *T. monococcum* have so far been conducted. QTL for PHS tolerance have been recognized utilizing the following parameters: PHS index, grain color, falling number, germination index, seed dormancy and alpha amylase activity (**Figure 4**).

Maximum numbers of QTL have been accounted for PHS index followed by seed dormancy, germination index, falling number, alpha amylase activity and grain color. About ~250 QTL were distinguished, among them just 29 QTL were major and stable across environments; these QTL are conveyed on 11 unique chromosomes (1B, 3A, 4A, 5A, 6A, 2B, 3B, 4B, 7B, 2D, 3D and 7D); the most noteworthy PV explained by an individual QTL range from 23% to 78.3%.

Chromosomes from homoeologous groups 3 and 4 together conveyed 17 of the 29 significant and stable QTL. The PHS and the germination index (a measure of dormancy) have regularly been utilized for estimation of tolerance against PHS. PHS indx is a simple to score parameter and reliable, with the goal that it has been widely used. The QTLs because of seed dormancy, which is characterized as the powerlessness of practical seeds to develop under conditions great for germination is additionally connected with PHS tolerance. The QTL for PHS tolerance, present on the long arms of chromosomes of homoeologous group 3, have regularly been accounted for to be related with genes for red grain color, which contributes to coat-imposed dormancy. A significant stable QTL for PHS (QPhs.ccsu-3A.1; 24.68–35.21% PV) was accounted [71–80]. The utilization of markers related with this QTL in MAS brought about significant level of PHS tolerance, which was tragically connected with red grain color.

In wheat markets, especially in Southeast Asia and Middle East, Africa and North America, there is a consumer preference for white grain. Along these lines, endeavors were later made to deliver white-grained PHS-tolerant wheat genotypes;

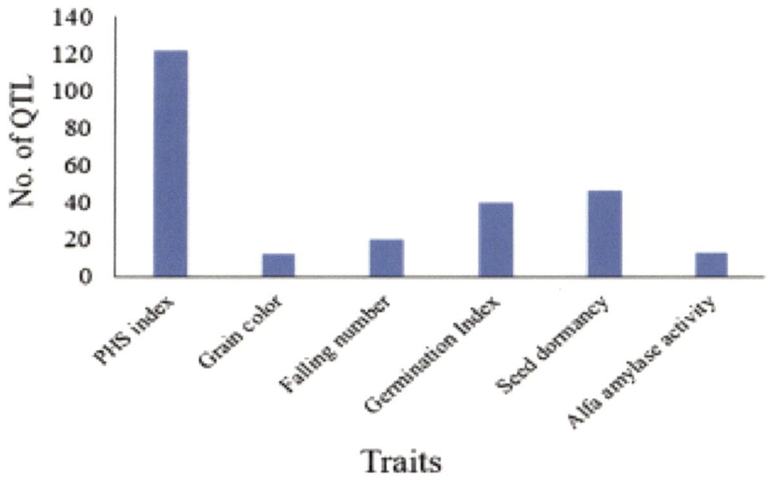

Figure 4.
Number of QTL for five different traits associated with pre-harvest sprouting tolerance reported in the 47 studies in wheat.

Sl. no.	Traits/QTL	PVE (%)	Linked marker	Physical position (Mbp)[c]	References
	FN/5A	26.4	Xpsr1194–Xpsr918b	ND	[81]
	α-AA/5A	30.0	Xpsr1194–Xpsr918b	ND	[81]
	SD/4AL (33–77.2)		Xcdo795/Xpsr115		[82]
	PHS/QPhs.ccsu-3A.1 (78.3)		Xwmc153–Xgwm155	701.7–702.9	[71]
	SD/QPhs.ocs-3A.1 (23.0–44.8)		Xbarc310/Xbcd907	7.1	
	GI/QGi.crc-3B	27.0	Xbarc77–Xwmc307	430.1–783.5	[83]
	SI/QSi.crc-3B	24.0	Xbarc77–Xwmc307	430.1–783.5	[83]
	FN/QFn.crc-3B	33.0	Xbarc77–Xwmc307	430.1–783.5,	[83]
	GI-14/QPhs. dpivic-3D.1	26.0–43.0	Red Grain Color RGC -wms1200	ND	[84]
	VI/QPhs.dpivic-4A.1	21.0	Xbarc170–Xgwm269c	605.7–607.8	[84]
11.	PHS/QPhs. pseru-3AS	31.26–44.96	Xbarc12–Xbarc321	11.7–15.4	[85]
	QPhs.dpi.vic.4A.2	27.78–39.84	Xgwm637–Xgwm937	617.4	
	PHS/2DS	25.73–27.50	Xgwm261–Xgwm484	19.6–48.1	[86]
	GI/QGI.crc-4B	28.2–66.6	Xwmc349	640.9	[87]
	PHS/QSI.crc-4B	6.2–26.9	Xwmc349	640.9	[87]
	PHS/QPhs.cnl-2B.1	24.0	Xbarc55–Xwmc474	133.5–172.6	—
	GC/QGc.ccsu-3B.1	15.28–40.42	Xgwm938–Xgwm980	ND	[88]
	PHS/QPhs.ccsu-6A.1	12.01–29.47	Xgwm1296–Xgwm1150	ND	[88]
	PHS/QPhs. caas-3AS.1	11.8–27.7	Xbarc294–Xbarc57	7.9–10.3	[89]
	GI/QGi.crc-4A	27.6–58.1	—	ND	[90]
	PHS(SI)/QSi.crc-4A	10.5–32.1	—	ND	[90]
	PHS(SI)/QSi.crc-7B	11.8–20.5	—	ND 1/2	[90]
	FN/QFn.crc-7D	13.2–20.6	—	ND	[90]
	PHS, SD/Qphs. pseru-4A	17.2–26.5	GBS_212432–GBS_109947	ND	[91]
	QPhs.spa-4B	35.0–60.0	Xwmc617b–Xwmc48a	15.7–98.7	[92]
	QPhs.spa-7D2	14.0–47.0	Xbarc76–Xcfa2257a	634.0	[92]
	GI/3AS	21.6–41.0	KASP-222	7.2	[93]
	qPHS.sicau-3D	8.65–42.47	AX-94415259	562.5–5	[94]

Table 5.
A summary of the major and stable QTL for pre-harvest sprouting/dormancy-related traits in wheat.

for this purpose, major and stable QTL on chromosomes of group 4 and different chromosomes were suggested. SSR markers are accessible for practically all major and stable QTL (**Table 5**); these SSR markers have been utilized for introgression of a QTL for PHS/dormancy to derive lines with high degree of PHS tolerance related with golden grains.

Author details

Sudhir Kumar[1]*, Shampa Purkyastha[2], Chandan Roy[1], Tushar Ranjan[3] and Rakesh Deo Ranjan[1]

1 Department of Plant Breeding and Genetics, Bihar Agricultural University, Bhagalpur, Bihar, India

2 Department of Genetics and Plant Breeding, Bidhan Chandra Krishi Vishwavidyalaya, Mohanpur, West Bengal, India

3 Department of Molecular Biology and Genetic Engineering, Bihar Agricultural University, Bhagalpur, Bihar, India

*Address all correspondence to: sudhir.hzb@gmail.com

References

[1] Oshino T, Miura S, Kikuchi S, Hamada K, Yano K, Watanabe M, et al. Auxin depletion in barley plants under high-temperature conditions represses DNA proliferation in organelles and nuclei via transcriptional alterations. Plant, Cell & Environment. 2011;**34**:284-290. DOI: 10.1111/j.1365-3040.2010.02242.x

[2] De Costa W. A review of the possible impacts of climate change on forests in the humid tropics. Journal of the National Science Foundation of Sri Lanka. 2011;**39**:281-302. DOI: 10.4038/jnsfsr.v39i4.3879

[3] Farooq M, Bramley H, Palta JA, Siddique KHM. Heat stress in wheat during reproductive and grain-filling phases. Critical Reviews in Plant Sciences. 2011;**30**:491-507. DOI: 10.1080/07352689.2011.615687

[4] Akter N, Islam R. Heat stress effects and management in wheat. A review. Agronomy for Sustainable Development. 2017;**37**:37

[5] Kaushal N, Bhandari K, Siddique KHM, Nayyar H. Food crops face rising temperatures: An overview of responses, adaptive mechanisms, and approaches to improve heat tolerance. Cogent Food & Agriculture. 2016;**2**:1134380

[6] IPCC (Intergovernmental Panel on Climate Change). Summary for policymakers. In: Stocker TF, Qin D, Plattner GK, Tignor M, Allen SK, Boschung J, Nauels A, Xia Y, Bex V, Midgley PM, editors. Climate Change 2013: The Physical Science Basis. Contribution of Working Group I to the Fifth Assessment Report of the Intergovernmental Panel on Climate Change. Cambridge: Cambridge University Press; 2014

[7] Wheeler T, Von Braun J. Climate change impacts on global food security. Science. 2013;**341**:508-513. DOI: 10.1126/science.1239402

[8] Suzuki N, Miller G, Morales J, Shulaev V, Torres MA. Respiratory burst oxidases: The engines of ROS signaling. Current Opinion in Plant Biology. 2011;**14**:691-699

[9] Ortiz R, Sayre KD, Govaerts B, Gupta R, Subbarao GV, Ban T, et al. Climate change: Can wheat beat the heat? Agriculture, Ecosystems & Environment. 2008;**126**:46-58. DOI: 10.1016/j.agee.2008.01.019

[10] Heffner EL, Sorrells ME, Jannink JL. Genomic selection for crop improvement. Crop Science. 2009;**49**:1-12. DOI: 10.2135/crop-sci2008.08.0512

[11] Pinto RS, Reynolds MP, Mathews KL, McIntyre CL, Olivares-Villegas J-J, Chapman SC. Heat and drought adaptive QTL in a wheat population designed to minimize confounding agronomic effects. Theoretical and Applied Genetics. 2010;**121**:1001-1021. DOI: 10.1007/s00122-010-1351-4

[12] Paliwal R, Röder MS, Kumar U, Srivastava JP, Joshi AK. QTL mapping of terminal heat tolerance in hexaploid wheat (*T. aestivum* L.). Theoretical and Applied Genetics. 2012;**125**:561-575. DOI: 10.1007/s00122-012-1853-3

[13] Liu C, Pinto F, Cossani CM, Sukumaran S, Reynolds MP. Spectral reflectance indices as proxies for yield potential and heat stress tolerance in spring wheat: Heritability estimates and marker-trait associations. Frontiers of Agricultural Science and Engineering. 2019;**6**:296-308

[14] Liu C, Sukumaran S, Claverie E, Sansaloni C, Dreisigacker S, Reynolds M. Genetic dissection of heat and drought stress QTLs in phenology controlled

synthetic-derived recombinant inbred lines in spring wheat. Molecular Breeding. 2019;**39**:34

[15] Mason RE, Mondal S, Beecher FW, Pacheco A, Jampala B, Ibrahim AMH, et al. QTL associated with heat susceptibility index in wheat (*Triticum aestivum* L.) under short-term reproductive stage heat stress. Euphytica. 2010;**174**:23-436. DOI: 10.1007/s10681-010-0151-x

[16] Sharma DK, Torp AM, Rosenqvist E, Ottosen CO, Andersen SB. QTLs and potential candidate genes for heat stress tolerance identified from the mapping populations specifically segregating for Fv/Fm in wheat. Frontiers in Plant Science. 2017;**8**:1668

[17] Chapman SC, Chakraborty S, Dreccer MF, Howden SC. Plant adaptation to climate change-opportunities and priorities in breeding. Crop & Pasture Science. 2012;**63**:251-268. DOI: 10.1071/CP11303

[18] Zheng B, Chenu K, Dreccer MF, Chapman SC. Breeding for the future: What are the potential impacts of future frost and heat events on sowing and flowering time requirements for Australian bread wheat (*Triticum aestivium*) varieties? Global Change Biology. 2012;**18**:2899-2914

[19] Fu J, Momclovic I, Prasad V. Molecular bases and improvement of heat tolerance in crop plants. In: Josipovic S, Ludwig E, editors. Heat Stress: Causes. Prevention and Treatments. USA: Nova Science; 2012. pp. 185-214

[20] Ristic Z, Bukovnik U, Momcilovic I, Fu J, Prasad PVV. Heat-induced accumulation of chloroplast protein synthesis elongation factor, EF-Tu, in winter wheat. Journal of Plant Physiology. 2008;**165**:192-202. DOI: 10.1016/j.jplph.2007.03.00

[21] Wang X, Cai J, Jiang D, Liu F, Dai T, Cao W. Pre-anthesis high-temperature acclimation alleviates damage to the flag leaf caused by post-anthesis heat stress in wheat. Journal of Plant Physiology. 2011;**168**:585-593. DOI: 10.1016/j.jplph.2010.09.016

[22] Sallam A, Alqudah A, Dawood MFA, Baenziger PS, Börner A. Drought stress tolerance in wheat and barley: Advances in physiology, breeding and genetics research. International Journal of Molecular Sciences. 2019;**20**:3137. DOI: 10.3390/ijms20133137

[23] Hasanuzzaman M, Nahar K, Alam MM, Roychowdhury R, Fujita M. Physiological, biochemical, and molecular mechanisms of heat stress tolerance in plants. International Journal of Molecular Sciences. 2013;**14**:9643-9684. DOI: 10.3390/ijms14059643

[24] Mathews KL, Malosetti M, Chapman S, McIntyre L, Reynolds M, Shorter R, et al. Multi-environment QTL mixed models for drought stress adaptation in wheat. Theoretical and Applied Genetics. 2008;**117**:1077-1091

[25] Von Kor M, Grando S, Del Greco A, This D, Baum M, Ceccarelli S. Quantitative trait loci associated with adaptation to Mediterranean dryland conditions in barley. Theoretical and Applied Genetics. 2008;**117**:653-669

[26] Quarrie S, Gulli M, Calestani C, Steed A, Marmiroli N. Location of a gene regulating drought-induced abscisic acid production on the long arm of chromosome 5A of wheat. Theoretical and Applied Genetics. 1994;**89**:794-800

[27] Obasa BT, Eglinton J, Coventry S, March T, Langridge P, Fleury D. Genetic analysis of developmental and adaptive traits in three doubled haploid populations of barley (*Hordeum vulgare* L.). Theoretical and Applied Genetics. 2016;**129**:1139-1151

[28] Kalladan R, Worch S, Rolletschek H, Harshavardhan VT, Kuntze L, Seiler C, et al. Identification of quantitative trait loci contributing to yield and seed quality parameters under terminal drought in barley advanced backcross lines. Molecular Breeding. 2013;**32**:71-90

[29] Maccaferri M, Sanguineti MC, Corneti S, Ortega JLA, Salem MB, Bort J, et al. Quantitative trait loci for grain yield and adaptation of durum wheat (*Triticum durum* Desf.) across a wide range of water availability. Genetics. 2008;**178**:489-511

[30] Tahmasebi S, Heidari B, Pakniyat H, McIntyre CL. Mapping QTLs associated with agronomic and physiological traits under terminal drought and heat stress conditions in wheat (*Triticum aestivum* L.). Genome. 2017;**60**:26-45

[31] Mora F, Quitral YA, Matus I, Russell J, Waugh R, Del Pozo A. SNP-based QTL mapping of 15 complex traits in barley under rain-fed and well-watered conditions by a mixed modeling approach. Frontiers in Plant Science. 2016;7:909

[32] Ogrodowicz P, Adamski T, Mikolajczak K, Kuczynska A, Surma M, Krajewski P, et al. QTLs for earliness and yield-forming traits in the Lubuski × Cam B barley RIL population under various water regimes. Journal of Applied Genetics. 2017;**58**:49-65

[33] Xu YF, Li SS, Li LH, Ma FF, Fu XY, Shi ZL, et al. QTL mapping for yield and photosynthetic related traits under different water regimes in wheat. Molecular Breeding. 2017;**37**:34

[34] Peleg ZVI, Fahima T, Krugman T, Abbo S, Yakir DAN, Korol AB, et al. Environment Genomic dissection of drought resistance in durum wheat × wild emmer wheat recombinant inbreed line population. Plant Cell Environ. 2009;**32**:758-779

[35] Kadam S, Singh K, Shukla S, Goel S, Vikram P, Pawar V, et al. Genomic associations for drought tolerance on the short arm of wheat chromosome 4B. Functional & Integrative Genomics. 2012;**12**:447-464

[36] Parent B, Shahinnia F, Maphosa L, Berger B, Rabie H, Chalmers K, et al. Combining field performance with controlled environment plant imaging to identify the genetic control of growth and transpiration underlying yield response to water-deficit stress in wheat. Journal of Experimental Botany. 2015;**66**:5481-5492

[37] Iehisa JC, Matsuura T, Mori IC, Takumi S. Identification of quantitative trait locus for abscisic acid responsiveness on chromosome 5A and association with dehydration tolerance in common wheat seedlings. Journal of Plant Physiology. 2014;**171**:25-34

[38] Castro AM, Tacaliti MS, Gimenez D, Tocho E, Dobrovolskaya O, Vasicek A, et al. Mapping quantitative trait loci for growth responses to exogenously applied stress induced hormones in wheat. Euphytica. 2008;**164**:719-727

[39] Acuna-Galindo MA, Mason RE, Subramanian NK, Hays DB. Meta-analysis of wheat QTL regions associated with adaptation to drought and heat stress. Crop Science. 2015;**55**:477-492

[40] Quarrie SA, Quarrie PS, Radosevic R, Rancic D, Kaminska A, Barnes JD, et al. Dissecting a wheat QTL for yield present in a range of environments: From the QTL to candidate genes. Journal of Experimental Botany. 2006;**57**:2627-2637

[41] Golabadi M, Arzani A, Mirmohammadi Maibody SAM, Tabatabaei BES, Mohammadi SA. Identification of microsatellite markers linked with yield components under

drought stress at terminal growth stages in durum wheat. Euphytica. 2011;**177**:207-221

[42] Lopes MS, Reynolds MP, McIntyre CL, Mathews KL, Jalal Kamali MR, Mossad M, et al. QTL for yield and associated traits in the Seri/Babax population grown across several environments in Mexico, in the West Asia, North Africa, and South Asia regions. Theoretical and Applied Genetics. 2013;**126**:971-984

[43] Liu B, Asseng S, Liu L, Tang L, Cao W, Zhu Y. Testing the responses of four wheat crop models to heat stress at anthesis and grain filling. Global Change Biology. 2016;**22**:1890-1903

[44] Chen L, Zhao J, Song J, Jameson PE. Cytokinin dehydrogenase: A genetic target for yield improvement in wheat. Plant Biotechnology Journal. 2019;**18**:614-630

[45] Salem KFM, Roder MS, Borner A. Identification and mapping quantitative trait loci for stem reserve mobilisation in wheat (*Triticum aestivum* L.). Cereal Research Communications. 2007;**35**:1367-1374

[46] Kumar S, Sehgal SK, Kumar U, Prasad PVV, Joshi AK, Gill BS. Genomic characterization of drought tolerance-related traits in spring wheat. Euphytica. 2012;**186**:265-276

[47] Acquaah G. Principles of Plant Genetics and Breeding. Oxford: Blackwell; 2007. p. 385

[48] FAO. High Level Expert Forum—How to Feed the World in 2050. Rome, Italy: Economic and Social Development, Food and Agricultural Organization of the United Nations; 2009

[49] Royo A, Abió D. Salt tolerance in durum wheat cultivars. Spanish Journal of Agricultural Research. 2003;**1**:27-35

[50] Munns R. Physiological processes limiting plant growth in saline soil: Some dogmas and hypotheses. Plant, Cell & Environment. 1993;**16**:15-24

[51] Munns R. Comparative physiology of salt and water stress. Plant, Cell & Environment. 2002;**25**:239-250

[52] Véry AA, Sentenac H. Molecular mechanisms and regulation of K^+ transport in higher plants. Annual Review of Plant Biology. 2003;**54**:575-603

[53] Platten JD, Cotsaftis O, Berthomieu P, Bohnert H, Davenport RJ, Fairbairn DJ. Nomenclature for HKT transporters, key determinants of plant salinity tolerance. Trends in Plant Science. 2006;**11**:372-374. DOI: 10.1016/j.tplants.2006.06.001

[54] Läuchli A. Salt exclusion: An adaptation of legumes for crops and pastures under saline conditions. In: Staples RC, editor. Salinity Tolerance in Plants: Strategies for Crop Improvement. New York, USA: Wiley; 1984. pp. 171-187

[55] Dubcovsky J, Sanata Maria G, Epstein E, Luo MC, Dvorak J. Mapping of K+/Na+ discrimination locus *Kna1* in wheat. Theoretical and Applied Genetics. 1996;**2**:448-454

[56] Lindsay MP, Lagudah ES, Hare RA, Munns R. A locus for sodium exclusion (*Nax1*), a trait for salt tolerance, mapped in durum wheat. Funct Plant Biol. 2004;**31**:1105-1114

[57] Ma L, Zhou E, Huo N, Zhou R, Wang G, Jia J. Genetic analysis of salt tolerance in a recombinant inbred population of wheat (*Triticum aestivum* L.). Euphytica. 2007;**153**:109-117

[58] Edwards J, Shavrukov Y, Ramsey C, Tester M, Langridge P, Schnurbusch T. Identification of a QTL on chromosome

7AS for sodium exclusion in bread wheat. In: Appels R, EastwoodR LE, Langridge P, Lynne MM, editors. Proceedings of 11th International Wheat Genetics Symposium. Australia: Sydney University Press; 2008

[59] De Leon JSL, Escoppinichi R, Geraldo N, Castellanos T, Mujeeb-Kazi A, Roder M. Quantitative trait loci associated with salinity tolerance in field grown bread wheat. Euphytica. 2011;**181**:371-383

[60] Hussain B, Lucas SJ, Ozturk L, Budak H. Mapping QTLs conferring salt tolerance and micronutrient concentrations at seedling stage in wheat. Sci Rep. 2017;7:1566

[61] Shamaya NJ, Shavrukov Y, Langridge P, Roy SJ, Tester M. Genetics of Na+ exclusion and salinity tolerance in Afghani durum wheat landraces. BMC Plant Biol. 2017;**17**:209

[62] Asif MA, Schilling RK, Tilbrook J, Brien C, Dowling K, Rabie H. Mapping of novel salt tolerance QTL in an Excalibur× Kukri doubled haploid wheat population. Theor Appl Genet. 2018;**131**:2179-2196

[63] Groos C, Gay G, Perretant MR, Gervais L, Bernard M, Dedryver F, et al. Study of the relationship between pre-harvest sprouting and grain color by quantitative trait loci analysis in a white × red grain bread-wheat cross. Theoretical and Applied Genetics. 2002;**104**(1):39-47

[64] Lan XJ, Zheng YL, Ren XB, Liu DC, Wei YM, Yan ZH. Utilization of pre harvest sprouting tolerance gene of synthetic wheat RSP. Journal of Plant Genetic Resources. 2005;**6**:204-209 (in Chinese)

[65] Lin RS, Horsley RD, Schwarz PB. Associations between caryopsis dormancy, α-amylase activity, and pre-harvest sprouting in barley. Journal of Cereal Science. 2008;**48**:446-456

[66] Yang JH, Yu YX, Cheng JS, Tan XL, Shen WP. Study on the pre harvest sprouting tolerance in *Triticum aestivum* ssp. yunnanense King. Journal of Triticeae Crops. 2011;**31**(4):747-752

[67] He ZT, Chen XL, Han YP. Progress on pre harvest sprouting resistance in white. Journal of Triticeae Crops. 2000;**20**:84-87

[68] Wilkinson MD, McKibbin RS, Bailey PC, Flintham JE, Gale MD, Lenton JR, et al. Use of comparative molecular genetics to study pre harvest sprouting in wheat. Euphytica. 2002;**126**:27-33

[69] Chang C, Zhang HP, Feng JM, Yin B, Si HQ, Ma CX. Identifying alleles of viviparous-1B associated with pre-harvest sprouting in micro-core collections of Chinese wheat germplasm. Molecular Breeding. 2010;**25**:481-490

[70] Divashuk M, Mayer N, Kroupin P, Rubets V, Pylnev V, Tkhi N, et al. The association between the allelic state of Vp-1B and pre-harvest sprouting tolerance in red-seeded hexaploid triticale. Open Journal of Genetics. 2012;**2**:51-55

[71] Kulwal PL, Kumar N, Gaur A, Khurana P, Khurana JP, Tyagi AK. Mapping of a major QTL for pre-harvest sprouting tolerance on chromosome 3A in bread wheat. Theoretical and Applied Genetics. 2005;**111**:1052-1059

[72] Mohan A, Kulwal P, Singh R, Kumar V, Mir RR, Kumar J. Genome-wide QTL analysis for pre-harvest sprouting tolerance in bread wheat. Euphytica. 2009;**168**:319-329

[73] Arifuzzaman M, Sayed MA, Muzammil S, Pillen K, Schumann H, Naz AA, et al. Detection and validation

of novel QTL for shoot and root traits in barley (*Hordeum vulgare* L.). Molecular Breeding. 2014;**34**:1373-1387

[74] Asthir B. Mechanisms of heat tolerance in crop plants. Biologia Plantarum. 2015;**59**:620-628. DOI: 10.1007/s10535-015-0539-5

[75] Muhammad Z, Waheed A, Muhammad Imran K, Shiraz A, Ali N, Amina B, et al. Breeding for pre-harvest sprouting resistance in bread wheat under rainfed conditions. Frontiers of Agricultural Science and Engineering. 2018;**5**(2):253-261

[76] Flintham JE. Different genetic components control coat-imposed and embryo-imposed dormancy in wheat. Seed Science Research. 2000;**10**(1):43-50

[77] Munns R, Tester M. Mechanism of salinity tolerance. Annual Review of Plant Biology. 2008;**59**:651-681

[78] Derera NF. Breeding for pre harvest sprouting in tolerance. In: Derera DF, editor. Preharvest Field Sprouting in Cereals. Boca Raton, Florida: CRC Press Incorporated; 1989. pp. 111-128

[79] Breiman A, Graur D. Wheat evolution. Israel Journal of Plant Sciences. 1995;**43**:85-98

[80] Caverzan A, Casassola A, Brammer SA. Antioxidant responses of wheat plants under stress. Genetics and Molecular Biology. 2016;**39**:1-6. DOI: 10.1590/1678-4685-GMB-2015-010

[81] Zanetti S, Winzeler M, Keller M, Keller B, Messmer M. Genetic analysis of pre-harvest sprouting resistance in a wheat × spelt cross. Crop Science. 2000;**40**:1406-1417

[82] Kato K, Nakamura W, Tabiki T, Miura H, Sawada S. Detection of loci controlling seed dormancy on group 4 chromosomes of wheat and comparative mapping with rice and barley genomes. Theoretical and Applied Genetics. 2001;**102**:980-985

[83] Fofana B, Humphreys DG, Rasul G, Cloutier S, Brûlé-Babel A, Woods S, et al. Mapping quantitative trait loci controlling pre-harvest sprouting resistance in a red × white seeded spring wheat cross. Euphytica. 2009;**165**:509-521

[84] Imtiaz M, Ogbonnaya FC, Oman J, Van Ginkel M. Characterization of quantitative trait loci controlling genetic variation for preharvest sprouting in synthetic backcross-derived wheat lines. Genetics. 2008;**178**:1725-1736

[85] Liu S, Cai S, Graybosch R, Chen C, Bai G. Quantitative trait loci for resistance to pre-harvest sprouting in US hard white winter wheat Rio Blanco. Theoretical and Applied Genetics. 2008;**117**:691-699

[86] Xiao-bo R, Xiu-jin L, Deng-cai L, Jia-li W, You-liang Z. Map-ping QTLs for pre-harvest sprouting tolerance on chromosome 2D in a synthetic hexaploid wheat × common wheat cross. Journal of Applied Genetics. 2008;**49**:333-341

[87] Rasul G, Humphreys DG, Brûlé-Babel A, McCartney CA, Knox RE, DePauw RM. Mapping QTLs for pre-harvest sprouting traits in the spring wheat cross 'RL4452/AC domain'. Euphytica. 2009;**168**:363-378

[88] Kumar A, Kumar J, Singh R, Garg T, Chhuneja P, Balyan HS. QTL analysis for grain colour and pre-harvest sprouting in bread wheat. Plant Science. 2009;**177**:114-122

[89] Miao XL, Zhang YJ, Xia XC, He ZH, Zhang Y, Yan J. Mapping quantitative trait loci for pre-harvest sprouting resistance in white-grained winter wheat line CA 0431. Crop & Pasture Science. 2013;**64**:573-579

[90] Cabral AL, Jordan MC, McCartney CA, You FM, Humphreys DG, Mac Lachlan R. Identification of candidate genes, regions and markers for pre-harvest sprouting resistance in wheat (*Triticum aestivum* L.). BMC Plant Biology. 2014;**14**:340

[91] Lin M, Cai S, Wang S, Liu S, Zhang G, Bai G. Genotyping-by-sequencing (GBS) identified SNP tightly linked to QTL for pre-harvest sprouting resistance. Theoretical and Applied Genetics. 2015;**128**:1385-1395

[92] Kumar S, Knox RE, Clarke FR, Pozniak CJ, DePauw RM, Cuthbert RD. Maximizing the identification of QTL for pre-harvest sprouting resistance using seed dormancy measures in a white-grained hexaploid wheat population. Euphytica. 2015;**205**:287-309

[93] Shao M, Bai G, Rife TW, Poland J, Lin M, Liu S. QTL mapping of pre-harvest sprouting resistance in a white wheat cultivar Danby. Theoretical and Applied Genetics. 2018;**131**:1683-1697

[94] Yang Y, Zhang CL, Chen XM, Wang DS, Xia LQ, Liu ZF. Identification and validatation of molecular markers for PHS tolerance in red-grained spring wheat. Journal of Triticeae Crops. 2011;**31**:54-59

Section 4

Morphological, Physio-Biochemical Mechanisms of Plants to Abiotic Stresses

Chapter 12

Morpho-Physiological Mechanisms of Maize for Drought Tolerance

Abu Sayeed Md. Hasibuzzaman, Farzana Akter, Shamim Ara Bagum, Nilima Hossain, Tahmina Akter and M. Shalim Uddin

Abstract

Maize is one of the mostly consumed grains in the world. It possesses a greater potentiality of being an alternative to rice and wheat in the near future. In field condition, maize encounters abiotic stresses like salinity, drought, water logging, cold, heat, etc. Physiology and production of maize are largely affected by drought. Drought has become a prime cause of agricultural disaster because of the major occurrence records of the last few decades. It leads to immense losses in plant growth (plant height and stem), water relations (relative water content), gas exchange (photosynthesis, stomatal conductance, and transpiration rate), and nutrient levels in maize. To mitigate the effect of stress, plant retreats by using multiple morphological, molecular, and physiological mechanisms. Maize alters its physiological processes like photosynthesis, oxidoreductase activities, carbohydrate metabolism, nutrient metabolism, and other drought-responsive pathways in response to drought. Synthesis of some chemicals like proline, abscisic acid (ABA), different phenolic compounds, etc. helps to fight against stress. Inoculation of plant growth-promoting rhizobacteria (PGPR) can result to the gene expression involved in the biosynthesis of abscisic acid which also helps to resist drought. Moreover, adaptation to drought and heat stress is positively influenced by the activity of chaperone proteins and proteases, protein that responds to ethylene and ripening. Some modifications generated by clustered regularly interspaced short palindromic repeat (CRISPR)-Cas9 are able to improve maize yield in drought. Forward and reverse genetics and functional and comparative genomics are being implemented now to overcome stress conditions like drought. Maize response to drought is a multifarious physiological and biochemical process. Applying data synthesis approach, this study aims toward better demonstration of its consequences to provide critical information on maize tolerance along with minimizing yield loss.

Keywords: abiotic stresses, osmotic adjustment, stomatal conductance, genomics, transpiration efficiency

1. Introduction

Maize being a main source of food, fodder, and fuel possesses great yield potentiality and nutritional value. It is the most widely produced cereal around the

IntechOpen

world which is an emerging crop throughout the world. Farmers are interested to its cultivation due to its higher benefit. Since agriculture is mainly reliant on nature/environment, like most of the cereal crops on which the world population depends, maize productivity is threatened by versatile stresses during its life cycle. Not only production of maize but also world's food security is at stake due to the adversities of abiotic stresses.

World food security is being challenged continuously by numerous abiotic stresses which are the consequences of exhausting climatic change in the recent decades. Abiotic stresses like salinity, drought, flooding, metal toxicity, nutrient deficiency, high temperature, and low temperature can limit the scope of crop choice as well as cause total productivity loss in severe cases. Among all other stresses, water scarcity or drought becomes an important restricting factor for crop production. According to Kramer and Boyer [1], around 28% of the world's land is too dry to support vegetation. For example, Bangladesh is one of the countries subjected to the detrimental influences of climate shifting, where 41 to 50% of land encounters a threat to experience drought each year with increased intensity (IPCC, 2013). To compete with the engrossing amount of drought stress, plants have developed a series of morphological, biochemical, physiological, cellular, and molecular mechanisms [2]. The mechanism involved in drought resistance in plants is either dehydration avoidance or dehydration tolerance [3]. In the first case, plants maintain an elevated water status during drought stress, whereas in the second case, plants function normally with a limited water condition [4].

In recent times, improved maize yield has been attained gradually, although its sensitivity to different abiotic stresses including drought has also increased. Drought being a major abiotic constraint to crop productivity as well as plant growth and development [5] can cause nearly 70% potential yield loss across the world largely because of changing climate [6]. Alike other crops, drought stress hampers maize plants in different biological, biochemical, and molecular aspects. Plants' answer to dehydration stress is somewhat very complex. Various elements which affect this response comprise environment, genotype, plant growth and development stage, and the severity and extent of the stress [7]. As maize is cultivated in over 170 million hectares in the world and is considered the second most important staple crop (FAO statistical database, http://faostat3.fao.org/home/E), it is very important to understand the mechanism behind drought adaptation in maize. Not only this but also the drought tolerance is also a prerequisite to sustain productivity of the plant. So, understanding the drought tolerance is very crucial for food security.

2. Effects of drought on growth and development of maize

Maize is prone to drought almost in every growth stage of its life cycle (**Table 1**), specifically during the reproductive stage [19]. Development and grain yield of maize are not only affected by the severity of drought stress but also by the growth stage of at which the plant is revealed to that stress; treatments of mild and severe drought stress can decrease final grain yield up to 63 and 85% [20]. At pre-anthesis and grain-filling stage, maize is more sensitive, but at seedling stage, drought can also become devastating [14]. Generally, late vegetative and anthesis growth stages have more water requirement than the seedling stage [21]. Drought condition during the seedling stage can cause harmful impact on the early crop establishment and grain yield potential due to premature tasseling which leads to extended anthesis to silk interval [13]. Even total biomass accumulation can be reduced in different developmental stages like 37% at silking stage, 34% at grain-filling, and 21% at maturity period (**Figure 1**) [22].

Feature	Effect of drought	Source
Morphological	Reduces leaf size, stem growth, and root proliferation	[8]
	Increased frequency of kernel abortion during pollination	[9, 10]
	Leaf rolling, stomata closure	[11]
	Reduced flowering, leaf number, biomass, and seed weight	[5]
	Delayed silking	[12]
	Premature flowering and longer anthesis-silk interval	[13]
Biological	Membrane damage and upset the activity of various enzymes	[14]
	Suppressed photosynthesis, carbohydrate metabolism, and energy metabolism	
Physiological	Unbalance plant-water relations and minimizes water-use efficiency	[14]
	Significantly inhibited photosynthetic rate	
	Reduced CO_2 assimilation by leaves is mainly due to stomatal closure, especially those of CO_2 fixation and adenosine triphosphate synthesis	
Biochemical	Malondialdehyde (MDA) accumulation as a biomarker for oxidative stress	[15]
	Antioxidant accumulation, reactive oxygen species (ROS) scavenging, and transcription activation	[16]
	Proline accumulation	[17]
	Higher activity of nonenzymatic antioxidants like β-carotene, ascorbate (ASC), α-tocopherol, reduced glutathione (GSH), carotenoid, enzymes include superoxide dismutase (SOD), peroxidase (POD), ascorbate peroxid (APX), catalase (CAT), polyphenol oxidase (PPO), and glutathione reductase (GR)	[18]

Table 1.
Negative impacts of drought in maize.

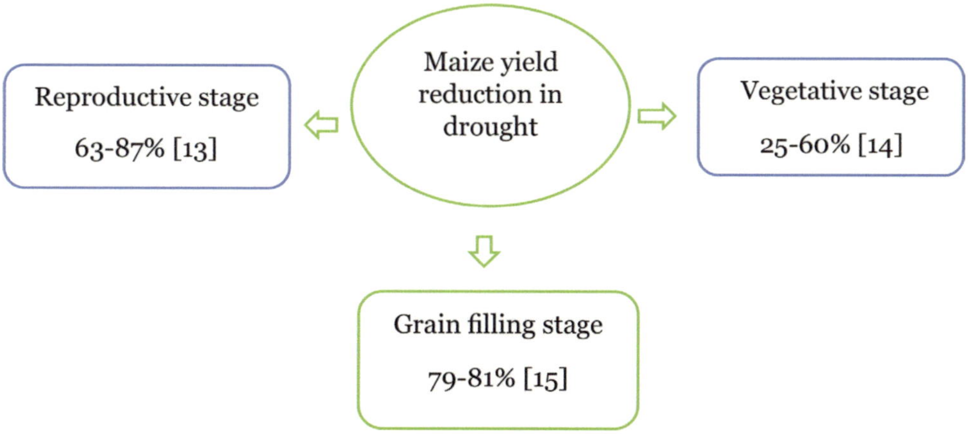

Figure 1.
Schematic representation of yield losses in maize due to drought [22–24].

3. Morpho-physiological mechanisms of drought tolerance

To fight against drought stress, plants have advanced multifarious adaptive mechanisms, comprising morphological, physiological, and biochemical adjustment, regulating transcription and gene expression, along with epigenetic plasticity [25, 26].

The most important mechanisms include enhanced water uptake along with prolific and deep root systems, partial water loss by increased diffusive resistance, and reducing leaf size making them succulent in order to decrease the transpirational loss. Potassium ions help in osmotic adjustment (OA); silicon improves the cell water balance and increases root endodermal silicification. Osmolytes with low molecular weight, like proline, glycinebetaine and other organic acids, amino acids, and polyols, are essential to maintain cellular functions while encountering drought. Different plant growth substances such as auxins, gibberellins, cytokinin, salicylic acid, and abscisic acid (ABA) regulate the plant responses toward drought [8].

3.1 Morphological mechanisms of drought tolerance

3.1.1 Leaf rolling and stomatal conductance

Water-deficient condition brings about harmful impacts on whole plant morphology. Cell and leaves undergo tremendous change in drought stress. Reduced leaves per plant, individual leaf size, and leaf area result in decreased photosynthesis [27]. Leaf senescence and rolling also indicate impacts of water deficiency in plants [28]. Thus, water stress adversely affects maize plants causing downgraded fresh and dry biomass production [29].

Stomatal resistance, relative water content (RWC), transpiration rate, leaf-water potential, and leaf and canopy temperature are important characteristics influencing plant-water relations. RWC highlights the water status in plant. It is the most useful index for drought tolerance in plants which reflects the metabolic activity in tissues [17]. Reduction in relative water content as a reply to drought stress has already been noticed in wide range of plants [30]. When plants are subjected to drought stress, they exhibit substantial decrease of leaf-water potential, relative water content, and transpiration rate [31].

3.1.2 Root characteristics

Maize root growth is not inhibited considerably in water stress [32]. Generally, in drought condition, the root and shoot ratio increases because of roots being less sensitive to growth retardation by low water potentials than shoots [33]. While undergoing drought condition, roots induce a signal flowing toward the shoots via xylem which causes physiological changes. Eventually this determines the level of plant adaptation to the stress. Drought induced root-to-leaf signaling flows through the transpiration stream resulting in the stomatal closure, which is an important alteration to reduced water supply [17]. Ethylene, cytokinin, abscisic acid, malate, and other factors are associated in the root-to-shoot signaling process. ABA endorses the K^+ ion from the guard cells, resulting in loss of turgor pressure which leads to stomatal closure. Loss of cell turgor due to dehydration of plants can increase ABA level up to 50 times [34].

3.2 Physiological mechanisms of drought tolerance

3.2.1 Osmotic adjustment, dehydration tolerance, and transpiration efficiency

Osmotic adjustment means the active accumulation of organic solutes in the plant tissue as a consequence of water-deficient condition or drought stress. This has been regarded as an important physiological adapting mechanism associated with drought tolerance which has gained significance in recent years.

Through osmotic adjustment, cytoplasmic and organelle activities occur at normal rate and assist plants to better growth, photosynthesis, and partitioning assimilates [17].

Maintaining cell turgor by reduced water potential, OA regulates photosynthesis and stomatal conductance at lower water potentials, improved root growth and water extraction, delayed leaf senescence and death, and reduced flower abortion which is crucial in dehydration tolerance [35].

3.2.2 Solute accumulation and storage sugar

Osmotic adjustment is involved in accumulating different solutes depending on the rate of drought stress. To maintain cell turgor, various organic and inorganic solutes are accumulated in the cytosol to lessen osmotic potential [36]. According to Hessini et al. [37], these accumulated compounds facilitate stressed cells either by functioning as cytoplasmic osmolytes which improve moisture uptake and preservation or by protecting and stabilizing biomolecules and their structures (i.e., proteins, membranes, chloroplasts, and liposomes) from damage caused by stress condition. Compatible solutes like amino acids such as proline or glycinebetaine, sugars, sugar alcohols (like mannitol and other low molecular weight metabolites), glycerol, and polyols can also be helpful in this process [38].

Balancing of leaf turgor pressure can be achieved by osmotic fine-tuning in drought through deposition of soluble carbohydrates, sucrose, proline, glycine-betaine, and other solutes in cytoplasm to improve water uptake. In water-deficit condition, proline accumulation is the first response of plants, which is the most widely studied due to its substantial significance in stress tolerance by reducing injury to cells. During signaling process, proline can modify mitochondrial functions, stimulate cell proliferation or cell death, and trigger some genic expression, which is ultimately essential for plant's recovery from drought stress [39]. It also helps to stabilize subcellular structures and take part in scavenging free radicals and buffering cellular redox potential during stress conditions [40].

3.2.3 Stay green (SG)/non-senescence

Stay green is an indicator of good plant health especially in drought conditions, which leads to reduced senescence, adaptability to stalk lodging, and post-flowering dehydration. This trait ensures superiority of plants compared to non-stay-green ones [41]. When green plant tissue area contributes more than the average of total plant area, and the grain moisture is lower or equal to the population average, that plants are regarded as stay green [42]. Swanckaert et al. [43] found that maize SG genotypes have higher photosynthetic capacity values with increased values for the proxies. Though an increased photosynthetic capability is not related with more assimilate accumulation in leaves, the stay-green trait was considered as a cosmetic SG. In maize this trait influenced N dynamics as the lower translocation of N from leaves resulted in low nitrogen concentration in the ear which consequently leaded to lower grain dry matter yield. SG trait generally causes changes in the partitioning of dry matter and nitrogen balance between both vegetative and reproductive tissues; the energy source also converts into cell wall material from starch (from ear source).

4. Functional genomics of drought tolerance

Different genes are translated and expressed in water-deficient condition or drought stress. The existing genotypes containing drought-inducible genes recommend the complex nature of plants in response to drought [26]. Studies executed to understand the molecular mechanisms of drought tolerance have identified some species-specific and conserved genes expressed in stress condition (**Table 2**) [55]. Additionally, transcription factors that regulate adaptive response in drought stress, such as myeloblastosis (MYB), dehydration responsive element binding (DREB), C-repeat binding factor (CBF), abscisic acid responsive elements binding factor (ABF), ABRE binding (AREB), (NAM, ATAF1/2, and CUC2 containing proteins) (NAC), WRKY, and SNF1-related kinase 2 (SnRK2), were also identified [56, 57]. In spite of these accomplishments, the gene network responsible for drought stress tolerance is still not fully revealed. Nowadays, bio-protective effects are being investigated in different ways to mitigate the crop losses caused due to drought. According to Cura et al. [53], plant growth-promoting rhizobacteria (PGPR), *A. brasilense*, strain SP-7, and *H. seropedicae*, strain Z-152, help plants to cope with the adverse effects of dehydration stress. Maize plants inoculated with these bacteria under drought condition resulted into higher carbon, nitrogen, and chlorophyll levels, higher biomass, and lower levels of ethylene and abscisic acid, plant growth-regulating hormones that affect the response toward stress. Less injuries to the cell membrane occur in the bacteria-inoculated plants than the non-inoculated plants in control condition, in the same levels of oxidative stress. Recently different genome editing tools are being applied like zinc-finger nucleases (ZFN), meganucleases, transcription activator-like effector nucleases (TALEN), and the clustered regularly interspaced short palindromic repeat (CRISPR)/CRISPR-associated nuclease protein (Cas) system which have provided the scope of modification to the desired genes in plants [58].

Gene for drought tolerance	Functions	Source
ZmVPP1	Enhanced activity Increased root growth and development	[44]
ZmNAC111	Expresses in transgenic maize Enhances drought tolerance	[45]
ZmPP2C-A10	Helps in abscisic acid signaling	[46]
ZmWRKY40	Regulate the stress-related genes as well as the reactive oxygen species (ROS) content in transgenic lines. ROS is reduced by enhancing the activities of peroxide dismutase (POD) and catalase (CAT)	[47]
MYB gene, ZmMYB3R	Elevated catalase (CAT), peroxidase (POD) and superoxide dismutase (SOD) enzyme activities, increased sensitivity to ABA, and regulation of the stomatal aperture	[48]
ASR gene ZmASR3	Lower MDA levels and higher relative water content and proline content	[49]
Maize glossy6 (gl6)	By involving GL6 protein in trafficking intracellular cuticular waxes, opening the door through which cuticular wax is transported from the site of biosynthesis to the plasma membrane	[50]
Late embryogenesis abundant (LEA) gene ZmLEA14tv	Enhanced the seed germination and survival rate	[51]
ARGOS8	Increased grain yield	[52]
ZmVP14	Biosynthesis of abscisic acid	[53]
ZmCIPK8	Involved in ABA and H2O2 signaling	[54]

Table 2.
Different genes which contribute in maize drought tolerance.

CRISPR-Cas9 system is easier to execute and very efficient. The system includes a Cas9 endonuclease which is derived from *Streptococcus pyogenes* and a chimeric single guide RNA. This leads Cas9 to a target sequence of DNA in the genome. CRISPR-Cas9 genome editing is undertaken by launching a DNA double-strand break via Cas9 in the target locus [52]. In order to resolve the molecular mechanisms underlying behind drought stress tolerance, comparative physiological and transcriptome analyses on dehydration-tolerant maize (*Zea mays* L.) are being done. Using an RNA sequencing (RNA-seq)-based approach, differentially expressed genes (DEGs) are being identified. From the critical sets of DEGs, specific drought-responsive genes are being mined out which are primarily associated with nitrogen metabolism, ribosome pathway, and amino acid biosynthesis pathways [11]. The tolerant genes can be associated with stress signal transduction; cellular redox homeostasis maintenance; MYB, NAC, WRKY, and PLATZ transcriptional factor modulated; carbohydrate synthesis and cell wall remodeling; amino acid biosynthesis; and protein ubiquitination processes.

5. Proteomics of drought tolerance

Alike some genes, several proteins are also translated in water-deficit condition; majority of the proteins are water soluble, which contribute to stress tolerance by hydrating various cellular structures [59]. In different studies, membrane-stabilizing proteins have been found to be drought responsive such as dehydrins, ABA-responsive proteins, phospholipase D, glyoxalase I or glutathione-S-transferase [60], and late embryogenic abundant proteins (LEA), increasing water-binding capacity of cells [55]. Transmembrane proteins like AQPs are found to be of vital significance to all stages of plant growth and development under osmotic stress induced by drought, through maintaining cell turgor pressure [61]. Proline accumulation has been found to be correlated with stress tolerance, which preserves the structure of complex proteins, maintains membrane integrity influencing protein solvation under dehydrated condition, and reduces oxidation of lipid membranes [62]. Heat shock proteins (HSPs) (**Table 3**) play a major role in stabilizing protein structure, which are chiefly involved in unwinding few folded proteins and averting protein denaturation under abiotic stress conditions [47, 68].

Maize CIPK gene ZmCIPK8, having a 1356 bp coding region, encodes a polypeptide of 451 amino acids. ZmCIPK8 protein contains two domains which are C-terminal regulatory domain and N-terminal protein kinase domain with a CBL-interacting NAF/FISL motif. They operate by cooperating with some

Protein	Function	Source
Hsp70	In chloroplast, mitochondria, endoplasmic reticulum, and cytosol, it prevents the aggregation and assists in refolding, protein import and translocation, signal transduction, and transcriptional activation	[63]
Chaperonin/Hsp60	In chloroplast, mitochondria, and cytosol, folding and assistance in refolding	[64]
Hsp90	In chloroplast, mitochondria, endoplasmic reticulum, and cytosol, it facilitates maturation of signaling molecules, genetic buffering	[65]
Hsp100/Clp	In cytosol and mitochondria, disaggregation, unfolding	[66]
sHsp	In chloroplast, mitochondria, endoplasmic reticulum, membrane, and cytosol, it prevents aggregation and stabilizes non-native proteins	[67]

Table 3.
Proteins related to drought tolerance in maize.

membrane-localized proteins as their targets resulting in higher activity of SOD, which is a major antioxidant enzyme, scavenging superoxide radicals [69] as well as lowering the levels of MDA under drought stress [54].

6. Breeding for drought tolerance

Breeding of drought-tolerant crops implies to fulfill the requirement of the expanding population around the world which will need more food, fodder, and fuel in a defensible way. Advancement in crops drought tolerance is eventually assessed by the increment of grain yield under water-deficient conditions. As it's difficult to predict the exact moment that drought will occur, effective and long-acting solutions must be brought up by the agriculturists. The metabolic systems and physiological activities responsible behind drought tolerance are very complex and often hard to allocate. The resistance to drought can be obtained by three different ways which are dehydration escape, dehydration avoidance, and dehydration tolerance [15]. Relevant morpho-physiological attributes include resistance to short anthesis-silking interval, plant wilting, deep root systems, rapid maturity, waxy cuticle, heavy glaucousness or dense pubescence, leaf-water retention, stay-green characteristics, osmotic adjustment, cellular membrane stability, and high harvest index, as well as biochemical traits like long-distance signals provided by plant hormones, abscisic acid, xylem sap pH, and inorganic ions that provide shoot water retention ability, etc. Traits can be improved through pedigree breeding, backcross breeding, bulk population breeding, and recurrent selection [70]. Strategies such as mass screening, marker-assisted selection, as well as genetic engineering can be adapted to attain drought resistance in plants. Maize breeding for drought resistance is prospective by utilizing the existing genetic resources and application of precise phenotyping and breeding informatics, introducing drought-resistance genes [71]. Moreover, for breeding drought-tolerant crops, CRISPR-Cas9 system is generating unique allelic variation [52].

7. Conclusion

Stress tolerance is a complex trait. Different plants adapt themselves differently to tolerate the adversity of stress. Maize as an important cereal exhibits some evidence of stress tolerance, especially drought stress tolerance. Inherently maize changes its morphological structures and physiological activities to mitigate the negative impact of drought. But, to do this the crop fails to produce yields with maximum potential. In this case, breeding for drought tolerance in maize opens a door to achieve higher yield under drought condition. So, more efforts are required to develop maize variety that can tolerate drought stress.

Author details

Abu Sayeed Md. Hasibuzzaman[1], Farzana Akter[1], Shamim Ara Bagum[2], Nilima Hossain[2], Tahmina Akter[3] and M. Shalim Uddin[2*]

1 Department of Genetics and Plant Breeding, Bangabandhu Sheikh Mujibur Rahman Agricultural University, Gazipur, Bangladesh

2 Bangladesh Agricultural Research Institute, Gazipur, Bangladesh

3 Biotechnology and Genetic Engineering Discipline, Khulna University, Khulna, Bangladesh

*Address all correspondence to: shalimuddin@yahoo.com

References

[1] Kramer PJ, Boyer JS. Water Relations of Plants and Soils. San Diego: Academic Press; 1995

[2] Fang Y, Xiong L. General mechanisms of drought response and their application in drought resistance improvement in plants. Cellular and Molecular Life Sciences. Switzerland. 2015;**72**(4):673-689

[3] Levitt J. Responses of Plants to Environmental Stress, Volume 1: Chilling, Freezing, and High Temperature Stresses. Cambridge, Massachusetts, USA: Academic Press; 1980

[4] Pires MV, de Castro EM, de Freitas BS, Lira JM, Magalhães PC, Pereira MP. Yield-related phenotypic traits of drought resistant maize genotypes. Environmental and Experimental Botany. 2019;**11**:103962

[5] Ghatak A, Chaturvedi P, Weckwerth W. Cereal crop proteomics: Systemic analysis of crop drought stress responses towards marker-assisted selection breeding. Frontiers in Plant Science. 2017;**8**:757

[6] Wu S, Ning F, Zhang Q, Wu X, Wang W. Enhancing omics research of crop responses to drought under field conditions. Frontiers in Plant Science. 2017;**8**:174

[7] Lata C, Muthamilarasan M, Prasad M. Drought stress responses and signal transduction in plants. In: Elucidation of Abiotic Stress Signaling in Plants. New York, NY: Springer; 2015. pp. 195-225

[8] Farooq M, Wahid A, Kobayashi N, Fujita DB, Basra SM. Plant drought stress: Effects, mechanisms and management. In: Sustainable Agriculture. Dordrecht: Springer; 2009. pp. 153-188

[9] Morgan PW. Effects of abiotic stresses on plant hormone systems. Plant Biology. USA. 1990;**12**:113-146

[10] Ober ES, Setter TL, Madison JT, Thompson JF, Shapiro PS. Influence of water deficit on maize endosperm development: Enzyme activities and RNA transcripts of starch and zein synthesis, abscisic acid, and cell division. Plant Physiology. 1991;**97**(1):154-164

[11] Zenda T, Liu S, Wang X, Liu G, Jin H, Dong A, et al. Key maize drought-responsive genes and pathways revealed by comparative transcriptome and physiological analyses of contrasting inbred lines. International Journal of Molecular Sciences. 2019;**20**(6):1268

[12] Cattivelli L, Rizza F, Badeck FW, Mazzucotelli E, Mastrangelo AM, Francia E, et al. Drought tolerance improvement in crop plants: An integrated view from breeding to genomics. Field Crops Research. 2008;**105**(1-2):1-4

[13] Li Z, Xu WJ, Xue BD, Cao P. Discuss on evaluating method to drought-resistance of maize in seedling stage. Journal of Maize Sciences. 2004;**2**:73-75

[14] Min H, Chen C, Wei S, Shang X, Sun M, Xia R, et al. Identification of drought tolerant mechanisms in maize seedlings based on transcriptome analysis of recombination inbred lines. Frontiers in Plant Science. 2016;**7**:1080

[15] Meena YK, Kaur N. Towards an understanding of physiological and biochemical mechanisms of drought tolerance in plant. Annual Research & Review in Biology. 2019;**25**:1-3

[16] Jogaiah S, Govind SR, Tran LS. Systems biology-based approaches toward understanding drought tolerance in food crops. Critical Reviews in Biotechnology. 2013;**33**(1):23-39

[17] Anjum SA, Xie XY, Wang LC, Saleem MF, Man C, Lei W.

Morphological, physiological and biochemical responses of plants to drought stress. African Journal of Agricultural Research. 2011;**6**(9):2026-2032

[18] Zhu JK. Epigenome sequencing comes of age. Cell. 2008;**133**(3):395-397

[19] Saseendran SA, Ahuja LR, Ma L, Nielsen DC, Trout TJ, Andales AA, et al. Enhancing the water stress factors for simulation of corn in RZWQM2. Agronomy Journal. 2014;**106**(1):81-94

[20] Ge T, Sui F, Bai L, Tong C, Sun N. Effects of water stress on growth, biomass partitioning, and water-use efficiency in summer maize (*Zea mays* L.) throughout the growth cycle. Acta Physiologiae Plantarum. 2012;**34**(3):1043-1053

[21] Maiti RK, Maiti LE, Maiti S, Maiti AM, Maiti M, Maiti H. Genotypic variability in maize cultivars (*Zea mays* L.) for resistance to drought and salinity at the seedling stage. Journal of Plant Physiology. 1996;**148**(6):741-744

[22] Kamara AY, Menkir A, Badu-Apraku B, Ibikunle O. The influence of drought stress on growth, yield and yield components of selected maize genotypes. The Journal of Agricultural Science. 2003;**141**(1):43-50

[23] Atteya AM. Alteration of water relations and yield of corn genotypes in response to drought stress. Bulgarian Journal of Plant Physiology. 2003;**29**(1-2):63-76

[24] Monneveux P, Sanchez C, Beck D, Edmeades GO. Drought tolerance improvement in tropical maize source populations. Crop Science. 2006;**46**(1):180-191

[25] Miao Z, Han Z, Zhang T, Chen S, Ma C. A systems approach to a spatio-temporal understanding of the drought stress response in maize. Scientific Reports. 2017;**7**(1):6590

[26] Zheng J, Fu J, Gou M, Huai J, Liu Y, Jian M, et al. Genome-wide transcriptome analysis of two maize inbred lines under drought stress. Plant Molecular Biology. 2010;**72**(4-5):407-421

[27] Rucker KS, Kvien CK, Holbrook CC, Hook JE. Identification of peanut genotypes with improved drought avoidance traits. Peanut Science. 1995;**22**(1):14-18

[28] Manivannan P, Jaleel CA, Kishorekumar A, Sankar B, Somasundaram R, Sridharan R, et al. Changes in antioxidant metabolism of *Vigna unguiculata* (L.) Walp. By propiconazole under water deficit stress. Colloids and Surfaces. B, Biointerfaces. 2007;**57**(1):69-74

[29] Zhao TJ, Sun S, Liu Y, Liu JM, Liu Q, Yan YB, et al. Regulating the drought-responsive element (DRE)-mediated signaling pathway by synergic functions of trans-active and trans-inactive DRE binding factors in Brassica napus. The Journal of Biological Chemistry. 2006;**281**(16):10752-10759

[30] Nayyar H, Gupta D. Differential sensitivity of C3 and C4 plants to water deficit stress: Association with oxidative stress and antioxidants. Environmental and Experimental Botany. 2006;**58**(1-3):106-113

[31] Siddique MR, Hamid AI, Islam MS. Drought stress effects on water relations of wheat. Botanical Bulletin of Academia Sinica. 2000;**41**:35-39

[32] Sacks MM, Silk WK, Burman P. Effect of water stress on cortical cell division rates within the apical meristem of primary roots of maize. Plant Physiology. 1997;**114**(2):519-527

[33] Wu Y, Cosgrove DJ. Adaptation of roots to low water potentials by changes

in cell wall extensibility and cell wall proteins. Journal of Experimental Botany. 2000;**51**(350):1543-1553

[34] Guerrero F, Mullet JE. Increased abscisic acid biosynthesis during plant dehydration requires transcription. Plant Physiology. 1986;**80**(2):588-591

[35] Taiz L, Zeiger E. Stress physiology. Plant Physiology. 2006;**4**:591-623

[36] Rhodes D, Samaras Y. Genetic control of osmoregulation in plants. Cellular and Molecular Physiology of Cell Volume Regulation. 1994;**416**

[37] Hessini K, Martínez JP, Gandour M, Albouchi A, Soltani A, Abdelly C. Effect of water stress on growth, osmotic adjustment, cell wall elasticity and water-use efficiency in Spartina alterniflora. Environmental and Experimental Botany. 2009;**67**(2):312-319

[38] Martınez-Ballesta MC, Martınez V, Carvajal M. Osmotic adjustment, water relations and gas exchange in pepper plants grown under NaCl or KCl. Environmental and Experimental Botany. 2004;**52**(2):161-174

[39] Szabados L, Savoure A. Proline: A multifunctional amino acid. Trends in Plant Science. 2010;**15**(2):89-97

[40] Ashraf MF, Foolad M. Roles of glycine betaine and proline in improving plant abiotic stress resistance. Environmental and Experimental Botany. 2007;**59**(2):206-216

[41] Bekavac G, Stojakovic M, Ivanovic M, Jockovic D, Vasic N, Purar B. Relationships of stay green trait in maize. Genetika. 2002;**34**(1):33-40

[42] Bekavac G, Stojaković M, Jocković D, Boćanski J, Purar B. Path analysis of stay-green trait in maize. Cereal Research Communications. 1998;**1**:161-167

[43] Swanckaert J, Pannecoucque J, Van Waes J, Steppe K, Van Labeke MC, Reheul D. Stay-green characterization in Belgian forage maize. The Journal of Agricultural Science. 2017;**155**(5):766-776

[44] Wang D, Yu C, Zuo T, Zhang J, Weber DF, Peterson T. Alternative transposition generates new chimeric genes and segmental duplications at the maize p1 locus. Genetics. 2015;**201**(3):925-935

[45] Mao H, Wang H, Liu S, Li Z, Yang X, Yan J, et al. A transposable element in a NAC gene is associated with drought tolerance in maize seedlings. Nature Communications. 2015;**6**:8326

[46] Xiang Y, Sun X, Gao S, Qin F, Dai M. Deletion of an endoplasmic reticulum stress response element in a ZmPP2C-A gene facilitates drought tolerance of maize seedlings. Molecular Plant. 2017;**10**(3):456-469

[47] Wang CT, Ru JN, Liu YW, Yang JF, Li M, Xu ZS, et al. The maize WRKY transcription factor ZmWRKY40 confers drought resistance in transgenic Arabidopsis. International Journal of Molecular Sciences. 2018;**19**(9):2580

[48] Wu J, Jiang Y, Liang Y, Chen L, Chen W, Cheng B. Expression of the maize MYB transcription factor ZmMYB3R enhances drought and salt stress tolerance in transgenic plants. Plant Physiology and Biochemistry. 2019;**137**:179-188

[49] Liang Y, Jiang Y, Du M, Li B, Chen L, Chen M, et al. ZmASR3 from the maize ASR gene family positively regulates drought tolerance in transgenic Arabidopsis. International Journal of Molecular Sciences. 2019;**20**(9):2278

[50] Li L, Du Y, He C, Dietrich CR, Li J, Ma X, et al. The maize glossy6

gene is involved in cuticular wax deposition and drought tolerance. Journal of Experimental Botany. 2019;**70**(12):3089-3099

[51] Minh BM, Linh NT, Hanh HH, Hien LT, Thang NX, Hai NV, et al. A LEA gene from a Vietnamese maize landrace can enhance the drought tolerance of transgenic maize and tobacco. Agronomy. 2019;**9**(2):62

[52] Shi J, Gao H, Wang H, Lafitte HR, Archibald RL, Yang M, et al. ARGOS 8 variants generated by CRISPR-Cas9 improve maize grain yield under field drought stress conditions. Plant Biotechnology Journal. 2017;**15**(2):207-216

[53] Curá J, Franz D, Filosofía J, Balestrasse K, Burgueño L. Inoculation with Azospirillum sp. and Herbaspirillum sp. bacteria increases the tolerance of maize to drought stress. Microorganisms. 2017;**5**(3):41

[54] Tai F, Wang Q, Yuan Z, Yuan Z, Li H, Wang W. Characterization of five CIPK genes expressions in maize under water stress. Acta Physiologiae Plantarum. 2013;**35**(5):1555-1564

[55] Kosová K, Vítámvás P, Prášil IT. Wheat and barley dehydrins under cold, drought, and salinity–what can LEA-II proteins tell us about plant stress response? Frontiers in Plant Science. 2014;**5**:343

[56] Singh D, Laxmi A. Transcriptional regulation of drought response: A tortuous network of transcriptional factors. Frontiers in Plant Science. 2015;**6**:895

[57] Mun BG, Lee SU, Park EJ, Kim HH, Hussain A, Imran QM, et al. Analysis of transcription factors among differentially expressed genes induced by drought stress in *Populus davidiana*. 3 Biotech. 2017;**7**(3):209

[58] Čermák T, Baltes NJ, Čegan R, Zhang Y, Voytas DF. High-frequency, precise modification of the tomato genome. Genome Biology. 2015;**16**(1):232

[59] Osakabe Y, Osakabe K, Shinozaki K, Tran LS. Response of plants to water stress. Frontiers in Plant Science. 2014;**5**:86

[60] Blein-Nicolas M, Negro SS, Balliau T, Welcker C, Cabrera-Bosquet L, Nicolas SD, et al. Integrating proteomics and genomics into systems genetics provides novel insights into the mechanisms of drought tolerance in maize. bioRxiv. 2019:636514

[61] Forrest KL, Bhave M. Major intrinsic proteins (MIPs) in plants: A complex gene family with major impacts on plant phenotype. Functional & Integrative Genomics. 2007;**7**(4):263

[62] Demiral T, Türkan I. Does exogenous glycinebetaine affect antioxidative system of rice seedlings under NaCl treatment? Journal of Plant Physiology. 2004;**161**(10):1089-1100

[63] Frydman J. Folding of newly translated proteins in vivo: The role of molecular chaperones. Annual Review of Biochemistry. 2001;**70**(1):603-647

[64] Bukau B, Horwich AL. The Hsp70 and Hsp60 chaperone machines. Cell. 1998;**92**(3):351-366

[65] Krishna P, Gloor G. The Hsp90 family of proteins in *Arabidopsis thaliana*. Cell Stress & Chaperones. 2001;**6**(3):238

[66] Goloubinoff P, Mogk A, Zvi AP, Tomoyasu T, Bukau B. Sequential mechanism of solubilization and refolding of stable protein aggregates by a bichaperone network. Proceedings of the National Academy of Sciences. 1999;**96**(24):13732-13737

[67] Lee GJ, Vierling E. A small heat shock protein cooperates with heat

shock protein 70 systems to reactivate a heat-denatured protein. Plant Physiology. 2000;**122**(1):189-198

[68] Zhang X, Lei L, Lai J, Zhao H, Song W. Effects of drought stress and water recovery on physiological responses and gene expression in maize seedlings. BMC Plant Biology. 2018;**18**(1):68

[69] Mittova V, Guy M, Tal M, Volokita M. Salinity up-regulates the antioxidative system in root mitochondria and peroxisomes of the wild salt-tolerant tomato species Lycopersicon pennellii. Journal of Experimental Botany. 2004;**55**(399):1105-1113

[70] Xiao YN, Li XH, George ML, Li MS, Zhang SH, Zheng YL. Quantitative trait locus analysis of drought tolerance and yield in maize in China. Plant Molecular Biology Reporter. 2005;**23**(2):155-165

[71] Luo L, Xia H, Lu BR. Editorial: Crop breeding for drought resistance. Frontiers in Plant Science. 2019;**10**:314

Chapter 13

Phytoremediation Strategies of Some Plants under Heavy Metal Stress

Momezul Haque, Karabi Biswas and Sankar Narayan Sinha

Abstract

Environments are polluted with heavy metals across the world because of increase in industrial garbage and sewage. Plants which are grow in polluted areas shows a reduction in growth, performance, productivity. Heavy metals affect physiological and biological process of plants. Heavy metals show metallic properties which are very harmful to the plants. Accumulation of heavy metals in plants through root are caused root malformation reduction in biomass and seed production, decrease in chlorophyll-aand carotenoid content. Phytoremediation is a natural biological process through which plants remove, detoxify or immobilise environmental heavy metals in a growth matrix.

Keywords: phytoremediation, heavymetal, pollution, sewage and detoxify

1. Introduction

Heavy metals are those elements which have density greater than 5 g cm^{-3} [1]. Some heavy metals namely, cobalt (Co), copper (Cu), molybdenum (Mo), manganese (Mn), nickel (Ni) iron (Fe), and zinc (Zn) are considered to be essential for plants. These heavy metal elements directly impact on plant growth, development, senescence and energy producing processes and other physiological process due to their high reactivity. The concentration of heavy metals in soil after the admissible limits is toxic to plants either provoke oxidative stress through free radicals or crumbling up the functions of enzymes by replacing metals and nutrients which are essential [2, 3]. Cell metabolism changes by the affect of heavy metals at first reduce the plant growth. However, toxicity of metals depends on various stage of their growth stage [4]. Maksymiec and Baszynski [5] reported that dicotyledonous plants like various beans and *Medicago sativa* were more resistant to heavy metals at the early growth stage [6]. So, the heavy metals toxicity on the plant physiology and metabolism are much more noticeable. Among the heavy metals, chromium and cadmium are of special concern due to their potential toxicity on plants even at low concentrations [7–9]. The various types of chromium toxicity in plant had described by [10], and the inhibition of enzymatic activity by vaeious types mutagenesis had also be described. The visible symptoms are reduction in growth, leaf chlorosis, stunting, and yield reduction [7, 11]. [12] has explain that Cadmium (Cd) is particularly is one of the most dangerous pollutant due to its high level of toxicity and much solubility in water. [13, 14], have reported that in some plant species Cd interacts with the absorption of metal nutrients such as Fe, Zn, Cu and

Mn, in addition to inducing a process named as peroxidation and breakdown of chlorophyll in plants, resulting in an enhanced production of reactive oxygen species (ROS) [15]. According to [16], Cadmium also inhibits the uptake of elements such as K, Ca, Mg, Fe because it uses the same transmembrane carriers. Cadmium acquisition in plants may also cause serious health hazard to human beings through food chain; however, it causes an extra risk to the children by direct ingestion of Cd-contaminated soil [17].

2. Origin and occurrence

Heavy metals remain in environment in various forms like colloidal, ionic, particulate and dissolved phases. The soluble forms of heavy metal elements are remain in environment as ionised or unionized organometallic chelates. According to [18], the metal concentrations of soil ranges from low to 100,000 mg kg^{-1} which depends on the location, area and the types of metals. [19], studied that among chemical elements, Cr is considered to be the seventh most abundant elements on earth and constitutes 0.1 to 0.3 mg kg − 1 of the crystal rocks. According to McGrath [20], In alloys and 15 percent in chemical industrial processes, mainly leather tanning, pigments, electroplating and wood preservation, about 60–70 percent of the total world production of Cr is used. Chromium has many oxidation states ranging from Cr^{2-} to Cr^{6+}; however, in a number of compounds, valences of I, II, IV and V have been shown to exist [21]. Cr (VI) is, however, considered the most toxic form of chromium and is also generally associated with oxygen as chromate (CrO_4^{2-}) or dichromate (CrO_4^{2-}) and dichromate ($Cr_2O_7^{2-}$) oxyanions. [22], observed that Cr (III) is less mobile and less toxic and is mainly bound to organic matter in soil and aquatic environments. According to [23], Cr present mostly in the form of Cr (III) in soil, and mineral environment. [24], has described that Cr and Fe $(OH)_3$ is a solid phase of Cr(III) having even lower solubility than $Cr(OH)_3$. Consequently, within the soil add up to solvent Cr(III) remains inside the allowable limits for drinking water for a wide extend of pH (4–12) due to precipitation of $Cr(OH)_3$, $Fe(OH)_3$[25, 26], moreover, major source of Cd is the parental fabric. Anthropogenic exercises have too been improved the sum of Cd in soil [27]. Overwhelming metals are regularly show at exceptionally moo concentrations in freshwaters [28], but the release of fluid squander from a wide assortment of businesses such as electroplating, metal wrapping up, calfskin tanning, chrome planning, generation of batteries, phosphate fertilizers, shades, stabilizers, and amalgams has solid impact in sea-going situations [29–31]. Cadmium pollution is also happened from rubber when car tires run over streets, and after a rain, the Cd is washed into sewage disposal systems and collected in the slush.

3. Mobility of heavy metals

Heavy metals are enter in environment are transported by water and air, also deposited in soil and sediments where they could be immobilized [32]. However, the bonding process of metals may take considerably long time. At the starting of the official handle the bio accessible division of metal components in soil is tall, but diminishes continuously in due course of time [33]. Metal dissolvability and bioavailability to plant is basically affected by the chemical properties of soil such as, soil pH, stacking rate, cation trade capacity, soil surface, redox potential, clay substance and natural matter [34–36]. For the most part, higher the slime or natural matter and soil pH, the metals will be relentlessly bound to soil with longer time and will be less organically accessible to the plants. Soil temperature is additionally an vital calculate for varieties in metal amassing by crops [37]. The bioavailability of metals is make greater in soil through several means, the secretion of phytosiderophores into the rhizosphere to chelate and solubilise metals that are soil bound [38]. Acidification of the rhizosphere and exudation of carboxylates are deliberated potential means to enhancing metal consumption.

4. Uptake of heavy metals

Heavy metals are taken through root cells of the vegetation after their mobilization inside the soil, and their improvement inside the soil relies upon in the main upon: (i) dissemination of steel additives alongside the attention attitude which has formed because of take-up of factors and ultimately inanition of the aspect inside the root region; (ii) interferences through roots, in which soil extent is uprooted through root extent after developing (iii) move of steel additives from enormous soil association down the water capacity slope [39]. Cell divider acts as a particle exchanger of relatively moo partiality and moo selectivity in which metals are first of all bound. From the mobileular divider, the shipping frameworks and intracellular high-affinity authoritative locations intercede and power the take-up of those metals over the plasma layer. A stable using power for the take-up of steel additives thru auxiliary transporters is made because of the layer capacity, that is bad at the indoors of the plasma movie and can exceed −200 mV in root epiderm. This is examined both in soil culture and in solution culture for Cd which might probably be due to low concentration of heavy metals per unit of absorption area [40, 41]. Both non-essential and essential metals are also preoccupied through leaves. Within the shape of gases, they input via thestomata withinside the leaves, while in ionic shape metals specifically input via theleaf cuticle [39, 42]. Hg in gaseous shape istaken up through stomata [43] and its uptake is recommended to bebetter in C3 than C4 flora [44]. The uptake of metals takes place viaectodesmata, non-plasmatic "channels" at a excessive level whichare much less dense elements of the cuticular layer which are located fundamental withinside theepidermal mobileular wall or cuticular membrane machine among shield cells andsubsidiary cells. Furthermore, the cuticle overlaying shield cells are oftenspecific to it overlaying everyday epidermal mobileular [39]. Most of the metallic factors are insoluble that won't capin an edge toflow freely withinside the vascular machine of flora and, as a result typically shapesulphate, phosphate or carbonate precipitates immobilizing them inextracellular booths i.e. apoplastic and intracellular compartment i.e. symplastic [45]. In the apoplastic pathway solute and also the water debris diffuse via mobileular membrane, consequently the pathway stays unregulated. The mobileularwall of the endodermal mobileular layer acts as an impediment for apoplastic diffusioninto the vascular machine. Generally, prior to the access of metallic ions withinside thexylem, solutes must be haunted through root symplasm [46]. Ifmetals are obsessed through the premise symplasm, their similarly motion from root tothe xylem is specifically ruled through 3 processes, including: (i) metallicsequestration arise into the premise symplasm, (ii) symplastic shipping ariseinto the stele, and (iii) launch of metals arise into the xylem. The ionshipping into the xylem is often occured through membrane shipping proteins. Metal factors which are not wished through the flora successfully compete thecritical heavy metals for his or her shipping the usage of the equal transmembranecarriers. Cr(III) uptake through the plant is specifically a passive process, whilst Cr(VI) shipping is mediated through sulphate carrier [47]. Inhibitors like, sodium azide and di nitrophenol inhibits the uptake of Cr(VI) through barley seedlings however this is not happened just in case of Cr(III) [47]. In keeping with [48], Group VI anions like SO_4^{-2} additionally inhibit the uptake of chromateswhile Ca^{2+} stimulates its shipping. This inhibition of chromate shipping is passed thanks to the aggressive inhibitiondue to the chemical similarity, whilst inspired shipping of Cr(VI) because of Ca is attributed to its critical position in flora for the receive and shipping of metallic factors [26, 49].

5. Accumulation of heavy metals

According to Kumar et al. [50], many plants species show an unusual capability to absorbe heavy metals through root system and accumulate of these heavy metals in their parts. Zayed and Terry [26] said that it seems a common tendency of all plant species to maintain Cr in their roots, but with quantitative differences. It is found that for the translocation of Cr to the plant tip, leafy vegetables such as spinach, turnip leaves that tend to acquire Fe appear to be the most effective [51]. While those leafy vegetables such as lettuce were considerably less effective for translocating Cr to their leaves, cabbage which accumulated relatively low Fe levels in their leaves. Zayed and Terry [26] have reported that some plant species attain substantially higher root or shoot concentration ratio than other species. However, a 'Soil–Plant Barrier' well protects the food chain from heavy metal toxicity, implying that, due to one or more of the following processes, heavy metal levels in edible plant tissues are reduced to safe levels for animals and humans: (i) prevention of metal element uptake due to soil insolubility, (ii) prevention of metal element translocation by making them immobile in roots, or (iii) prevention of metal element translocation for animals and humans to the permissible level [52]. Within plant tissues, some elements such as B, Mo, Cd, Mn, Se, and Zn are readily absorbed and translocated, while others such as Al, Ag, Cr, Fe, Hg, and Pb are less mobile because of their strong binding to soil components or root cell walls. However, at certain concentrations, all of these elements are mobilised, even against a concentration gradient, within the transport system of the plant. Kinetic data show, for instance, that essential Cu^{2+}, Ni^{2+} and Zn^{2+} and non-essential Cd^{2+} compete for their transport with the same transmembrane carrier [53]. As is the case of phytosiderophore such as Fe-transport in graminaceous species, metal chelate complexes can be transported by plasma membrane [54]. Among the most important parameter the most influencing factor of heavy metal accumulation in plants is soil pH [55–58]. At higher soil pH, metal elements in soil solution decrease their bioavailability, and at lower soil pH metalelements in soil solution increase their bioavailability to plants [59].

6. Effect on growth and development

Heavy metals mitigate the growth and development of the plant [60, 61]. The plant parts which are associated with the heavy metals polluted soils normally the roots express rapid and sensorial changes in their growth and development [62]. It is well observed that the very significant effects of a number of metals (Cd, Al, Cu, Fe, Ni, Pb, Hg, Cr, Zn,) on the growth of above ground plant parts vary [63]. Through the formation of free radicals and reactive oxygen species (ROS), heavy metals mainly affect plant growth, which causes constant oxidative damage by decreasing important cellular components. [64, 65]. For example, rice seedlings irradiated to Cd or Ni [66] and runner bean plants treated with Cd and Cu have shown an increase in carbohydrate content and a decrease in photosynthesis process, resulting in growth inhibition [67]. Similarly, in cucumber plants, Cu limits K uptake by leaf and inhibits the photosynthesis via sugar acquisition resulting into the inhibition of cell expansion [68]. Limped leaves, growth inhibition, progressive chlorosis in certain leaves and leaf sheaths and browned root systems, especially the root tips, are the symptoms of Cd toxicity in rice plants [7, 69]. Moreover, plant growth has also been retarded in maize (*Zea mays*) Cd [70, 71]. Some phenotypic abnormalities such as stunted growth, less branching and less fruiting are also shown by tomato plants irrigated with polluted water. However, acquisition of heavy metals is much more appears in stems, roots, and leaves as compared to fruits [72].

6.1 Germination

Seed germination is the breaking of seed dormancy which is inhibited by heavy metals. Germination of seeds and growth of seedling may sensitive towards environmental conditions [59]. So as per [73], the performance of germination, breaking of seed dormancy and seedlings growth rates are therefore often used to assess the abilities of plant tolerance to metal elementsIn comparison to control, higher concentrations such as 1 μM, 5 μM and 10 μM of heavy metals such as Cu, Zn, Mg and Na significantly inhibit seed germination and early growth of rice, barley, wheat and maize seedlings [74]. The ability of a seed to germinate in a moderate containing any metal element like Cr would be a direct indication of its level of tolerance to this metal, but seed germination is the first physiological process affected by toxic elements [73]. At 200 μM of Cr treatment, the seed germination of *Echinochloa colona*is decreased to 25 percent [75], and high levels (500 ppm) of Cr (VI) in soil decreased *Phaseolus vulgaris* germination by up to 48 percent [76]. Jain et al. [77] observed reductions in sugarcane bud germination of up to 35 per cent and 60 per cent at 20 and 80 ppm Cr application, respectively. In another study by Peralta et al. [73], at 40 ppm Cr (VI) treatment, *Medicago sativa*cv germination was reduced to 23 percent.

6.2 Root

Among the plant parts, roots are firstly come into contact with toxic elements and they usually absorbed more metals by root hair through absorbption process but shoots are not that [78–80]. The inhibition or retard of root elongation appears to be the first visible effect of metal toxicity. Elongations of root are reduced by the inhibition of cell division, the decrease of cell expansion, decrease of cell size in the elongation zone [81]. So the first visible effect of metal toxicity is the inhibition of root elongation, the root length can be used as most important tolerance index [82–85]. *Medicago sativa* plants grown in solid media watered with 20 mg L^{-1} of Cr (VI) in another [73] study, the ratio of Cr in shoots to Cr in roots was approximately 43 percent. This is an indication that in the roots, 50 percent of the absorbed Cr is held. The response of roots to heavy metals in both herbaceous plant species and trees has been extensively studied. [86–89]. After the work of numerous researchers [86, 87, 89, 90]. The main morphological and structural effects of metal root toxicity can be summarised as: (i) decrease in root elongation, (ii) decrease in biomass, (iii) decrease in vessel diameter, (iv) damage to tip, (v) collapse of root hair or decrease in number of roots, (vi) increase or decrease in lateral root formation, (vii) enhancement of suberification, (viii) enhancement of lignifications, (ix) translocation process become hampered. The research work of [91], revealed that Cr affects the root length than the other parts of plant as compared to other heavy metals. Mokgalaka-Matlala et al. [92], have observed that when increasing concentrations of As (V) and As (III) in *Prosopis juliflora,* the root elongation decreased significantly. It is reported that when Cr has applied on *Salix viminalisis* then the root length is affected more than by Cd and Pb [91]. In fact, the inhibition effect of Cr on the growth of the *Salix alba* root is similar to that of Hg and stronger than that of Cd and Pb, whereas the root length of Ni decreased less than Cr [93, 94]. In Salix viminalisis, the order of metal toxicity to the new root rimordial was reported to be Cd > Cr > Pb [91].

6.3 Stem

The heavy metal elements highly affect the plant height as well as shoot growth [95]. Cr transport to the various part of the plant have a direct impact on cellular metabolism as a result shoots contributing affected so plant height ultimately reduces [61]. It is observed that reduction of 11, 22 and 41% respectively compared to control in oat plants at 2, 10 and 25 ppm of Cr content in nutrient solutions in sand cultures [96]. Joseph et al. [97] observed a similar reduction in the height of Curcumas sativus, *Lactuca sativa* and *Panicum miliaceum* due to Cr (VI). Shoot growth in *Medicago sativa* is inhibited by Cr (III) [98]. In a glasshouse experiment after 32 and 96 days, Sharma and Sharma [99] noted a significant decrease in the height of *Triticum aestivum* when sown in sand with 0.5 μM sodium dichromate. A significant reduction in height of *Sinapsis albaat* a level of 200 or 400 mg kg^{-1} of Cr in soil along with N, P, K and S fertilizers was reported by Hanus and Tomas [100]. Very recently, it is found that a reduction in stem height at various concentrations (10, 20, 40 and 80 ppm) of Cd and Cr have been reported in *Dalbergia sisso* seedlings compared to the control [101].

6.4 Leaf

The heavy metal elements severely affect the leaf height as well as leaf growth. Metal elements like Cd induce morphological changes such as drying of older leaves, wilt, and chlorosis and necrosis of younger leaves. *Datura innoxia, D. metel,* plants grown in a contaminated environment with Cr(VI) exhibited toxic symptoms at 0.1 mM to 0.2 mM of Cr(VI) in the form of leaf fall and wilting of leaves at 0.4 to 0.5 mM Cr(VI) in soil [97, 102]. A similar reduction in the height of Curcumas sativus, *Lactuca sativa* and *Panicum miliaceum* due to Cr(VI) was observed (1995). In *Medicago sativa*, shoot growth is inhibited by Cr(III) [98]. Sharma and Sharma [99] noted a significant drop in the height of *Triticum aestivum* when sown in sand with 0.5 μM sodium dichromate in a glasshouse experiment after 32 and 96 days [103]. In *Zea mays*, Acacia holosericeaOryza sativa, and *Leucaena leucocephala* plants treated with tannery effluent of varying concentrations, leaf dry weight and leaf area slowly decreases [104]. The effect of Cr(III) and Cr(VI) on the *Spinacia oleracea* plant was found in a study. Singh [105] reported that Cr applied to soil at a rate of 60 mg kg^{-1} and higher levels decreased the size of the leaves, causing leaf foliage, leaf tips or margins to burn, and slowed the rate of leaf growth.

7. Effect on physiological process of plant

The physiological process of the plant is severely affected by heavy metal elements. In reaction to heavy metal stress, plants show morphological, physiological, biochemical and metabolic changes which are thought to be adaptive responses [106]. Cd not only inhibits growth, for example, but also changes different physiological and biochemical features such as water balance, nutrient uptake, photosynthesis, breathing, mineral, nutrition and ion uptake, translocation, plant hormone [107–109] and Photosynthetic electron transport around PS I and PS II photosystems [110–112]. Likewise, Cr inhibits electron transport, decreases

CO2 fixation, malformation of chloroplast [113–115], decreases water potential, increases transpiration rate, decreases diffusive resistance, and causes a reduction intercalary meristem [116].

7.1 Photosynthesis

The photosynthetic mechanism is significantly impacted by the heavy metal elements. The photosynthetic apparatus tends to be very susceptible to the toxicity of heavy metals, which directly or indirectly affect the photosynthetic process by inhibiting the enzyme activities of the Calvin cycle and CO2 deficiency in the plant body due to stomatal closure [59, 117, 118]. Cr has a well-cited detrimental effect on the photosynthic process in terrestrial plants. The influence of Cr on the PS I was more conspicuous than on the PS II operation in isolated chloroplasts of Pisumsativum plant [119] according to different reports. Photo inhibition in the leaves of *Lolium perenne* due to the influence of 250 μM Cr on the primary photochemistry of PS II, according to the Vernay et al. [120] report and A decrease in the overall photochemical efficiency of plant PS II at 500 μM of Cr was noted. Shanker et al. [61] argued that Cr triggered oxidative stress in plants because, due to the loss of molecular oxygen, Cr improves alternate sinks for the electrons. The ultimate influence of Cr ions on photosynthesis and conversion of excitation energy will be attributed to Cr-induced anomalies such as thylakoid expansion and reduction in the amount of grana in the ultrastructure of the chloroplast [121]. The impact of Cr on photosynthesis in higher plants is widely known [122, 123], it is not well known to what degree Cr induces photosynthesis inhibition either because of ultra-structure chloroplast malformation and the influence of Cr on the Calvin cycle enzymes or because of electron transport inhibition [116]. Krupa and Baszynski explained in 1995 that some theories applied to all photosynthesis pathways of heavy metal toxicity and introduced a list of primary photosynthetic carbon reduction enzymes that inhibited mainly cereal and legume crops in heavy metal treated plants. The 40 percent inhibition of whole plant photosynthesis in 52-day-old *Pisum sativum* seedlings at 0.1 mM Cr(VI) was further increased to 65 and 95 percent after 76 and 89 days of growth respectively [119]. A potential explanation of Cr mediated reduction rate of photosynthetic is a malformation of the chloroplast ultra structure and inhibition or returdation of electron transport processes due to Cr and a diversion of electrons from the electron donation side of PS-I to Cr (VI). It is likely that, as demonstrated by the low photosynthetic rate of the Cr stressed plants, electrons generated by the photo chemical process are not generally used for carbon fixation. According to [124–126], bioaccumulation of Cr and its toxicity to photosynthetic pigments in various crops and trees has been investigated. [127]; has extensively studied the effect of Cr present in tannery effluent sludge which directly get into chloroplast pigment content in *Vigna radiata* and reported that irrespective of Cr concentration, chlorophyll a, chlorophyll b, chlorophyll d and total chlorophyll decreased in 6 days old seedlings as compared to control. Chatterjee and Chatterjee [128] have reported that a dramatic decrease in chlorophylls a, b and d in leaves was recorded in *Brassica oleracea* grown in distilled sand with full nutrition with control and Co, Cr and Cu at 0.5 mM each. The stress order was Co > Cu > Cr. Conversely, a broad analysis on the tolerance of Cr and Ni in *Echinochloa colona* found that in terms of survival under elevated Cr concentration, the chlorophyll content was high in resistant calluses [129]. Chromium (VI) at 1 and 2 mg L^{-1} significantly decreased chlorophylls a, b and d and carotenoid concentrations in *Salvinia minima* [130]. The decrease in the chlorophyll a/b ratio brought about by Cr indicates that Cr toxicity possibly reduces the size of the peripheral part of the antenna complex [114]. It has been hypothesized that the decrease in chlorophyll b due to Cr could be due to the destabilization and degradation of the proteins of the peripheral part [61]. The interaction of heavy metals with the functional SH groups of proteins according to Van Assche and Clijsters [131, 132] is a possible mechanism of action for heavy metals.

7.2 Water relation

Every physiological process is directly linked to water's chemical potential. Water's chemical potential is a quantitative expression of water-related energy. In plant growth regulation, water can be considered as the most important factor because it affects all growth processes directly or indirectly [133]. Plants grown in contaminated heavy metal soils often suffer from drought stress due primarily to poor physicochemical properties of the soil and shallow root system; researchers are interested in investigations on plant water relation under heavy metal stress. According to Barcelo et al. [134], Selection of drought resistance species can be considered to be an important trait in phytoremediation of soils polluted with heavy metals. The heavy metal stress can induce stress in plants through a series of events leading to decreased water loss like enhanced water conservation, decrease in number and size of leaves, decrease in root hair, malformation of parenchymatous cells stomatal size, number and diameter of xylem vessels, increased stomatal resistance, enhancement of leaf rolling and leaf abscission, higher degree of root suberization [90]. It has been suggested that through various mechanisms operating on the apoplastic and/or the symplastic pathway, heavy metals may influence root hydraulic conductivity. Reduced cell expansion can occur in the growth medium at relatively low concentrations without damaging the integrity of the cells. In bean plants, for instance, leaf expansion growth was inhibited after 48 h in bean plants exposed to 3 uM Cd. The most significant higher toxic effect of Cr (VI) is to degenerate the stomatal conductance that could damage the cells and membranes of stomatal guard cells. In this way, the relationship between water and many plant species has been affected.

8. Mechanism of metal tolerance

Complex processes has used by plants to adjust their metabolism to rapidly changing environment. These processes include transduction, transcription, perception, and transmission of stress stimuli [135–137]. During stressing conditions plants adopt various process likes mechanisms of resistance and tolerance, later involves the immobilization of a metal in roots and in cell walls [138]. The plants adopt a series of mechanisms to avoid heavy metal toxicity which include: (i) Through auto oxidation and Fenton reaction plant produce reactive oxygen, (ii) blocking of main functional group, and (iii) from biomolecules displacement of metal ions, [139]. Plants are capable of growing in polluted soils because; (i) plants avoid metal absorption by aerial components or sustain low metal concentrations over a wide range of metal concentrations in soil by trapping metals in their roots [140]; (ii) plants deliberately absorb metals in their epidermal tissues due to the development of metal binding chelators (iii) they storing metals in non-sensitive parts by alter metal compartmentalisation pattern that is called metal indicators, and (iv) by the process of hyperaccumulators i.e. they can accumulate metals at much higher levels than soil in their aerial components [141, 142]. The processes used for hyperaccumulation are still unclear. Plants that can accumulate either As, Cu, Cr, Ni, Pb, or Co > 1000 mg kg^{-1} or zinc >10,000 mg kg^{-1} in their shot dry matter ([141, 143–145]; Baker and Reeves 2000) or Mo > 1500 mg kg^{-1} [146] are the standard for classifying plants as hyperaccumulators. (ii) Plants that absorb metals 10–500 times higher than average amounts in shoots [147], (iii) plants that accumulate metal components more in shoots than in roots [141]. Very few higher plant species have adaptations that enable them to live and replicate with Zn, Cu, Pb, Cd, Ni, and As highly polluted soils. [148, 149]. The tree roots of these plants can deliberately forage towards less polluted soil areas [150] and can "rest and wait" for optimal growth conditions even with highly reduced growth [151].

9. Conclusion

For the biological, biochemical and physiological functions of plants, various types of heavy metal elements are very important, including protein biosynthesis, lipids, nucleic acids, growth substances, hormones, chlorophyll and secondary metabolism synthesis, stress tolerance, morphological, structural and functional integrity of different membranes and other cellular compounds. These metal components, however, become poisonous in nature, above allowable limits, depending on the types of plants and the nature of the metal. Metal toxicity can inhibit the transport chain of electrons, reduce CO2 fixation, decrease the production of biomass, and cause chloroplast malformation. It can also affect plant growth by generating free radicals and ROS and other substances, which, by decreasing important cellular components, pose a threat to continuous oxidative damage. In addition, heavy metal stress can induce many events in plants leading to decrease in number and size of leaves, enhancement of leaf rolling and leaf abscission, leave erosion, changes in stomatal size, guard cell size, and stomatal resistance, and higher degree of root ligninization, suberization. Symptoms that are visible in plant by the affect of heavy metal toxicity include drying of older leaves, chlorosis, and necrosis of young leaves, stunting, wilting, canker, colour changes, blotch wrinkling and yield reduction. However, plants use complex processes (perception, transduction, and transmission of stress stimuli) and several non enzymatic and enzymatic mechanisms such as CAT, SOD, POD, and APX that activate the cell for their metabolism to heavy metal stress.

Author details

Momezul Haque, Karabi Biswas and Sankar Narayan Sinha*
Environmental Microbiology Research Laboratory, Department of Botany, University of Kalyani, Kalyani 741235, West Bengal, India

*Address all correspondence to: sinhasn62@yahoo.co.in

References

[1] Adriano DC (2001) Trace elements in terrestrial environments. Biochemistry, Alburry, Australia, pp. 1-16

[2] Henry JR (2000) In an overview of phytoremediation of lead and mercury. NNEMS ReportWashington, pp 3-9

[3] Prasad MNV (2008) Trace Elements as Contaminants and Nutrients: Consequences in Ecosystemsand Human Health. Wiley, New York

[4] Skórzyńska-Polit E, Baszynski T (1997) Difference in sensitivity of the photosynthetic apparatusin Cd-stressed runner bean plants in relation to their age. Plant Science 128:11-21

[5] Maksymiec W, Baszynski T (1996) Different susceptibility of runner bean plants to excess copperas a function of growth stages of primary leaves. Journal of Plant Physiology 149:217-221

[6] Peralta-Videa JR, de la Rosa G, Gonzalez JH, Gardea-Torresdey JL 2004. Effect of the growthstage on the heavy metal tolerance of alfalfa plants. Advances in Environmental Research 8:679-685

[7] Das P, Samantaray S, Rout GR (1997) Studies on cadmium toxicity in plants: a review. EnvironPollut 98:29-36

[8] Sharma DC, Chaterjee C, Sharma CP (1995) Chromium accumulation and its effects on wheat(Triticum aestivumL. cv. DH220) metabolism. Plant Science 111:145-151

[9] Shukla OP, Dubey S, Rai UN (2007) preferential accumulation of cadmium and chromium: Toxicity inBacopa monnieriL. Under mixed metal treatments. B Environ Contam Toxicol78:252-257

[10] Barcelo J, Poschenrieder C, Vazquez MD, Gunse B, Vernet JP (1993) Beneficial and toxic effectsof chromium in plants: Solution culture, pot and field studies. Studies in Environmental ScienceNo. 55, Paper Presented at the 5th International Conference on Environmental Contamination.Morges, Switzerland

[11] Boonyapookana B, Upatham ES, Kruatrachue M, Pokethitiyook P, Singhakaew S (2002)Phytoaccumulation and phytotoxicity of cadmium and chromium in duckweedWolffia globosa. Int J Phytoremed 4:87-100

[12] Pinto AP, Mota AM, de Varennes A, Pinto FC (2004) Influence of organic matter on the uptake ofcadmium, zinc, copper and iron by sorghum plants. Sci Tot Environ 326:239-247

[13] Wu FB, Zhang GP (2002) Genotypic variation in kernel heavy metal concentrations in barley andas affected by soil factors. Journal of Plant Nutrition 25:1163-1173

[14] Zhang GP, Fukami M, Sekimoto H (2002) Influence of cadmium on mineral concentration andyield components in wheat genotypes differing in Cd tolerance at seedling stage. Field CropRes 4079:1-7

[15] Hegedüs A, Erdei S, Janda T, Toth E, Horvath G, Dubits D (2004) Transgenic tobacco plantsover producing alfafa aldose/aldehyde reductase show higher tolerance to low temperature andcadmium stress. Plant Science 166:1329-1333

[16] Rivetta A, Negrini N, Cocucci M (1997) Involvement of Ca+– calmodulin in Cd2+toxicity dur-ing the early phases of radish (Raphanus sativusL.) seed germination. Plant, Cell & Environment 20:600-608

[17] Nordberg G (2003) Cadmium and human health: a perspective based on

recent studies in China.J Trace Elem Exp Med 16:307-319

[18] Blaylock JM, Huang JW (2000) Phytoextraction of metals; In: Raskin I, Ensley BD (eds) Phytoremediation of toxic metals: Using plants to clean up the environment. Wiley, New York

[19] Cervantes C, Campos-Garcia J, Devars S, Gutiérrez-Corona F, Loza-TaveraH,Torres-GuzmànJC,Moreno-Sànchez R (2001) Interactions of chromium with microorganisms and plants. FEMSMicrobiol Rev 25:335-347

[20] McGrath SP (1995) Chromium and nickel. In: Alloway BJ (ed) Heavy metal in soils, 2nd edn.Chapman and Hall, Great Britain, pp 152-178

[21] Krishnamurthy S, Wilkens MM (1994) Environmental chemistry of Cr. Northeastern Geol16 (1):14-17

[22] Becquer T, Quantin C, Sicot M, Boudot JP (2003) Chromium availability in ultramafic soils fromNew Caledonia. Sci Total Environ 301:251-261

[23] Adriano DC (1986) Trace elements in the terrestrial environment. Springer-Verlag, New York, pp. 105-123

[24] Rai D, Sass BM, Moore DA (1987) Cr(III) hydrolysis constants and solubility of Cr(III) hydroxide.Inorganic Chemistry 26:345-349

[25] Rai D, Eary LE, Zachara JM (1989) Environmental chemistry of chromium. Sci Total Environ86:15-23

[26] Zayed AM, Terry N (2003) Chromium in the environment: factors affecting biological remediation.Plant Soil 249:139-156

[27] Kabata-Pendias A, Pendias H (2001) Trace elements in soils and plants. CRC Press, Boca Raton

[28] Le Faucheur S, Schildknecht F, Behra R, Sigg L (2006) Thiols inScenedesmus vacuolatusuponexposure to metals and metalloids. Aquatic Toxicology 80:355-361

[29] Booth B (2005) The added danger of counterfeit cigarettes. Environmental Science & Technology 39:34

[30] El-Nady FE Atta MM (1996) Toxicity and bioaccumulation of heavy metals to some marine biotafrom the Egyptian coastal waters. Journal of Environmental Science and Health, Part A 31(7):1529-1545

[31] Stephens WE, Calder A (2005) Source and health implications of high toxic metal concentrationsin illicit tobacco products. Environmental Science & Technology 39:479-488

[32] Ozturk M, Yucel E, Gucel S, Sakcali S, Aksoy A (2008) Plants as biomonitors of trace ele-ments pollution in soil. In: Prasad MNV (eds) Trace elements: environmental contamination, nutritional benefits and health implications, Chap. 28, Wiley, New York, pp 723-744

[33] Martin HW, Kaplan DI (1998) Temporal changes in cadmium, thallium and vanadium mobility insoil and phytoavailability under field conditions. Water, Air, and Soil Pollution 101:399-410

[34] Logan TJ, Chaney RL (1983) Metals. In: Page AL (ed) Utilization of municipal wastewater andsludge on land. University of California, Riverside, pp. 235-326

[35] Verloo M, Eeckhout M (1990) Metal species transformations in soil: an analytical approach. Int JEnviron Anal Chem 39:170-186

[36] Williams DE, Vlamis J, Purkite AH, Corey JE (1980) Trace element accumulation movementand distribution in the soil profile from

massive applications of sewage sludge. Soil Science 1292:119-132

[37] Chang AC, Page AL, Warneke JE (1987) Long-term sludge application on cadmium and zincaccumulation in Swiss chard and radish. Journal of Environmental Quality 16:217-221

[38] Kinnersely AM (1993) The role of Phytochelates in plant growth and productivity. Plant GrowRegul 12:207-217

[39] Marschner H (1995) Mineral nutrition of higher plants. Academic Press, Cambridge

[40] Greger M (1997) Willow as phytoremediator of heavy metal contaminated soil. Proceedings of the2nd international conference on element cycling in the environment. Warsaw, pp 167-172

[41] Greger M, Brammer E, Lindberg S, Larson G, Ildestan-Almquist J (1991) Uptake and physiolog-ical effects of cadmium in sugar beet (Beta vulgaris) related to mineral provision. J Exp Bot42:729-737

[42] Lindberg SE, Meyers TP, Taylor Jr GE, Turner RR, Schroeder WH (1992) Atmosphere-surfaceexchange of mercury in a forest: results of modeling and gradient approached. J Geophys Res97:2519-2528

[43] Cavallini A, Natali L, Durante M, Maserti B (1999) Mercury uptake, distribution and DNA affinityin durum wheat (Triticum durumDesf.) plants. Sci Total Environ 243/244:119-127

[44] Du ShH, Fang ShC (1982) Uptake of elemental mercury vapour by C3and C4species. EnvironExp Bot 22:437-443

[45] Raskin I, Smith RD, Salt DE (1997) Phytoremediation of metals: using plants to remove pollutantsfrom the environment. Current Opinion in Biotechnology 8:221-226

[46] Tester M, Leigh RA (2001) Partitioning of nutrient transport processes in roots. Journal of Experimental Botany 52:445-457

[47] Skeffington RA, Shewry PR, Peterson PJ (1976) Chromium uptake and transport in barleyseedlings (Hordeum vulgareL.). Planta 132:209-214

[48] Shewry PR, Peterson PJ (1974) The uptake and transport of chromium by barley seedlings (Hordeum vulgareL.). Journal of Experimental Botany 25:785-797

[49] Montes-Holguin MO, Peralta-VideaJR, MeitznerG, MartinezA, Rosa G, Castillo-Michel H,Gardea-Torresdey JL (2006) Biochemical and spectroscopic studies of the response ofConvolvulus arvensisL. to chromium (III) and chromium (VI) stress. Environ Toxicol Chem25(1):220-226

[50] Kumar P, Dushenkov V, Motto H, Raskin I (1995) Phytoextraction: the use of plants to removeheavy metals from soils. Environmental Science & Technology 29:1232-1238

[51] Cary EE, Allaway WH, Olson OE (1977) Control of Cr concentrations in food plants. 1. Absorptionand translocation of Cr by plants. Journal of Agricultural and Food Chemistry 25(2):300-304

[52] Chaney RL (1980) Health risks associated with toxic metals in municipal sludge. In: Britton G (ed) Sludge: health risks of land application. Ann Arbor Science Publications, Ann Arbor, Michigan,pp. 58-83

[53] Crowley DE, Wang YC, Reid CP, Szaniszlo PJ (1991) Mechanisms of iron acquisition fromsiderophores by

microorganisms and plants. Plant and Soil 130:179-198

[54] Cunningham SD, Berti WR (1993) Remediation of contaminated soils with green plants:An overview. In Vitro Cellular & Developmental Biology 29P:207-212

[55] Deng H, Ye ZH ZH, Wong MH (2006) Lead and zinc accumulation and tolerance in populationsof six wetland plants. Environmental Pollution 141:69-80

[56] Kirkham MB (2006) Cadmium in plants on polluted soils: effects of soil factors, hyperaccumula-tion, and amendments. Geoderma 137:19-32

[57] Piechalak A, Tomaszewska B, Baralkiewicz D (2003) Enhancing phytoremediative ability ofPisum sativumby EDTA application. Phytochemistry 4:1239-1251

[58] Ramos I, Esteban E, Lucena JJ Garate A (2002) Cadmium uptake and subcellular distribution inplants ofLactucasp. Cd–Mn interaction. Plant Science 162:761-767

[59] Seregin IV, Ivanov VB (2001) Physiological aspects of cadmium and lead toxic effects on higherplants. Russian J Plant Physiol 4:523-544

[60] Shafiq M, Iqbal MZ (2005) Tolerance ofPeltophorum pterocarpumD. C. Baker Ex K. Heyneseedlings to lead and cadmium treatment. J New Seeds 7:83-94

[61] Shanker AK, Cervantes C, Loza-Tavera H, Avudainayagam S (2005) Chromiumtoxicityinplants.Environment International 31:739-751

[62] Baker AJM, Walker PL (1989) Physiological responses of plants to heavy metals and thequantitificatioin of tolerance and toxicity. Chem Spec Biovail 1:7-17

[63] Wong MH, Bradshaw AD (1982) A comparison of the toxicity of heavy metals, using rootelongation of rye grass,*Lolium perenne*. The New Phytologist 91:255-261

[64] Pandey V, Dixit V, Shyam R (2005) Antioxidative responses in elation to growth of mustard (Brassica junceacv. Pusa Jai Kisan) plants exposed to hexavalent chromium. Chemosphere61:40-47

[65] Qureshi MI, Israr M, Abdin MZ Iqbal M (2005) Responses ofArtemisia annuaL. to lead and saltinduced oxidative stress. Environmental and Experimental Botany 53:185-193

[66] Moya JL, Ros R, Picazo I (1993) Influence of cadmium and nickel on growth, net photosynthesisand carbohydrate distribution on rice plants. Photosynthesis Research 36:75-80

[67] Skórzyńska-Polit E, Tukendorf A, Selstam E, Baszyński T (1998) Calcium modifies Cd effect onrunner bean plants. Environmental and Experimental Botany 40:275-286

[68] Alaoui-Sosse B, Genet P, Vinit-Dunand F, Toussaint ML, Epron D, Badot PM (2004) Effect ofcopper on growth in cucumber plants (Cucumis sativus) and its relationships with carbohydrateaccumulation and changes in ion contents. Plant Science 166:1213-1218

[69] Chugh LK, Sawhney SK (1999) Photosynthetic activities ofPisum sativumseedlings grown in thepresence of cadmium. Plant Physiology and Biochemistry 37(4):297-303

[70] Liu DH, Jiang WS, Gao XZ (2003/2004). Effects of cadmium on root growth, cell division andnucleoli in root tip cells of garlic. Biologia Plantarum 47(1):79-83

[71] Talanova VV, Titov AF, Boeva NP (2001) Effect of increasing concentrations of heavy metals onthe growth of barley and wheat seedlings. Russian J Plant Physiol 48:100-103

[72] Gupta S, Nayek S, Saha N, Satpati S (2008) Assessment of heavy metal accumulation in macro-phyte, agricultural soil and crop plants adjacent to discharge zone of sponge iron factory. Environ Geol 55:731-739

[73] Peralta JR, Torresdey JLG, Tiemann KJ, Gomez E, Arteaga S, Rascon E (2001) Uptake and effectsof five heavy metals on seed germination and plant growth in alfalfa (Medicago sativa)L.B Environ Contam Toxicol 66:727-734

[74] Mahmood T, Islam KR, Muhammad S (2007) Toxic effects of heavy metals on early growth andtolerance of cereal crops. Pakistan Journal of Botany 39(2):451-462

[75] Rout GR, Sanghamitra S, and Das P (2000) Effects of chromium and nickel on germination and growthin tolerant and non-tolerant populations' ofEchinochloa colona (L). Chemosphere 40:855-859

[76] Parr PD, Taylor FG Jr. (1982) Germination and growth affects of hexavalent chromium in OrocolTL (a corrosion inhibitor) onPhaseolus vulgaris. Environment International 7:197-202

[77] Jain R, Srivastava S, Madan VK, Jain R (2000) Influence of chromium on growth and cell divisionof sugarcane. Indian Journal of Plant Physiology 5:228-231

[78] Rout GR, Samantaray S, Das P (2001) Differential lead tolerance of rice and black gram genotypesin hydroponic culture. Rost. Výroba (Praha) 47:541-548

[79] Salt DE, Prince RC, Pickering IJ, Raskin I (1995) Mechanisms of cadmium mobility andaccumulation in Indian mustard. Plant Physiology 109:1427-1433

[80] Wójcik M, Tukiendorf A (1999) Cd-tolerance of maize, rye and wheat seedlings. Acta PhysiolPlant 21:99-107

[81] Fiskesjo G (1997) Aliumtest for screening chemicals; evaluation of cytological parameters.In; Wang W, Gorsuch JW, Hughes JS (eds) Plants for environmental studies. Lewis Publ., BocaRaton, pp 307-333

[82] Belimov AA, Safronova VI, TsyganovVE,BorisovAY,KozhemyakovAP, Stepanok VV,Martenson AM, Gianinazzi-Pearson V, Tikhonovich IA (2003) Genetic variability in tolerance to cadmium and accumulation of heavy metals in pea (Pisum sativumL.). Euphytica131 (1):25-35

[83] Han YL, Yuan HY, Huang SZ, Guo Z, Xia B, Gu J (2007) Cadmium tolerance and accumulationby two species of Iris. Ecotoxicology 16:557-563

[84] Odjegba VJ, Fasidi IO (2004) Accumulation of trace elements byPistia stratiotes: Implicationsfor phytoremediation. Ecotoxicology 13:637-646

[85] Piechalak A, Tomaszewaska B, Baralkiewisz D (2002) Accumulation and detoxification of leadion in legumes. Phytochemistry 60:153-162

[86] Hagemeyer J, Breckle SW (1996) Growth under trace element stress. In: Waisel Y, Eshel A, Kafkafi U (eds) Plant root: the hidden half, 2nd edn. Dekker, New York, pp. 415-433

[87] Hagemeyer J, Breckle SW (2002) Trace element stresses in roots. In: Waisel Y, Eshel A, Kafkafi U(eds) Plant root: the hidden half, 3rd edn. Decker, New York, pp. 763-785

[88] Khale H (1993) Response of roots of trees to heavy metals. Environmental and Experimental Botany 33:99-119

[89] Punz WF Sieghardt H (1993) the response of roots of herbaceous plant species to heavy metals.Environmental and Experimental Botany 33:85-86

[90] Barcelo J, Poschenrieder CH (1990) Plant water relations as affected by heavy metal stress: areview. Journal of Plant Nutrition 13:1-37

[91] Prasad MNV, Greger M, Landberg T (2001) Acacia niloticaL. Bark removes toxic elements fromsolution: corroboration from toxicity bioassay usingSalix viminalisL.In hydroponic system.Int J Phytoremed 3:289-300

[92] Mokgalaka-Matlala NS, Flores-Tavizön E, Castillo-Michel H, Peralta-Videa JR, Gardea-TorresdeyJL (2008) Toxicity of arsenic (III) and (V) on plant growth, element uptake, and total amylolyticactivity of mesquite (Prosopis juliflorax p. velutina). Int J Phytoremed 10:47-60

[93] Farga^svá A (1994) Effect of Pb, Cd, Hg, As, and Cr on germination and root growth ofSinapisalbaseeds. Bulletin of Environmental Contamination and Toxicology 52:452-456

[94] Farga^svá A (1998) Root growth inhibition, photosynthetic pigments production, and metal accu-mulation inSinapis albaas the parameters for trace metals effect determination. Bull EnvironContam Toxicol 61:762-769

[95] Rout GR, Samantaray S, Das P (1997) Differential chromium tolerance among eight mungbeancultivars grown in nutrient culture. Journal of Plant Nutrition 20:473-483

[96] Anderson AJ, Meyer DR, Mayer FK (1972) Heavy metal toxicities: Levels of nickel, cobalt andchromium in the soil and plants associated with visual symptoms and variation in growth of anoat crop. Australian Journal of Agricultural Research 24:557-71

[97] Joseph GW, Merrilee RA, Raymond E (1995) Comparative toxicities of six heavy metals usingroot elongation and shoot growth in three plant species. The symposium on environmentaltoxicology and risk assessment, Atlanta, pp. 26-9

[98] Barton LL, Johnson GV, O'Nan AG, Wagener BM (2000) Inhibition of ferric chelate reductase inalfalfa roots by cobalt, nickel, chromium, and copper. Journal of Plant Nutrition 23:1833-1845

[99] Sharma DC, Sharma CP (1993) Chromium uptake and its effects on growth and biological yield ofwheat. Cereal Research Communications 21:317-321

[100] Hanus J, Tomas J (1993) An investigation of chromium content and its uptake from soil in whitemustard. Acta Fytotech 48:39-47

[101] Shah FR, Ahmad N, Masood KR, Zahid DM (2008) The influence of Cd and Cr on the biomassproduction of Shisham (Dalbergia sissooRoxb.) seedlings. Pakistan Journal of Botany 40(4):1341-1348

[102] Vernay P, Gauthier-Moussard C, Jean L, Bordas F, Faure O, Ledoigt G, Hitmi A (2008) Effectof chromium species on phytochemical and physiological parameters In*datura innoxia*chemosphere 72:763-771

[103] Wallace A, Soufi SM, Cha JW, Romney EM (1976) Some effects of chromium toxicity on bushbean plants grown in soil. Plant and Soil 44:471-473

[104] Karunyal S, Renuga G, Paliwal K (1994) Effects of tannery effluent on seed germination, leaf area, biomass and mineral content of some plants. Bioresource Technology 47:215-218

[105] Singh AK (2001) Effect of trivalent and hexavalent chromium on spinach (Spinacea oleraceaL).Environment and Ecology 19:807-810

[106] Singh S, Sinha S (2004) Scanning electron microscopic studies and growth response of the plantsofHelianthus annuusL. grown on tannery sludge amended soil. Environment International 30:389-395

[107] Drazic G, Mihailovic N, Lojic M (2006) Cadmium accumulation in Medicago sativa seedlings treated with salicylic acid. Biologia Plantarum 50:239-244

[108] Scebba F, Arduini I, Ercoli L, Sebastiani L (2006) Cadmium effects on growth and antioxidantenzymes activities inMiscanthus sinensis. Biologia Plantarum 50:688-692

[109] Vassilev A, Yordanov I, Tsonev T (1997) Effects of Cd2+on the physiological state andphotosynthetic activity of young barley plants. Photosynthetica 34:293-302

[110] Siedlecka A, Baszynski T (1993) Inhibition of electron transport flow around photosystem I inchloroplasts of Cd-treated maize plants is due to Cd-induced iron deficiency. Physiol Plant87:199-202

[111] Skórzyńska-Polit E, Baszynski T (1995) Photochemical activity of primary leaves in cadmiumstressedPhaseolus coccineusdepends on their growth stages. Acta Societatis Botanicorum Poloniae 64:273-279

[112] Vassilev A, Lidon F, Scotti P, Da Graca M, Yordanov I (2004) Cadmium-induced changes inchloroplast lipids and photosystem activities in barley plants. Biologia Plantarum 48:153-156

[113] Davies FT, Puryear JD, Newton RJ, Egilla JN, Grossi JAS (2002) Mycorrhizal fungi increasechromium uptake by sunflower plants: influence on tissue mineral concentration, growth, andgas exchange. Journal of Plant Nutrition 25:2389-407

[114] Shanker AK (2003) Physiological, biochemical and molecular aspects of chromium toxicity andtolerance in selected crops and tree species. PhD Thesis, Tamil Nadu Agricultural University, Coimbatore, India

[115] Zeid IM (2001) Responses ofPhaseolus vulgaristo chromium and cobalt treatments. Biol Plant44:111-115

[116] Vazques MD, Poschenrieder C, Barcelo J (1987) Chromium (VI) induced structural changes inbush bean plants. Annals of Botany 59:427-438

[117] Bertrand M, Poirier I (2005) Photosynthetic organisms and excess of metals. Photosynthetica43 (3):345-353

[118] Linger P, Ostwald A, Haensler J (2005) Cannabis sativaL. growing on heavy metal contaminatedsoil: growth, cadmium uptake and photosynthesis. Biologia Plantarum 49(4):567-576

[119] Bishnoi NR, Chugh LK, Sawhney SK (1993a) Effect of chromium on photosynthesis, respirationand nitrogen fixation in pea (Pisum sativumL) seedlings. Journal of Plant Physiology 142:25-30

[120] Vernay P, Gauthier-Moussard C, Hitmi A (2007) Interaction of bioaccumulation of heavy metalchromium with water relation, mineral nutrition and photosynthesis in developed leaves of Lolium perenneL. Chemosphere 68:1563-1575

[121] Rocchetta I, Mazzuca M, Conforti V, Ruiz L, Balzaretti V, Ríos deMolina MC (2006) Effect of chromium on the fatty acid composition of two strains of Euglena gracilis. Environ Poll141:353-358

[122] Foy CD, Chaney RL, White MC (1978) The physiology of metal toxicity in plants. Ann Rev PlantPhysiol 29:511

[123] Van Assche F, Clijsters H (1983) Multiple effects of heavy metals on photosynthesis. In: Marcelle R (ed) Effects of Stress on Photosynthesis. The Hague: Nijhoff/Junk. pp. 371-382

[124] Barcelo J, Poschenrieder C, Gunse B (1986) Water relations of chromium VI treated bush beanplants (Phaseolus vulgarisL. cv. Contender) under both normal and water stress conditions.J Exp Bot 37:178-187

[125] Sharma DC, Sharma CP (1996) Chromium uptake and toxicity effects on growth and metabolicactivities in wheat, Triticum aestivumL. cv. UP 2003. Indian Journal of Experimental Biology 34:689-691

[126] Vajpayee P, Sharma SC, Tripathi RD, Rai UN, Yunus M (1999) Bioaccumulation of chromium andtoxicity to photosynthetic pigments, nitrate reductase activity and protein content ofNelumbonuciferaGaertn. Chemosphere 39:2159-2169

[127] Bera AK, Kanta-Bokaria AK, Bokaria K (1999) Effect of tannery effluent on seed germination, seedling growth and chloroplast pigment content in mungbean (Vigna radiataL. Wilczek). Environment and Ecology 17(4):958-961

[128] Chatterjee J, Chatterjee C (2000) Phytotoxicity of cobalt, chromium and copper in cauliflower.Environmental Pollution 109:69-74

[129] Samantaray S, Rout GR, Das P (2001) Induction, selection and characterization of Cr andNi-tolerant cell lines ofEchinochloa colona (L) in vitro. Journal of Plant Physiology 158:1281-1290

[130] Nichols PB, Couch JD, Al Hamdani SH (2000) Selected physiological responses ofSalviniaminimato different chromium concentrations. Aquatic Botany 68:313-319

[131] Van Assche F, Clijsters H (1990a) Effect of metals on enzyme activity in plants. Plant Cell Environ13:195-206

[132] Van Assche F, Clijsters H (1990b) Effects of metals on enzyme activity in plants. Plant Cell Environ13:195-206

[133] Kramer PJ, Boyer JS (1995) Water relations of plants and soils. Academic Press, San Diego,p 495

[134] Barcelo J, Poschenrieder C, Lombini A, Llugany M, Bech J, Dinelli E (2001) Mediterranean plantspecies for phytoremediation. In: Abstracts costs action 837 WG2 workshop on phytoremedi-ation of trace elements in contaminated soils and waters (with special emphasis on Zn, Cd, Pband As), Madrid, 5-7 April, p 23

[135] Kopyra M, Gwozdz EA (2004) The role of nitric oxide in plant growth regulation and responses toabiotic stresses. Acta Physiologiae Plantarum 26:459-472

[136] Turner JG, Ch E, Devoto A (2002) The jasmonate signal pathway. Plant Cell 14 (Suppl):153-164

[137] Xiong L, Schumaker KS, Zhu JK (2002) Cell signaling during cold, drought and salt stress. PlantCell 14(Suppl):165-183

[138] Garbisu C, Alkorta I (2001) Phytoextraction: a cost-effective plant-based technology for theremoval of metals from the environment. Bioresource Technology 77:229-236

[139] Clemens S (2006) Toxic metal accumulation, responses to exposure and mechanisms of tolerancein plants. Biochimie 88:1707-1719

[140] Cunningham SD (1995) In proceedings/abstracts of the fourteenth annual symposium, currenttopics in

plant biochemistry, physiology, and molecular biology columbia, April 19-22,pp 47-48

[141] Baker AJM, Reeves RD, Hajar ASM (1994) Heavy metal accumulation and tolerance in Britishpopulation of the metallophyteThalaspi caerulesensJ. and C. Presl (Brassicaeae). New Phytol127:61-68

[142] Raskin I, Kumar PBAN, Dushenkov S, Salt DE (1994) Bioconcentration of heavy metals by plants.Current Opinion in Biotechnology 5:285-290

[143] Brown SL, Chaney RL, Angle JS, Baker AJM (1994) Phytoremediation potential ofThlaspicaerulescensandBladder campionfor zinc- and cadmium contaminated soil. J Environ Qual23:1151-1157

[144] Brooks RR (1998) Plants that hyperaccumulate heavy metals. Cambridge University Press, New York

[145] Ma LQ, Komar KM, Kennelley ED (2001) Methods for removing pollutants from contaminatedsoil materials with a fern plant. Document type and number: United States Patent 6280500.http://www.freepatentsonline.com/6280500.html

[146] Lombi E, Zhao FJ, Dunham SJ, McGrath SP (2001) Phytoremediation of heavy metal, contami-nated soils, natural hyperaccumulation versus chemically enhanced phytoextraction. J EnvironQual 30:1919-1926

[147] Shen ZG, Liu YL (1998) Progress in the study on the plants that hyperaccumulate heavy metal.Plant Physiol Commun 34:133-139

[148] Dahmani-Muller H, van Oort F, Gelie B, Balabane M (2000) Strategies of heavy metal uptake bythree plant species growing near a metal smelter. Environmental Pollution 109:231-238

[149] Pulford ID, Watson C (2003) Phytoremediation of heavy metal-contaminated land by trees- areview. Environment International 29:529-540

[150] Turner AP, Dickinson NM (1993) Survival ofAcer pseudoplatanusL. (Sycamore) seedlings onmetalliferous soils, The New Phytologist 123:509

[151] Watmough SA (1994) Adaptation to pollution stress in trees: metal tolerance traits, Ph.D. thesis,Liverpool John Moore University, Liverpool

Chapter 14

The Impact of Changing Climate on the Cambial Activity during Radial Growth in Some *Citrus* Species

Moin Ahmad Khan and M. Badruzzaman Siddiqui

Abstract

This study on radial growth in the stem of *Citrus* was carried out with an aim to notice the behavior of vascular cambium with respect to climatic and age effects. The fusiform initials vary in length from 137 to 363 μm in *C. limon*, 100 to 463 μm in *C. paradisi*, 137 to 413 μm in *C. reticulata* var. *kinnow*, and 137 to 375 μm in *C. sinensis*. The length rises with age, followed by decline and then again increase in *C. limon*. In *C. paradisi*, there is increase up to maximum and after decline is soon followed by constancy. In *C. reticulata* var. *kinnow*, increase in length from top to base in *C. sinensis*, increase up to maximum followed by a decline. Swelling of cambial cells occurs in the third week of March in *C. limon*, last week of March in *C. paradisi*, third week of April in *C. reticulata* var. *kinnow*, and second week of April in *C. sinensis*. The cambium turns dormant in early October in *C. limon*, late December in *C. paradisi*, early December in *C. reticulata* var. *kinnow*, and early November in *C. sinensis*. Thus, the cambium remains active for about 6 months in *C. limon* and *C. sinensis*, 9 months in *C. paradisi*, and 7 months in *C. reticulata* var. *kinnow*.

Keywords: cambium, radial growth, fusiform initials, ray initials

1. Introduction

In most dicotyledons and gymnosperms, a layer of procambial cells between the primary phloem and primary xylem matures into fascicular cambium, while the cells of pith or medullary rays which lie in between the edges of the fascicular cambium divide accordingly to form a new layer of cambium across the medullary rays, known as interfascicular cambium, resulting in the formation of a complete ring of cambium. In this way, a new lateral meristem, the vascular cambium, which is responsible for the "growth in thickness by the formation of secondary vascular tissues (radial growth)," is formed and adds secondary phloem toward the outer side and secondary xylem toward the inner side.

2. Cambium

In three-dimensional view, the cambium is a continuous cylindrical sheath about the xylem. In most of the plants, the vascular cambium is reported to exhibit successive active and dormant phases during a calendar year. This behavior of cambium is believed to be regulated by several internal and external factors that include heredity constitution, physiological phenomenon, and environmental conditions of the habitat [1]. Therefore, there is further need to investigate the influence of different physical and climatic factors on cambial makeup and its activity and then to suggest measures for the maintenance of desirable growth pattern to ensure a vigorous production of derivative tissues and their content, although in the past, several workers have conducted such type of studies in different species growing in tropical and subtropical regions [2–12].

Butterfield [13] defines **cambium** as a "multiseriate zone of periclinally dividing cells lying between the differentiating secondary xylem and phloem, with distinct initials capable of both periclinal and anticlinal divisions lying somewhere within each radial file of cells." The same terminology has been adopted for describing cambium in the present study.

In spite of the fact that Indian subcontinent is one of the richest tropical tree flora on earth, the studies on the radial growth of these trees, that is, the activity of cambium, its structure, and behavior are still meager. Much, therefore, remains to be known about the growth phenomenon of Indian tropical trees, particularly the vascular cambium and its derivative tissues, xylem, and phloem, their cellular organization with age and varying climatic conditions.

The tropical trees in general exhibit a continuous growth unlike temperate ones where the growth phenomenon is sharply rhythmic. A majority of tropical trees grow in multiple flushes or in an intermittent manner due to the prolonged favorable climatic conditions that prevail in the tropical belt. Keeping in view the aforesaid fact, the present anatomical studies are an attempt to elaborate the structure and behavior of vascular cambium and its derivative tissues in some **Rutaceae** members in relation to various seasonal conditions of the study site and age of the trees. My study includes the following aspects:

1. Structure and activity of vascular cambium.

2. The effect of climate on the activity and structure of vascular cambium.

3. The effect of age on the activity and structure of vascular cambium.

In fact, no information is available with regard to the cambial activity and formation of its derivative tissues in *Citrus* species of Rutaceae family. It is noteworthy that *Citrus* is of immense medicinal importance as well as economic value.

2.1 *Citrus*

A genus of evergreen, usually armed, aromatic shrubs or small trees distributed in the Indo-Malayan region, South-east Asia, and China but cultivated throughout the tropical and temperate regions for fruits. Currently, *Citrus* is commercially grown primarily between the latitudes 40°N to 40°S.

Four species of genus *Citrus*, available in and around district Aligarh, Uttar Pradesh, India, have been selected for a comparative anatomical study on the aspects as described earlier.

1. *Citrus limon* (Linn.) Burm.f.

Classification [14]
Class: Dicotyledons

Series: Lignosae
Order: Rutales
Family: Rutaceae
Genus: *Citrus*
Species: *limon*
Hindi: Baranibu, Jambira, Paharikaghzi, Paharinimbu, Kinnanibu

2. *Citrus paradisi* Macf.

Classification [14]
Class: Dicotyledons
Series: Lignosae
Order: Rutales
Family: Rutaceae
Genus: *Citrus*
Species: *paradisi*
Hindi: Chakotra

3. *Citrus reticulata* var. *kinnow*

Classification [14]
Class: Dicotyledons
Series: Lignosae
Order: Rutales
Family: Rutaceae
Genus: *Citrus*
Species: *reticulata* var. *kinnow*
Hindi: Kinnow

4. *Citrus sinensis* (Linn.) Osbeck

Classification [14]
Class: Dicotyledons
Series: Lignosae
Order: Rutales
Family: Rutaceae
Genus: *Citrus*
Species: *sinensis*
Hindi: Mosammi, Malta

3. The vascular cambium: structure

The term cambium was coined by Grew [15] who presented the classification of plant tissues. In the year 1863, Sanio [16] recognized vascular cambium, its origin and function as a lateral meristem. The vascular cambium in all the species investigated forms a continuous cylinder between the xylem and phloem. Vascular cambium as a rule is made up of exclusively mononucleate elongated spindle-shaped elements with long tapering end walls, the fusiform initials, and almost isodiametric or rectangular ray initials [17]. In the presently investigated species, the arrangement of end walls of the adjacent cambial initials overlaps to a considerable extent depicting a clear non-storied (non-stratified) structure in all species investigated (**Figure 1(i)**).

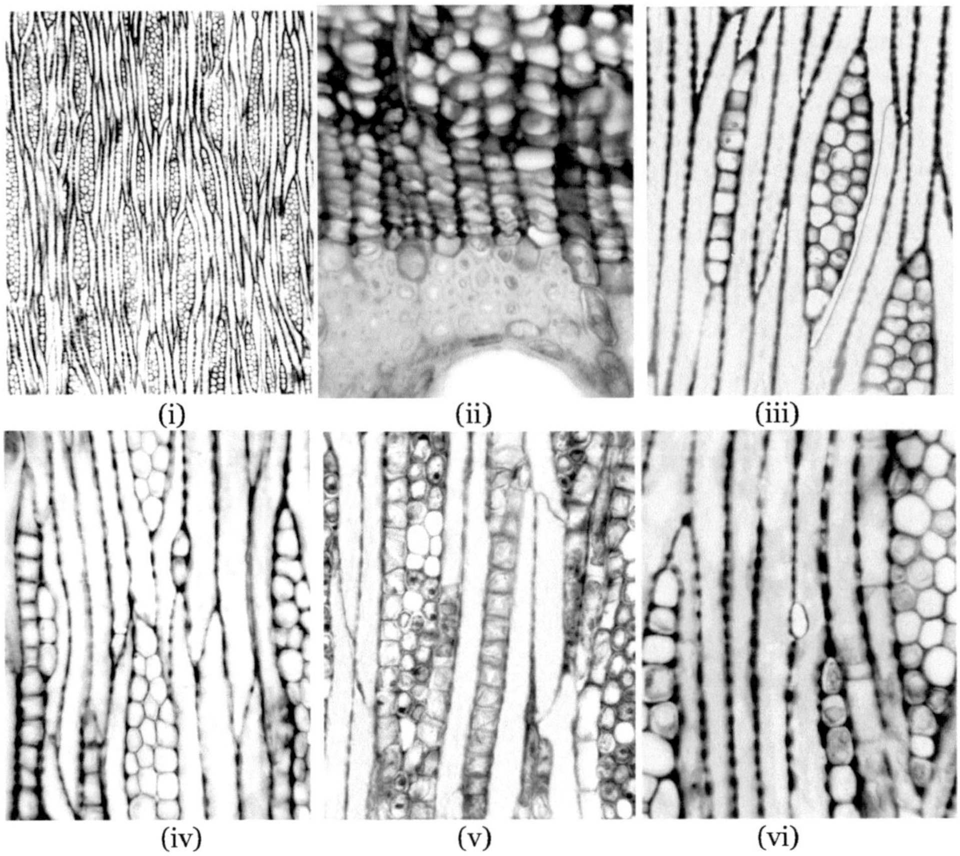

Figure 1.
(i) C. sinensis *T.L.S. through dormant cambium showing arrangement of various components of cambium, (ii)* C. limon *T.S. showing dormant cambial zone, (iii)* C. paradisi *T.L.S. through active cambium showing pseudotransverse wall in dividing fusiform initial, (iv)* C. sinensis *T.L.S. through active cambium showing two-celled newly formed ray, terminally cut ray cell and lateral fusion of rays, (v)* C. sinensis *T.L.S. through active cambium showing terminal and lateral fusion of rays and transverse septation of fusiform initial, (vi)* C. paradisi *T.L.S. through active cambium showing laterally cut ray cell.*

After measuring cambial initials of a wide variety of tropical as well as temperate trees Bailey [18] concluded that the fusiform initials vary in length from 460 to 4400 μm showing non-stratified cambium. The observations regarding this aspect indicate that in the presently investigated species, the length of fusiform initials ranges from 137 to 363 μm in *Citrus limon*, 100 to 463 μm in *C. paradisi*, 137 to 413 μm in *C. reticulata* var. *kinnow*, and 137 to 375 μm in *C. sinensis*, which is contrary to the findings of Bailey [18]. But the present findings are in agreement with the results of some earlier workers like Ghouse and Iqbal [19] in some arid zone species of *Acacia* and *Prosopis*, Kojs [20] in selected woody species, Khan [10] in *Jacaranda mimosifolia*, *Pterospermum acerifolium*, and *Terminalia arjuna* who have found fusiform initial length to fall shorter than Bailey's [18] reported limit for non-storied cambium.

Among the species investigated in this study, *C. reticulata* var. *kinnow* has been found to possess comparatively short fusiform initials while *C. limon* having the longest with the other two species falling in between these. If the size of fusiform initials is taken as a criterion for phylogenetic advancement, then obviously *C. reticulata* var. *kinnow* appears to be the most evolved form among the presently investigated species.

The walls of the fusiform initials bear primary pit fields and have distinct plasmodesmata connections with the contiguous elements, especially with the ray initials. The radial walls of fusiform initials have been observed to be usually thicker than tangential walls in the present study in all the species investigated. Especially during dormancy, the primary pit fields appear deeply depressed in tangential longitudinal view giving a beaded look to the radial walls. Similar situation has been noticed by Iqbal [21], Khan et al. [22], and Khan and Siddiqui [12].

The cambial initials have been reported to undergo anticlinal and periclinal divisions periodically [23]. Anticlinal division which is also known as multiplicative division increases the cambial population, whereas the periclinal or additive division increases the number of cambial derivatives emanating new phloem and xylem elements [24]. The anticlinal division in the cambial initials has been noted to be pseudotransverse wall formation takes place running askew intersecting the two radial walls at two different levels [12, 22, 23, 25, 26].

The pseudotransverse wall formation observed in this study varies in the length from short to long in all the species investigated. Sometimes, the dividing wall almost extending from one end of the cell to the other, as it has been reported earlier by Khan and Siddiqui [12] in *Alstonia* species.

The ray initials may arise primarily as a single cell, which is cut at the ends of fusiform initials as terminal segments [27] or as lateral segments [20, 27] or they may arise by transverse segmentation of fusiform initials [20, 27]. In the presently investigated species, the first and last types of ray development are found to be more frequent than the lateral segments. Once the ray initials get established, they continue to undergo multiplication resulting in expansion of rays in height and width [27]. The rays also increase in height and width by fusion of two or more vertically and radially aligned rays. These fusions result due to the transverse segmentation of the intervening fusiform initials or by multiplication of already existing ray initials of the adjacent panel of rays [20, 27]. Apart from the above fact, some long and broad rays get split into smaller units by intrusion of adjacent fusiform initials in all the species investigated as has already been reported by [20, 27–29].

The ray initials form an integral part of the cambial cylinder in all the species investigated. The relative proportion of ray initials to that of fusiform initials has been found to vary from species to species. A maximum of 23% has been observed in *C. paradisi* and *C. reticulata* var. *kinnow* and minimum of 18% in *C. limon*, whereas in *C. sinensis*, the ray initials constitute about 22% of tangential area of cambial cylinder in adult trees [30–34].

4. The vascular cambium: developmental changes in the structure

In transverse sections of the young shoots, the cambial zone consists of three to five layers of cells in all the species, whereas the number of cell layers in the cambial zone of adult trees varies from three to nine (**Figure 1(ii)**). It is evident that the fusiform cambial initial experience considerable length variation as the tree grows in thickness in all the species investigated in the present research. Average length of fusiform initials has been noticed to vary from 158.78 to 259.71 μm in *C. limon*, 228.53 to 257.59 μm in *C. paradisi*, 118.00 to 246.70 μm in *C. reticulata* var. *kinnow*, and 156.00 to 232.43 μm in *C. sinensis* in different age group samples (**Table 1**). On close observations, it is seen that the average length rises with age from 158.78 to 232.46 μm that is followed by a decline to 220.31 μm and then again increase in the old stem to 259.71 μm in case of *C. limon* which coincides with the findings of Cumbie [35] and Ajmal et al. [36]. In *C. paradisi*, there is a gradual increase up to

Circumference of axis in mm	Length (μm) *C. limon*				Circumference of axis in mm	Length (μm) *C. paradisi*			
	Range	**Mean ± SE**	**SD**	**CV%**		**Range**	**Mean ± SE**	**SD**	**CV%**
20	200–325	158.78 ± 1.81	36.25	22.83	20	137–325	228.53 ± 2.43	48.63	21.27
40	225–363	184.21 ± 1.94	38.85	21.09	40	137–338	234.68 ± 2.68	53.61	22.84
65	250–400	210.28 ± 2.02	40.46	19.24	65	112–350	249.65 ± 2.85	57.04	22.84
95	275–400	232.46 ± 1.80	36.19	15.56	95	175–375	257.59 ± 2.17	43.50	16.89
150	262–400	220.31 ± 1.79	35.91	16.29	164	175–363	238.56 ± 2.51	50.35	21.10
195	250–363	259.71 ± 1.80	36.01	13.86	204	125–363	239.00 ± 2.79	55.83	23.35
LSD at 5%		= 44.60			LSD at 5%		= 73.33		
LSD at 1%		= 59.73			LSD at 1%		= 98.19		
Circumference of axis in mm	Length (μm) *C. reticulata* var. *kinnow*				Circumference of axis in mm	Length (μm) *C. sinensis*			
	Range	Mean ± SE	SD	CV%		Range	Mean ± SE	SD	CV%
20	50–200	118.00 ± 2.00	40.14	34.01	20	100–213	156.00 ± 1.79	35.94	23.03
45	75–263	124.18 ± 2.04	40.94	32.96	45	87–250	178.15 ± 2.03	40.60	22.78
67	100–250	129.50 ± 1.95	39.19	30.26	67	137–288	204.81 ± 1.92	38.45	18.77
96	112–263	132.62 ± 1.96	39.26	29.60	96	150–313	232.43 ± 1.96	39.36	16.93
147	187–338	235.54 ± 2.01	40.25	17.08	143	150–288	219.57 ± 1.88	37.64	17.14
199	125–400	246.70 ± 3.59	71.92	29.15	198	137–263	205.12 ± 1.84	36.94	18.00
LSD at 5%		= 68.48			LSD at 5%		= 73.33		
LSD at 1%		= 91.70			LSD at 1%		= 98.19		

Table 1.
Changes in the length size of fusiform initials (as observed in tangential longitudinal section) of cambial zone along tree axis of varying girth.

maximum from 228.53 to 257.59 μm and after slight decline to 238.56 μm is soon followed by constancy in the adult region which coincides with the report of Evert [37, 38] and Ghouse and Yunus [39]. In *C. reticulata* var. *kinnow*, a significant impact of age is seen on the vascular cambium as the fusiform initials increase in length from top to base, that is, from 118.00 to 246.70 μm with an increase in girth of the axis, which coincides with the report of Khan [40], Khan [10], and Mahmood [11]. Whereas in *C. sinensis*, there is an initial increase up to maximum from 156.00 to 232.43 μm with the advancing age of stem axis which is followed by a decline toward the basal region which goes in agreement with the results of Ghouse and Iqbal [41], Khan et al. [42], Ajmal [43], and Mahmood [11].

In general, fusiform initials are found longer and broader in stouter axes than in the slender ones. The rate of increase happens to be high in young shoots and low in older ones. It appears, therefore, that the ability of newly formed initials to elongate in size depends on the age of the meristem. The younger is the cambium, greater is the ability of the initials to elongate, and conversely, the older is the meristem, the lesser is the ability of the initials to undergo expansion. A similar comparative analysis of the data obtained on the width of the fusiform initials has revealed that they do not show any significant change with the increase in the circumference of the stem axes.

A similar analysis of the ray initials of the investigated species (**Table 2**) shows a slight initial increase from 12.44/9.86 to 15.98/13.80 μm which is followed by constancy in anticlinal and periclinal diameters of *C. limon* as has already been reported by Khan [10] in *Terminalia arjuna* and slight increase from 8.97/7.82 to 11.96/10.13 μm with the increasing diameter of axis in *C. reticulata* var. *kinnow* which coincides with the findings of Khan [10] in *Jacaranda mimosifolia*, whereas a slight initial increase followed by a decrease in the basal region is seen in anticlinal diameter of *C. sinensis* as has been reported by Khan [10] in *Pterospermum acerifolium*. The ray initials do not show any appreciable change in their dimension in relation to age of the axis in *C. paradisi* as has been reported by earlier researchers like Ajmal [43]. However, they undergo multiplication to become multiseriate in older axis [11, 41]. As a consequence of various developmental changes in cambial zone, relative proportion of fusiform and ray initials also varies with age of stem axis. The ray initials occupy a relatively greater area in the cambial cylinder in the old axis as compared to younger ones. The ray initials occupy 14–17% in *C. limon*, 7–25% in *C. paradisi*, 14–23% in *C. reticulata* var. *kinnow*, and 12–18% in *C. sinensis* of the total tangential area of the cambial cylinder.

With the growing girth of the axis, the cambial cylinder also expands by adding more cells. The fusiform initials undergo pseudotransverse divisions and give rise to sister initials (**Figure 1(iii)**). Similarly, the ray initials also divide and give rise to new ray initials (**Figure 1(iv)**). All this happens in order to cope with the expansion of the axis. The new ray initials are also produced by the fusiform initials and this happens either by the transverse septation of the fusiform cells (**Figure 1(v)**) or by the formation of new initials as terminal (**Figure 1(vi)**) or lateral segments (**Figure 1(vi)**). Occasionally, the rays are seen fusing with one another to form tall and wide bodies (**Figure 1(v)**). This is brought about by the conversion of the intervening fusiform initials into a group of ray initials, which forms the bridge between the two already existing groups of ray initials. The newly produced rays having a limited height in the beginning grow into tall structures by the divisions of the existing initials. At times, the fusiform initials are found to intrude into a panel of ray initials, resulting in the division of a broad or tall ray into a number of smaller entities (**Figure 2(i)**).

Vascular cambium, therefore, constantly undergoes changes in its composition, as an accommodative measure to meet the increasing circumference of the vascular

Circumference of axis in mm	Anticlinal diameter (μm)				Periclinal diameter (μm)			
	Range	Mean ± SE	SD	CV%	Range	Mean ± SE	SD	CV%
				In *Citrus limon*				
20	6–17	12.44 ± 0.11	2.31	18.56	6–14	9.86 ± 0.10	2.17	22.00
40	10–17	14.41 ± 0.12	2.50	17.34	10–17	13.12 ± 0.12	2.45	18.67
65	10–21	15.30 ± 0.17	3.48	22.74	10–17	13.60 ± 0.11	2.35	17.27
95	10–21	15.98 ± 0.18	3.74	23.40	10–17	13.80 ± 0.11	2.29	16.59
150	10–17	13.46 ± 0.11	2.35	17.45	10–17	11.49 ± 0.08	1.78	15.49
195	10–21	14.75 ± 0.16	3.30	22.37	10–17	12.17 ± 0.09	1.81	14.87
LSD at 5%		= 4.46				= 3.10		
LSD at 1%		= 5.97				= 4.15		
				In *Citrus paradisi*				
20	6–14	8.63 ± 0.12	2.47	28.62	6–14	7.95 ± 0.11	2.21	27.79
40	6–17	10.47 ± 0.13	2.62	25.02	6–14	8.50 ± 0.11	2.38	28.00
65	6–14	10.60 ± 0.12	2.42	22.83	6–14	8.84 ± 0.12	2.54	28.73
95	6–14	10.20 ± 0.14	2.80	27.45	6–14	8.36 ± 0.91	1.82	21.77
164	6–14	9.92 ± 0.11	2.24	22.58	6–14	9.18 ± 0.09	1.95	21.24
204	6–14	9.72 ± 0.11	2.25	23.14	6–14	8.97 ± 0.13	2.61	29.09
LSD at 5%		= 3.55				= 3.04		
LSD at 1%		= 4.75				= 4.07		
				In *Citrus reticulata* var. *kinnow*				
20	6–14	8.97 ± 0.13	2.70	30.10	6–14	7.82 ± 0.10	2.17	27.74
40	6–14	9.52 ± 0.10	2.04	21.42	6–14	7.68 ± 0.09	1.90	24.73
67	6–17	9.99 ± 0.10	2.08	20.82	6–14	8.43 ± 0.11	2.38	28.23
96	6–14	10.54 ± 0.11	2.38	22.58	6–14	8.77 ± 0.13	2.64	30.10
147	6–14	10.33 ± 0.12	2.45	23.71	6–14	8.29 ± 0.09	1.82	21.95
199	6–17	11.96 ± 0.18	3.68	30.76	6–14	10.13 ± 0.13	2.76	27.24
LSD at 5%		= 3.73				= 3.32		
LSD at 1%		= 4.99				= 4.45		
				In *Citrus sinensis*				
20	10–17	12.92 ± 0.12	2.54	19.65	6–14	8.84 ± 0.13	2.72	30.76
45	10–21	14.96 ± 0.17	3.47	23.19	6–14	9.65 ± 0.11	2.29	23.73
67	10–21	14.82 ± 0.16	3.38	22.80	6–14	10.88 ± 0.12	2.54	23.34
96	10–17	14.28 ± 0.12	2.54	17.78	6–14	11.22 ± 0.10	2.17	19.34
143	10–17	13.26 ± 0.11	2.38	17.94	6–14	10.20 ± 0.14	2.80	27.45
198	6–17	11.83 ± 0.18	3.74	31.61	6–14	9.79 ± 0.09	1.88	19.20
LSD at 5%		= 4.23				= 3.42		
LSD at 1%		= 5.67				= 4.58		

Table 2.
Changes in the cell size of ray initials (as observed in tangential longitudinal section) of cambial zone along tree axis of varying girth.

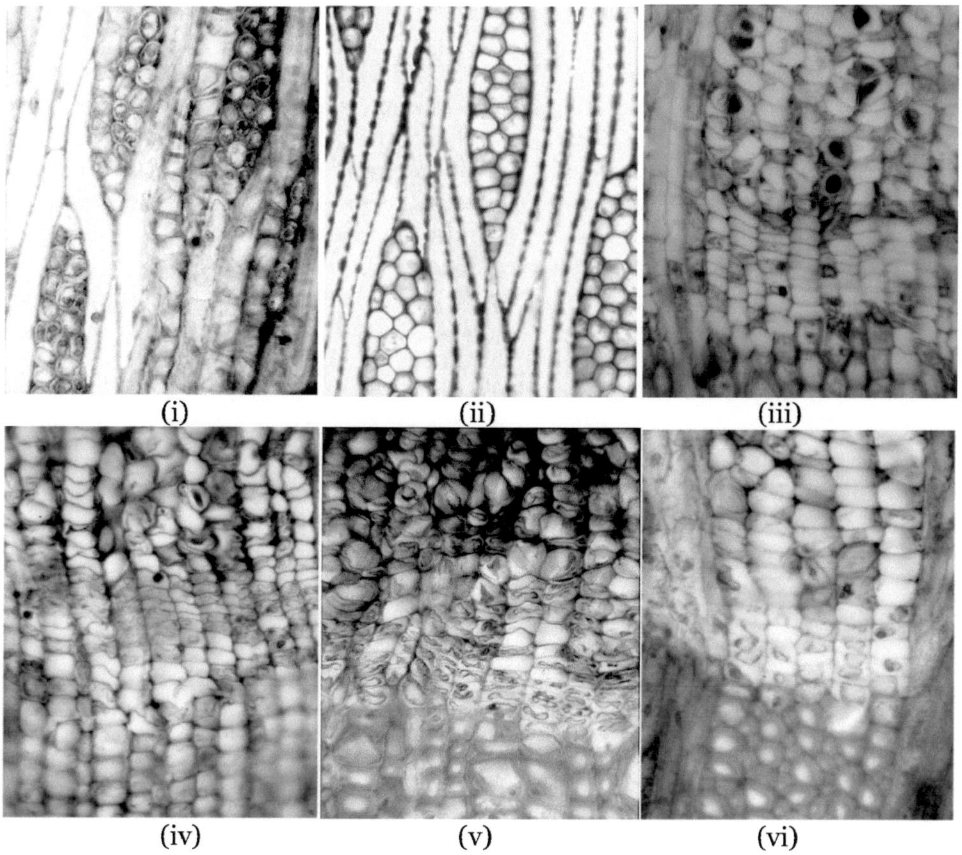

Figure 2.
(i) C. reticulata *var.* kinnow *T.L.S. through active cambium showing transverse septation of fusiform initial and splitting of ray, (ii)* C. paradisi *T.L.S. through dormant cambium showing beaded radial walls, (iii)* C. limon *T.S. showing swelling of cambial zone, (iv)* C. paradisi *T.S. showing swelling of cambial zone, (v)* C. reticulata *var.* kinnow *T.S. showing swelling of cambial zone, (vi)* C. sinensis *T.S. showing swelling of cambial zone.*

cylinder. This usually resulted in a considerable change in the corresponding volume of the different initials. Thus, in the young shoots, the fusiform initials have been found to occupy 86% of the total area of the cambial cylinder in *C. limon*, 93% in *C. paradisi*, 86% in *C. reticulata* var. *kinnow*, and 88% in *C. sinensis*, while the corresponding area of fusiform initials in the cambial cylinder gets reduced toward the mature stem.

5. The vascular cambium: seasonal changes in the structure

As far as the impact of seasonal changes on the dimensions of fusiform initials is concerned, it has been noted that length, width, and tapering ends averages of fusiform initials as well as magnitude of ray initials vary to some extent depending on the time of development of new cambial initials and the period of their growth. Analysis of the data obtained during three consecutive years has revealed that both the structure and the contents of the cambial initials vary from season to season. Short fusiform initials with narrow width and comparatively tapering ends coincided with the activity of the cambium in all the species investigated [44]. Also, the size of ray initials show smaller diameter during the height of cambial activity as has earlier been reported by Catesson [3], Kitin et al. [8], Espinosa et al. [45], Gricar et al. [46], and Begum et al. [47].

In *C. limon*, the averages of length and width of fusiform initials varies from 233.72 to 278.00 and 17.47 to 21.28 μm, respectively, while the average size of their tapering ends ranges from 76.75 to 90.75 μm. Comparatively shorter fusiform initials occur in May to August and then rest of the months (**Table 3**). The size of ray initials also shows minor variation in different seasons. The mean value of the anticlinal and periclinal diameters ranges from 12.64/11.22 to 16.18/14.82 μm during a calendar year (**Table 4**). In *C. paradisi*, the averages of length and width of fusiform initials range from 217.74 to 284.00 and 16.38 to 21.48 μm, respectively and the end walls range from 72.75 to 91.50 μm. Comparatively shorter fusiform initials are found from June to November (**Table 5**). Similarly, anticlinal and periclinal diameters of ray initials range from 8.63/8.43 to 15.50/14.96 μm in different seasons (**Table 6**). In *C. reticulata* var. *kinnow*, the averages of length and width of fusiform initials range from 211.64 to 256.03 and 16.25 to 18.97 μm, respectively. The size of their tapering ends varies from 72.25 to 82.25 μm. Comparatively, shorter fusiform initials are found from June to October (**Table 7**). The mean value of anticlinal and periclinal diameters of ray initials varies from 10.47/9.24 to 15.23/13.32 μm (**Table 8**). In *C. sinensis*, the averages of length and width of fusiform initials range from 205.36 to 270.50 and 15.98 to 19.51 μm, respectively. The size of their tapering ends varies from 71.25 to 84.25 μm. Comparatively, shorter fusiform initials occur from May to September (**Table 9**). The mean value of anticlinal and periclinal diameters of ray initials varies from 11.76/10.67 to 15.16/12.44 μm (**Table 10**).

The frequency of ray types, when studied in fortnightly collections has revealed that the size as well as the formation of their development happens to be highly influenced by the seasonal conditions. The short and medium sized rays are more frequent during the activity of the cambium in all the species investigated [44]. The distribution of uniseriate to multiseriate rays is found to be influenced by the weather conditions. Frequency of uniseriate and short cambial rays has been found higher in active period than in the inactive phase of the cambium in the species investigated presently. In *C. limon*, the multiseriate rays are dominant in number and constitute 45–65% of the cambial zone in different months lowest being in June and highest in January. Following this, the uniseriate rays generally vary from 23 to 34% and biseriate rays generally vary from 12 to 21% of the total rays of the cambial zone. In *C. paradisi*, multiseriate and uniseriate rays occur more frequently than the biseriate ones. The uniseriate rays are more frequent, that is, 22–30%, whereas the biseriate ones are from 16 to 28%. The highest number of multiseriate rays noticed is 62% (February) and lowest is 42% (June). In *C. reticulata* var. *kinnow*, uniseriate rays vary from 4 to 14%, whereas the biseriate ones are from 13 to 23%. Multiseriate rays are very common in this species and occur frequently within the range of 63–83% minimum being found in May and maximum in February. In *C. sinensis*, uniseriate rays vary from 22 to 30%, biseriate from 16 to 28% and multiseriate from 42 to 62% maximum in the month of March and minimum in June. Earlier workers have also reported such changes both in size and magnitude of different types of cambial initials in tropical trees [48–53].

The amount of ray and fusiform initials shows some minor fluctuations in different months of a calendar year. In *C. limon*, the percentage area occupied by ray initials varies from 14 to 22%, the maximum being in November, while the minimum occurring in May. In *C. paradisi*, it is found to vary from 18 to 27% with lowest being in May and highest in January. In *C. reticulata* var. *kinnow*, 19 to 28% with minimum in June and maximum in December and in *C. sinensis* from 18 to 31% with highest value occurring in January and lowest in July. Thus, *C. sinensis* shows

Months	Length (μm)				Width (μm)				Tapering ends (μm)			
	Range	Mean ± SE	SD	CV%	Range	Mean ± SE	SD	CV%	Range	Mean ± SE	SD	CV%
January	200–313	265.29 ± 1.81	36.22	13.65	17–21	18.63 ± 0.08	1.70	9.12	75–88	81.00 ± 0.31	6.25	7.71
February	200–338	264.00 ± 1.78	35.65	13.50	17–24	19.65 ± 0.11	2.28	11.60	75–100	84.75 ± 0.42	8.41	9.92
March	200–350	269.67 ± 2.01	40.35	14.96	17–24	20.06 ± 0.12	2.47	12.31	75–100	86.25 ± 0.45	9.11	10.56
April	212–325	264.55 ± 1.65	33.07	12.50	17–24	18.97 ± 0.09	1.81	9.54	75–100	82.25 ± 0.33	6.66	8.09
May	175–300	238.32 ± 1.63	32.67	13.70	13–21	17.74 ± 0.10	2.07	11.66	62–88	77.75 ± 0.38	7.62	9.80
June	137–313	233.72 ± 2.02	40.58	17.36	13–21	17.47 ± 0.12	2.45	14.02	62–88	76.75 ± 0.45	9.02	11.75
July	150–313	241.40 ± 1.93	38.66	16.01	13–21	17.88 ± 0.11	2.23	12.47	62–88	78.25 ± 0.41	8.22	10.50
August	150–325	247.00 ± 1.85	37.09	15.01	13–24	18.36 ± 0.10	2.15	11.71	62–100	80.00 ± 0.39	7.91	9.88
September	187–338	263.44 ± 1.79	35.92	13.63	13–24	18.70 ± 0.10	2.07	11.06	62–100	81.25 ± 0.38	7.61	9.36
October	200–338	269.34 ± 1.89	37.87	14.06	17–24	19.44 ± 0.11	2.26	11.62	75–100	84.00 ± 0.41	8.31	9.89
November	250–263	274.84 ± 1.79	35.97	13.08	17–24	21.28 ± 0.11	2.23	10.47	75–100	90.75 ± 0.41	8.22	9.05
December	225–350	278.00 ± 1.66	33.39	12.01	17–24	20.40 ± 0.11	2.25	11.02	75–100	87.50 ± 0.41	8.30	9.48
LSD at 5%		= 50.48				= 3.03				= 11.14		
LSD at 1%		= 67.60				= 4.05				= 14.92		

Table 3.
Changes in the cell size of fusiform initials (as observed in tangential longitudinal section) of cambial zone during various months of a calendar year in Citrus limon.

Months	Anticlinal diameter (μm)				Periclinal diameter (μm)			
	Range	Mean ± SE	SD	CV%	Range	Mean ± SE	SD	CV%
January	10–21	15.23 ± 0.17	3.42	22.45	10–21	14.62 ± 0.16	3.21	21.95
February	10–21	15.16 ± 0.16	3.35	22.09	10–17	12.78 ± 0.12	2.41	18.85
March	10–21	15.43 ± 0.17	3.55	23.00	10–21	14.82 ± 0.16	3.38	22.80
April	10–21	13.46 ± 0.13	2.63	19.53	10–14	11.22 ± 0.07	1.56	13.90
May	10–17	12.64 ± 0.10	2.15	17.00	10–14	11.35 ± 0.08	1.61	14.18
June	10–17	12.85 ± 0.12	2.48	19.29	6–14	11.69 ± 0.09	1.82	15.56
July	10–17	12.92 ± 0.12	2.54	19.65	10–17	12.10 ± 0.10	2.17	17.93
August	10–17	14.48 ± 0.12	2.43	16.78	10–17	12.30 ± 0.10	2.03	16.50
September	10–17	13.46 ± 0.11	2.35	17.45	10–17	12.17 ± 0.09	1.81	14.87
October	10–21	15.57 ± 0.16	3.34	21.45	10–17	13.60 ± 0.11	2.35	17.27
November	10–21	16.18 ± 0.19	3.82	23.60	10–17	14.55 ± 0.11	2.36	16.21
December	10–21	15.84 ± 0.18	3.64	22.97	10–21	13.94 ± 0.16	3.21	23.02
LSD at 5%		= 4.34				= 3.38		
LSD at 1%		= 5.81				= 4.53		

Table 4.
Changes in the cell size of ray initials (as observed in tangential longitudinal section) of cambial zone during various months of a calendar year in Citrus limon.

maximum fluctuation as compared to others [44]. The ray initials multiply considerably to become multiseriate in older axis in all the investigated species as has been reported by earlier workers [10, 11, 52].

6. Periodicity of the vascular cambium

The activity of vascular cambium is not uniform but shows great variation depending on the genetic constitution of plants and difference in the internal and external environment [1]. There are plants whose cambium is active throughout the entire life of the plant, that is, the cells of cambium divide continuously and the resulting cells undergo gradual differentiation to form xylem and phloem. Such type of activity usually occurs in plants growing in tropical regions [54]. However, not all tropical trees exhibit continuous cambial activity [55–58].

In the present study, it has been observed that the vascular cambium of all three species shows a periodic activity rather than a continuous growth as reported in other tropical species of Indian subcontinent [10, 11, 52, 57–61]. During the dormant stage, the cambial zone is represented by a narrow zone of tangentially flattened cells constituting of three to seven layers in *C. limon*, seven to nine layers in *C. paradisi*, four to six layers in *C. reticulata* var. *kinnow*, and five to seven layers in *C. sinensis* (**Figure 1(ii)**). The radial walls of cambial cells during dormant stage are found comparatively thicker than what they are during the active phase. In tangential view, the radial walls are found prominently beaded during the resting period (**Figure 2(ii)**) due to the alternatively thickened areas and the deeply depressed primary pit fields, through which they communicate by plasmodesmata connections with the contiguous elements. The fusiform cambial cells during their active phase possess relatively thin and almost smooth radial walls due to the absence of thickened areas, alternating with the primary pit fields (**Figure 1(v)**).

Months	Length (μm)				Width (μm)				Tapering ends (μm)			
	Range	Mean ± SE	SD	CV%	Range	Mean ± SE	SD	CV%	Range	Mean ± SE	SD	CV%
January	187–438	284.00 ± 3.18	63.72	22.43	13–28	20.12 ± 0.16	3.25	16.15	62–113	86.50 ± 0.59	11.96	13.82
February	187–463	281.35 ± 3.12	62.53	22.22	13–31	21.48 ± 0.18	3.63	16.89	62–125	91.50 ± 0.66	13.34	14.57
March	125–450	272.87 ± 3.15	63.18	23.15	10–31	19.65 ± 0.18	3.74	19.03	50–125	84.75 ± 0.68	13.77	16.24
April	237–343	270.73 ± 1.76	35.35	13.05	17–24	20.53 ± 0.12	2.45	11.93	75–100	88.00 ± 0.45	9.01	10.23
May	137–400	266.37 ± 2.90	58.17	21.83	13–24	19.38 ± 0.15	3.13	16.15	62–100	83.75 ± 0.57	11.53	13.76
June	100–425	217.74 ± 2.98	59.65	27.39	10–28	16.38 ± 0.15	3.02	18.43	50–113	72.75 ± 0.55	11.10	15.25
July	125–313	220.32 ± 2.26	45.28	20.55	10–21	16.86 ± 0.13	2.72	16.13	50–88	74.50 ± 0.50	10.00	13.42
August	150–375	227.06 ± 2.31	46.32	20.39	13–24	16.86 ± 0.12	2.45	14.53	62–100	74.50 ± 0.45	9.01	12.09
September	125–363	239.00 ± 2.79	55.83	23.35	10–24	17.61 ± 0.16	3.31	18.79	50–100	77.25 ± 0.60	12.18	15.76
October	125–363	243.58 ± 2.70	54.08	22.20	10–24	17.74 ± 0.16	3.28	18.48	50–100	77.75 ± 0.60	12.08	15.53
November	137–338	244.09 ± 2.56	51.35	21.03	13–24	18.36 ± 0.14	2.96	16.12	62–100	80.00 ± 0.54	10.91	13.63
December	150–450	267.41 ± 3.19	63.95	23.91	13–28	19.24 ± 0.18	3.64	18.91	62–113	83.25 ± 0.66	13.38	16.07
LSD at 5%		= 81.05				= 4.61				= 16.95		
LSD at 1%		= 108.53				= 6.17				= 22.70		

Table 5.
Changes in the cell size of fusiform initials (as observed in tangential longitudinal section) of cambial zone during various months of a calendar year in Citrus paradisi.

Months	Anticlinal diameter (μm)				Periclinal diameter (μm)			
	Range	Mean ± SE	SD	CV%	Range	Mean ± SE	SD	CV%
January	10–21	15.36 ± 0.16	3.35	21.80	10–21	14.96 ± 0.16	3.33	22.25
February	10–17	14.14 ± 0.12	2.48	17.53	6–14	10.13 ± 0.13	2.76	27.24
March	10–21	14.55 ± 0.12	2.55	17.52	10–17	12.85 ± 0.12	2.48	19.29
April	10–17	14.21 ± 0.12	2.42	17.03	10–17	12.51 ± 0.10	2.19	17.50
May	6–17	10.54 ± 0.14	2.90	27.51	6–14	9.45 ± 0.10	2.18	23.06
June	6–14	9.72 ± 0.11	2.25	23.14	6–14	8.97 ± 0.13	2.61	29.09
July	6–14	8.63 ± 0.12	2.47	28.62	6–14	8.43 ± 0.09	1.83	21.70
August	6–14	8.77 ± 0.13	2.64	30.10	6–14	8.56 ± 0.11	2.38	27.80
September	10–17	13.46 ± 0.12	2.45	18.20	6–14	10.33 ± 0.11	2.25	21.78
October	10–21	15.43 ± 0.16	3.35	21.71	10–17	14.55 ± 0.12	2.45	16.83
November	10–17	14.34 ± 0.12	2.48	17.29	10–21	13.87 ± 0.15	3.18	22.92
December	10–21	15.50 ± 0.17	3.48	22.45	10–21	14.82 ± 0.16	3.31	22.33
LSD at 5%		= 3.77				= 3.77		
LSD at 1%		= 5.05				= 5.06		

Table 6.
Changes in the cell size of ray initials (as observed in tangential longitudinal section) of cambial zone during various months of a calendar year in Citrus paradisi.

The beaded nature of radial walls, if at all present during the active period, is not as prominent as in the dormant period. The cambial zone as a whole during the active phase takes light stain due to the absence of colored contents and loss of chromaticity of protoplasm (**Figure 1(v)**).

The vascular cambium in all the species appears to undergo activation once in a year, after undergoing a definite period of rest. The first sign of activity appears in April in *C. limon* and *C. paradisi* and in May in *C. reticulata* var. *kinnow* and *C. sinensis*. The cells in the cambial zone undergo radial expansion in the third week of March in *C. limon* (**Figure 2(iii)**) and in the last week of March in *C. paradisi* (**Figure 2(iv)**), in the third week of April in *C. reticulata* var. *kinnow* (**Figure 2(v)**), and in the second week of April in *C. sinensis* (**Figure 2(vi)**). As a result of this enlargement, the cambial zone swells up from 32 to 40 μm in *C. limon*, 51 to 79 μm in *C. paradisi*, 17 to 64 μm in *C. reticulata* var. *kinnow*, and 34 to 40 μm in *C. sinensis*. Several criteria have been employed in the past to judge the initiation and the duration of cambial activity in tropical as well as in the different temperate species. Firstly, Priestly et al. [62] demonstrated the case with which the bark separate itself from wood of a tree trunk during the active period, a phenomenon what they named as "slippage of the bark." Subsequent workers later employed several other criteria to recognize the reactivation of cambium after its winter dormancy. The important finding in this connection is Frankenstein et al. [63]. In the present study, however, a number of criteria have been used in combination while studying periodicity of cambium. The initiation of cambial reactivation has been taken from the time of radial expansion of cambial initials, but the activity of cambium has been counted from the actual cell division and not from the date of histochemical changes or physical expansion of initials. The ceassation of activity has been taken by stopping of cell division which normally proceeds to the histochemical change in the initials.

Months	Length (μm)				Width (μm)				Tapering ends (μm)			
	Range	Mean ± SE	SD	CV%	Range	Mean ± SE	SD	CV%	Range	Mean ± SE	SD	CV%
January	162–325	252.51 ± 2.02	40.43	16.01	13–21	18.42 ± 0.10	2.05	11.12	62–88	80.25 ± 0.37	7.54	9.39
February	150–325	251.13 ± 1.68	33.61	13.38	17–24	18.56 ± 0.09	1.82	9.80	75–100	80.75 ± 0.33	6.72	8.32
March	150–300	248.00 ± 1.97	39.57	15.95	13–21	17.88 ± 0.11	2.33	13.03	62–88	78.25 ± 0.43	8.60	10.99
April	137–400	246.61 ± 3.60	72.00	29.19	13–28	18.36 ± 0.19	3.97	21.62	62–113	80.00 ± 0.72	14.59	18.23
May	200–325	242.00 ± 1.66	33.29	13.75	17–24	18.97 ± 0.09	1.81	9.54	75–100	82.25 ± 0.33	6.66	8.09
June	150–288	230.42 ± 1.88	37.60	16.31	13–21	17.40 ± 0.11	2.32	13.33	62–88	76.50 ± 0.42	8.54	11.16
July	150–413	226.38 ± 2.08	41.67	18.40	13–28	17.06 ± 0.12	2.59	15.18	62–113	75.25 ± 0.47	9.52	12.65
August	137–338	211.64 ± 1.88	37.73	17.82	13–24	16.25 ± 0.11	2.28	14.03	62–100	72.25 ± 0.42	8.41	11.64
September	137–275	216.52 ± 1.95	39.11	18.06	13–21	16.59 ± 0.12	2.42	14.58	62–88	73.50 ± 0.44	8.89	12.09
October	175–300	237.00 ± 1.69	33.85	14.28	13–21	17.68 ± 0.11	2.25	12.72	62–88	77.50 ± 0.41	8.30	10.70
November	175–300	245.69 ± 1.62	32.56	13.25	13–21	18.22 ± 0.94	1.89	10.37	62–88	79.50 ± 0.34	6.97	8.76
December	150–338	256.03 ± 2.01	40.33	15.75	13–24	18.83 ± 0.10	2.06	10.93	62–100	81.75 ± 0.37	7.59	9.28
LSD at 5%		= 58.27				= 3.37				= 12.40		
LSD at 1%		= 78.03				= 4.51				= 16.61		

Table 7.
Changes in the cell size of fusiform initials (as observed in tangential longitudinal section) of cambial zone during various months of a calendar year in Citrus reticulata *var.* kinnow.

Months	Anticlinal diameter (μm)				Periclinal diameter (μm)			
	Range	Mean ± SE	SD	CV%	Range	Mean ± SE	SD	CV%
January	10–21	15.23 ± 0.17	3.48	22.84	10–17	13.32 ± 0.12	2.53	18.99
February	10–21	14.68 ± 0.12	2.58	17.57	10–17	13.26 ± 0.11	2.38	17.94
March	10–17	12.44 ± 0.10	2.11	16.96	6–14	10.94 ± 0.12	2.48	22.66
April	10–21	13.53 ± 0.12	2.59	19.14	6–17	11.56 ± 0.15	3.12	26.98
May	10–17	13.73 ± 0.12	2.45	17.84	10–14	11.28 ± 0.07	1.58	14.00
June	6–14	10.47 ± 0.11	2.34	22.34	6–14	9.38 ± 0.11	2.21	23.56
July	6–17	11.69 ± 0.14	2.98	25.49	6–14	9.45 ± 0.09	1.96	20.74
August	6–17	10.74 ± 0.14	2.83	26.35	6–14	9.24 ± 0.10	2.04	22.07
September	6–17	11.96 ± 0.18	3.68	30.76	6–14	10.13 ± 0.13	2.76	27.24
October	10–17	13.80 ± 0.12	2.49	18.04	10–14	11.15 ± 0.07	1.52	13.63
November	10–21	14.75 ± 0.16	3.23	21.89	10–17	11.76 ± 0.09	1.95	16.58
December	10–17	13.94 ± 0.11	2.38	17.07	10–17	12.51 ± 0.11	2.30	18.38
LSD at 5%		= 3.91				= 3.26		
LSD at 1%		= 5.24				= 4.36		

Table 8.
Changes in the cell size of ray initials (as observed in tangential longitudinal section) of cambial zone during various months of a calendar year in Citrus reticulata *var.* kinnow.

In all the species investigated, the reactivation of vascular cambium has been indicated by radial expansion of cambial initials which has been described as "swelling" of cambial cell by earlier workers [52, 60, 64]. This phenomenon has been observed in the present study to occur a few days before the cells start dividing to produce new derivatives in all the species presently investigated. The present study shows swelling of the cambial zone in March in *C. limon* and *C. paradisi* and in April in *C. reticulata* var. *kinnow* and *C. sinensis*. However, the extent of radial expansion was found varying in different species.

After swelling, the cell divisions start within a week or two in the cambial zone which in turn is followed by a number of histochemical changes in the initials. A decrease in the density of cell protoplast coupled with the loss of chromaticity and the leaning of cell wall as a result of reduction in wall thickening and in the size of beads of radial walls. More or less similar changes in the nature and structure of cambial initials have been described by Oribe et al. [65–67]. The initiation of cell division during hot weather conditions shows that this phenomenon depends upon high temperature and low humidity as has been reported earlier by Mellerowicz et al. [68] in Abies balsamea, Barnett and Miller [69] in Picea sitchensis, Oribe and Kubo [70] in Conifers, and Espinosa et al. [45] in 83 tropical trees.

In *C. limon*, the cells start dividing in the first week of April which causes an increase in the layers of cells up to nine layers. In *C. paradisi*, cambial cell division begins from the second week of April, increasing the number of cambial layers up to 12. Similarly, in *C. reticulata* var. *kinnow* and in *C. sinensis*, the cells start dividing in the first week of May, causing an increase in the layers of cambium up to 10 in both the species. The newly produced derivatives differentiate first into xylary elements in all the species investigated as a result of which new xylem is being added in *C. limon* and *C. paradisi* in the month of April, in *C. reticulata* var. *kinnow* and *C. sinensis* in May. The phloem production, out of the newly produced cambial derivatives, is observed in two flushes first in May, June, and then again in September in

Months	Length (μm)				Width (μm)				Tapering ends (μm)			
	Range	Mean ± SE	SD	CV%	Range	Mean ± SE	SD	CV%	Range	Mean ± SE	SD	CV%
January	175–325	264.75 ± 1.95	39.06	14.75	13–24	18.08 ± 0.11	2.30	12.72	62–100	79.00 ± 0.42	8.46	10.70
February	137–300	261.00 ± 2.02	40.53	15.52	13–21	17.40 ± 0.12	2.42	13.90	62–88	76.50 ± 0.44	8.89	11.62
March	200–338	263.63 ± 1.72	34.45	13.06	17–24	19.04 ± 0.09	1.92	10.08	75–100	82.50 ± 0.35	7.07	8.56
April	150–363	260.39 ± 2.18	43.79	16.81	13–24	18.83 ± 0.12	2.47	13.11	62–100	81.75 ± 0.45	9.09	11.11
May	175–350	235.33 ± 2.04	40.84	17.35	13–24	17.54 ± 0.12	2.48	14.13	62–100	77.00 ± 0.45	9.14	11.87
June	137–275	215.45 ± 1.93	38.74	17.98	13–21	16.52 ± 0.12	2.45	14.83	62–88	73.25 ± 0.45	9.02	12.31
July	137–300	205.36 ± 1.84	36.96	17.99	13–21	15.98 ± 0.10	2.17	13.57	62–88	71.25 ± 0.40	8.01	11.24
August	137–300	227.04 ± 2.10	42.07	18.52	13–21	17.27 ± 0.12	2.43	14.07	62–88	76.00 ± 0.44	8.96	11.78
September	162–325	238.58 ± 1.68	33.75	14.14	13–21	17.81 ± 0.10	2.10	11.79	62–88	78.00 ± 0.38	7.73	9.91
October	187–325	256.00 ± 1.88	37.73	14.73	13–24	18.70 ± 0.09	1.95	10.42	62–100	81.25 ± 0.35	7.18	8.83
November	150–325	262.66 ± 2.07	41.40	15.76	13–24	18.02 ± 0.11	2.38	13.20	62–100	78.75 ± 0.43	8.76	11.12
December	200–375	270.50 ± 2.05	41.04	15.17	17–24	19.51 ± 0.11	2.33	11.94	75–100	84.25 ± 0.43	8.60	10.20
LSD at 5%		= 56.77				= 3.24				= 11.91		
LSD at 1%		= 76.03				= 4.34				= 15.95		

Table 9.
Changes in the cell size of fusiform initials (as observed in tangential longitudinal section) of cambial zone during various months of a calendar year in Citrus sinensis.

Months	Anticlinal diameter (μm)				Periclinal diameter (μm)			
	Range	Mean ± SE	SD	CV%	Range	Mean ± SE	SD	CV%
January	10–17	14.14 ± 0.12	2.48	17.53	10–17	12.30 ± 0.11	2.24	18.21
February	10–17	12.85 ± 0.12	2.48	19.29	10–14	11.15 ± 0.07	1.52	13.63
March	10–17	13.32 ± 0.11	2.34	17.56	10–14	11.56 ± 0.08	1.66	14.35
April	10–17	13.19 ± 0.10	2.11	15.99	10–17	12.24 ± 0.11	2.25	18.38
May	10–17	12.78 ± 0.09	1.87	14.63	6–14	10.88 ± 0.12	2.54	23.34
June	6–17	11.76 ± 0.16	3.20	27.21	6–17	10.67 ± 0.15	3.04	28.49
July	10–14	12.58 ± 0.07	1.56	12.40	6–14	10.74 ± 0.09	1.84	17.13
August	10–17	12.64 ± 0.10	2.15	17.00	6–14	10.94 ± 0.11	2.28	20.84
September	10–17	13.26 ± 0.11	2.38	17.94	6–14	11.22 ± 0.10	2.17	19.34
October	6–21	13.53 ± 0.19	3.88	28.67	6–14	11.90 ± 0.11	2.28	19.15
November	10–17	14.34 ± 0.12	2.48	17.29	10–14	11.49 ± 0.08	1.65	14.36
December	10–21	15.16 ± 0.16	3.28	21.63	10–17	12.44 ± 0.12	2.41	19.37
LSD at 5%		= 3.77				= 3.16		
LSD at 1%		= 5.05				= 4.24		

Table 10.
Changes in the cell size of ray initials (as observed in tangential longitudinal section) of cambial zone during various months of a calendar year in Citrus sinensis.

C. limon. In *C. paradisi*, new phloem is added in the last week of October followed by another addition of new phloem in November and December. In *C. reticulata* var. *kinnow*, new phloem is added in October and November. In *C. sinensis*, first flush of new phloem is added in June and July. The second flush of phloem differentiates in October.

The cessation of cambial activity occurs in early October in *C. limon*, early November in *C. sinensis*, early December in *C. reticulata* var. *kinnow*, and late December in *C. paradisi*. Thus, it appears that in the investigated species, extreme fall in temperature brings down the dormancy as reported earlier by Paliwal and Paliwal [71] in *Rhododendron arboreum*, Khan [10] in *Jacaranda mimosifolia*, *Pterospermum acerifolium*, *Terminalia arjuna*, and Mahmood [11] in *Alstonia scholaris*, *Emblica officinalis*, *Putranjiva roxburghii*.

Thus, in the presently investigated species, the cambium remains active for about 6 months in *C. limon* and *C. sinensis*, 9 months in *C. paradisi*, and 7 months in *C. reticulata* var. *kinnow*. More or less similar prolonged tends of duration of 5–9 months of radial growth has been reported earlier by Fahn [72], Zhang et al. [57], Rao and Rajput [60], Khan [10], and Mahmood [11].

Author details

Moin Ahmad Khan[1*] and M. Badruzzaman Siddiqui[2]

1 Department of Biology, Ballsbridge University, Dominica, West Indies

2 Department of Botany, Aligarh Muslim University, Aligarh, India

*Address all correspondence to: moin_a_khan11@yahoo.co.in

References

[1] Philipson WR, Ward JM, Butterfield BG. The Vascular Cambium, Its Development and Activity. London: Chapmen & Hall; 1971. ISBN: 0412104008

[2] Iqbal M, Ghouse AKM. Cambial concept and organization. In: Iqbal M, editor. The Vascular Cambium. Tauton, New York; Tauton, Somerset, England: John Wiley & Sons Inc; Research Studies Press Ltd.; 1990. pp. 1–36. ISBN: 0-86380-095-5

[3] Catesson AM. Cambial ultrastructure and biochemistry: changes in relation to vascular tissue differentiation and the seasonal cycle. International Journal of Plant Sciences. 1994;**155**:251–261. DOI: 10.1086/297165

[4] Larson PR. The Vascular Cambium. Development and Structure. Berlin: Springer-Verlag; 1994. eBook ISBN 978-3-642-78466-8. DOI: 10.1007/978-3-642-78466-8

[5] Rao KS, Rajput KS, Srinivas T. Comparative structure of vascular cambium and its derivatives in some species of Sterculia. IAWA Journal. 1996;**17**:311–318. DOI: 10.1163/22941932-90001581

[6] Chaffey NJ. Cambium: old challenges —New opportunities. Trees. 1999;**13**: 138–151. DOI: 10.1007/PL00009745

[7] Lachaund S, Catesson AM, Bonnemain JL. Structure and functions of the vascular cambium. Comptes Rendus de l' Academie des Sciences. Sciences de la Vie (Life Sciences). 1999; **322**:633–650. DOI: 10.1016/s0764-4469(99)80103-6

[8] Kitin P, Funada R, Sano Y, Beeckman H, Ohtani J. Variations in the lengths of fusiform cambial cells and vessel elements in *Kalopanax pictus*. Annals of Botany. 1999;**84**:621–632. DOI: 10.1006/anbo.1999.0957

[9] Kitin P, Funada R, Sano Y, Ohtani J. Analysis by confocal microscopy of the structure of cambium in the hardwood *Kalopanax pictus*. Annals of Botany. 2000;**86**:1109–1117. DOI: 10.1006/anbo.2000.1281

[10] Khan HA. Studies on the seasonal variation of phloem and xylem production in some tropical trees [thesis]. Aligarh Muslim University; 2001

[11] Mahmood A. Studies on growth activities of some tropical trees [thesis]. Aligarh Muslim University; 2001

[12] Khan MA, Siddiqui MB. Size variations in the vascular cambium and its derivatives in two *Alstonia* species. Acta Botânica Brasílica. 2007;**21**(3): 531–538. DOI: 10.1590/S0102-33062007000300003

[13] Butterfield BG. Terminology used for describing the cambium. IAWA Bulletin. 1975;**1**:13–14. Available from: https://agris.fao.org/agris-search/search.do?recordID=US201303005043

[14] Hutchinson J. Families of flowering plants. In: 1 Dicotyledons: 1-510; 2 Monocotyledons: 511-792. 2nd ed. Oxford: Clarendon press; 1959

[15] Grew N. The Anatomy of Plants. London: W. Rawlings; 1682. Available from: http://hdl.loc.gov/loc.rbc/General.06649.1

[16] Sanio C. Vergleichede untersuchungen uber die zusammensetzung. Des Holzkörper Botany. 1863;**21**:357-363, 369-375, 377-385, 389-399, 401-412

[17] Kitin P, Sano Y, Funada R. Fusiform cells in the cambium of Kalopanax pictus are exclusively mononucleate. Journal of Experimental Botany. 2002;

53:483–488. DOI: 10.1093/jexbot/53.368.483

[18] Bailey IW. The cambium and its derivative tissues: II. Size variation of cambial initials in gymnosperms and angiosperms. American Journal of Botany. 1920;7:355–367. DOI: 10.1002/j.1537-2197.1920.tb05590.x

[19] Ghouse AKM, Iqbal M. A comparative study on the cambial structure of some arid zone species of *Acacia* and *Prosopis*. Botaniska Notiser. 1975;**128**:327–331

[20] Kojs P. The mechanisms of cell rearrangement in storied cambium of selected woody species (in Polish) [thesis]. Katowice: Silesian University; 2000

[21] Iqbal M. The Vascular Cambium. Taunton, Somerset, England: Research Studies Press ltd; 1990. ISBN: 0-86380-095-5

[22] Khan MA, Siddiqui MB, Bhat S, Shahab D. Pollution effects on the cambial structure of *Citrus reticulata* var. tangelos. International Journal of Botany. 2007;**3**(4):342–350. DOI: 10.3923/ijb.2007.342.350

[23] Esau K. Anatomy of Seed Plants. 2nd ed. Singapore: John Wiley & Sons (Asia) Pvt. Ltd.; 2002. ISBN: 978-0-471-24520-9

[24] Bannan MW. A survey of cell length and frequency of multiplicative divisions in the cambium of conifers. Canadian Journal of Botany. 1970;**48**: 1585–1589. DOI: 10.1139/b70-234

[25] Han KS, Woong YS. Developmental changes of cambial initials and their derivative cell in the trunk of *Diospyrus khaki* Thumb and Firmiana simplex W. F. Wight in relation to girth increase. Korean Journal of Botany. 1991;**34**: 191–199

[26] Iqbal M. Structural and operational specializations of the vascular cambium of seed plants. In: Iqbal M, editor. Growth Patterns in Vascular Plants. Portland, OR; Oregon, USA: Dioscorides Press; 1994. pp. 211–271. ISBN: 0-931146-26-7

[27] Khan MA, Siddiqui MB. Development of rays in some *Citrus* species. Vegetos. 2007;**20**(2):9–11. Available from: http://vegetosindia.org/journal/Vegetos-20(2)2007/Development-of-Rays-in-Some-Citrus-Species.html

[28] Wloch W, Polap E. The intrusive growth of initial cells in re-arrangement of cells in cambium of Tilia cordata Mill. Acta Societatis Botanicorum Poloniae. 1994;**63**:109–116. Available from: https://rebus.us.edu.pl/bitstream/20.500.12128/7079/1/Wloch_The_intrusive_growth_of_initial_cells.pdf

[29] Jura J, Kojs P, Iqbal M, Szymanowska-Pulka J, Wloch W. Apical intrusive growth of cambial fusiform initials along the tangential walls of adjacent fusiform initials: Evidence for a new concept. Australian Journal of Botany. 2006;**54**:493–504. DOI: 10.1071/BT05130

[30] Khan MA, Siddiqui MB. Ratio of fusiform and ray initials in *Citrus sinensis*. Indian Journal of Applied and Pure Biology. 2007;**22**(1):161–164

[31] Khan MA, Siddiqui MB. Ratio of fusiform and ray initials in *Citrus limon* (Linn.) Burm.f. Indian Journal of Applied and Pure Biology. 2007;**22**(1): 175–178

[32] Khan MA, Siddiqui MB. Ratio of fusiform and ray initials in *Citrus reticulata* var. tangelos. Indian Journal of Applied and Pure Biology. 2007;**22**(2): 275–278

[33] Khan MA, Siddiqui MB. Ratio of fusiform and ray initials in *Citrus*

paradisi. Indian Journal of Applied and Pure Biology. 2007;**22**(2):287–290

[34] Khan MA, Khan MIH, Siddiqui MB, Bhat S. Ratio of ray and fusiform initals in some *Citrus* species. Vegetos. 2005;**18** (1&2):99–103

[35] Cumbie BG. Development changes in the vascular cambium of *Leitneria floridana*. American Journal of Botany. 1967;**54**:414–424. DOI: 10.2307/2440830

[36] Ajmal S, Khan R, Iqbal M. Cambial structure of *Holoptelea integrifolia* Planch. in relation to age. Flora. 1986; **178**:197–202. DOI: 10.1016/S0367-2530(17)31492-5

[37] Evert RF. Some aspects of cambial development in *Pyrus communis*. American Journal of Botany. 1961;**48**: 479–488. DOI: 10.1002/j.1537-2197.1961.tb11672.x

[38] Evert RF. The cambium and seasonal development of phloem in *Pyrus malus*. American Journal of Botany. 1963;**50**:149–159. DOI: 10.1002/j.1537-2197.1963.tb07190.x

[39] Ghouse AKM, Yunus M. Some aspects of cambial development in the shoots of *Dalbergia sissoo* Roxb. Flora. 1973;**162**:549–558. DOI: 10.1016/S0367-2530(17)31736-X

[40] Khan S. Studies on seasonal activity of vascular cambium and secondary phloem in some Myrtaceae [thesis]. Aligarh Muslim University; 1980

[41] Ghouse AKM, Iqbal M. Variation trends in the cambial structure of *Prosopis spicigera* L. in relation to the girth of the tree axis. Bulletin of the Torrery Botanical Club. 1977;**104**(3): 197–201. DOI: 10.2307/2484298

[42] Khan MIH, Siddiqui TO, Khan S. Ontogenetic changes in the cambial structure of *Citrus sinensis* L. Flora. 1983; **173**:151–158. DOI: 10.1016/S0367-2530(17)31994-1

[43] Ajmal S. Studies on vascular cambium and its derivatives in some arborescent Moraceae [thesis]. Aligarh Muslim University; 1985

[44] Khan MA. Anatomical studies on the activity of vascular cambium and production of vascular tissues in some tropical trees [thesis]. Aligarh Muslim University; 2009

[45] Espinosa LY, Terrazas T, Lopez-Mata L. Integrated analysis of tropical trees growth: A multivariate approach. Annals of Botany. 2006;**98**:637–645. DOI: 10.1093/aob/mcl142

[46] Gričar J, Zupančič M, Čufar K, Koch G, Schmitt U, Oven P. Effect of local heating and cooling on cambial activity and cell differentiation in the stem of Norway spruce (*Picea abies*). Annals of Botany. 2006;**97**:943–951. DOI: 10.1093/aob/mcl050

[47] Begum S, Nakaba S, Oribe Y, Kubo T, Funada R. Induction of cambial reactivation by localized heating in a deciduous hardwood hybrid Popular (*Populus sieboldii* X P. grandidentata). Annals of Botany. 2007;**100**:439–447. DOI: 10.1093/aob/mcm130

[48] Creber GT, Chaloner WG. Environmental influences on cambial activity. In: Iqbal M, editor. The Vascular Cambium. Tauton, Somerset, U.K.: Research Studies Press; 1990. pp. 159–199. ISBN: 0-86380-095-5

[49] Fahn A, Werker E. Seasonal cambial activity. In: Iqbal M, editor. The Vascular Cambium. Vol. 1990. Taunton Somerset. England: Research Studies Press; 1990. pp. 139–158. ISBN: 0-86380-095-5

[50] Blanche CA, Jr L, Sommers RA, Hodges JD, Nebeker TE. Seasonal

cambial growth and development of loblolly pine: Xylem formation, inner bark chemistry, resin ducts and resin flow. Forest Ecology and Management. 1992;**49**:151–165. DOI: 10.1016/0378-1127(92)90167-8

[51] Rao KS, Srinivas T, Rajput KS. Seasonal anatomy of vascular cambium in young branches of *Bombax ceiba* Brum. Acta Botanica Indica. 1996;**24**(1): 17–20

[52] Rajput KS. Seasonal cambial activity and wood development in some timber trees growing in different forest region of Gujarat state [thesis]. M.S. University of Baroda; 1997

[53] Borchert R. Climatic periodicity, phenology and cambium activity in tropical dry forest trees. IAWA Journal. 1999;**20**:239–247. DOI: 10.1163/22941932-90000687

[54] Fahn A. Plant Anatomy. 3rd ed. Oxford: Pergamon press; 1982. DOI: 10.1111/j.1756-1051.1983.tb01458.x

[55] Lu CY, Chiang SHT. Seasonal activity of the cambium in the young branch of *Liquidambar formosana* Hance. Taiwania. 1975;**20**:32–47

[56] Fahn A, Burley J, Longman KA, Mariaux A, Tomlinson PB. Possible contributions of wood anatomy to the determination of the age of tropical trees. In: Bormann FH, Berlyn GP, editors. Age and Growth of Tropical Trees. School of Forestry and Environmental Studies. Vol. 94. New Haven, USA: Bull|Yale University; 1981. pp. 31–54. Available from: https://elischolar.library.yale.edu/yale_fes_bulletin/6

[57] Zhang ZJ, Lin J, Chen Z, Zhang YT. Periodicity of cambial activity and seasonal changes of the secondary phloem in 4 tannic plants. Acta Botanica Yunnanica. 1997;**19**(3):271–274. Available from: http://europepmc.org/article/CBA/305492

[58] Priya PB, Bhat KM. Influence of rainfall, irrigation and age on the growth, periodicity and wood structure in teak (Tectona grandis). IAWA Journal. 1999;**20**(2):181–192. Available from: https://brill.com/downloadpdf/journals/iawa/20/2/article-p181_9.pdf

[59] Eshete G, Stahl G. Tree rings as indicators of growth periodicity of acacias in the Rift Valley of Ethiopia. Forest Ecology and Management. 1999; **116**:107–117. DOI: 10.1016/s0378-1127(98)00442-3

[60] Rao KS, Rajput KS. Seasonal behavior of vascular cambium in teak (*Tectona grandis*) growing in moist deciduous and dry deciduous forests. IAWA Journal. 1999;**20**:85–93. Avaialble from: https://agris.fao.org/agris-search/search.do?recordID=US201900438848

[61] Rao KS, Rajput KS. Relationship between seasonal cambial activity, development of xylem and phenology in *Azadirachta indica* growing in different forests of Gujarat state. Annals of Forests Science. 2001;**58**:691–698. DOI: 10.1051/forest:2001156

[62] Priestley JH, Scott LT, Mellins NE. A new method of studying cambial activity. Proceedings of the Leeds Philosophical and Literary Society. 1933; **2**:365–374

[63] Frankenstein C, Eckstein D, Schmitt U. The onset of cambium activity—A matter of agreement? Dendrochronologia. 2005;**23**:57–62. DOI: 10.1016/j.dendro.2005.07.007

[64] Fujita M. Three dimensional analyses of cambial activity and xylem differentiation. In: 15th International Botanical Congress, Yokohama, Japan, 28 August 3 September, 1993. Abstracts; 1993. p. 88

[65] Oribe Y, Funada R, Shibagaki M, Kubo T. Cambial reactivation in locally

heated stems of the evergreen conifer *Abies sachalinensis* (Schmidt) Masters. Planta. 2001;**212**:684–691. DOI: 10.1007/s004250000430

[66] Oribe Y, Funada R, Kubo T. Relationships between cambial activity, cell differentiation and the localization of starch in storage tissues around the cambium in locally heated stems of *Abies sachalinensis* (Schmidt) Masters. Trees. 2003;**17**:185–192. DOI: 10.1007/s00468-002-0231-1

[67] Oribe Y, Funada R, Kubo T. Cambial activity in locally heated stems of evergreen and deciduous conifers during winter cambial dormancy. In: Baas P, Barnett JR, Carcaillet C, et al., editors. Proceedings of the International Symposium on Wood Science Organized by IAWA. Montpelier, France: IAWS and CIRAD; 2004. p. 47

[68] Mellerowicz EJ, Coleman WK, Riding RT, Little CHA. Periodicity of cambial activity in *Abies balsamea* L. Effects of temperature and photoperiod on cambial dormancy and frost hardiness. Physiologia Plantarum. 1992; **85**:515–525. DOI: 10.1111/j.1399-3054.1992.tb05820.x

[69] Barnett JR, Miller H. The effect of applied heat on graft union formation in dormant Picea sitchensis (Bong.) Carr. Journal of Experimental Botany. 1994; **45**:135–143. Available from: https://www.jstor.org/stable/23694726

[70] Oribe Y, Kubo T. Effect of heat on cambial reactivation during winter dormancy in evergreen and deciduous conifers. Tree Physiology. 1997;**17**:81–87. DOI: 10.1093/treephys/17.2.81

[71] Paliwal SP, Paliwal GS. Influence of climatic variations on seasonal behaviour of vascular cambium in some Himalayan trees. III. *Rhododendron arboreum* Smith. Phytomorphology. 1992;**40**(3–4):257–271

[72] Fahn A. Plant Anatomy. 2nd ed. Oxford: Pargamon Press; 1974. ISBN 10: 0080119433 ISBN 13: 9780080119434

Chapter 15

Understanding the Adaptive Mechanisms of Plant in Low Phosphorous Soil

Mehtab Muhammad Aslam, Kashif Akhtar, Joseph K. Karanja, Noor-ul-Ain and Fasih Ullah Haider

Abstract

With the rapidly increasing world population and escalating food demand in the face of changing weather patterns, it is imperative to improve our understanding of how root functional traits enhance water acquisition and nutrient foraging for improved crop yields. Phosphorous (P) is poorly bioavailable element and essential for plant growth and development. Natural P reserves are very limited, and its availability is greatly influenced by several environmental factors, e.g., due to finite natural resources, soil pH, organic matter, and soluble complexes with cations (Al, Fe, and Ca); therefore, P limitation is a major factor that adversely affects crop production. To ensure an efficient and stable agricultural system, the establishment of P efficient crop production is inevitable. Plants have evolved different adaptability mechanisms to overcome these nutrient stresses. Low P adapted responses in plants are considered as an important trait for developing new lines with improved P acquisition, water uptake efficiency, and eventually protect roots from physical impedance. Previous studies showed that, modification in root architecture is potentially correlated with water, nutrient and phosphorus uptake. During P deficit condition, plant root undergoes several phenotypic (root hair density, cluster root, and lateral root) and biochemical modifications (citrate, malate, and acid phosphates secretion) leading to the solubilization and acquisition of unavailable P complexes in soil. This chapter reveals the biochemical, physiological, and molecular mechanisms of plant adaptive responses to low P availability. Moreover, this chapter proposes how plant competes with various abiotic stresses such as P deficiency, drought, and salinity. Screening of plants with superior root hair traits would be an important approach toward the development of P efficient crop varieties.

Keywords: phosphorous deficiency, plant adaptability mechanisms, P uptake, modification in root traits, sustainable crop production

1. Introduction

Phosphorous (P) is an indispensable limiting factor for plant growth and development [1]. Agricultural land comprised on low P availability is about 67% to sustain a better crop production [2]. P is mostly absorbed by diffusion through root absorption by creating gradients force. Very little (0.05 g^{-1}) of phosphate concentration is soil moved to the roots through capillary water movement. The value of P extracted is low by P concentration at the root-soil surface, and wheat roots have to grow to come into contact with new soil from which can extract phosphate. Thus, the length of root is a major factors of absorbing surface area [3]. Organic P is not directly amenable for plant to capture to make it easily accessible for plant uptake, conversion of organic P into inorganic Pi ($H_2PO_4^-$, PO_4^{3-}, and HPO_4^{2-}) is a prerequisite [4]. Plants have adapted a range of strategies to improve Pi availability such as microbial symbiotic association [5], modification in root system architecture (RSA) [6], cluster root (CR) formation [7], organic acid exudation [8], H^+ secretion, and genetic modification [9, 10] (**Figure 1**). For instance, white lupin (*Lupinus albus*) has developed extreme tolerance to low Pi condition through forming specialized dense root structures known as cluster root [10]. Cluster root secretes large number of organic acids, protons, and acid phosphatases into the soil, that increases Pi availability [19, 20].

Another important strategy for improving Pi availability and uptake under P limited region is the exudation of organic compounds and acid phosphatase by plant roots into the rhizospheric zone. Cluster roots of white lupin are known as exudate organic acid such as citrate, malate, malonate, carboxylate, and acid phosphatase into the soil [7, 21, 22]. Several other distantly related plant families have the ability to form cluster root, and is commonly found in proteaceae family [23]. It is not mandatory that, every genus of plant family produce CR root, like some member of other families can form CR (Restionaceae, Moraceae, Myricaceae, Elaeagnaceae, Fabaceae, Casuarinaceae, Cyperaceae, Cucurbitaceae, and Betulaceae) [24].

Previous studies have shown that the exudation of PAP (purple acid phosphatase) may facilitate the use of organic P for plants [25, 26]. Membrane localized high affinity transporters (PHT1) also exhibit great contribution in improving P uptake, and have been recognized in soybean, rice, and wheat roots [27–30]. Arbuscular mycorrhizal fungi (AMF) symbiotic association plays vital role in improving plant

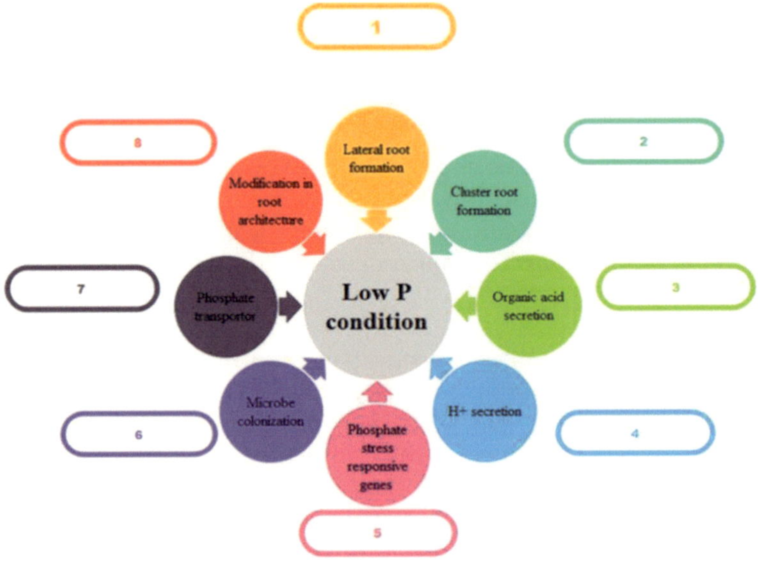

Figure 1.
Under P stress condition plant evolved multiple adaptive responses to improve Pi uptake, recycling and transportation [11–18].

ability to acquire inorganic P from rhizosphere [5]. Additionally, AMF symbiotic process activates expression of PSI genes (Pi starvation inducible), involving phosphate transporters, ATPases, and acid phosphatases, [28, 31, 32], which increases the ability of Pi acquisition in plants. Further, studies on identifying the whole genetic mechanisms underlying P adaptability mechanisms would provide a better understanding in producing modern P efficient agricultural crops, that will not only reduce fertilizer cost but also improves plant production.

2. Phosphorous concentration in soil

Most of the soils have a large reservoirs of total phosphorus, while available P is at low level [33], and it is further reported that soil total P is about 100 times higher than available P to crops plants. Phosphorous is a key determinant factor in regulating plant cell metabolism, and is a major constituent of nucleic acid, phospholipid, ATP and NADPH. It is not amenable for plant to uptake like other growth nutrient due to its high reactivity [34]. Freely available Pi can form complexes with Al and Fe under acidic and with Mg and Ca under alkaline/neutral soil, rendering the Pi inaccessible for plant to uptake [35]. Furthermore, phytic acid bounds with 60–80% of agricultural Pi and restricts its availability, that requires mineralization of Pi before assimilated by plant root [36]. This problem of Pi starvation can be solved by applying phosphate fertilizer [37]. But due to the limited availability of phosphate resources it is not a permanent solution to rely on it for future agricultural production, however, it becomes a major threating bulletin towards future agriculture system [38]. However, a deep understanding of plant adaptability and respond mechanism to low P condition would help in establishing modern strategy for efficient utilization of Pi by plants.

The whole agriculture system relies on the use of fertilizer to increase yield, and maintaining plant growth. Some ecological and economical drawbacks have provoked the interest to explore alternative approaches to fulfill the demand of global food supply [10, 39–41]. To determine the mechanism that facilitates plant growth on poor nutrient soil, scientists are learning from those plants that are extremely tolerance to nutrient deficiency condition, such as cluster root forming plant species.

2.1 Uptake vs. utilization efficiency of phosphorus

Phosphorus utilization (grain yield per unit P in the plant) is dependent on the plants P requirement. The P utilization efficiency can improve due to the increase in harvest index, P harvest index, and low P concentration in grain. Moreover, the strategy for reducing P content in grain has some limits. Therefore, in a P deficits soils, excessively low values of P concentration in grain affects seed vigor [42, 43]. To improve P utilization efficiency that selection of wheat genotypes is important, which removes small amount of P from soil due to their low P concentration in grains contributes in soil sustainability [44].

2.2 P-solubilizing microbes improves plant growth

The availability of soluble P uptake by plant is due to PS microbes, and the release of important nutrients can also improve growth and development of plants [45]. Therefore, due to symbiotic and asymbiotic the change in the concentration of phytohormones, e.g., indole acetic acid also gave the positive results about the increase in growth and development of plants [46, 47]. This mechanism is active at different growth stages; however, PS microbes have the ability for synthesizing plant growth promoting nutrients at different climatic conditions [48].

3. Plant low P adaptability mechanisms

Naturally, plants have evolved several different mechanisms to cope with nutrient limiting (Pi stress) conditions, either by acquiring more phosphate from soil or by maintain Pi homeostasis within plant body. These adaptive mechanisms could be appearing as biochemical, physiological, or molecular responses to low P conditions.

3.1 Biochemical

In a Pi stress condition, the plant roots undergo a range of phosphate stress responses, involving modification in root system architecture (RSA), increasing/inducing expression of Pi transporters, secretion of large amount of organic acid and acid phosphatases. Root exudates are below ground substances released by the plant root which plays multiple role in plant defense and nutrient uptake such as attractants, stimulator, signaling molecules, and also as an inhibitor against toxic pathogen. Root exudates are continuing source of fixed carbon to carry out plant's photosynthetic activity. Major differences in the root exudation type, exudation levels, and root architecture system distinctly varies from plant species to species. It is speculated that, nutrient influx and efflux by plant root is heterogeneous among time and space [49]. Mucilage exuded by the roots, with its high water holding capacity [50], may increase water holding capacity of the rhizosphere (area around plant root).

Mucilage has positive effects on root water and nutrient uptake, it has the potential to increase the capability of young root segments to capture water from soils, particularly under drought condition. Such characteristics potentially help plants to use soil resources and survive drought spells [51]. However, the role of root exudates and the rhizosphere on nutrient uptake and drought tolerance has not yet been demonstrated and remains largely hypothetical. Plant roots exude several compounds such as phenolic, amino acids, sugars, and organic acids [52]. Major organic acids e.g. citrate, malate, and oxalate are implicating in regulating nutrient acquisition, and stimulating toxic metal detoxification mechanisms [53–55].

There is an evidence, indicating direct role of organic acid in mobilizing phosphorous for plant uptake and detoxification of Al^{3+}, Fe, and Mn^{2+} [56, 57]. It is noteworthy to mention that, from total P fraction exist in soil only Pi (inorganic P) and is directly available to capture by plant root [58]. A number of plant species respond to low P condition by secreting large amount of organic acids such as *Lupinus albus, Glycine max, Zea mays, Triticum aestivum, Cajanus cajan, Phaseolus vulgaris; Cassia tora, Hordeum vulgare* and *Solanum tuberosum* [55, 59–65]. For example, *Lupinus albus* cluster root forming plant were shown to secrete citric acid, and proposed that citrate greatly improved Pi acquisition by forming ferric-hydroxy-phosphate compound diffused to the root and release Pi in to the rhizosphere [59]. Similarly, *Cajanus cajan* exudates malonic and piscidic acid that solubilized fixed P to directly available Pi form [63, 66].

3.2 Physiological

Plants survive in heterogeneous environment, are exposed to various abiotic factors such as; high temperature, salinity, drought, and nutrient deficiency etc. Phosphorous deficient soil is one of the major abiotic factors compromising plant growth status, particularly by reducing crop yield. Drought is a major stress on plants that partially limit nutrient availability, acquisition and remobilization [67]. Under low P availability plants adapted various physiological responses such as anthocyanin accumulation [68, 69], inhibition of primary root elongation, massive production of lateral and cluster root development [68]. Root tip serves as entry point for P sensing, modification in root system potentially contribute to nutrient uptake for maintaining plant survival under P starvation [70–72]. It is suggested that well developed root architecture is an important adaptive strategy for plants to acquire more Pi from soil. It has been revealed that, *Phaseolus vulgaris* genotype having highly branched root architecture showed efficient P acquisition ability [73].

Root hairs are also quite important for the uptake of poorly mobile growth factor such as P by improving soil exploration. It was reported that, under P deficient condition root hairs regulates almost 63% of the total P uptake [17]. Therefore, different plant species or genotypes with different root hairs/length may exhibit different P uptake efficiency [74]. Cluster root excretes large number of citrate, malonate, and phosphatases, that help in solubilization of fixed P to available form that is easily accessible for plant to capture [75]. Many studies elaborated that root hairs exhibit primary role in P acquisition under low P soil [17, 68]. It is concluded that root hairs showed strong correlation in phosphorus acquisition [76].

3.3 Molecular

Generally, plants employ a range of molecular mechanisms to confer resistance against multiple abiotic and biotic stresses that influence nutrient availability, uptake, and recycling. The ability of plant to sense and transduce signals is regulated by multiple genes or transcription factor. A growing body of evidence from mammals and yeast proposes that role of chromatic structure governs by metabolic signals [77], while the identification of molecular players involved in crosstalk of signal transduction pathways remains largely unknown [78]. Understanding the molecular mechanism behind belowground root traits would help to identify genetic markers to improve abiotic/biotic stress tolerance and environmental variability. Plants exposed to P starvation conditions evolved different adapted responses controlled by phosphorous starvation and root development related genes. For example, AtPHR1 and OsPTF1 genes are considered to be central regulator for P starvation responses [79, 80], upregulation of these genes may improve P availability, which is important for plant root growth, and development. This is indirect evidence that, root hairs and length are major key determinant and positively correlates with nutrient uptake.

A clear understanding of molecular mechanism of root system architecture (RSA) is necessary to improve nutrient acquisition, and plant productivity. OsFH1 plays critical role in root hair development and elongation [80, 81]. Phosphorous is an essential macronutrient for plant survival, due to its limited reservoirs the establishment of phosphorous efficient crops is needed. Pup1 phosphorous deficiency tolerance locus has been identified in rice (Kasalath variety). Pup1 is protein kinase gene later named as phosphorous starvation tolerance-1 (PSTOL1) (**Table 1**) [28]. The overexpression of PSTOL1 gene in rice which naturally lacks PSTOL1 showed greatly increased grain yield in P deficient soil. It also triggered root growth initiation, and resulting the nutrients and P uptake ability from soil. Thereby, PSTOL1 confer tolerance to drought and P deficient soil [82].

Collectively it is suggested that, all root development/elongation related genes play critical role in increasing P acquisition and bioavailability. However, to understand candidate genes involved in development of root would enable farmers and breeders to screen out cultivars with better adapted root system through marker assisted selection tool.

No.	Genes/transcription factors	Plants	Function	Reference
1.	*OsPTF1*	*Oryzae sativa*	Contribute to P availability	[79]
2.	*OsPSTOL1*	*Oryzae sativa*	Confer tolerance to drought and increased crop yield	[82]
3.	*OsFH1*	*Oryzae sativa*	Improves root hairs growth and elongation	[81]
4.	*DRO1*	*Oryzae sativa*	Develop deeper root system	[83]
5.	*OsEXPA17*	*Oryzae sativa*	Involved in root elongation	[84]
6.	*OsSNDP1*	*Oryzae sativa*	Promotes root hair elongation	[85]
7.	*OsSAPK10*	*Oryzae sativa*	Increases root hair length	[86]
8.	*AtPHR1*	*Arabidopsis thaliana*	Contribute to P availability, important role in regulating PSRs	[79, 80]
9.	*PHO1 and AVP1*	*Arabidopsis thaliana*	Improved resistance to drought, and maintain Pi homeostasis, plant productivity	[87, 88]
10.	AVP1	*Solanum lycopersicum*	Increased Pi transport and root/ shoot dry weight, resistant to P deficient soil	[89]

Table 1.
The phosphorous starvation induced genes and transporters involved in promoting plant growth and development.

4. "Omics" approaches contribute to Pi adaptation mechanism

The prime objective for future crop production is the development of well adapted lines to Pi starvation condition. Identification of key genes are upregulated under Pi deficient soil could be a useful tool for understanding plant development responses, and use as marker selection for crop improvement, and reported in various plant species transcriptomic and metabolomics approaches had identified bunch of genes and metabolites involved in regulating plant developmental responses and cluster root formation, and provides deep insight in identifying Pi acquisition pathway and network [90, 91]. Genetic engineering has great potential

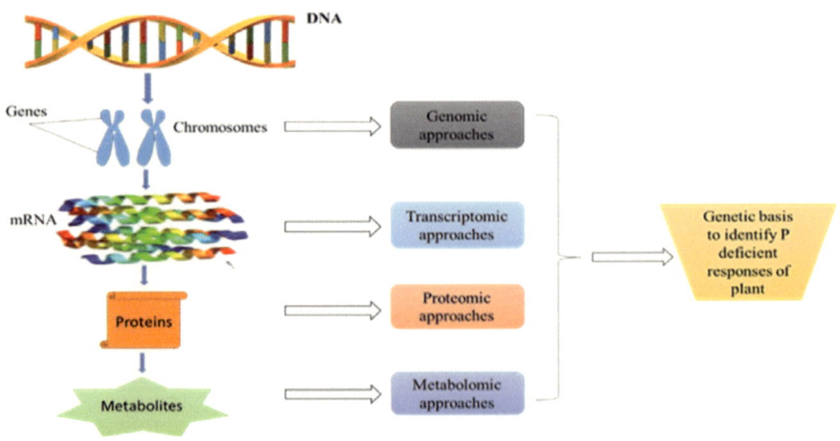

Figure 2.
Omics approaches can reveal molecular basis of plant developmental adaptation to poor nutrient soil.

to revolutionize functional analysis of gene (**Figure 2**), particularly in those plants which have developed stable transformation method.

5. Generation of phosphorous efficient crops

Molecular engineering is a useful approach for breeding and production of transgenic, efficient P uptake plants. It has been shown in rice and Arabidopsis studies that, overexpression of PSTOL1 in rice increases P uptake efficiency under low P availability condition [82], and overexpression of AVP1 also improves P uptake in Arabidopsis and several other plant species [92], suggesting that molecular approaches can significantly improves P uptake efficiency.

Overexpression strategy has also been reported to change exudation rate of acid phosphatase and H^+ secretion in tomato root, that promotes the solubilization of soil fixed P to Pi form [93]. Contrastingly, knockout approaches can also be used for altering Pi homeostasis, for example, OsPHT1.8 and OsPHF1 reduces P uptake and translocation [94, 95].

6. Concluding remarks

P deficiency is an important limiting factor in terms of plant nutrition and growth in cultivated soils. Although the exogenous application of chemical P fertilizer is extensively exploited to fulfill crop nutrition demands. The overuse of chemical fertilizer is not a permanent solution due to finite P reserves and imposes serious threats to environment safety. The excessive use of P fertilizer adversely affects soil biota (microbes, earthworms) and its physical or mechanical properties, eventually reduces crop productivity. As a consequence, soil compactness serves as a major constraint that restricts root growth and elongation. Despite of reduction in root length, root hairs endure as a unique trait for enhancing P acquisition ability under highly compacted low P soil. An efficient uptake of nutrients is a cornerstone towards crop improvement and productivity. Improved phosphorous use efficiency will be arising as a demanding approach in the future to achieve higher crop productivity. Root hairs and density significantly contribute to improve P availability under diverse soil constraints. In future, a clear understanding of molecular mechanism underlying root system architecture (RSA) is necessary to improve nutrient acquisition, and plant yield. More studies in a wide range of plants at the genetic level would provide breeders with molecular markers useful for improving nutrient uptake in plants growing in soils having heterogeneous P levels. The recurring theme is that potential importance of P efficient crops in improving agricultural yield under limited resources is still poorly identified. Such studies will provide important clues for potential targets that can be utilized to engineer biofertilizers which can increase phosphorus use efficiency by changes root trait modification in poor nutrient availability soil.

Acknowledgement

We would like to acknowledge Prof. Xu Weifeng for his support and guidance.

Conflict of interest

There is no conflict of interest exist to declare.

Author's contribution

MA conceived the first idea and prepared the first draft, KA helped in improving writing. MA, KA, and JK critically reviewed the final draft. All author(s) read and approved the final draft.

Author details

Mehtab Muhammad Aslam[1*], Kashif Akhtar[2], Joseph K. Karanja[1], Noor-ul-Ain[1] and Fasih Ullah Haider[3]

1 Center for Plant Water-Use and Nutrition Regulation and College of Life Sciences, Joint International Research Laboratory of Water and Nutrient in Cops, Fujian Agriculture and Forestry University, Fuzhou, China

2 Institute of Nuclear Agricultural Sciences, Key Laboratory of Nuclear Agricultural Sciences of Ministry of Agriculture and Zhejiang Province, Zhejiang University, Hangzhou, China

3 College of Resource and Environmental Sciences, Gansu Agricultural University, Lanzhou, China

*Address all correspondence to: 2171916002@fafu.edu.cn

References

[1] Ågren GI, Wetterstedt JM, Billberger MF. Nutrient limitation on terrestrial plant growth–modeling the interaction between nitrogen and phosphorus. New Phytologist. 2012;**194**:953-960

[2] Batjes NA. World dataset of derived soil properties by FAO–UNESCO soil unit for global modelling. Soil Use and Management. 1997;**13**:9-16

[3] Ortiz-Monasterio J, Manske G, Van Ginkel M. Nitrogen and phosphorus use efficiency. In: Reynolds MP, Ortiz-Monasterio JI, McNab A, editors. Application of Physiology in Wheat Breeding. Mexico: CIMMYT Wheat Program; 2001. pp. 200-207

[4] Withers PJ, Rodrigues M, Soltangheisi A, De Carvalho TS, Guilherme LR, Benites VM, et al. Transitions to sustainable management of phosphorus in Brazilian agriculture. Scientific Reports. 2018;**8**:1-13

[5] Aslam MM, Karanja J, Bello SK. *Piriformospora* indica colonization reprograms plants to improved P-uptake, enhanced crop performance, and biotic/abiotic stress tolerance. Physiological and Molecular Plant Pathology. 2019;**106**:232-237

[6] Heppell J, Talboys P, Payvandi S, Zygalakis K, Fliege J, Withers P, et al. How changing root system architecture can help tackle a reduction in soil phosphate (P) levels for better plant P acquisition. Plant, Cell & Environment. 2015;**38**:118-128

[7] Cheng L, Bucciarelli B, Liu J, Zinn K, Miller S, Patton-Vogt J, et al. White lupin cluster root acclimation to phosphorus deficiency and root hair development involve unique glycerophosphodiester phosphodiesterases. Plant Physiology. 2011;**156**:1131-1148

[8] Qin R, Hirano Y, Brunner I. Exudation of organic acid anions from poplar roots after exposure to Al, Cu and Zn. Tree Physiology. 2007;**27**:313-320

[9] Guo W, Zhao J, Li X, Qin L, Yan X, Liao H. A soybean β-expansin gene gmexpb2 intrinsically involved in root system architecture responses to abiotic stresses. The Plant Journal. 2011;**66**:541-552

[10] Vance CP, Uhde-Stone C, Allan DL. Phosphorus acquisition and use: Critical adaptations by plants for securing a nonrenewable resource. New Phytologist. 2003;**157**:423-447

[11] Abel S, Ticconi CA, Delatorre CA. Phosphate sensing in higher plants. Physiologia Plantarum. 2002;**115**:1-8

[12] Calderon-Vazquez C, Ibarra-Laclette E, Caballero-Perez J, Herrera-Estrella L. Transcript profiling of *Zea mays* roots reveals gene responses to phosphate deficiency at the plant-and species-specific levels. Journal of Experimental Botany. 2008;**59**:2479-2497

[13] Kretzschmar T, Kohlen W, Sasse J, Borghi L, Schlegel M, Bachelier JB, et al. A petunia abc protein controls strigolactone-dependent symbiotic signalling and branching. Nature. 2012;**483**:341-344

[14] Mayzlish-Gati E, De-Cuyper C, Goormachtig S, Beeckman T, Vuylsteke M, Brewer PB, et al. Strigolactones are involved in root response to low phosphate conditions in arabidopsis. Plant Physiology. 2012;**160**:1329-1341

[15] Lynch JP. Root phenes for enhanced soil exploration and phosphorus acquisition: Tools for future crops. Plant Physiology. 2011;**156**:1041-1049

[16] Zhao J, Fu J, Liao H, He Y, Nian H, Hu Y, et al. Characterization of root architecture in an applied core collection for phosphorus efficiency of soybean germplasm. Chinese Science Bulletin. 2004;**49**:1611-1620

[17] Gahoonia TS, Nielsen NE. Barley genotypes with long root hairs sustain high grain yields in low-p field. Plant and Soil. 2004;**262**:55-62

[18] Haling RE, Brown LK, Bengough AG, Young IM, Hallett PD, White PJ, et al. Root hairs improve root penetration, root-soil contact, and phosphorus acquisition in soils of different strength. Journal of Experimental Botany. 2013;**64**:3711-3721

[19] Shen J, Li H, Neumann G, Zhang F. Nutrient uptake, cluster root formation and exudation of protons and citrate in lupinus albus as affected by localized supply of phosphorus in a split-root system. Plant Science. 2005;**168**:837-845

[20] Wasaki J, Yamamura T, Shinano T, Osaki M. Secreted acid phosphatase is expressed in cluster roots of lupin in response to phosphorus deficiency. Plant and Soil. 2003;**248**:129-136

[21] Chen J, Liu Y, Ni J, Wang Y, Bai Y, Shi J, et al. Osphf1 regulates the plasma membrane localization of low-and high-affinity pi transporters and determines pi uptake and translocation in rice. Plant Physiology. 2011;**157**(1):269-278

[22] Lambers H, Finnegan PM, Laliberté E, Pearse SJ, Ryan MH, Shane MW, et al. Phosphorus nutrition of proteaceae in severely phosphorus-impoverished soils: Are there lessons to be learned for future crops? Plant Physiology. 2011;**156**:1058-1066

[23] Dinkelaker B, Hengeler C, Marschner H. Distribution and function of proteoid roots and other root clusters. Botanica Acta. 1995;**108**:183-200

[24] Neumann G, Martinoia E. Cluster roots—An underground adaptation for survival in extreme environments. Trends in Plant Science. 2002;7:162-167

[25] Robinson WD, Park J, Tran HT, Del Vecchio HA, Ying S, Zins JL, et al. The secreted purple acid phosphatase isozymes AtPAP12 and AtPAP26 play a pivotal role in extracellular phosphate-scavenging by *Arabidopsis thaliana*. Journal of Experimental Botany. 2012;**63**(18):6531-6542

[26] Wang X, Wang Y, Tian J, Lim BL, Yan X, Liao H. Overexpressing atpap15 enhances phosphorus efficiency in soybean. Plant Physiology. 2009;**151**:233-240

[27] Ai P, Sun S, Zhao J, Fan X, Xin W, Guo Q, et al. Two rice phosphate transporters, osphtl; 2 and osphtl; 6, have different functions and kinetic properties in uptake and translocation. The Plant Journal. 2009;**57**:798-809

[28] Chen A, Gu M, Sun S, Zhu L, Hong S, Xu G. Identification of two conserved cis-acting elements, mycs and p1bs, involved in the regulation of mycorrhiza-activated phosphate transporters in *Eudicot* species. New Phytologist. 2011;**189**:1157-1169

[29] Miao J, Sun J, Liu D, Li B, Zhang A, Li Z, et al. Characterization of the promoter of phosphate transporter tapht1.2 differentially expressed in wheat varieties. Journal of Genetics and Genomics. 2009;**36**:455-466

[30] Qin L, Zhao J, Tian J, Chen L, Sun Z, Guo Y, et al. The high-affinity phosphate transporter gmpt5 regulates phosphate transport to nodules and nodulation in soybean. Plant Physiology. 2012;**159**:1634-1643

[31] Li C, Gui S, Yang T, Walk T, Wang X, Liao H. Identification of soybean purple acid phosphatase genes and their expression responses to phosphorus

availability and symbiosis. Annals of Botany. 2012;**109**:275-285

[32] Xu G-H, Chague V, Melamed-Bessudo C, Kapulnik Y, Jain A, Raghothama KG, et al. Functional characterization of lept4: A phosphate transporter in tomato with mycorrhiza-enhanced expression. Journal of Experimental Botany. 2007;**58**:2491-2501

[33] Al-Abbas A, Barber S. A soil test for phosphorus based upon fractionation of soil phosphorus: I. Correlation of soil phosphorus fractions with plant-available phosphorus. Soil Science Society of America Journal. 1964;**28**:218-221

[34] Shen J, Yuan L, Zhang J, Li H, Bai Z, Chen X, et al. Focus issue on phosphorus plant physiology: Phosphorus dynamics: From soil to plant. Plant Physiology. 2011;**156**(3):997-1005

[35] Hinsinger P. Bioavailability of soil inorganic P in the rhizosphere as affected by root-induced chemical changes: A review. Plant and Soil. 2001;**237**:173-195

[36] Runge-Metzger A. Closing the cycle: Obstacles to efficient P management for improved global food security. Scope-Scientific Committee on Problems of the Environment International Council of Scientific Unions. 1995;**54**:27-42

[37] Rosemarin A, Schroder J, Dagerskog L, Cordell D, Smit A. Future supply of phosphorus in agriculture and the need to maximise efficiency of use and reuse. International Fertiliser Society. Proceedings. 2011;**685**:1-28

[38] Vickers NJ. Animal communication: When i'm calling you, will you answer too? Current Biology. 2017;**27**:713-715

[39] Cordell D, Drangert J-O, White S. The story of phosphorus: Global food security and food for thought. Global Environmental Change. 2009;**19**:292-305

[40] Cramer MD. Phosphate as a limiting resource: Introduction. Plant and Soil. 2010;**334**:1-10

[41] López-Grimau V, Gutierrez M. Decolourisation of simulated reactive dyebath effluents by electrochemical oxidation assisted by uv light. Chemosphere. 2006;**62**:106-112

[42] Batten GD. A review of phosphorus efficiency in wheat. Plant and Soil. 1992;**146**:163-168

[43] Jones G, Blair G, Jessop R. Phosphorus efficiency in wheat—A useful selection criterion? Field Crops Research. 1989;**21**:257-264

[44] Schulthess U, Feil B, Jutzi SC. Yield-independent variation in grain nitrogen and phosphorus concentration among ethiopian wheats. Agronomy Journal. 1997;**89**:497-506

[45] Thomas GV, Shantaram M, Saraswathy N. Occurrence and activity of phosphate-solubilizing fungi from coconut plantation soils. Plant and Soil. 1985;**87**:357-364

[46] Wani P, Khan M, Zaidi A. Co-inoculation of nitrogen-fixing and phosphate-solubilizing bacteria to promote growth, yield and nutrient uptake in chickpea. Acta Agronomica Hungarica. 2007;**55**:315-323

[47] Wani PA, Khan MS, Zaidi A. Chromium reduction, plant growth–promoting potentials, and metal solubilizatrion by *Bacillus* sp. isolated from alluvial soil. Current Microbiology. 2007;**54**:237-243

[48] Khan MS, Zaidi A, Wani PA, Oves M. Role of plant growth promoting rhizobacteria in the remediation of metal contaminated

soils. Environmental Chemistry Letters. 2009;**7**:1-19

[49] Rubio G, Walk T, Ge Z, Yan X, Liao H, Lynch JP. Root gravitropism and below-ground competition among neighbouring plants: A modelling approach. Annals of Botany. 2001;**88**:929-940

[50] McCully M, Boyer J. The expansion of maize root-cap mucilage during hydration. 3. Changes in water potential and water content. Physiologia Plantarum. 1997;**99**:169-177

[51] Ahmed MA, Kroener E, Holz M, Zarebanadkouki M, Carminati A. Mucilage exudation facilitates root water uptake in dry soils. Functional Plant Biology. 2014;**41**:1129-1137

[52] Sharma S, Kumar V, Tripathi RB. Isolation of phosphate solubilizing microorganism (PSMs) from soil. Journal of Microbiology and Biotechnology Research. 2011;**1**:90-95

[53] Andersson H, Bergström L, Djodjic F, Ulén B, Kirchmann H. Topsoil and subsoil properties influence phosphorus leaching from four agricultural soils. Journal of Environmental Quality. 2013;**42**:455-463

[54] Hinsinger P, Gobran GR, Gregory PJ, Wenzel WW. Rhizosphere geometry and heterogeneity arising from root-mediated physical and chemical processes. New Phytologist. 2005;**168**:293-303

[55] Pineros MA, Shaff JE, Manslank HS, Alves VMC, Kochian LV. Aluminum resistance in maize cannot be solely explained by root organic acid exudation. A comparative physiological study. Plant Physiology. 2005;**137**:231-241

[56] Jones DL. Organic acids in the rhizosphere—A critical review. Plant and Soil. 1998;**205**:25-44

[57] Neumann G, Romheld V. The release of root exudates as affected by the plant's physiological status. In: The Rhizosphere. CRC Press, Taylor & Francis Group; 2000. pp. 57-110

[58] Park KH, Lee CY, Son HJ. Mechanism of insoluble phosphate solubilization by pseudomonas fluorescens raf15 isolated from ginseng rhizosphere and its plant growth-promoting activities. Letters in Applied Microbiology. 2009;**49**:222-228

[59] Bar-Yosef B. Root excretion and their environmental effects: Influence on availability of phosphorus. In: Waisel Y et al., editors. Plant Roots, the Hidden Half Plant Roots: The Hidden Half. New York: Marcel Dekker; 1991. pp. 529-557

[60] Dechassa N, Schenk MK. Exudation of organic anions by roots of cabbage, carrot, and potato as influenced by environmental factors and plant age. Journal of Plant Nutrition and Soil Science. 2004;**167**:623-629

[61] Dong D, Peng X, Yan X. Organic acid exudation induced by phosphorus deficiency and/or aluminium toxicity in two contrasting soybean genotypes. Physiologia Plantarum. 2004;**122**:190-199

[62] Sasaki T, Yamamoto Y, Ezaki B, Katsuhara M, Ahn SJ, Ryan PR, et al. A wheat gene encoding an aluminum-activated malate transporter. The Plant Journal. 2004;**37**:645-653

[63] Subbarao GV, Ae N, Otani T. Genotypic variation in iron-, and aluminum-phosphate solubilizing activity of pigeonpea root exudates under P deficient conditions. Soil Science and Plant Nutrition. 1997;**43**:295-305

[64] Yang ZM, Sivaguru M, Horst WJ, Matsumoto H. Aluminium tolerance is achieved by exudation of citric

acid from roots of soybean (*Glycine max*). Physiologia Plantarum. 2000;**110**:72-77

[65] Zheng SJ, Yang JL, He YF, Yu XH, Zhang L, You JF, et al. Immobilization of aluminum with phosphorus in roots is associated with high aluminum resistance in buckwheat. Plant Physiology. 2005;**138**:297-303

[66] Gerke J, Römer W, Jungk A. The excretion of citric and malic acid by proteoid roots of *Lupinus albus* l.; effects on soil solution concentrations of phosphate, iron, and aluminum in the proteoid rhizosphere in samples of an oxisol and a luvisol. Zeitschrift für Pflanzenernährung und Bodenkunde. 1994;**157**:289-294

[67] He M, Dijkstra FA, Zhang K, Li X, Tan H, Gao Y, et al. Leaf nitrogen and phosphorus of temperate desert plants in response to climate and soil nutrient availability. Scientific Reports. 2014;**4**:6932

[68] Brown L, George T, Thompson J, Wright G, Lyon J, Dupuy L, et al. What are the implications of variation in root hair length on tolerance to phosphorus deficiency in combination with water stress in barley (*Hordeum vulgare*)? Annals of Botany. 2012;**110**:319-328

[69] Bustos R, Castrillo G, Linhares F, Puga MI, Rubio V, Pérez-Pérez J, et al. A central regulatory system largely controls transcriptional activation and repression responses to phosphate starvation in arabidopsis. PLoS Genetics. 2010;**6**(9):e1001102. DOI: 10.1371/journal.pgen.1001102

[70] Chiou T-J, Aung K, Lin S-I, Wu C-C, Chiang S-F, Su C-l. Regulation of phosphate homeostasis by microrna in arabidopsis. The Plant Cell. 2006;**18**:412-421

[71] Niu YF, Chai RS, Jin GL, Wang H, Tang CX, Zhang YS. Responses of root architecture development to low phosphorus availability: A review. Annals of Botany. 2013;**112**:391-408

[72] Sato A, Miura K. Root architecture remodeling induced by phosphate starvation. Plant Signaling & Behavior. 2011;**6**:1122-1126

[73] Lynch JP, Brown KM. Topsoil foraging—An architectural adaptation of plants to low phosphorus availability. Plant and Soil. 2001;**237**:225-237

[74] Eticha D, Schenk M. Phosphorus efficiency of cabbage varieties. In: Horst WJ et al., editors. Plant Nutrition. Developments in Plant and Soil Sciences. Vol. **92**. Dordrecht: Springer; 2001. pp. 542-543

[75] Bista DR, Heckathorn SA, Jayawardena DM, Mishra S, Boldt JK. Effects of drought on nutrient uptake and the levels of nutrient-uptake proteins in roots of drought-sensitive and-tolerant grasses. Plants. 2018;**7**(2):28

[76] Delhaize E, Taylor P, Hocking PJ, Simpson RJ, Ryan PR, Richardson AE. Transgenic barley (*Hordeum vulgare* l.) expressing the wheat aluminium resistance gene (taalmt1) shows enhanced phosphorus nutrition and grain production when grown on an acid soil. Plant Biotechnology Journal. 2009;**7**:391-400

[77] Lu C, Thompson CB. Metabolic regulation of epigenetics. Cell Metabolism. 2012;**16**:9-17

[78] Sakr S, Wang M, Dédaldéchamp F, Perez-Garcia M-D, Ogé L, Hamama L, et al. The sugar-signaling hub: Overview of regulators and interaction with the hormonal and metabolic network. International Journal of Molecular Sciences. 2018;**19**:2506

[79] Gu M, Chen A, Sun S, Xu G. Complex regulation of plant phosphate transporters and the gap between molecular mechanisms and practical application: What is missing? Molecular Plant. 2016;**9**:396-416

[80] Yang XJ, Finnegan PM. Regulation of phosphate starvation responses in higher plants. Annals of Botany. 2010;**105**:513-526

[81] Zhang Z, Zhang Y, Tan H, Wang Y, Li G, Liang W, et al. Rice morphology determinant encodes the type ii formin fh5 and regulates rice morphogenesis. The Plant Cell. 2011;**23**:681-700

[82] Gamuyao R, Chin JH, Pariasca-Tanaka J, Pesaresi P, Catausan S, Dalid C, et al. The protein kinase pstol1 from traditional rice confers tolerance of phosphorus deficiency. Nature. 2012;**488**:535-539

[83] Wang L, Guo M, Li Y, Ruan W, Mo X, Wu Z, et al. Large root angle1, encoding ospin2, is involved in root system architecture in rice. Journal of Experimental Botany. 2018;**69**:385-397

[84] Yu ZM, Bo K, He XW, Lv SL, Bai YH, Ding WN, et al. Root hair-specific expansins modulate root hair elongation in rice. The Plant Journal. 2011;**66**:725-734

[85] Huang Y, Zeng Y, Yu Z, Zhang J, Feng H, Lin X. In silico and experimental methods revealed highly diverse bacteria with quorum sensing and aromatics biodegradation systems—A potential broad application on bioremediation. Bioresource Technology. 2013;**148**:311-316

[86] Wang T, Li C, Wu Z, Jia Y, Wang H, Sun S, et al. Abscisic acid regulates auxin homeostasis in rice root tips to promote root hair elongation. Frontiers in Plant Science. 2017;**8**:1121

[87] Pizzio GA, Paez-Valencia J, Khadilkar AS, Regmi K, Patron-Soberano A, Zhang S, et al. Arabidopsis type I proton-pumping pyrophosphatase expresses strongly in phloem, where it is required for pyrophosphate metabolism and photosynthate partitioning. Plant Physiology. 2015;**167**:1541-1553

[88] Regmi P, Holgate B, Fredericks D, Miller MW, Wett B, Murthy S, et al. Optimization of a mainstream nitritation-denitritation process and anammox polishing. Water Science and Technology. 2015;**72**:632-642

[89] Yang H, Zhang X, Gaxiola RA, Xu G, Peer WA, Murphy AS. Over-expression of the arabidopsis proton-pyrophosphatase avp1 enhances transplant survival, root mass, and fruit development under limiting phosphorus conditions. Journal of Experimental Botany. 2014;**65**:3045-3053

[90] Lan P, Li W, Schmidt W. 'Omics' approaches towards understanding plant phosphorus acquisition and use. Annual Plant Reviews online. 2018:65-97

[91] Rai A, Saito K, Yamazaki M. Integrated omics analysis of specialized metabolism in medicinal plants. The Plant Journal. 2017;**90**:764-787

[92] Yang H, Knapp J, Koirala P, Rajagopal D, Peer WA, Silbart LK, et al. Enhanced phosphorus nutrition in monocots and dicots over-expressing a phosphorus-responsive type I H^+-pyrophosphatase. Plant Biotechnology Journal. 2007;**5**:735-745

[93] Gao N, Su Y, Min J, Shen W, Shi W. Transgenic tomato overexpressing ath-mir399d has enhanced phosphorus accumulation through increased acid phosphatase and proton secretion as well as phosphate transporters. Plant and Soil. 2010;**334**:123-136

[94] Jia H, Ren H, Gu M, Zhao J, Sun S, Zhang X, et al. The phosphate transporter gene ospht1; 8 is involved in phosphate homeostasis in rice. Plant Physiology. 2011;**156**:1164-1175

[95] Wu Z, Ren H, McGrath SP, Wu P, Zhao F-J. Investigating the contribution of the phosphate transport pathway to arsenic accumulation in rice. Plant Physiology. 2011;**157**:498-508

Chapter 16

Role of Plant Carbonic Anhydrases under Stress Conditions

Natalia N. Rudenko, Maria M. Borisova-Mubarakshina, Lyudmila K. Ignatova, Tatiana P. Fedorchuk, Elena M. Nadeeva-Zhurikova and Boris N. Ivanov

Abstract

Carbonic anhydrases (CAs) are enzymes catalyzing the reversible hydration of carbon dioxide with the generation of protons and bicarbonate. The components of the reaction are involved in almost all metabolic processes in higher plants and algae, maintaining the balance of electrolytes and pH, gluconeogenesis, lipogenesis, ethylene synthesis, and others. The CAs may take part in transmitting signals to activate cascades of protective response genes. Our findings reveal significant changes in the content of carbonic anhydrase gene transcripts in response to changes in environmental conditions. Here we discuss the functions of CAs located in the plasma membrane, chloroplast envelope, chloroplast stroma, and in thylakoids in plant protection under stress conditions, such as high illumination, low and high concentration of carbon dioxide in the environment, drought, and salinity.

Keywords: carbonic anhydrase, plants, chloroplasts, thylakoids, high illumination, carbon dioxide

1. Introduction

Carbonic anhydrases (CAs) are the group of Zn-containing enzymes that are the biological catalysts accelerating both the carbon dioxide hydration reaction and the bicarbonate dehydration reaction:

$$CO_2 + H_2O \leftrightarrow HCO_3^- + H^+$$

In the absence of CA, these reactions proceed relatively slowly to ensure the physiological needs of the cell. The CAs were found in cells of all living organisms: prokaryotes, fungi, plants, and animals. Notably, both binding and release of protons in this reaction are important for many biochemical processes in cellular metabolism. All metabolic processes in higher plants and algae, including the electrolyte balance and pH, gluconeogenesis, lipogenesis, ethylene synthesis, and others depend on/require the components of this reaction. In this review, we focus on the functioning of plant CAs, including algae and angiosperms with different types of CO_2 fixation: on C3 and C4 organic matter as well as under normal and stress conditions respectively.

CO_2 is the main source of carbon in higher plants and algae cells. The compounds with a common name “inorganic carbon” (Ci) were found in nature both in the form of an unhydrated or hydrated carbon dioxide molecule and in the form of a bicarbonate or carbonate ion. The content of various forms of Ci in solution is pH-dependent. CO_2 prevails at pH value lower than 6.4, HCO_3^- at pH between 6.4 and 10.3, and CO_3^{2-} at a pH of 10.3 and higher [1]. The substrate for the enzyme ribulose-1,5-bisphosphate carboxylase/oxygenase (Rubisco), which carries out the first Ci fixation reaction in the Benson-Calvin cycle, is the CO_2 molecule. This poses difficulties for aquatic photosynthetic organisms, since at pH of 8.0–8.3 in seawater about 95% of Ci is in the form of HCO_3^-, whereas the dissolved CO_2 represents only a small part of Ci. Thus, the main stress that the aquatic photosynthesizers, namely cyanobacteria and algae, especially in the habitats with alkaline water in seas and oceans, have to face is a low CO_2 content in their environment. Therefore, these organisms have evolved the mechanisms of Ci concentrating to increase CO_2 content near Rubisco. The main role in supplying Ci and converting it from HCO_3^- to CO_2 is played by CAs located in various cellular compartments of cells in aqueous photosynthesizers [2].

The terrestrial higher plants have adapted to survival under constant stress conditions, such as high temperature, high illumination, and low soil moisture, leading to the closure of stomata. They do not suffer from CO_2 deficiency since another type of Ci concentrating mechanism, C4 type of photosynthesis, has evolved. The first reaction that ensures the concentrating of Ci in plants of C4 species is converting CO_2 to bicarbonate, which is used by phosphoenolpyruvate carboxylase in forming a four-carbon product. It has been established that the CA, which is mainly located in cytoplasm in C4 plants, plays a key role in this transformation. In C4 species, the suppression of the synthesis of these CAs leads to dramatic effects in these plants [3, 4].

In C3 plant leaves, the existence of a Ci-concentrating mechanism has not been stated. At the same time, the story is much more complicated since C3 plants have a large number of genes encoding CAs, the physiological role of which has not been established yet. It is still not entirely clear whether all these genes are expressed, since the expression of a number of CA genes is induced under certain conditions, such as low CO_2 [5] and high illumination [6] and osmotic stress [7].

In chloroplasts, performing C3 photosynthesis, at least six CAs have been discovered to date, both soluble and membrane-bound [8], and belonging to different CA families. In the review, we discuss the possible role of two soluble CAs in photosynthesis, one of them was known for a long time [9] and another one discovered rather recently [10]. Both of them are situated in the chloroplast stroma, where carbohydrates from Ci are formed. Thus, here we discuss the functional relationship of these CAs with Rubisco and their response to stress factors. Recent studies show that the CAs are important not only for photosynthesis and for a number of metabolic pathways, including the interconversion of Ci forms, but also they are necessary under certain stress conditions. In this case, changes (fluctuations) in the CA activity can orchestrate the intensity of certain metabolic processes including the rate of photosynthesis [11–13].

Recently, a functional relationship between cytoplasmic and chloroplast CAs and aquaporins has been found [14–16]. The last ones are the protein channels that facilitate the transport not only of water, but also of CO_2, and even of H_2O_2 [16], known as a signal molecule in retrograde signaling.

The enzymatic activity and the content of thylakoid CAs also change under stress conditions. CA activity in thylakoids of higher plants was detected in granal thylakoid membranes, enriched with PSII, as well as in lamellar membranes, enriched with PSI and ATPase in thylakoid lumen [17–20]. Incorporation of these CAs in plant defense systems of higher plants under changing environmental conditions is reviewed.

2. Modern classification of carbonic anhydrases

Based on the conservative nucleotide sequences in the genes encoding CAs, they could be classified into nine evolutionarily independent families (α, β, γ, δ, ζ, ε, η, θ and the recently discovered ι-CAs) [21–24]. Some researchers do not ascribe ε-CAs to a separate family, since these CAs are highly modified β-CAs [25]. Although these enzymes have completely different primary, tertiary, and quaternary structures and differ in the organization of the active center, they all are called CAs because they catalyze the same reaction using similar catalysis mechanisms.

Representatives of rare and small δ-, ζ-, θ-, and ι-CA families in eukaryotes were found in diatoms and some other unicellular microalgae, whereas α-, β-, and γ-CAs were discovered in most algae species and in all higher plants, including mosses and Lycopodium.

2.1 α-CAs

Since all CAs in human cells belong to the α-CA family, this family is the most studied and widespread. Representatives of this family were found in eubacteria [26, 27], ascomycetes [28], algae [29, 30], higher plants [5, 31], and animals [32]. In green algae *Chlamydomonas reinhardtii*, there are three α-CAs: CAH1, CAH2 [30, 33] in the periplasm, and CAH3 found on the luminal side of thylakoid membranes [34]. Eight genes encoding α-CAs were found in *Arabidopsis thaliana* genome. α-CA1 was discovered in chloroplast stroma [10], and α-CA4 was found during proteomic studies among the proteins of the thylakoid membranes [35, 36]. Evidence suggests that α-CA2 is present in thylakoid membranes [37]. α-CA3 was found in flowers and pods [5], as well as during proteomic analysis of mature pollen proteins [38, 39].

All functionally active α-CAs contain three residues of histidine, which are conservative in all active α-CAs. These histidines are the ligands of Zn atom [40, 41] located at the bottom of the conical cavity of the active center. Histidines in α-CAs' structure enhance the net positive charge of the metal ion, which is essential to achieve effective catalysis [42]. Most α-CAs are the monomers with molecular mass of about 30 kDa, although the periplasmic CA in *Chlamydomonas reinhardtii*, CAH1, is a heterotetramer with two 27-kDa subunits and two 4-kDa subunits connected by disulfide bridges [43].

2.2 β-CAs

The representatives of β-CAs were found in archaebacteria [44], cyanobacteria [45], eubacteria [46], chemoautotrophic bacteria [47], fungi [48], algae [49], higher plants with C3- and C4-type photosynthesis, dicotyledons and monocotyledons [50, 51]. In *Chlamydomonas reinhardtii* cells, there are six β-CAs located in mitochondria, chloroplast stroma, and cytoplasm [2]. The analysis of the Arabidopsis genome has shown the presence of six genes encoding β-CA; moreover, as it was found by Fabre et al. [5], all of these genes are expressed. The same group of researchers using the method of gene fusion with a green fluorescent protein gene confirmed that the two most active CAs are located in stroma and cytoplasm. These CAs were discovered by Atkins et al. in 1972 [9] in the soluble fractions of higher plant leaves of different species. Fabre et al. [5] have found these CAs in the chloroplast stroma and in the cytoplasm. They were called β-CA1 and β-CA2, correspondingly. Other β-CAs were also located in Arabidopsis cells: β-CA3, in the cytoplasm, β-CA4 in plasma membranes, β-CA5 in chloroplasts, and β-CA6 in the mitochondrial matrix.

The enzymes of this family are as effective in catalysis as α-CAs are. One histidine and two cysteine residues are zinc ligands in the active center of β-CAs [52, 53]. The active site in β-CAs' structure is a tight pocket and the only access to it from the bulk solvent is via a bottleneck between Gln, Gly, Asp, and Tyr residues. This bottleneck is too narrow for anything larger than a water molecule, meaning that to implement the catalytic cycle some rearrangement should take place [54]. The study of the structure of β-CAs revealed the presence of the so-called non-catalytic bicarbonate binding site [55], which represents a "pocket" located 8 Å from the active center zinc. A network of at least seven molecules linked together by hydrogen bonds is capable of holding a bicarbonate molecule. Studies have shown that this is not just an anion-binding site, the addition of bicarbonate leads to the reorganization in the protein molecule and in the active center optimizing its functioning [56]. The hydrogen bonds of the Trp, Arg, and Tyr residues in this site are well organized to recognize the bicarbonate ion such anions as acetate or nitrate are not able to carry out all the necessary interactions with the hydrogen bonds of these amino acid residues. Tyr residues in the active site and bicarbonate binding site structures were recently found to be the key amino acids for the reversible modification by phosphorylation and nitration in response to abiotic and biotic stresses [11, 12]. Such modifications allow for blocking the passage of substrate and inhibiting the activity of β-CAs under stress conditions.

2.3 γ-CAs

γ-CAs are present in cells of bacteria, green algae, diatoms [2], and higher plants [57]. In *C. reinhardtii,* three γ-CAs are situated in mitochondria [58]. In *A. thaliana*, the CA domain consisting of five γ-CAs was discovered as a part of mitochondrial complex l [57]. Each of these CAs is encoded by a separate gene [59]. Three CAs of the domain, γ-CA1, γ-CA2, and γ-CA3, are close in structure to the first discovered γ-CA from *Methanosarcina thermophila* (Cam). Two more CAs were called γ-CAL1 and γ-CAL2 ("gamma carbonic anhydrase like") since they had sequences less similar to those of Cam for readability [60].

γ-CAs function as trimers consisting of identical subunits [61], which contain one zinc atom per subunit; however, unlike α and β-CAs, the active center is located between the subunits. In the active center of γ-CAs, there are three histidine and one H_2O residues coordinating Zn atom, like in α-CAs, but these are histidines of two opposite subunits. The mechanism of catalysis is similar to that of α-CAs [62].

3. The participation of CAs of aquatic photosynthesizers in the CO_2-concentrating mechanism

To increase the CO_2 content near Rubisco during the evolutionary adaptation to growth conditions, some groups of plants, that is, aquatic photosynthesizers, developed CO_2 concentrating mechanisms (CCMs) in their cells. All currently known CCM pathways require at least one or, more often, several CAs.

In cyanobacteria, negatively charged HCO_3^- penetrates the cells through specialized transporters on the plasma membrane, utilizing the energy of ATP and NADPH [63]. The dissolved Ci is captured with the help of extracellular CA, supplying HCO_3^- to bicarbonate transporters and preventing leakage of Ci from the cell [45]. The other CAs situated in cyanobacterial carboxysomes accelerate the formation of CO_2 for Rubisco. The last one is also placed in carboxysomes, which serve as a barrier to CO_2 leakage.

In algae cells, Ci should cross the cell wall, plasma membrane, and chloroplast membrane to reach the place of CO_2 fixation. Like in cyanobacteria, CAs in algae cells play a crucial role in CCM, which is important under conditions of low CO_2 in water. Periplasmic CA in *C. reinhardtii*, the so-called CAH1, is the most important CA in CCM [2]. This enzyme is intensively expressed at a low concentration of CO_2 in water. If CAH1 activity is inhibited, photosynthesis in *C. reinhardtii* cells is also suppressed at high pH in media, when most Ci is in the form of bicarbonates [64]. This result suggests that CAH1 facilitates the CO_2 entry into the cell. Other CAs are also involved in the delivery of carbon dioxide into the cell, namely, CAH8 and CAH9. CAH8 appears to be localized in the plasma membrane and this protein might also facilitate the diffusion of CO_2 across the plasma membrane by facilitating the conversion of HCO_3^- to CO_2 at the cell surface [65]. Likewise, CAH9 might mediate the movement of CO_2 across the cytoplasmic region to the chloroplast. One more CA, CAH3, located in thylakoid membranes of chloroplasts in *C. reinhardtii* cells is also important for algae survival under conditions of low carbon dioxide content in water. The fact that *C. reinhardtii* cells with inhibited synthesis of CAH3 were not able to grow at a normal CO_2 concentration indicates that CAH3 is involved in photosynthesis [66, 67]. However, the expression of CAH3 is constitutive; the concentration of CO_2 had no effect on the transcription level of the gene encoding this enzyme and the role of CAH3 in CCM is still not accepted.

4. CAs in higher plants

4.1 The role of CAs in photosynthesis of C4 higher plants

The plants constantly growing under stress conditions, such as drought, salinity, or high temperature, have to keep stomata closed most of the day to reduce water loss as a result of transpiration. At the same time, to avoid starvation in the absence of CO_2 in leaves, a C4 type of photosynthesis has evolved. It includes an additional carbon conversion cycle, called the Hatch-Slack cycle. CCM exists in the form of the so-called four-carbon (C4) photosynthesis with the spatial separation of the primary carboxylation reactions and the Calvin cycle. The Hatch-Slack cycle allows Ci to be concentrated in leaf tissues, carrying out the primary fixation of carbon dioxide through the carboxylation of phosphoenolpyruvate using the enzyme phosphoenolpyruvate carboxylase (PEPC). In the cytosol of the mesophyll cells of C4 plants, there is a high amount of CA [68]. This enzyme plays a decisive role in C4 photosynthesis. CA catalyzes the first reaction of C4 pathway, increasing its rate by 10^4 times, due to providing bicarbonate to PEPC [3]. After that, four-carbon acids produced as a result of PEPC activity are decarboxylated in bundle-sheath cells. This leads to an increase in carbon dioxide concentration around Rubisco, which is located in the chloroplasts of the bundle-sheath cells in C4 plants [69]. Interestingly, the Genomic Southern analysis by Tetu et al. in 2007 [70] revealed the presence of two forms of CA from the β-family in the cytoplasm in the cells of *Flaveria bidentis* leaves. The only one of them, the abundant CA, located in the cytoplasm, plays the described role in C4 photosynthesis. The other CA, which is also a cytosolic CA isoform, is not directly involved in C4 photosynthesis. This CA was suggested to be the housekeeping form of the enzyme supplying bicarbonate for such anaplerotic processes, as replenishment of tricarboxylic acid cycle intermediates, carbon for amino acid biosynthesis, seed maturation, and pH balance [71, 72]. The transformation of the gene encoding the abundant cytosolic CA in *Flaveria bidentis* cells with an antisense construct confirmed the important role this enzyme plays in the C4 photosynthetic pathway. Some of the primary transformants had impaired CO_2

assimilation rates and required a high level of CO_2 for growth. In the mutants with the CA activity, less than 10% of the WT CO_2 assimilation rate was very low and these transformants grew poorly at ambient CO_2 in the atmosphere. Reduced CA activity also increased the partial pressure of carbon dioxide required to saturate the assimilation rate of CO_2 [4].

4.2 The role of CAs in photosynthesis of C3 higher plants and in the plant defense systems under stress conditions

4.2.1 CA in the plasma membrane and envelope chloroplast membrane in C3 plants

The mechanisms of CO_2 penetration inside leaf cells have been studied for decades. Chemically, CO_2 is a lipophilic compound and it should easily diffuse through membranes [73, 74]. However, biological membranes appear to have a low CO_2 permeability [75] due to a high level of protein and sterol content [74, 76]. Recently, using plasma membrane vesicles of pea leaves it has been demonstrated that plasma membrane aquaporins could facilitate CO_2 transport [77]. The same data were earlier presented for tobacco, Arabidopsis, fava bean, and others [78–80]. The plant aquaporins are the plasma membrane intrinsic proteins (PIPs). They are mainly represented by two groups: PIP1 and PIP2, which possess some structural differences in the N and C terminal end and are subdivided into several subgroups [81, 82]. Each aquaporin has different transport properties for H_2O, CO_2, and solutes; however, the functional interaction between PIP1 and PIP2 was proposed to occur with the emphasis on their coupling under stress conditions [82]. Using the inhibitor analysis, the presence of CA located in the plasma membrane of photosynthetic pea leaf cells was proved [83] and the importance of the CA functioning in the plasma membrane for Ci transport into higher plant leaves was confirmed [84]. Later, the presence of β-CA4 in the plasma membrane was shown [5].

Wang et al. in 2016 [14] identified β-CA4 as an interactor with aquaporin PIP2;1 in Arabidopsis. This connection allows CO_2 permeability across the plasma membrane to be facilitated. Further, the decrease in the CA concentration was shown to lead to a lower CO_2 permeability in plasma membrane vesicles [77]. Therefore, it can be proposed that the functioning of plasma membrane aquaporins depends on the CA activity under both normal and stress conditions. The roles of aquaporins and CA in plant defense responses against biotic and abiotic stress factors are broadly discussed [85]. Aquaporins should interact with CAs at the membrane-liquid phase interfaces, and a term that describes their combined functioning as "cooporin" was proposed [86]. The extent of water and CO_2 permeability via aquaporins of the plasma membrane is consistent with the expression levels of both PIPs and CA when exposed to environmental challenges. Experimental evidence suggests that aquaporins and CA are mutually involved in the regulation of CO_2 in stomatal and mesophyll conductance in leaves of higher plants [15] that is crucial for both processes: acclimation to stress and recovery after stress. A number of studies conducted using genetic transformation methods showed that changes in the content of aquaporins per unit area of leaves lead to the corresponding changes in mesophyll conductance [87–89].

In tobacco plants, aquaporin NtAQP1 has been identified as a membrane CO_2 pore not only in the plasma membrane but also in the inner chloroplast envelope membranes [79]. Therefore, it was proposed that aquaporins may be involved in CO_2 transfer in both the plasma membrane and in the chloroplast envelope [75]. Along with PIP, various forms of tonoplast intrinsic proteins (TIPs) and other intrinsic proteins were described in the Arabidopsis envelope fraction [90]. We have previously provided data showing the presence of the CA activity, associated with the isolated chloroplast envelope [16]. CA activity was also detected in chloroplast envelope membranes of *Chlamydomonas reinhardtii* and shown to be induced under conditions of low inorganic carbon concentrations [91]. Perez-Martin et al. [15] have studied the relationship between the functioning of envelope aquaporins and CA of the β-family (Olea_CA) in leaves of *Olea europaea*, which is a drought-tolerant plant. By studying the change in the expression intensity of genes encoding these proteins in the leaves of olive plants grown under conditions of sufficient moisture, drought, and during the recovery period after drought, the authors made the conclusion that these proteins can function together by supplying carbon dioxide to the stroma chloroplast. In this case, the authors presume that Olea_CA takes part in the joint functioning with envelope aquaporins, proceeding from the assumption that this CA corresponds to β-CA1. Moreover, they also demonstrated that the nucleotide sequence of the Olea_CA gene corresponds to encoding the other CA in *A. thaliana*, β-CA5. This CA was discovered by Fabre et al. in 2007 [5] in Arabidopsis chloroplasts, but the exact location of this enzyme in the chloroplast is still unknown. What is important, β-CA5 gene is the only CA encoding gene, the knockout of which leads to death when growing under ambient CO_2. The seeds of these mutants could germinate only at high CO_2 concentrations, although they were much worse developed as compared to WT plants (J. Moroney, personal communication). Based on the above, it seems more likely that not the stromal β-CA1 but β-CA5 (or one of the isoforms of β-CA5) located in the chloroplast envelope membrane may function mutually with aquaporins.

CA protects the cells from H_2O_2-induced apoptosis [92]. It is known that among various chloroplast signals H_2O_2 plays a major role in various signaling pathways under stress conditions [93, 94]. An essential factor to implement the retrograde signal (the signal from the organelle to the nucleus) is the ability of H_2O_2 to diffuse over long distances from the place of formation to the place of signaling. Earlier, we demonstrated that H_2O_2 that was produced inside chloroplasts diffused from a chloroplast to cytoplasm through chloroplast envelope membranes and the amount of H_2O_2 outside the chloroplasts increased under the conditions of ascorbate peroxidase inhibition [95]. Using acetazolamide (AZA), which is known as CA inhibitor, as a non-specific aquaporin inhibitor, we have shown that aquaporins facilitate the diffusion of H_2O_2 molecules through the chloroplast membrane [16]. AZA was established to be an efficient inhibitor of aquaporins through interacting with the guanidyl group of Arg, backbone carbonyl of Gly, carboxyl of Asp, Ser, His, Ile and Asn of aquaporins [96–98]. Taking into consideration that AZA also inhibits the activity of CA, we cannot exclude that inhibition of H_2O_2 diffusion through the envelope membrane in the presence of AZA was the only result of blocking of the aquaporins. This inhibition could also result from inhibiting the envelope CA, especially if this CA is attached to aquaporins in the envelope. If this is the case, the inhibitory effect of AZA on the H_2O_2 diffusion could be a consequence of AZA binding to envelope CA, leading to the conformational changes of CA with subsequent conformational changes of aquaporin proteins and therefore blocking the envelope aquaporins. Thus, the data in [16] can represent the evidence of the joint functioning of the envelope CA with aquaporins in diffusing hydrogen peroxide through the chloroplast envelope. Considering the presence of CA of β-family in the chloroplast envelope (see above) and the facilitation of not only CO_2, but also H_2O_2 diffusion through the envelope by the functioning of envelope aquaporins, several propositions on the mechanisms of the CA incorporation in signaling under stress conditions can be made. One of them is that CA is involved in the cascades of mitogen-activated protein kinases (MAPKs), which are serine-threonine kinases mediating intracellular signaling through changes in the redox state of their cysteine residues. Since cysteine residues are the main target of hydrogen peroxide in MAPK cascades [99], and two cysteine residues are located in the β-CAs active center, we can assume that H_2O_2 changes the redox status of cysteine residues of β-CAs, resulting in incorporation of CA in the global MAPK signaling network in plant cells.

It was quantified that the amount of water in chloroplast stroma and thylakoid lumen was much lower than it is enough to perform photosynthetic water oxidation per day. Therefore, the existence of aquaporins in the thylakoid membrane was also suggested [100]. This question is still under debate, some researches have detected some forms of TIPs in the thylakoid membranes of *Arabidopsis thaliana* [101, 102]. However, there is still little direct experimental evidence demonstrating the existence of aquaporins in thylakoid membranes and this question needs to be clarified in the future.

4.2.2 CA in mitochondria and the possible role of CA in chloroplast-mitochondria communication for activation of the CO_2-concentrating-like mechanism

C3 higher plants are believed to lack any mechanisms of CO_2 concentrating. The carbon dioxide content in a liquid phase in leaves is not very high, about 11 μM at the current concentration of CO_2 in the atmosphere. Under the illumination pH value of the chloroplast stroma, where Rubisco is located in higher plant cells, increases up to 7.7. The bicarbonate content there reaches 230 μM at 25°C. This effect can be considered as inorganic carbon concentrating.

For higher plants, especially when they grow under stress conditions, such as high light or high temperature, Zabaleta et al. [103] supposed the existence of CMM as in plants in such conditions stomata are closed leading to the decrease of Ci content in chloroplasts. At the same time, a large amount of CO_2 is released in mitochondria due to the reactions of both the tricarboxylic acid cycle and photorespiration. The authors supposed that CO_2 from mitochondria can be used by chloroplasts in the Calvin cycle. Possibly, CO_2 may transfer into chloroplast by diffusion, harnessing some mechanisms of an active HCO_3^- transport would be more efficient. In the genome of *A. thaliana*, five genes encoding γ-CA, which form part of the mitochondrial complex I, were found [57, 104]. The CAs of the three subunits of the complex have a fully active center (γ-CA1, γ-CA2, and γ-CA3), and the CAs of two other subunits, in which some conservative amino acid residues forming the active center are absent, are called CA-like proteins named γ-CA4 and γ-CA5. The domain consisting of five γ-CAs attached to mitochondrial complex I probably plays a role in the conversion of CO_2 to HCO_3^- and/or even in a transfer of HCO_3^- from mitochondria to cytosol. Possibly, one more CA, β-CA6, located in the mitochondrial matrix, can also participate in the formation of HCO_3^-. The last one can be subsequently transferred from mitochondria to chloroplasts by a putative bicarbonate carrier in the chloroplast membrane or C4-like pathway; however, the mechanism of such transfer is yet unknown [103].

Interestingly, in double mutants without γ-CA4 and γ-CA5, light-dependent activation of the key enzyme for the synthesis of anthocyanins, chalcone synthase, was observed [105]. The authors suggested that both these CAs play an important role in the growth and development of *A. thaliana* in conditions of high illumination.

4.2.3 Carbonic anhydrases in chloroplasts of C3 plants

The most known and the most studied CA in chloroplasts of C3 higher plants is the stromal CA, which belongs to β-family. This CA was named β-CA1 by Fabre et al. in 2007 [5] in the study of CAs of β-family in Arabidopsis leaves. Later, this name started to be used to designate the corresponding stromal CAs in leaves of other plants [12, 15]. This enzyme is the second after Rubisco in terms of the amount of protein in the cell (0.5–2% of the total) [106]. The rate of the spontaneous interconversion of Ci forms is low, which implies the role of CA in accelerating the supply of CO_2 to carboxylation centers; however, direct data on the role of the enzyme in this process are lacking.

In some studies, no association of CA activity with photosynthesis was found. Growing plants under conditions of zinc deficiency showed that the rate of photosynthesis in these plants remained almost unchanged with a sharp decrease in CA activity [107]. In transgenic plants of *Nicotiana tabacum* containing 10% or less soluble β-CA activity compared to WT plants, there were no significant differences in Rubisco activity, chlorophyll content, stomatal conductivity, dry weight per unit leaf area, and in the ratio of the partial pressure of intracellular and external CO_2 in comparison with WT plants [108]. However, in these mutants, the carbon isotopic composition of the leaf dry matter was changed.

Another group of researchers has found that plants compensated for a decrease in CA activity by increasing the permeability of stomata, which, however, has led to a higher rate of water loss [109]. Plant growth under conditions of nitrogen deficiency in the soil led to a significant decrease in the activity of soluble CA in leaves of sugar beet plants, and upon restoration of nitrogen nutrition, a gradual reverse increase in CA activity by 80% of the initial values has been observed (Novichkova, personal comunication). Importantly, many researchers have discovered the convincing evidence of the functional relationship of Rubisco and stromal β-CA. In *Phaseolus vulgaris* plants grown under the increased CO_2 content in the air, a significant decrease in the activities of both CA and Rubisco has been observed [110]. The activity of these enzymes and the content of transcripts of the genes encoding them were reduced in *Pisum sativum* plants grown at 1000 ppm CO_2, compared with plants grown at atmospheric carbon dioxide concentration; the transfer of pea plants grown at a carbon dioxide concentration of 1000 ppm to the conditions of normal CO_2 content in the atmosphere led to a rapid increase in the expression level of CA and Rubisco genes, following which the activity of the corresponding enzymes in the leaves increased [50]. Immunocytolocalization experiments indicated that CA is a neighbor of Rubisco in the stroma of pea leaves' chloroplast [111]. One CA in the stroma of chloroplasts was later discovered to be associated with Rubisco on the outside of the thylakoid membrane [112]. In the early stages of the adaptation of sugar beet plants to a high concentration of carbon dioxide, of 700 ppm, a decrease in both the activity of the soluble and of the membrane-bound CAs was observed. However, the activity of Rubisco was the same in plants grown in conditions of high CO_2 concentration and in the ambient carbon dioxide content [113].

In the study of the drought tolerance mechanisms in *Brassica napus* [12], β-CA1 was identified using mass-spectrometry analysis as a protein interacting with isoforms of the large Rubisco subunit in several protein spots obtained by 2D electrophoresis of the proteins of the rapeseed plant leaves. The content of β-CA1 in these spots was higher in the samples from plants after drought treatment in comparison to those from the control plants. RT-PCR analysis of the expression level of the gene encoding β-CA1 has shown a similar trend at the transcriptional and translational levels meaning that both the expression level and the protein level increased under drought stress in leaves of *Brassica napus* plants.

Nevertheless, the activity of CA in leaves of plants exposed to water deficit was lower than in the control plants grown under normal conditions [12] due to the phosphorylation of several amino acid residues in the active site and substrate-binding site of β-CA1. The same mechanism of inhibition of CA activity was observed by studying the action of high-temperature stress in the leaves of *Helianthus annum* [11]. The key suppression mechanism of this activity was the nitration of tyrosine residues in the structure of β-CA. Nitration, as well as phosphorylation of Tyr, blocks the passage of substrate to active site cavity.

It is not entirely clear why the content of CA protein increases with the simultaneous suppression of CA activity. It can be assumed that plants apply this mechanism to prepare for normal moisture conditions. The answer to this question is very important to determine the physiological role of β-CA1 stroma in higher plant metabolism. Although the participation of β-CA1 in photosynthesis is not considered to be proved, the described facts support this suggestion. The fact that it turned out to be one of the main differentially abundant proteins in response to drought testifies to the assumption that β-CA1 supplies CO_2 for Rubisco activity by catalyzing the reversible reaction of bicarbonate to carbon dioxide and thus regulating the rate of photosynthesis under stress condition. Water deficit is known to cause stomatal closure with a reduction in plant photosynthetic efficiency and inhibition of Rubisco activity [114]. These responses often result in a change in photosynthetic and energy metabolism-associated protein accumulation. The inhibition of the activity of β-CA1 under stress by phosphorylation and nitration [11, 12] may play an important role in regulating the photosynthetic process in response to stress.

Herewith Yu et al. [7] showed that in rice seedlings the expression of the gene encoding CA of β-family (OsCA1) as well as the total CA activity were upregulated by osmotic stress, in particular, by salt stress. Moreover, the same group showed that *Arabidopsis thaliana* mutant plants over-expressing OsCA1 had a greater salt tolerance. These data imply that the OsCA1 has an important role in the response of plants to environmental stress conditions. In the rice genome among salt tolerance genes, the *LOC_Os09g28910* gene encoding CA has been recently identified [115]. This gene, as well as the gene encoding OsCA1, had the chloroplast precursor sequence, which implies the location of both CAs in the chloroplasts in rice.

The experimental data presented, however, show that the role of β-CA1 in photosynthesis, as well as the participation of this enzyme in regulating the rate of photosynthetic processes under stress conditions, is still controversial. This contradiction can be explained by the suggestion that two stromal CAs, β-CA1 and α-CA1, participate in CO_2 supply to Rubisco. The latter was detected in the chloroplast stroma in the study of the pathway of newly synthesized proteins into the chloroplast through the endo-membrane system of the Golgi apparatus [10]. The data on the decrease in photosynthetic activity, as well as the ability to accumulate starch in plants with a knockout of the gene encoding α-CA1 [116], may indicate the role of α-CA1 in photosynthesis and the possibility that this CA is involved in CO_2 supply to Rubisco. Using primers designed for possible alternative splicing of the template RNA of the *At3g01500* gene encoding β-CA1, two isoforms were found to be present in the leaves of Arabidopsis [6]. An increase in plant illumination led to a change in the content of transcripts of most genes encoding CA of chloroplasts, and a significant difference was observed in the expression intensity of α-CA1 genes and two forms of β-CA1. The opposite effect of increasing the light intensity on the content of transcripts of two RNA forms of a gene encoding β-CA1, an increase in the content of transcripts of one form and a decrease in the other, can indicate different functions of the proteins encoded by them. An increase in the expression of α-CA1 and one of the forms of RNA of the *β-ca1* gene with increased illumination suggests their cooperation in the CO_2 supply to Rubisco [6]. In general, it can be assumed, on the basis of the available data, that plant CAs, as a rule, jointly control one or another metabolic process.

Studies using double mutants showed that the reduced synthesis of several CAs had an effect on photosynthesis. Plants with knocked-out genes encoding β-CA1 and β-CA4, which are located in the stroma and plasma membrane, respectively, showed a higher stomata conductivity compared to WT plants [117]. Plants are known to activate anion channels in response to environmental signals such as drought, high levels of carbon dioxide, and bacterial invasion. Recently, two new gene families encoding major groups of anion channels have been identified. SLAC/SLAH channels are the representatives of this group characterized by slow voltage-dependent activation (S-type) [118]. Xue et al. in 2011 [119] and Tian et al. in 2015 [120] showed that β-CA4 and β-CA1 take part in the regulation of gas exchange between the atmosphere and leaves by opening/closing the stomatal aperture. Intracellular bicarbonate generated by β-CA4 and β-CA1 acts as a second messenger and activates S-type anion channels in guard cells.

Evidence suggests that β-CA1 may be important not only for photosynthetic processes but, for example, for the synthesis of ethylene during plant germination, a process that is also dependent on CO_2. The seeds of the *A. thaliana* mutant with knocked-out gene encoding β-CA1 showed a significant decrease in germination on sterile artificial media at ambient CO_2 concentration in air [121]. The germination ability of these mutants was restored to that of WT when grown at a high CO_2 content (1500 μl/L) or after adding sucrose to the medium [122].

To understand the role of β-CA1 in plants, we should take into account that this CA located in the stroma of the chloroplast not only exhibits CA activity but can also bind salicylic acid (SA) [123]. As it is known, SA plays a role in signaling cascades that activate the synthesis of the proteins involved in protecting plants from oxidative stress at both the transcription and translation levels. The intensity of stromal CAs' gene expression was shown to respond to signals associated with activation of biotic stress protection systems: infection with the late blight of potato plants [124], treatment of tomatoes with mycotoxin fusicoccin [125] and treatment of Arabidopsis plants with methyl jasmonates [126]. *Nicotiana benthamiana* mutant plants with knocked-out gene encoding chloroplast soluble CA showed high susceptibility to late blight [124]. Medina-Puche et al. [127] showed that in *A. thaliana* plants inoculated with the phytopathogenic bacterium *Pseudomonas syringae*, the expression level of *β-ca1*, *β-ca2*, and *β-ca4* genes was repressed, with the induction of the expression of the gene encoding the other CA, β-CA6, located in the mitochondrial matrix.

There is also a hypothesis that the stromal β-CA can ensure protection against the stress through the biosynthesis of fatty acids. The lower content of stromal β-CA was shown [128] to suppress the fatty acid synthesis leading to a lower expression of genes regulated by jasmonic acid, another signaling molecule that triggers an alternative gene cascade of a protective response. Omega-6 fatty acids are the intermediate compounds of biosynthesis of fatty acids that are synthesized in the stroma of chloroplasts.

4.2.4 CAs in lamellar thylakoid membranes

In the early 2000s, the information on the presence of more than one carrier of CA activity in the thylakoid membranes began to appear [17, 129]. The CA activity was identified in preparations of lamellar thylakoid membranes enriched with PSI and ATP synthase complexes isolated from pea and Arabidopsis plants [18, 20]. The activity of this CA differed in properties from the CA activity of granular thylakoid membranes enriched with PSII complexes. The CAs in lamellar and granal thylakoid membranes were different in their effect on the activity of inhibitors and detergents, their molecular masses were different. The CAs' activity of lamellar thylakoid membranes was inhibited equally by both ethoxyzolamide, which is able to penetrate into membranes and acetazolamide, poorly penetrating into lipid membranes. These data show that this CA is situated at the stromal surface of the thylakoid membrane, where it is accessible to both inhibitors. Recently [130, 131] the stimulating effect of adding bicarbonate on the rate of phosphorylation in isolated thylakoids that is known since the 60th of the last century was explained by the presence of CAs in lamellar thylakoid membranes. We have shown that mafenide, a water-soluble inhibitor of CAs, suppressed the phosphorylation stimulation by HCO_3^- without inhibiting the rate of electron transport and had no effect on phosphorylation rate in the absence of bicarbonate [132].

We suggested a hypothesis of the mechanism of this CA's involvement in the increase of photophosphorylation in thylakoid membranes. In the reaction of bicarbonate dehydration catalyzed by membrane-bound CAs, CO_2 molecules are formed near or even in the surface layer of the thylakoid membrane. With an increase in their local concentration, a stream of CO_2 in the thylakoid lumen is possible. In the lumen, CO_2 is hydrated by the luminal CAs (see below) with proton release. As a result, an increase in proton concentration in lumen may cause an intensification in ATP production, that is, acceleration of photophosphorylation. Since the uncoupling effect of ammonium salts on ATP synthesis in thylakoids resulted mainly from proton bonding in the thylakoid lumen by NH_3 molecules penetrating there, the facts that the addition of bicarbonate reduced the uncoupling effect of these salts [132, 133] is in line with the proposed hypothesis.

Thus, CAs located in lamellar thylakoid membranes, that is, directed immediately to the stromal phase, may operate as one of the suppliers of CO_2 for Rubisco, and in the case of low effectiveness of CO_2 fixation under some stressful conditions can be involved in the extra production of ATP that is especially important under such conditions.

4.2.5 CA in granal thylakoid membranes

In 2004, α-CA4 was detected among the proteins of thylakoid membranes [35]. We have shown that the fresh weight of the plant leaves with knocked-out *α-ca4* gene was 10% higher, the starch and H_2O_2 content was significantly higher, and the rate of CO_2 assimilation in leaves was lower in comparison to WT plants. The effective quantum yield of photosynthetic electron transport at saturating light intensity and CO_2 concentration was higher in mutants than in the WT, while nonphotochemical chlorophyll fluorescence quenching (NPQ) was lower [13, 37, 134, 135]. The content of transcripts of the *At4g20990* gene encoding α-CA4 was two times higher under high (400 μmol quanta $m^{-2}\ s^{-1}$) than under low illumination (80 μmol quanta $m^{-2}\ s^{-1}$) when grown under "short-day" (8 hours day/16 hours night) and 16 times higher under "long-day" conditions (16 hours day/8 hours night) [6]. Thus, the above-described effects of the *At4g20990* gene knockout are consistent with a significant increase in the expression level of this gene in plants under high illumination; these are the conditions where the activation of NPQ, which protects PSII from photoinhibition, is important. Since the energy-dependent quenching (the part of the NPQ that is associated with the accumulation of protons in the thylakoid lumen) is the main process, which had the effect on changes in NPQ, we suggested that α-CA4 is involved in the protonation of either PsbS protein or violaxantin deepoxidase. The role of α-CA4 in NPQ in PSII antenna was confirmed by the fact that the knockout of the *At4g20990* gene affected both the protein content of the light-harvesting complex PSII and the expression level of the genes encoding these proteins [135]. The content of the major PSII antenna proteins, Lhcb1, and Lhcb2 was lower in α-CA4 knockouts than in the WT plants. We have also found that the mutant plants grown under high illumination compensated for the absence of α-CA4 and for reduced NPQ by the increase in the contents of both PsbS protein and the violaxanthin cycle components, the latter accompanied by an increase in violaxanthin deepoxidase activity [13]. In addition, *α-ca4* gene was found among the genes that change the expression level under osmotic stress [136]. The content of *α-ca4* gene transcripts became an order of magnitude higher in conditions of drought stress (Rudenko et al., in preparation).

Our research has provided evidence that one more CA, α-CA2, is present in higher plant thylakoids. The knockout of *α-ca2* gene has led to the opposite changes in the properties of plants compared with those resulting from the knockout of *α-ca4* gene. Fresh leaf weight, chlorophyll a/chlorophyll b ratio, and starch and H_2O_2 content in leaves of α-CA2 mutants were lower than in the WT plants, while CO_2 assimilation rate was higher [37, 134, 135]. In α-CA2 knockouts, the effective quantum yield of photosynthetic electron transport was lower than in WT, while NPQ was higher, also due to the energy-dependent component. The set of physiological effects that occur when the α-CA2 and α-CA4 genes are turned off suggests that not only α-CA4, but also α-CA2, is present in chloroplasts, participating in the functioning of the opposite "regulatory pathways" that respond to changes in external conditions. This phenomenon of regulation of metabolic states, when there are two enzymes that oppositely affect particular conditions, for example, kinase and phosphatase on protein phosphorylation, is well known. The detection of this phenomenon in the regulation of NPQ is very important. It shows that the change in NPQ involved in the protection of the photosynthetic apparatus from photoinhibition is under operational control, which allows the fast increase in the thermal dissipation of solar energy with its excess. In the opposite situation, in conditions of the low illumination, the above-described mechanism could reduce this dissipation when it could decrease the useful incoming energy, which is necessary for photosynthesis.

Since the regulation of the energy-dependent component of NPQ taking place with the participation of α-CA4 and α-CA2 is based on a change in proton concentration in the lumen, these CAs are also able to regulate the redox state of plastoquinone pool by the regulation of plastohydroquinone oxidation rate by cytochrome complex in the photosynthetic electron transport chain. According to a number of studies [137–139], it regulates the adaptive response of plants to environmental conditions' changes. In addition, the concentration of protons is important for the functional activity of enzymes such as thioredoxins and kinases, whose activities are dependent on the state of sulfhydryl groups.

4.2.6 CA in the thylakoid lumen

The CA activity in thylakoid membranes of higher plants was discovered in the early 1980s. A number of properties distinguish the CA activity of thylakoid membranes from the activity of soluble stromal CA. In particular, the dehydration activity of thylakoid CA depends on pH with a maximum at 6.8–7.0, and the activity of soluble CA does not depend on pH [140]. Antibodies against soluble CA from spinach have also shown a strong cross-reaction with soluble CA from pea chloroplasts, but not with thylakoids exhibiting similar CA activity [141]. In our studies, a soluble CA that belongs to β-family located in the thylakoid lumen has been discovered [19, 142], and it was suggested that this is β-CA5, previously detected in chloroplasts from Arabidopsis [5]. The exact position of β-CA5 in Arabidopsis chloroplasts is unknown, but it cannot be excluded that there are two forms of this enzyme with one located in the chloroplast envelope (see above) and the other situated in the thylakoid lumen.

The expression level of the *At4g33580* gene encoding β-CA5 is 2–3 orders lower than that of, for example, the *At3g01500* gene encoding stromal β-CA1 in plants grown at atmospheric CO_2 concentration in low-light conditions [135]. The transcription intensity of the gene encoding β-CA5 was higher when illumination decreased both in short-day and long-day conditions. CA in thylakoid lumen may enable more free diffusion of protons to ATP-synthase channel together with CO_2/HCO_3^- buffer, and the value of such diffusion should decrease at low light intensity when proton inflow into the lumen is low and they may be "lost" on the way to the ATP-synthase.

5. Conclusion

The role of CAs in mechanisms of stress tolerance has been studied in plants possessing CCM, that is, in algae and C4 higher plants. The lack of understanding of the functions of individual CAs in higher plants with C3 type of photosynthesis may be explained by an involvement of more than one of CAs in one biochemical pathway. The effect of the absence of CAs in mutants is more obvious under various adverse conditions, which are comfortable for plants. These are the conditions when CO_2 supply and fixation are not limited to the general plant metabolism under optimal light intensity, temperature, mineral nutrition, etc. Moreover, the effect of the absence of CAs in mutants is more obvious in various adverse conditions. That means that the functioning of CAs in plants is the most important under stress.

Funding information

The preparation of the manuscript was supported by the Russian Science Foundation (project no. 17-14-01371) and by the Ministry of Science and Higher Education of the Russian Federation, State Scientific Program, theme no. AAAA-A17-117030110135-1.

Author details

Natalia N. Rudenko*, Maria M. Borisova-Mubarakshina, Lyudmila K. Ignatova, Tatiana P. Fedorchuk, Elena M. Nadeeva-Zhurikova and Boris N. Ivanov
Institute of Basic Biological Problems, Federal Research Center–Pushchino Scientific Center for Biological Research of the Russian Academy of Sciences, Pushchino, Russia

*Address all correspondence to: nataliacherry413@gmail.com

References

[1] Manahan S. Environmental Chemistry. Boston: Willard Grant Press; 1984

[2] Moroney J, Ma Y, Frey W, Fusilier K, Pham T, Simms T, et al. The carbonic anhydrase isoforms of *Chlamydomonas reinhardtii*: Intracellular location, expression, and physiological roles. Photosynthesis Research. 2011;**109**:133-149. DOI: 10.1371/journal.pone.0001426

[3] Hatch M, Burnell J. Carbonic anhydrase activity in leaves and its role in the first step of C4 photosynthesis. Plant Physiology. 1990;**93**:825-828. DOI: 10.1104/pp.93.2.825

[4] von Caemmerer S, Quinn V, Hancock N, Price G, Furbank T, Ludwig M. Carbonic anhydrase and C4 photosynthesis: A transgenic analysis. Plant, Cell and Environment. 2004;**27**:697-703. DOI: 10.1111/j.1365-3040.2003.01157.x

[5] Fabre N, Reiter I, Becuwe-Linka N, Genty B, Rumeau D. Characterization and expression analysis of genes encoding alpha and beta carbonic anhydrases in Arabidopsis. Plant, Cell & Environment. 2007;**30**:617-629. DOI: 10.1111/j.1365-3040.2007.01651.x

[6] Rudenko N, Vetoshkina D, Fedorchuk T, Ivanov B. Effect of light intensity under different photoperiods on expression level of carbonic anhydrase genes of the α- and β-families in *Arabidopsis thaliana* leaves. Biochemistry (Moscow). 2017;**82**(9):1025-1035. DOI: 10.1134/S000629791709005X

[7] Yu S, Zhang X, Guan Q, Takano T, Liu S. Expression of a carbonic anhydrase gene is induced by environmental stresses in Rice (*Oryza sativa* L.). Biotechnology Letters. 2007;**29**:89-94. DOI: 10.1007/s10529-006-9199-z

[8] Rudenko N, Ignatova L, Fedorchuk T, Ivanov B. Carbonic anhydrases in photosynthetic cells of higher plants. Biochemistry (Moscow). 2015;**80**(6):674-687. DOI: 10.1134/S0006297915060048

[9] Atkins C, Patterson B, Graham D. Plant carbonic anhydrases. II. Preparation and some properties of monocotyledon and dicotyledon enzyme types. Plant Physiology. 1972;**50**:218-223. DOI: 10.1104/pp.50.2.218

[10] Villarejo A, Buren S, Larsson S, Dejardin A, Monne M, Rudhe C, et al. Evidence for a protein transported through the secretory pathway en route to the higher plant chloroplast. Nature Cell Biology. 2005;**7**:1224-1231. DOI: 10.1038/ncb1330

[11] Chaki M, Carreras A, Lypez-Jaramillo J, Begara-Morales J, Sάnchez-Calvo B, Valderrama R, et al. Tyrosine nitration provokes inhibition of sunflower carbonic anhydrase (β-CA) activity under high temperature stress. Nitric Oxide. 2013;**29**:30-33. DOI: 10.1111/j.1399-3054.1992

[12] Wang L, Jin X, Li Q, Wang X, Li Z, Wu X. Comparative proteomics reveals that phosphorylation of β carbonic anhydrase 1 might be important for adaptation to drought stress in *Brassica napus*. Scientific Reports. 2016;**6**:39024. DOI: 10.1038/srep39024

[13] Rudenko N, Fedorchuk T, Terentyev V, Dymova O, Naydov I, Golovko T, et al. The role of carbonic anhydrase α-CA4 in the adaptive reactions of photosynthetic apparatus. The study with α-CA4 knockout plants. Protoplasma. 2020;**257**:489-499. DOI: 10.1007/s00709-019-01456-1

[14] Wang C, Hu H, Qin X, et al. Reconstitution of CO_2 regulation of

SLAC1 anion channel and function of CO_2-permeable PIP2;1 aquaporin as carbonic anhydrase4 interactor. The Plant Cell. 2016;**28**(2):568-582. DOI: 10.1105/tpc.15.00637

[15] Perez-Martin A, Michelazzo C, Torres-Ruiz J, Flexas J, Fernández J, Sebastiani L, et al. Regulation of photosynthesis and stomatal and mesophyll conductance under water stress and recovery in olive trees: Correlation with gene expression of carbonic anhydrase and aquaporins. Journal of Experimental Botany. 2014;**65**:3143-3156. DOI: 10.1093/jxb/eru160

[16] Borisova (Mubarakshina) M, Kozuleva M, Rudenko N, Naydov I, Klenina I, Ivanov B. Photosynthetic electron flow to oxygen and diffusion of hydrogen peroxide through the chloroplast envelope via aquaporins. BBA – Bioenegetics. 2012;**1817**:1314-1321. DOI: 10.1016/j.bbabio.2012.02.036

[17] Lu Y, Stemler A. Extrinsic photosystem II carbonic anhydrase in maize mesophyll chloroplasts. Plant Physiology. 2002;**128**:643-649. DOI: 10.1104/pp.010643

[18] Ignatova L, Rudenko N, Khristin M, Ivanov B. Heterogeneous origin of carbonic anhydrase activity of thylakoid membranes. Biochemistry (Moscow). 2006;**71**:525-532. DOI: 10.1134/S0006297906050099

[19] Rudenko N, Ignatova L, Ivanov B. Multiple sources of carbonic anhydrase activity in pea thylakoids: Soluble and membrane-bound forms. Photosynthesis Research. 2007;**91**(1):81-89. DOI: 10.1007/s11120-007-9148-2

[20] Ignatova L, Rudenko N, Mudrik V, Fedorchuk T, Ivanov B. Carbonic anhydrase activity in *Arabidopsis thaliana* thylakoid membrane and fragments enriched with PSI or PSII. Photosynthesis Research. 2011;**110**:89-98. DOI: 10.1007/s11120-011-9699-0

[21] Hewett-Emmett D, Tashian R. Functional diversity, conservation, and convergence in the evolution of the α, β, and γ-carbonic anhydrase gene families. Molecular Phylogenetics and Evolution. 1996;**65**:50-77

[22] Liljas A, Laurberg M. A wheel invented three times. The molecular structures of the three carbonic anhydrases. EMBO Reports. 2000;**1**(1):16-17. DOI: 10.1006/mpev.1996.0006

[23] De Simone G, Di Fiore A, Capasso C, Supuran C. The zinc coordination pattern in the η-carbonic anhydrase from plasmodium falciparum is different from all other carbonic anhydrase genetic families. Bioorganic & Medicinal Chemistry Letters. 2015;**25**:1385-1389. DOI: 10.1016/j.bmcl.2015.02.046

[24] Jensen E, Clement R, Kosta A, Maberly S, Gontero B. A new widespread subclass of carbonic anhydrase in marine phytoplankton. The ISME Journal. 2019;**13**:2094-2106. DOI: 10.1038/s41396-019-0426-8

[25] DiMario R, Clayton H, Mukherjee A, Ludwig M, Moroney J. Plant carbonic anhydrases: Structures, locations, evolution, and physiological roles. Molecular Plant. 2017;**10**:30-46. DOI: 10.1016/j.molp.2016.09.001

[26] Soltes-Rak E, Mulligan M, Coleman J. Identification and characterization of a gene encoding a vertebrate-type carbonic anhydrase in cyanobacteria. Journal of Bacteriology. 1997;**179**:769-774. DOI: 10.1128/jb.179.3.769-774

[27] Elleby B, Chirica L, Tu C, Zeppezauer M, Lindskog S. Characterization of carbonic anhydrase from *Neisseria gonorrhoeae*.

European Journal of Biochemistry. 2001;**268**:1613-1619. DOI: 10.1046/j.1432-1327.2001.02031.x

[28] Elleuche S, Poggeler S. Evolution of carbonic anhydrases in fungi. Current Genetics. 2009;**55**:211-222. DOI: 10.1007/s00294-009-0238-x

[29] Fukuzawa H, Fujiwara S, Yamamoto Y, Dionisto-Sese M, Miyachi S. cDNA cloning, sequence, and expression of carbonic anhydrase in *Chlamydomonas reinhardtii*: Regulation by environmental CO_2 concentration. PNAS. 1990;**87**: 4383-4387. DOI: 10.1073/pnas.87.11.4383

[30] Fujiwara S, Fukuzawa H, Tachiki A, Miyachi S. Structure and differential expression of two genes encoding carbonic anhydrase in *Chlamydomonas reinhardtii*. PNAS. 1990;**87**:9779-9783. DOI: 10.1073/pnas.87.24.9779

[31] Tuskan G, Difazio S, Jansson S, et al. The genome of black cottonwood, *Populus trichocarpa* (Torr., Gray). Science. 2006;**313**(5793):1596-1604. DOI: 10.1126/science.1128691

[32] Meldrum N, Roughton F. Carbonic anhydrase: Its preparation and properties. Nature. 1933;**80**:113-142. DOI: 10.1113/jphysiol.1933.sp003077

[33] Rawat M, Moroney J. Partial characterization of a new isoenzyme of carbonic anhydrase isolated from *Chlamydomonas reinhardtii*. The Journal of Biological Chemistry. 1991;**266**:9719-9723. DOI: 10.1371/journal.pone.0079909

[34] Karlsson J, Hiltonen T, Husic H, Ramazanov Z, Samuelsson G. Intracellular carbonic anhydrase of *Chlamydomonas reinhardtii*. Plant Physiology. 1995;**109**:533-539. DOI: 10.1007/s11120-011-9635-3

[35] Friso G, Giacomelli L, Ytterberg A, Peltier J, Rudella A, Sun Q, et al. In-depth analysis of the thylakoid membrane proteome of *Arabidopsis thaliana* chloroplasts: New proteins, new functions, and a plastid proteome database. The Plant Cell. 2004;**16**: 478-499. DOI: 10.1105/tpc.017814

[36] Sun Q, Zybailov B, Majeran W, Friso G, Olinares P, van Wijk K. PPDB, the plant proteomics database at Cornell. Nucleic Acids Research. 2009;**37**:969-974. DOI: 10.1093/nar/gkn654

[37] Zhurikova E, Ignatova L, Rudenko N, Mudrik V, Vetoshkina D, Ivanov B. The participation of two carbonic anhydrases of alpha family in photosynthetic reactions in Arabidopsis thaliana. Biochemistry (Moscow). 2016;**81**(10):1182-1187. DOI: 10.1134/S0006297916100151

[38] Holmes-Davis R, Tanaka C, Vensel W, Hurkman W, McCormick S. Proteome mapping of mature pollen of *Arabidopsis thaliana*. Proteomics. 2005;**5**:4864-4884. DOI: 10.1002/pmic.200402011

[39] Noir S, Brautigam A, Colby T, Schmidt J, Panstruga R. A reference map of the *Arabidopsis thaliana* mature pollen proteome. Biochemical and Biophysical Research Communications. 2005;**337**:1257-1266. DOI: 10.1016/j.bbrc.2005.09.185

[40] Eriksson A, Jones T, Liljas A. Refined structure of human carbonic anhydrase II at 2.0 angstrom resolution. Proteins: Structure, Function, and Genetics. 1988;**4**:274-282. DOI: 10.1002/prot.340160104

[41] Christianson D, Cox J. Catalysis by metal-activated hydroxide in zinc and manganese metalloenzymes. Annual Review of Biochemistry. 1999;**68**:33-57. DOI: 10.1146/annurev.biochem.68.1.33

[42] Christianson D, Fierke C. Carbonic anhydrase: Evolution of the zinc binding site by nature and by design. Accounts

of Chemical Research. 1996;**29**:331-339. DOI: 10.1021/ar9501232

[43] Kamo T, Shimogawara K, Fukuzawa H, Muto S, Miyachi S. Subunit constitution of carbonic anhydrase from *Chlamydomonas reinhardtii*. European Journal of Biochemistry. 1990;**192**: 557-562. DOI: 10.1111/j.1432-1033.1990.tb19261.x

[44] Smith K, Ferry J. A plant-type (beta-class) carbonic anhydrase in the thermophilic metanoarchaeon Methanobacterium thermoautotrophicum. Journal of Bacteriology. 1999;**181**:6247-6253

[45] So A, Espie G. Cloning, characterization and expression of carbonic anhydrase from the cyanobacterium Synechocystis PCC6803. Plant Molecular Biology. 1998;**37**:205-215. DOI: 10.1023/A:1005959200390

[46] Smith K, Jakubzick C, Whittam T, Ferry J. Carbonic anhydrase is an ancient enzyme widespread in prokaryotes. PNAS. 1999;**96**(26):15184-15189. DOI: 10.1073/pnas.96.26.15184

[47] Sawaya M, Cannon G, Heinhorst S, Tanaka S, Williams E, Yeates T, et al. The structure of beta-carbonic anhydrase from the carboxysomal shell reveals a distinct subclass with one active site for the price of two. The Journal of Biological Chemistry. 2006;**281**:7546-7555. DOI: 10.1074/jbc

[48] Gotz R, Gnann A, Zimmermann F. Deletion of the carbonic anhydrase-like gene NCE103 of the yeast Saccharomyces cerevisiae causes an oxygen-sensitive growth defect. Yeast. 1999;**15**:855-864. DOI: 10.1002/(SICI)1097-0061(199907)15:10A<855::AID-YEA425>3.0.CO;2-C

[49] Eriksson M, Karlsson J, Ramazanov Z, Gardestrom P, Samuelsson G. Discovery of an algal mitochondrial carbonic anhydrase. Molecular cloning and characterization of a low-CO_2-induced polypeptide in *Chlamydomonas reinhardtii*. PNAS. 1996;**93**(21):12031-12034. DOI: 10.1073/pnas.93.21.12031

[50] Majeau N, Coleman J. Effect of CO_2 concentration on carbonic anhydrase and ribulose-1,5-bisphosphate carboxylase/oxygenase expression in pea. Plant Physiology. 1996;**112**(2):569-574

[51] Fett J, Coleman J. Characterization and expression of two cDNAs encoding carbonic anhydrase in Arabidopsis thaliana. Journal of Plant Physiology. 1994;**105**:707-713. DOI: 10.1104/pp.112.2.569

[52] Provart N, Majeau N, Coleman J. Characterization of pea chloroplastic carbonic anhydrase. Expression in Escherichia coli and site-directed mutagenesis. Plant Molecular Biology. 1993;**22**:937-943. DOI: 10.1007/BF00028967

[53] Rowlett R, Chance M, Wirt M, Sidelinger D, Royal J, Woodroffe M, et al. Kinetic and structural characterization of spinach carbonic-anhydrase. Biochemistry. 1994;**33**:13967-13976. DOI: 10.1021/bi00251a003

[54] Kimber M, Pai E. The active site architecture of Pisum sativum β-carbonic anhydrase is a mirror image of that of α-carbonic anhydrases. The EMBO Journal. 2000;**19**:1407-1418. DOI: 10.1093/emboj/19.7.1407

[55] Cronk J, Rowlett R, Zhang K, Tu C, Endrizzi J, Lee J, et al. Identification of a novel noncatalytic bicarbonate binding site in eubacterial beta-carbonic anhydrase. Biochemistry. 2006;**45**:4351-4361. DOI: 10.1021/bi052272q

[56] Rowlett R. Structure and catalic mechanism of the β-carbonic anhydrase. BBA. 1804;**2010**:362-373. DOI: 10.1007/978-94-007-7359-2_4

[57] Parisi G, Perales M, Fornasari M, Gonztalez-Schain A, Gomez-Casati D, Zimmermann S, et al. Gamma carbonic anhydrases in plant mitochondria. Plant Molecular Biology. 2004;**55**:193-207

[58] Cardol P, Vanrobaeys F, Devreese B, Van Beeumen J, Matagne R, Remacle C. Higher plant-like subunit composition of mitochondrial complex I from *Chlamydomonas reinhardtii*: 31 conserved components among eukaryotes. BBA. 1658;**2004**:212-224. DOI: 10.1007/s11103-004-0149-7

[59] Sunderhaus S, Dudkina N, Jansch L, Klodmann J, Heinemeyer J, Perales M, et al. Carbonic anhydrase subunits form a matrix-exposed domain attached to the membrane arm of mitochondrial complex I in plants. The Journal of Biological Chemistry. 2006;**281**(10):6482-6488. DOI: 10.1074/jbc

[60] Perales M, Eubel H, Heinemeyer J, Colaneri A, Zabaleta E, Braun H. Disruption of a nuclear gene encoding a mitochondrial gamma carbonic anhydrase reduces complex I and supercomplex I + III2 levels and alters mitochondrial physiology in Arabidopsis. Journal of Molecular Biology. 2005;**350**:263-277. DOI: 10.1016/j.jmb.2005.04.062

[61] Kisker C, Schindelin H, Alber B, Ferry J, Rees D. A left-handed beta-helix revealed by the crystal structure of a carbonic anhydrase from the archaeon *Methanosarcina thermophila*. The EMBO Journal. 1996;**15**:2323-2330. DOI: 10.1002/j.1460-2075.1996.tb00588.x

[62] Smith K, Ferry J. Prokaryotic carbonic anhydrases. FEMS Microbiology Reviews. 2000;**24**:335-366. DOI: 10.1111/j.1574-6976.2000.tb00546.x

[63] Price G. Inorganic carbon transporters of the cyanobacterial CO_2 concentrating mechanism. Photosynthesis Research. 2011;**109**:33-45. DOI: 10.1007/s11120-010-9608-y

[64] Moroney J, Husic H, Tolbert N. Effect of carbonic anhydrase inhibitors on inorganic carbon accumulation by *Chlamydomonas reinhardtii*. Plant Physiology. 1985;**79**:177-183. DOI: 10.1104/pp.79.1.177

[65] Ynalvez R, Xiao Y, Ward A, Cunnusamy K, Moroney J. Identification and characterization of two closely related beta carbonic anhydrases from *Chlamydomonas reinhardtii*. Physiologia Plantarum. 2008;**133**:15-26. DOI: 10.1111/j.1399-3054.2007.01043.x

[66] Spalding M, Spreitzer R, Ogren W. Carbonic anhydrase deficient mutant of *Chlamydomonas reinhardii* requires elevated carbon-dioxide concentration for photoautotrophic growth. Plant Physiology. 1983;**73**:268-272. DOI: 10.1104/pp.73.2.268

[67] Moroney J, Tolbert N, Sears B. Complementation analysis of the inorganic carbon concentrating mechanism of *Chlamydomonas reinhardtii*. Molecular & General Genetics. 1986;**204**:199-203. DOI: 10.1007/BF00425498

[68] Ku M, Kano-Murakami Y, Matsouka M. Evolution and expression of C4 photosynthesis genes. Plant Physiology. 1996;**111**:949-957. DOI: 10.1104/pp.111.4.949

[69] von Caemmerer S, Furbank R. The C4 pathway: An efficient CO_2 pump. Photosynthesis Research. 2003;**77**:191-207. DOI: 10.1016/B978-0-12-675408-7.50012-4

[70] Tetu S, Tanz S, Vella N, Burnell J, Ludwig M. The *Flaveria bidentis* β-carbonic anhydrase gene family encodes cytosolic and chloroplastic isoforms demonstrating distinct organ-specific expression patterns. Plant Physiology.

2007;**144**(3):1316-1327. DOI: 10.1104/pp.107.098152

[71] Raven J, Newman J. Requirement for carbonic anhydrase activity. Plant, Cell & Environment. 1994;**17**:123-130. DOI: 10.1111/j.1365-3040.1994.tb00275.x

[72] Chollet R, Vidal J, O'Leary M. Phosphoenolpyruvate carboxylase: A ubiquitous, highly regulated enzyme in plants. Annual Review of Plant Physiology and Plant Molecular Biology. 1996;**47**:273-298. DOI: 10.1146/annurev.arplant.47.1.273

[73] Missner A, Pohl P. 110 years of the Meyer–Overton rule: Predicting membrane permeability of gases and other small compounds. ChemPhysChem. 2009;**10**:1405-1414. DOI: 10.1002/cphc.200900270

[74] Boron W, Endeward V, Gros G, Musa-Aziz R, Pohl P. Intrinsic CO_2 permeability of cell membranes and potential biological relevance of CO_2 channels. ChemPhysChem. 2011;**12**:1017-1019. DOI: 10.1002/cphc.201100034

[75] Uehlein N, Otto B, Hanson D, Fischer M, McDowell N, Kaldenhoff R. Function of *Nicotiana tabacum* aquaporins as chloroplast gas pores challenges the concept of membrane CO_2 permeability. The Plant Cell. 2008;**20**:648-665. DOI: 10.1105/tpc.107.054023

[76] Endeward V, Al-Samir S, Itel F, Gros G. How does carbon dioxide permeate cell membranes? A discussion of concepts, results and methods. Frontiers in Physiology. 2014;**4**:21. DOI: 10.3389/fphys.2013.00382

[77] Zhao M, Tan HT, Scharwies J, Levin K, Evans JR, Tyerman S. Association between water and carbon dioxide transport in leaf plasma membranes: Assessing the role of aquaporins. Plant, Cell & Environment. 2017;**40**(6):789-801. DOI: 10.1111/pce.12830

[78] Terashima I, Ono K. Effects of $HgCl_2$ on CO_2 dependence of leaf photosynthesis: Evidence indicating involvement of aquaporins in CO_2 diffusion across the plasma membrane. Plant & Cell Physiology. 2002;**43**:70-78. DOI: 10.1093/pcp/pcf001

[79] Uehlein N, Lovisolo C, Siefritz F, Kaldenhoff R. The tobacco aquaporin NtAQP1 is a membrane CO_2 pore with physiological functions. Nature. 2003;**425**:734-737. DOI: 10.1038/nature02027

[80] Uehlein N, Sperling H, Heckwolf M, Kaldenhoff R. The Arabidopsis aquaporin PIP1;2 rules cellular CO_2 uptake. Plant, Cell & Environment. 2012;**35**:1077-1083. DOI: 10.1111/j.1365-3040.2011.02473.x

[81] Soto G, Alleva K, Amodeo G, Muschietti J, Ayub N. New insight into the evolution of aquaporins from flowering plants and vertebrates: Orthologous identification and functional transfer is possible. Gene. 2012;**503**:165-176. DOI: 10.1016/j.gene.2012.04.021

[82] Yaneff A, Sigaut L, Marquez M, Alleva K, Pietrasanta L, Amodeo G. Heteromerization of PIP aquaporins affects their intrinsic permeability. PNAS 2014;111(1):231-236. DOI: 10.1073/pnas.1316537111.

[83] Ignatova L, Romanova A. Participation of carbonic anhydrase in inhibition of photosynthesis in pea protoplasts by CO_2 excess. Russian Journal of Plant Physiology. 1992;**39**:82-88

[84] Ignatova L, Moskvin O, Ivanov B, Romanova A. The effect of CO_2 uptake by pea protoplasts on O_2 evolution rate and parameters of chlorophyll fluorescence quenching. Journal of

Plant Biochemistry & Physiology. 1993;**31**:295-301

[85] Afzal Z, Howton T, Sun Y, Mukhtar M. The roles of aquaporins in plant stress responses. Journal of Developmental Biology. 2016;**4**(1):9. DOI: 10.3390/jdb4010009

[86] Terashima I, Hanba Y, Tazoe Y, Vyas P, Yano S. Irradiance and phenotype: Comparative eco-development of sun and shade leaves in relation to photosynthetic CO_2 diffusion. Journal of Experimental Botany. 2006;**57**:343-354. DOI: 10.1093/jxb/erj014

[87] Hanba Y, Shibasaka M, Hayashi Y, Hayakawa T, Kasamo K, Terashima I, et al. Overexpression of the barley aquaporin HvPIP2;1 increases internal CO_2 conductance and CO_2 assimilation in the leaves of transgenic rice plants. Plant & Cell Physiology. 2004;**45**:521-529. DOI: 10.1093/pcp/pch070

[88] Flexas J, Ribas-Carbo M, Hanson D, Bota J, Otto B, Cifre J, et al. Tobacco aquaporin NtAQP1 is involved in mesophyll conductance to CO_2 in vivo. The Plant Journal. 2006;**48**:427-439. DOI: 10.1111/j.1365-313X.2006.02879.x

[89] Katsuhara M, Hanba YT. Barley plasma membrane intrinsic proteins (PIP aquaporins) as water and CO_2 transporters. Pflügers Archiv. 2008;**456**:687-691. DOI: 10.1007/s00424-007-0434-9

[90] Simm S, Papasotiriou D, Ibrahim M, Leisegang M, Müller B, Schorge T, et al. Defining the core proteome of the chloroplast envelope membranes. Frontiers in Plant Science. 2013;**4**:11. DOI: 10.3389/fpls.2013.00011

[91] Villarejo A, Rolland N, Martínez F, Sültemeyer D. A new chloroplast envelope carbonic anhydrase activity is induced during acclimation to low inorganic carbon concentrations in *Chlamydomonas reinhardtii*. Planta. 2001;**213**:286-295. DOI: 10.1007/s004250000508

[92] Raisanen S, Lehenkari P, Tasanen M, Rahkila P, Harkonen P, Vaananen H. Carbonic anhydrase III protects cells from hydrogen peroxide-induced apoptosis. The FASEB Journal. 1999;**13**:513-522. DOI: 10.1096/fasebj.13.3.513

[93] Gadjev I, Vanderauwera S, Gechev T, Laloi C, Minkov I, Shulaev V, et al. Transcriptomic footprints disclose specificity of reactive oxygen species signaling in Arabidopsis. Plant Physiology. 2006;**141**:436-445. DOI: 10.1104/pp.106.078717

[94] Ivanov B, Kozuleva M, Mubarakshina M. Oxygen metabolism in chloroplast. In: Babulya P, editor. Cell Metabolism - Cell Homeostasis and Stress Response. Rijeka: InTechOpen; 2012. pp. 39-73. DOI: 10.1134/S1990747814060026

[95] Mubarakshina MM, Ivanov BN, Naydov IA, Hillier W, Badger MR, Krieger-Liszkay A. Production and diffusion of chloroplastic H_2O_2 and its implication to signalling. Journal of Experimental Botany. 2010;**61**:3577-3587. DOI: 10.1093/jxb/erq171

[96] Gao J, Wang X, Chang Y, Zhang J, Song Q, Yu H, et al. Acetazolamide inhibits osmotic water permeability by interaction with aquaporin-1. Analytical Biochemistry. 2006;**350**:165-170. DOI: 10.1016/j.ab.2006.01.003

[97] Haddoub R, Rutzler M, Robin A, Flitsch S. Design, synthesis and assaying of potential aquaporin inhibitors. In: Beitz E, editor. Handbook of Experimental Pharmacology: Aquaporins. Berlin: Springer; 2009. pp. 385-402. DOI: 10.1007/978-3-540-79885-9_19

[98] Huber V, Tsujita M, Yamazaki M, Sakimura K, Nakada T. Identification of arylsulfonamides as aquaporin 4 inhibitors. Bioorganic & Medicinal Chemistry Letters. 2007;**17**:1270-1273. DOI: 10.1016/j.bmcl.2006.12.010

[99] Kim J, Yoon H, Kwon K, Lee S, Rhee S. Identification of proteins containing cysteine residues that are sensitive to oxidation by hydrogen peroxide at neutral pH. Analytical Biochemistry. 2000;**283**(2):214-221. DOI: 10.1006/abio.2000.4623

[100] Beebo A, Mathai J, Schoefs B, Cornelia S. Assessment of the requirement for aquaporins in the thylakoid membrane of plant chloroplasts to sustain photosynthetic water oxidation. FEBS Letters. 2013;**587**: 2083-2089. DOI: 10.1016/j.febslet.2013.05.046

[101] Zybailov B, Rutschow H, Friso G, et al. Sorting signals, N-terminal modifications and abundance of the chloroplast proteome. PLoS One. 2008;**3**(4):e1994. DOI: 10.1371/journal.pone.0001994

[102] Gao L, Lu Z, Ding L, Guo J, Wang M, Ling N, et al. Role of aquaporins in determining carbon and nitrogen status in higher plants. International Journal of Molecular Sciences. 2018;**19**(1):35-47. DOI: 10.3390/ijms19010035

[103] Zabaleta E, Hans V, Braun P. A basal carbon concentrating mechanism in plants? Plant Science. 2012;**187**:97-104. DOI: 10.1016/j.plantsci.2012.02.001

[104] Millar A, Sweetlove L, Giege P, Leaver C. Analysis of the Arabidopsis mitochondrial proteome. Plant Physiology. 2001;**127**:1711-1727. DOI: 10.1104/pp.010387

[105] Wang Q, Fristedt R, Yu X, Chen Z, Liu H, Lee Y, et al. The γ-carbonic anhydrase subcomplex of mitochondrial complex I is essential for development and important for photomorphogenesis of Arabidopsis. Plant Physiology. 2012;**160**(3):1373-1383. DOI: 10.1104/pp.112.204339

[106] Badger M, Price G. The role of carbonic anhydrase in photosynthesis. Annual Review of Plant Physiology and Plant Molecular Biology. 1994;**45**:369-392. DOI: 10.1146/annurev.pp.45.060194.002101

[107] Edwards G, Mohamed A. Reduction of carbonic anhydrase activity in zinc deficient leaves of *Phaseolus vulgaris* L. Crop Science. 1973;**13**:351-354. DOI: 10.2135/cropsci1973.0011183X001300030018x

[108] Price G, von Caemmerer S, Evans J, Yu J, Lloyd J, Oja V, et al. Specific reduction of chloroplast carbonic anhydrase activity by anti-sense RNA in transgenic tobacco plants has a minor effect on photosynthetic CO_2 assimilation. Planta. 1994;**193**:331-340. DOI: 10.1007/BF00201810

[109] Majeau N, Arnoldo M, Coleman J. Modification of carbonic anhydrase activity by antisense and over-expression constructs in transgenic tobacco. Plant Molecular Biology. 1994;**25**:377-385. DOI: 10.1007/BF00043867

[110] Porter M, Grodzinski B. Acclimation to high CO_2 in bean carbonic anhydrase and ribulose bisphosphate carboxylase. Plant Physiology. 1984;**74**:413-416. DOI: 10.1104/pp.74.2.413

[111] Anderson L, Carol A. Enzyme co-localization with rubisco in pea leaf chloroplasts. Photosynthesis Research. 2004;**82**:49-58. DOI: 10.1023/B:PRES.0000040443.92346.37

[112] Lazova G, Stemler A. A 160 kDa protein with carbonic anhydrase activity is compexed with rubisco on the outer

surface of thylakoids. Cell Biology International. 2008;**32**:646-653

[113] Ignatova L, Novichkova N, Mudrik V, Lyubimov V, Ivanov B, Romanova A. Growth, photosynthesis, and metabolism of sugar beet at an early stage of exposure to elevated CO_2. Russian Journal of Plant Physiology. 2005;**52**(2):158-164

[114] Alam M, Nahar K, Hasanuzzaman M, Fujita M. Exogenous jasmonic acid modulates the physiology, antioxidant defense and glyoxalase systems in imparting drought stress tolerance in different Brassica species. Plant Biotechnology Reports. 2014;**8**:279-293

[115] Liu C, Chen K, Zhao X, Wang X, Shen C, Zhu Y, et al. Identification of genes for salt tolerance and yield-related traits in rice plants grown hydroponically and under saline field conditions by genome-wide association study. Rice. 2019;**12**(1):88-100. DOI: 10.1186/s12284-019-0349-z

[116] Buren S. Targeting and function of CAH1-Characterisation of a novel protein pathway to the plant cell chloroplast [thesis]. Umea: Umea University; 2010

[117] Hu H, Boisson-Dernier A, Israelsson-Nordstrom M, Bohmer M, Xue S, Ries A, et al. Carbonic anhydrases are upstream regulators of CO_2-controlled stomatal movements in guard cells. Nature Cell Biology. 2010;**12**:87-93. DOI: 10.1038/ncb2009

[118] Roelfsema MRG, Hedrich R, Geiger D. Anion channels: Master switches of stress responses. Trends in Plant Science. 2012;**17**(4):221-229. DOI: 10.1016/j.tplants.2012.01.009

[119] Xue S, Hu H, Ries A, Merilo E, Kollist H, Schroeder J. Central functions of bicarbonate in S-type anion channel activation and OST1 protein kinase in CO_2 signal transduction in guard cell. The EMBO Journal. 2011;**30**(8): 1645-1658. DOI: 10.1038/emboj.2011.68

[120] Tian W, Hou C, Ren Z, Pan Y, Jia J, Zhang H, et al. A molecular pathway for CO_2 response in Arabidopsis guard cells. Nature Communications. 2015;**6**:6057. DOI: 10.1038/ncomms7057

[121] Kende H. Ethylene biosynthesis. Plant Molecular Biology. 1993;**44**: 283-307. DOI: 10.1146/annurev.pp.44.060193.001435

[122] Ferreira F, Guo C, Coleman J. Reduction of plastid-localized carbonic anhydrase activity results in reduced Arabidopsis seedling survivorship. Plant Physiology. 2008;**147**:585-594. DOI: 10.1104/pp.108.118661

[123] Slaymaker D, Navarre D, Clark D, del Pozo O, Martin G, Klessig D. The tobacco salicylic acid binding protein 3 (SABP3) is the chloroplast carbonic anhydrase which exhibits antioxidant activity and plays a role in the hypersensitive defense response. PNAS. 2002;**99**(10):11640-11645. DOI: 10.1073/pnas.182427699

[124] Restrepo S, Myers K, del Pozo O, Martin G, Hart A, Buell C, et al. Gene profiling of a compatible interaction between *Phytophthora infestans* and *Solanum tuberosum* suggests a role for carbonic anhydrase. Molecular Plant Microbe Interactions Journal. 2005;**18**(9):913-922. DOI: 10.1094/MPMI-18-0913

[125] Frick U, Schaller A. cDNA microarray analysis of fusicoccin-induced changes in gene expression in tomato plants. Planta. 2002;**216**:83-94. DOI: 10.1007/s00425-002-0887-1

[126] Schenk P, Kazan K, Wilson I, Anderson J, Richmond T, Somerville S, et al. Coordinated plant defense responses in Arabidopsis revealed by microarray analysis. PNAS. 2000;**97**:11655-11660. DOI: 10.1073/pnas.97.21.11655

[127] Medina-Puche L, Castelló M, Canet J, Lamilla J, Colombo M, Tornero P. β-Carbonic anhydrases play a role in salicylic acid perception in Arabidopsis. PLoS One. 2017;**12**(7):e0181820. DOI: 10.1371/journal.pone.0181820

[128] Hoang C, Chapman K. Biochemical and molecular inhibition of plastidial carbonic anhydrase reduces the incorporation of acetate into lipids in cotton embryos and tobacco cell suspensions and leaves. Plant Physiology. 2002;**128**:1417-1427. DOI: 10.1104/pp.010879

[129] Pronina N, Allakhverdiev S, Kupriyanova E, Klyachko-Gurvich G, Klimov V. Carbonic anhydrase in subchloroplast particles of pea plants. Russian Journal of Plant Physiology. 2002;**49**(3):303-310. DOI: 10.1023/A:1015589215862

[130] Punnett T, Iyer R. The enhancement of photophosphorylation and the hill reaction by carbon dioxide. The Journal of Biological Chemistry. 1964;**239**:2335-2339. DOI: 10.1104/pp.40.6.1074

[131] Cohen W, Jagendorf T. Inhibition of energy-linked reactions in chloroplasts by polygalacturonate. Archives of Biochemistry and Biophysics. 1972;**150**:235-243. DOI: 10.1016/0003-9861(72)90031-8

[132] Fedorchuk T, Opanasenko V, Rudenko N, Ivanov B. Bicarbonate-induced stimulation of photophosphorylation in isolated thylakoids: Effects of carbonic anhydrase inhibitors. Biological Membranes. 2018;**35**:34-41. DOI: 10.7868/S0233475518010048

[133] Cohen W, MacPeek W. A proposed mechanism for the stimulatory effect of bicarbonate ions on ATP synthesis in isolated chloroplasts. Plant Physiology. 1980;**66**(2):242-245. DOI: 10.1104/pp.66.2.242

[134] Zhurikova E, Ignatova L, Semenova G, Rudenko N, Mudrik V, Ivanov B. Effect of knockout of α-carbonic anhydrase 4 gene on photosynthetic characteristics and starch accumulation in leaves of *Arabidopsis thaliana*. Russian Journal of Plant Physiology. 2015;**62**:564-569. DOI: 10.1134/S1021443715040214

[135] Rudenko N, Fedorchuk T, Vetoshkina D, Zhurikova E, Ignatova L, Ivanov B. Influence of knockout of At4g20990 gene encoding α-CA4 on photosystem II light-harvesting antenna in plants grown under different light intensities and day lengths. Protoplasma. 2018;**255**(1):69-78. DOI: 10.1007/s00709-017-1133-9

[136] Li Y, Zhu Y, Shu Y, Meng F, Lu Y, Bai X, et al. Genome-wild identification of osmotic stress response in *Arabidopsis thaliana*. Genomics. 2008;**92**:488-493. DOI: 10.1016/j.ygeno.2008.08.011

[137] Allen J. Plastoquinone redox control of chloroplast thylakoid protein phosphorylation and distribution of excitation energy between photosystems: Discovery, background, implications. Photosynthesis Research. 2002;**73**:139-148. DOI: 10.1007/1-4020-3324-9_17

[138] Maciejewska U, Polkowska-Kowalczyk L, Swiezewska E, Szkopinskae A. Plastoquinone: possible involvement in plant disease resistance. Acta Biochimica Polonica. 2002;**49**:775-780. DOI: 024903775

[139] Frigerio S, Campoli C, Zorzan S, Fantoni L, Crosatti C, Drepper F, et al. Photosynthetic antenna size in higher plants is controlled by the plastoquinone redox state at the post-transcriptional rather than transcriptional level. The Journal of Biological Chemistry.

2007;**282**:29457-29469. DOI: 10.1074/jbc.M705132200

[140] Ignatova L, Moskvin O, Romanova A, Ivanov B. Carbonic anhydrases in the C3-plant leaf cell. Australian Journal of Plant Physiology. 1998;**25**:673-677. DOI: 10.1071/PP97137

[141] Moskvin O, Shutova T, Khristin M, Ignatova L, Villarejo A, Samuelsson G, et al. Carbonic anhydrase activities in pea thylakoids. Photosynthesis Research. 2004;**79**:93-100. DOI: 10.1023/B:PRES.0000011925.93313.db

[142] Fedorchuk T, Rudenko N, Ignatova L, Ivanov B. The presence of soluble carbonic anhydrase in the thylakoid lumen of chloroplasts from Arabidopsis leaves. Journal of Plant Physiology. 2014;**171**(11):903-906. DOI: 10.1016/j.jplph.2014.02.009

Chapter 17

Active Deformation in the Tunic of *Halocynthia roretzi*: How the Tissue Composed of Cellulose Responds to Stimuli and Deforms

Yoko Kato

Abstract

Halocynthia roretzi, belonging to class Ascidiacea, has highly pure and crystalline cellulose Iβ, and sulfated chitin in its tunic. Cells, including hemocytes in the open circulatory system, are scattered in the tunic. The tunic, which maintains its thickness by continuous proliferation and removal, can be classified into active tissues. Recently, it has been reported that various stimuli, such as mechanical stimuli and changes in the mechanical environment, could cause active deformations of the tunic without changes in the characteristics of the tissue structure, which would be associated with influx and efflux of water. In this chapter, the system associated with active deformation, tissue structure and flux of water in the tunic is shown, with reference to the previous reports.

Keywords: cellulose, sulfated chitin, active deformation, water, stimuli, adaptation, tissue structure

1. Introduction

Halocynthia roretzi, which is a solitary ascidian and of the class Ascidiacea (the subphylum Tunicata and the phylum Chordata) in marine habitats, is entirely covered with the tissue called tunic. An example of *Halocynthia roretzi* is shown in **Figure 1A**. The tunic, where blood vessels and various cells including hemocytes have been observed [1–3], shows the system to keep its thickness by continuous removal and secretion [1] and defense system by the secreted substances of the hemocytes [4–11]. While it has been reported that the species in Tunicata has cellulose in its tunic [12], whose elastic modulus is 143 GPa [13], cellulose Iβ in the tunic of *Halocynthia roretzi* shows pure and highly crystalline form [14]. Also, sulfated chitin, which is biocompatible as well as biodegradable [15], has been observed in the tunic [16, 17]. In addition to the aforementioned components, α-smooth muscle actin and elastic fiber, which are expected to directly influence the mechanical properties of the tunic, and nervous systems, have been observed [18]. In the meantime, the active deformation in the tunic of *Halocynthia roretzi*, caused by acetylcholine (neurotransmitter) [18], mechanical stimuli [18, 19], electric stimuli [20] and enzyme (α-chymotrypsin) [20], has been reported. The active deformation responding to the mechanical environment has been associated with change in mass

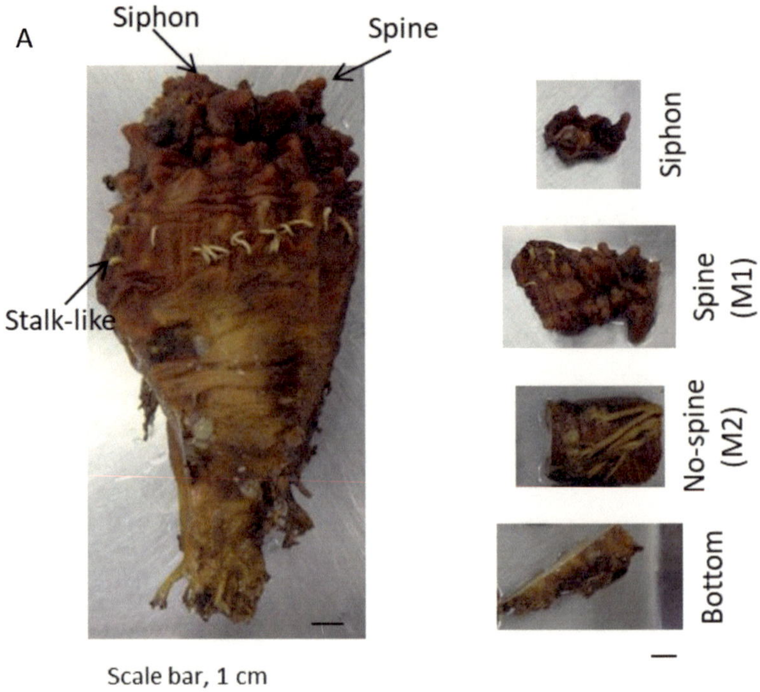

Figure 1.
Sample of Halocynthia roretzi. *A, entire image; B, the tunic sample in each category (siphon, M1 (tunic with spines), M2 (tunic without a spine) and bottom (thickest part)).*

of the tunic [21]. Because the change in mass of the tunic agreed with that in water content of the tunic, influx and efflux of water would be involved with the tunic deformation [21]. When the tunic sample was put into the seawater, the absorbance at 220 nm and 250–350 nm [22–27], which is influenced by the concentrations of nitrate and dissolved organic matter, was changed so that the substances released from the tunic would be added to the seawater [21, 28].

As **Figure 1** shows, the tunic tissue can be categorized by characteristics in shape: siphon, tubular parts where seawater is passing through; M1, tunic with spines; M2, tunic without a spine; and bottom, thickest part. While the mechanical stimuli caused a decrease in mass in every category, the tunic in the seawater at 5°C indicated an increase in the mass of the tunic, which became smaller as the position was closer to bottom [21]. While the outer layer and collapse of blood vessels could cause the difference in change of mass [21], the cells extracted from the tunic by centrifugation, kept in the seawater at 5°C for 10 days, showed motility [28] so that these cells would also influence change in mass. While the absorbance at 220 nm and 250–350 nm in the seawater used for keeping the tunic at 5°C was decreased after the removal of the tunic samples [28], the influence of the tunic category has been barely examined. Also, whether or not the cells in the tunic are obtained from all the tunic categories by centrifugation at the same degree has not been clear. If the effect of centrifugation on separating the cells from the tunic tissue is dependent on the tunic category, the characteristics of the tunic structure would be diverse and influence mass transfer.

In this chapter, why the tunic category, composed of siphon, M1, M2 and bottom, could influence the active deformation was examined. The absorbance of the seawater, which kept the tunic sample in each category separately, was evaluated by spectroscopic analysis in order to examine the change in the components of the seawater. The seawater after removing the tunic sample was also evaluated in the same way. In the meantime, the hemocytes in each category of the tunic, which would secrete halocyamines (antimicrobial substance) [5] and hemagglutinin [10], were obtained by centrifugation to examine the influence of the tissue category on separating the cells from the tunic.

2. The following materials and methods have been followed to find out the findings

2.1 Change in mass and components from the tunic

The samples of *Halocynthia roretzi* were obtained from Yamanaka Inc. and Marutaki Suisan (Miyagi, Japan) (n = 3). The tunic was removed from other organs and cut into samples in each category (siphon, M1, M2 and bottom) by tweezers and trimming blades (feather trimming blade; Feather Safety Razor, Co. Ltd., Osaka, Japan) as **Figure 1B** shows. The sample in each category was put into the artificial seawater (Reef Crystals, Aquarium Systems, Sarrebourg, France) separately, and kept at 5°C for 10 days (Day 10) or 15 days (Day 15). The mass of the tunic, which was wrapped by paper (Kimwipe; Nippon Paper Crecia, Tokyo, Japan) for 10 s to remove water on the surface, was measured with the balance (UW420S; Shimadzu Corporation, Kyoto, Japan), in order to check whether or not the change in mass of the tunic sample agreed with that in the previous report [21]. After removing the tunic sample, two types of the seawater samples, filtrated (1001-150 (Whatman); GE Healthcare Japan, Tokyo, Japan) and not filtrated, were prepared. The two types of seawater samples were kept at 5°C for 10 days, 17 days or 30 days. The absorbance of the seawater at 190–1100 nm was measured by the spectrometer (UV-1280; Shimadzu Corporation, Kyoto, Japan) before and after removing the tunic sample. The absorbance at 220 nm and mean absorbance at 250–350 nm, which are influenced by the concentrations of nitrate and dissolved organic matter [21–27], and the peak absorbance around 970 nm, which was clearly observed, were used to evaluate the characteristics of the seawater. For the evaluation in the shape of the absorbance curve at 220–350 nm, the standard deviation of the absorbance at 250–350 nm, divided by the mean absorbance at the same range, and the mean absorbance at 250–350 nm, divided by the absorbance at 220 nm, which is named shape index, were used. Shape index was also used for estimating the change in the component ratio of the seawater.

2.2 Hemocytes

While there are several types of hemocytes in *Halocynthia roretzi* [11], the hemocyte secreting halocyamines and hemagglutinin could be obtained by the centrifugation (1000 G, 7 min) of hemolymph [5, 10]. Considering that effect of centrifugal force on separating the hemocyte from the tunic could be a parameter to evaluate the characteristics of the tunic structure, the tunic samples in each category were centrifuged in the previous report [5, 10] (n = 5). During the centrifugation, the tunic sample was put into the artificial seawater (Suprema21; Tomy, Tokyo, Japan). After removing the supernatant and tunic sample, the cells were obtained. Because the cells seemed damaged during counting the number by hemocytemeter, the number of the obtained cells was estimated by observation under the microscope (CX41-31PHP; Olympus, Tokyo, Japan).

3. Outcomes of the present study

3.1 Change in the mass of the tunic and components from the tunic

An example of a change in mass of the tunic sample is shown in **Figure 2**. The tunic bottom underwent smaller changes than those in other categories. The tendency, which was observed in all the samples, agreed with that in the previous report [21].

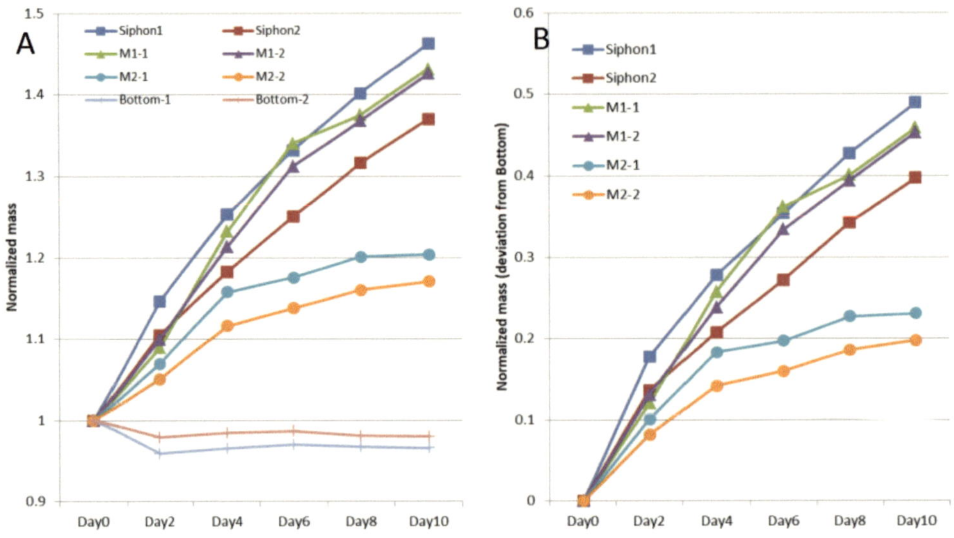

Figure 2.
Change in the mass of the tunic sample kept in the seawater at 5°C up to 10 days (day 10). A, normalized by the mass before the immersion; B, deviation from the normalized mass in bottom. All the samples indicated the same tendency.

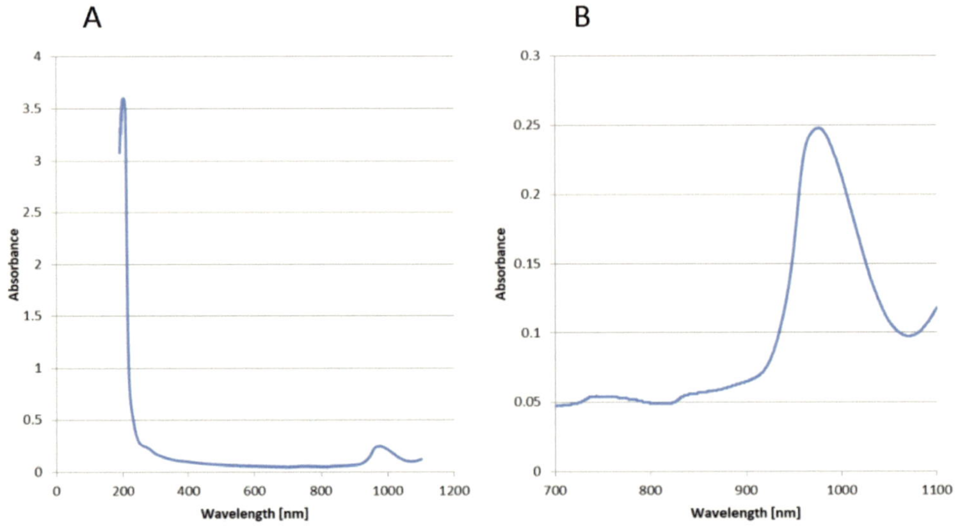

Figure 3.
Absorbance for the seawater containing the tunic sample (siphon) for 10 days at 5°C (190–1100 nm). This absorbance at 190–1100 nm was one of the results. A, entire range; B, around 1000 nm.

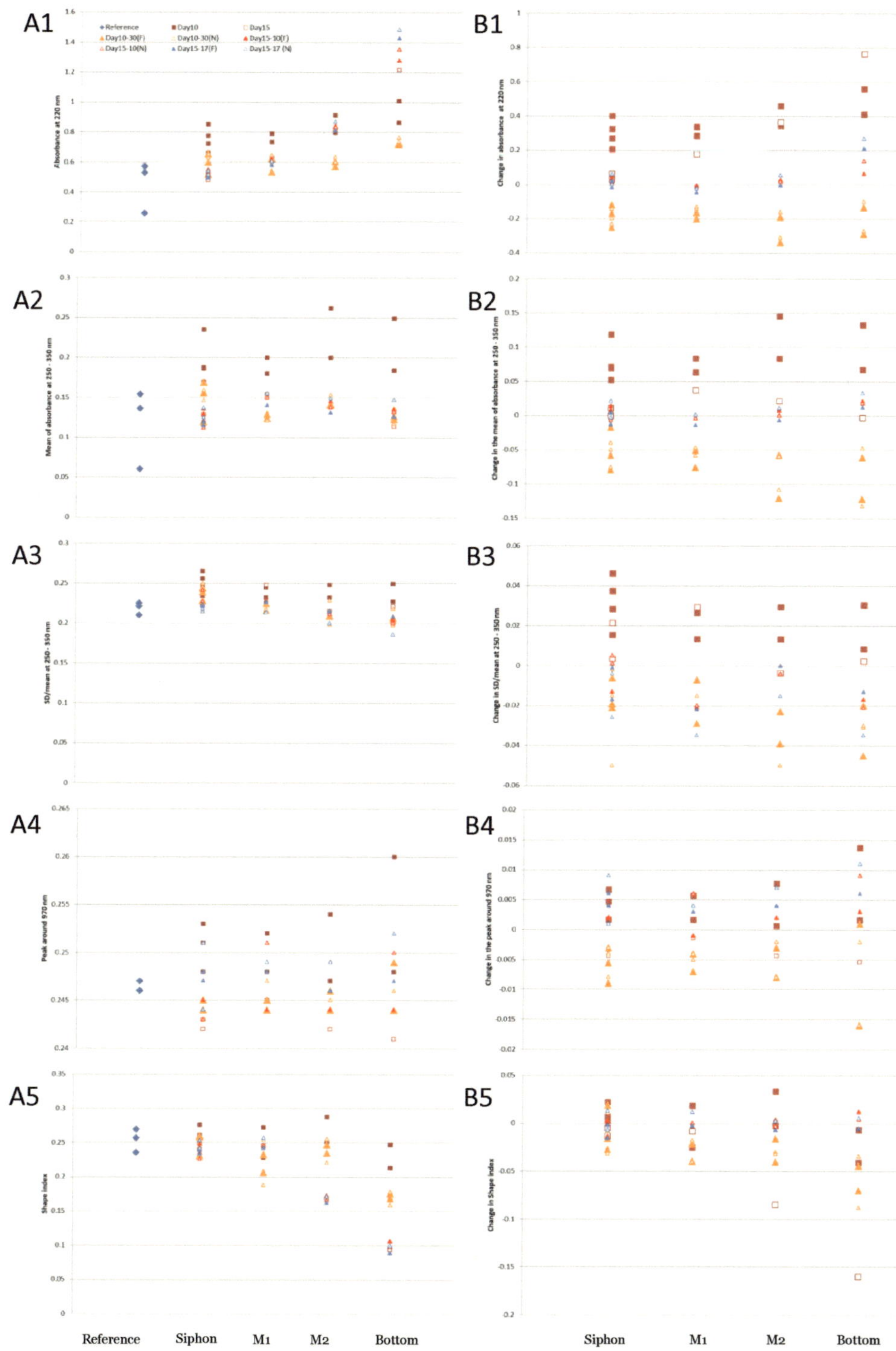

Figure 4.
Absorbance at the characteristic wavelength and related parameter. The absorbance at each wavelength and related parameter (A1–A5, left), and their change between the adjacent processes (B1–B5, right): before adding, keeping and removing the tunic samples in the seawater. The seawater samples labelled as follows: reference, without usage; day i *(*i = 10, 15*), keeping the tunic sample at 5°C for* i *days; day* i–j *(F or N) (*i = 10, 15, j = 10, 17, 30*), keeping the tunic sample at 5°C for* i *days and kept at 5°C after removing the tunic samples for* j *days with filtration (F), or without filtration (N).*

An example of the absorbance at 190–1100 nm is shown in **Figure 3**. The shape of the absorbance curve is almost the same in all the samples. The absorbance at the characteristic wavelength and related parameter, shape index, and their changes, caused by the adjacent process, in each seawater sample are shown in **Figure 4**. Considering the influences of the tunic sample categories (siphon, M1, M2 and bottom) on the absorbance, the absorbance and related parameter are indicated in each sample category. The mean value and change between the adjacent processes and their ranges through all the processes are indicated in **Figures 5** and **6**, respectively. As **Figures 4–6** show, the absorbance values at the characteristic wavelength and related parameters were changed by the tunic category as well as the presence and removal of the tunic samples. While the change in shape index between the adjacent processes was zero or less, other absorbance values and parameters increased before the removal of the tunic samples, and decreased after the removal, in all the tunic categories, as **Figure 5B** shows. Because the presence and absence of the tunic samples in the seawater directly influenced these parameters, and change in the component ratio of the seawater was kept through the processes, the substances released from the tunic sample would be partially degradable with progress in the change of the component ratio in the seawater. But the influences of the tunic category and process in other results were so complicated that they could be hardly explained in such a simple way. These results indicated that each category might have different systems to control its active deformation.

3.2 Cells

Figure 7 shows the cells from M1 by centrifugation (1000 G, 7 min). The cells were also obtained from the tunic samples of siphon and M2, but barely from bottom. Considering blood vessels in bottom and open circulation in the entire body, few cells in bottom would be hardly expected. Hence, there might be the

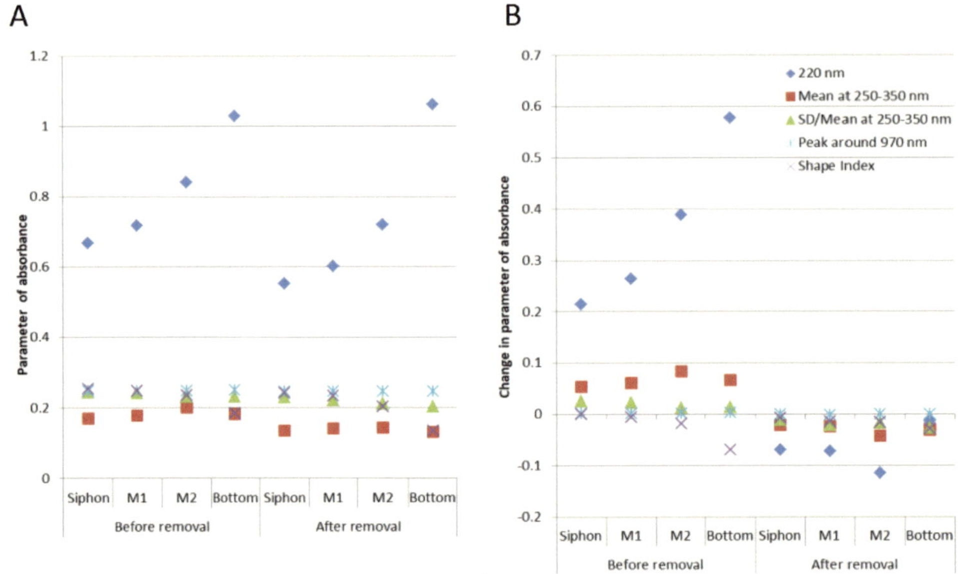

Figure 5.
Mean absorbance and related parameter. The parameter of absorbance (A) and its change between the adjacent processes (B), before and after the removal of the tunic samples in the seawater, are shown.

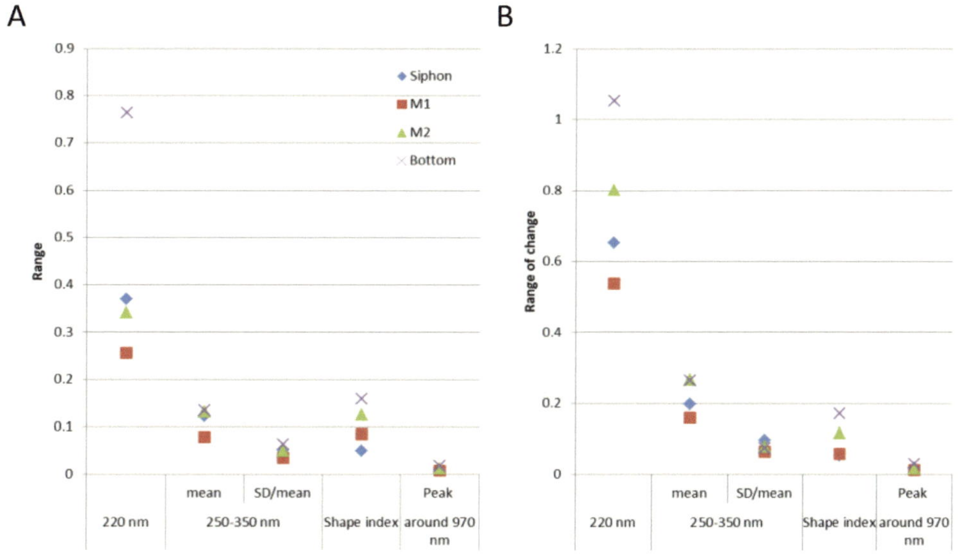

Figure 6.
Range of the parameter and change through all the processes. The range of the parameter (A) and change through all the processes from reference (B) are shown.

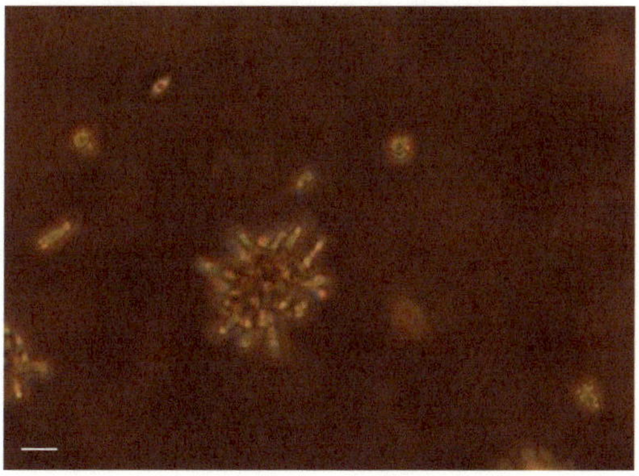

Figure 7.
Cells from M1. These cells were obtained by centrifugation (1000 G, 7 min). Scale bar, 20 μm.

characteristics of the tissue structure in bottom, which would cause cells to be hardly separated by an external force, but not in other categories of the tunic, siphon, M1 and M2.

4. Discussion of the findings as compared to earlier studies

In this chapter, the difference in the tunic categories, which are siphon, M1, M2 and bottom, was investigated to examine the system for active deformation in the tunic. Considering that influx and efflux from the tunic, which are associated with the active deformation of the tunic, would bring some components to the seawater, change in the components of the seawater was evaluated by the absorbance at the characteristic wavelength and related parameters. In all the tunic categories, these parameters, except shape index, which continuously decreased, were increased by keeping the tunic in the seawater and decreased by removing them. These results indicated that the substances, released from the tunic, would disappear without continuous supply and keep the change in the component ratio of the seawater. The released substances would be degradable partially as well as reactive, associated with the change of the component ratio of the seawater. In the meantime, the influence of each tunic category on these parameters was complicated. Hence, the active deformation would be controlled by two types of substances, which would be in every category of the tunic sample, and specific in each category. The details of the substances will be investigated in the future.

In the meantime, the cells were obtained from siphon, M1 and M2 by centrifugation, but not from bottom. Considering the open circulation system and blood vessels in bottom, bottom would have cells, which would be hardly separated from the surrounding by centrifugation because of the characteristics in the tissue structure of bottom, different from those in other tunic categories. The result that change in mass of the tunic was smallest at bottom would agree with this unique feature of bottom. Why the cells in bottom are hardly obtained by centrifugation and how the cells in bottom can be obtained will be investigated in the future.

5. Conclusion

In this chapter, the active deformation of the tunic in *Halocynthia roretzi*, a solitary ascidian, was investigated by the substances released from the tunic, and cells obtained from the tunic by centrifugation. The absorbance at the characteristic wavelength and related parameter, except shape index, in the seawater were enhanced by keeping the tunic samples and decreased by removing them while shape index was continuously decreased. Hence, the substances released from all the tunic categories would be partially degradable, and reactive enough to stable change in the component ratio of the seawater. The difference in the influences of the tunic category on these parameters, which was complicated, would contribute to a difference in the active deformation in each tunic category. The cells in bottom were hardly obtained by centrifugation although those in other categories were successfully obtained. Hence, bottom would have the specific characteristics in the tissue structure that would keep the cells in the tunic firmly. Also, these characteristics in bottom would prevent change in mass of the tunic at bottom.

Author details

Yoko Kato
Faculty of Engineering, Tohoku Gakuin University, Tagajo, Japan

*Address all correspondence to: ykato@mail.tohoku-gakuin.ac.jp

References

[1] Berill NJ. The Tunicata with an Account of British Species. Pisces Conservation Ltd: Lymington; 1950. pp. 365

[2] Goodbody I. The physiology of ascidians. In: Russel FS, Yonge M, editors. Advances in Marine Biology. Vol. 12. Amsterdam: Elsevier; 1974. pp. 1-149. DOI: 10.1016/S0065-2881(08)60457-5

[3] Das SM. On the structure and function of the ascidian test. Journal of Morphology. 1936;**59**(3):589-600. DOI: 10.1002/jmor.1050590308

[4] Yokosawa H, Sawada H, Abe Y, Numakunai T, Ishii S. Galactose-specific lectin in the hemolymph of solitary ascidian, *Halocynthia roretzi*: Isolation and characterization. Biochemical and Biophysical Research Communications. 1982;**107**(2):451-457. DOI: 10.1016/0006-291X(82)91512-1

[5] Azumi K, Yokosawa H, Ishii S. Halocyamines: Novel antimicrobial tetrapeptide-like substances isolated from the hemocytes of the solitary ascidian *Halocynthia roretzi*. Biochemistry. 1990;**29**(1):159-165. DOI: 10.1021/bi00453a021

[6] Yokosawa H, Harada K, Igarashi K, Abe Y, Takahashi K, Ishii S. Galactose-specific lectin in the hemolymph of solitary ascidian, *Halocynthia roretzi*. Molecular, binding and functional properties. Biochimica & Biophysica Acta. 1986;**870**(2):242-247. DOI: 10.1016/0167-4838(86)90228-1

[7] Harada-Azumi K, Yokosawa H, Ishii S. N-acetyl-galactosamine-specific lectin, a novel lectin in the hemolymph of the ascidian *Halocynthia roretzi*: Isolation, characterization and comparison with galactose-specific lectin. Comparative Biochemistry and Physiology. B. 1987;**88B**(1):375-381. DOI: 10.1016/0305-0491(87)90130-1

[8] Azumi K, Yoshimizu S, Suzuki S, Ezura Y, Yokosawa H. Inhibitory effect of halocyamine, an antimicrobial substance from ascidian hemocytes, on the growth of fish viruses and marine bacteria. Experientia. 1990;**46**(10):1066-1068. DOI: 10.1007/BF01940675

[9] Azumi K, Yokosawa H, Ishii S. Lipopolysaccharide induces release of a metallo-protease from hemocytes of the ascidian, *Halocynthia roretzi*. Developmental and Comparative Immunology. 1991;**15**(1-2):1-7. DOI: 10.1016/0145-305X(91)90041-V

[10] Azumi K, Ozeki S, Yokosawa H, Ishii S. A novel lipopolysaccharide-binding hemagglutinin isolated from hemocytes of the solitary ascidian, *Halocynthia roretzi*: It can agglutinate bacteria. Developmental and Comparative Immunology. 1991;**15** (1-2):9-16. DOI: 10.1002/jez.1402650312

[11] Azumi K, Satoh N, Yokosawa H. Functional and structural characterization of hemocytes of the solitary ascidian, *Halocynthia roretzi*. Journal of Experimental Zoology. 1993;**265**:309-316. DOI: 10.1002/jez.1402650312

[12] Zhao Y, Li J. Excellent chemical and material cellulose from tunicates: Diversity in cellulose production yield and chemical and morphological structures from different tunicate species. Cellulose. 2014;**21**(5):3427-3441. DOI: 10.1007/s10570-014-0348-6

[13] Štrucova A, Davies GR, Eichhorn SJ. Elastic modulus and stress-transfer properties of tunicate cellulose whiskers. Biomacromolecules.

2005;**6**(2):1055-1061. DOI: 10.1021/bm049291k

[14] Nishiyama Y, Langan P, Chanzy H. Crystal structure and hydrogen-bonding system in cellulose Iβ from synchrotron X-ray and neutron fiber diffraction. Journal of the American Chemical Society. 2002;**124**(31):9074-9082. DOI: 10.1021/ja0257319

[15] Jayakumar R, New N, Tokura S, Tamura H. Sulfated chitin and chitosan as novel biomaterials. International Journal of Biological Macromolecules. 2007;**40**(3):175-181. DOI: 10.1016/j.ijbiomac.2006.06.021

[16] Anno K, Otsuka K, Seno N. A chitin sulfate-like polysaccharide from the test of the tunicate *Halocynthia roretzi*. Biochimica et Biophysica Acta. 1974;**362**(1):215-219. DOI: 10.1016/0304-4165(74)90043-9

[17] Wagner GP. Evolution and multi-functionality of the chitin system. EXS. 1994;**69**:559-577. DOI: 10.1007/978-3-0348-7527-1_33

[18] Kato Y. Active movement of the tunic in *Halocynthia roretzi*. Journal of Biomechanical Science and Engineering. 2010;**5**(2):163-174. DOI: 10.1299/jbse.5.163

[19] Kato Y. Mechanical senses and the tunic structure in *Halocynthia roretzi*. In: Proceedings of ISOPE-2011: 19-23 June 2011; Maui, California. Volume II. 2011. pp. 250-253

[20] Kato Y. The role of protein as a deformation controller in cellulose tissue. In: Proceedings of ASME 2012 International Mechanical Engineering Congress and Exposition; 9-15 Nov 2012; Houston. New York: ASME. 2012. pp. 607-613

[21] Kato Y. Deformation control and mass transfer in the tunic of *Halocynthia roretzi*. Open Chemistry Journal. 2018;**5**:1-17. DOI: 10.2174/1874842201805010001

[22] Armstrong FAJ, Boalch GT. Volatile organic matter in algal culture media and sea water. Nature. 1960;**185**:761-762. DOI: 10.1038/185761b0

[23] Armstrong FAJ, Boalch GT. The ultra-violet absorption of sea water. Journal of the Marine Biological Association of the United Kingdom. 1961;**41**:591-597. DOI: 10.1017/S0025315400016179

[24] Ogura N, Hanya T. Nature of ultra-violet absorption of sea water. Nature. 1966;**212**:758. DOI: 10.1038/212758a0

[25] Ogura N, Hanya T. Ultraviolet absorption of the sea water, in relation to organic and inorganic matters. International Journal of Oceanology and Limnology. 1967;**1**(2):91-102

[26] Foster P, Morris AW. The use of ultra-violet absorption measurements for the estimation of organic pollution in inshore sea water. Water Research. 1971;**5**:19-27. DOI: 10.1016/0043-1354(71)90059-5

[27] Collos Y, Mornet F, Sciandra A, Waser N, Larson A, Harrison PJ. An optical method for the rapid measurement of micromolar concentrations of nitrate in marine phytoplankton cultures. Journal of Applied Phycology. 1999;**11**:179-184. DOI: 10.1023/A:1008046023487

[28] Kato Y. Maintenance of the cell motility in the tunic of *Halocynthia roretzi*. In: Proceedings of 2018 Sustainable Industrial Processing Summit & Exhibition; Volume 6: New and Advanced Materials and Technologies; 4-7 November 2018; Rio de Janeiro. Quebec: FLOGEN. 2018. pp. 207-214

Chapter 18

Amelioration of Drought Stress on Plants under Biostimulant Sources

Ana Carolina Feitosa de Vasconcelos

Abstract

Water stress is one of the most important environmental factors inducing physiological changes in plants, such as decreasing water potential of the cells and the stomatal closure, resulting in reduced CO_2 availability for the plants and inhibiting photosynthesis. One common feature of these stress conditions is the development of oxidative processes mediated by reactive oxygen species (ROS). ROS accumulate in the cells and cause damage in important cellular components, such as thylakoids and chloroplasts. Plants have antioxidant defense systems to cope with ROS. Antioxidants enzymes superoxide dismutase (SOD), catalase (CAT), and ascorbate peroxidase (APX) are efficient scavengers of ROS: superoxide, hydroxyl radicals, and singlet oxygen. The activities of antioxidant enzymes in plants are normally favored when plants are subjected to some kind of improvement in the conditions in which they are grown. In this sense, biostimulants cause changes in vital and structural processes in order to influence plant growth through improved tolerance to abiotic stresses by increasing the antioxidant activity in plants.

Keywords: reactive oxygen species, antioxidant enzymes, plant drought resistance, humic substances, seaweed extracts, hormones

1. Introduction

Water availability is one of the most important environmental factors for plant growth and development. The water deficit caused by drought or salinity in soils is one of the most serious environmental problems that limit agricultural production in various regions of the world. According to [1], water deficit occurs when all water content in the cell is below the highest water content displayed in the state of greatest hydration.

Plants experience a water deficit when water supply to the roots becomes difficult or when the rate of evapotranspiration becomes very high. These two conditions generally coincide in regions with an arid and semiarid climate and affect plants to a greater or lesser extent according to the tolerance that species have [2].

Plant response to biotic and abiotic stresses is a complex network of reactions, which involves different physiological pathways of the primary and secondary metabolism. At the cellular level, membranes and proteins can be damaged by a reduction in hydration and an increase in reactive oxygen species (ROS) [3]. ROS derive from oxidative processes such as photosynthesis and respiration, and, in normal conditions, they are produced in low concentration without any negative consequences for the plants. In stressful conditions (biotic or abiotic), ROS levels increase as an index of the oxidative burst induced by the stress agent [4]. When ROS become toxic, they can result in a series of damages to plant metabolism, such as deterioration of photosynthetic components, inactivation of proteins and enzymes, and destruction of the structure and permeability of the cell membrane by lipid peroxidation [5, 6].

Antioxidants and their role in the plant defense system have received a lot of attention in scientific research. Many results suggest that the effects of environmental stresses, such as salinity, drought, low temperatures, and herbicide residues, damage plants directly or indirectly by increasing endogenous ROS [7].

Plant cells are protected against the damaging effects of ROS by a complex antioxidant system composed of enzymatic antioxidants, such as superoxide dismutase (SOD), catalase (CAT), and ascorbate peroxidase (APX) [8]. The close relationship between antioxidant activity and stress tolerance has been identified in many crops such as maize (*Zea mays* L.) [7], tobacco (*Nicotiana tabacum*) [9], and grasses [10].

Biostimulants are extracts obtained from organic raw materials containing bioactive compounds. The most common components of the biostimulants are mineral elements, humic substances (HSs), vitamins, and amino acids [6]. Seaweed extracts have been used in agriculture as soil conditioners or as plant stimulators. They are applied as foliar spray and enhance plant growth; freezing, drought, and salt tolerance; photosynthetic activity; and resistance to fungi, bacteria, and virus, improving the yield and productivity of many crops [11, 12]. Seaweeds used for biostimulant production contain cytokinins and auxins or other hormone-like substances [13]. From a legal point of view, the biostimulants can contain traces of natural plant hormones, but their biological action should not be ascribed to them; otherwise they should be registered as plant growth regulators [6].

Humic acids have been used in the composition of many commercial products because they have phytohormones [14] that favor protection against oxidative damage in plants caused by environmental stresses. Thus, the use of biostimulants in agriculture has been emphasized, which are products that contain active ingredient or organic agent free of pesticides, capable of acting, directly or indirectly, on all or part of the cultivated plants, increasing their productivity [15].

The components of biostimulants can change the hormonal status of the plant and have a great influence on its development and health. Seaweed, humic acids, and vitamins are commonly present in biostimulants and are important in improving plant development and hormonal activity [16]. In addition, these products increase the antioxidant activity in plants, especially when they are under water stress, severe temperatures, and herbicide action, among others [7].

Several studies have shown results in improving the resistance of plants to water stress when subjected to the application of biostimulants. The activity levels of the antioxidant enzymes superoxide dismutase (SOD), ascorbate peroxidase (APX), and catalase (CAT) have been determined. In general, increases in these antioxidant enzymes have been observed with the use of biostimulants [16]. Another parameter that has been improved in the plant with the application of biostimulants is the photochemical efficiency [17].

Thus, the objective of this chapter was to approach the role of biostimulants in plants submitted to water supply deficit, by affecting the activities of enzymatic antioxidants.

2. Use of biostimulants in plants

Biostimulants are components that produce responses in plant growth by improving tolerance to abiotic stresses. Many of the effects of these products are based on their ability to influence the hormonal activity of plants. Phytohormones are chemical messengers that regulate the normal development of plants by growing roots and shoots, in addition to regulating responses to the environment where they are located [18].

Many statements about biostimulants also refer to the improvements they provide in the tolerance of plants to water stress, a limiting factor in the management of the crops. Water stress affects many metabolic functions in plants, specifically photosynthesis. The application of biostimulants increases the defense system of the plant by increasing its level of antioxidant enzymes [15].

The components of biostimulants can alter the plant's hormonal status and have a major influence on its growth and health. Seaweed, humic acids and vitamins are commonly present in biostimulants and are important in improving plant development and hormonal activity [19]. In addition, these products increase the antioxidant activity in plants, especially when they are under water stress, severe temperatures and herbicide action, among others [20].

However, the composition of biostimulants is partly unknown; the complexity of the extracts and the wide range of molecules contained in the solution make it very difficult to understand which the most active compounds are. Moreover, the isolation and study of a single component present in a biostimulant can produce unreliable results because the effects on plants are often due to the combination and synergistic action of different compounds. In addition, the mechanisms activated by biostimulants are difficult to identify and still under investigation [6].

Plants usually thrive when the environment is favorable. Under these conditions, the effects of biostimulants may not be easily identified. However, when plants are stressed and undergo treatment with biostimulants, they develop better, as their defense system becomes more efficient due to the increase in their levels of antioxidants [20]. Besides, many of the active substances of biostimulants can be present in very low concentrations, sometimes below the levels detectable with commonly available technologies, but can provide strong biological effects [6].

Biostimulants and humic substances have shown an influence on many metabolic processes in plants, such as respiration, photosynthesis, synthesis of nucleic acids, and ion absorption. Within the cell, humic substances can increase the chlorophyll content resulting in greener leaves and reduction of some problems in plants, such as leaf chlorosis, since humic substances improve the capacity of nutrient uptake by the roots [21]. Beyond humic substances, various raw materials have been used in biostimulant compositions, such as hormones, algae extracts, and plant growth-promoting bacteria [22].

3. Water stress in plants

Water availability is one of the most limiting environmental factors that affect crop productivity. In the semiarid tropics, the occurrence of drought or water deficit in the soil is quite common, despite the fact that crops in regions of tropical and temperate climate suffer seasonal periods of water deficit, especially during the summer [23].

Drought is a prevalent stress factor especially in arid and semiarid areas and can affect different aspects of plant growth, development, and metabolism. Drought is a multidimensional stress factor, and hence its effects on plants are complex. Its effects on plants can occur on a molecular level up to a whole-plant level. There are several reasons for drought in nature, including low rainfall, salinity, high temperature, and high intensity of light, among others [24].

Some of the plants' first responses to stress appear to be mediated by biophysical events, rather than changes in chemical reactions resulting from dehydration. The closing of stomata, the reduction of photosynthesis, and osmotic adjustments are the responses of some plants to the first stage of water deficit [25]. As the water content of the plant decreases, the cells shrink, and the cell walls relax. With this, the solutes increase their concentration in the cells, and the plasma membrane becomes thicker and more compressed, as it covered a smaller area than before [1]. Cell expansion occurs when the turgor pressure is greater than the growth of the cell wall. Water stress greatly decreases cell expansion and plant growth due to low turgor pressure [26].

Stomata provide the main mechanism for controlling the rate of water loss. However, the site of water loss is also the site of carbon gain by the plant, so a reduction in water loss by stomatal control also results in a reduction in assimilation with consequent effects on productivity and the accumulation of reactive oxygen species [27]. These responses hinder the supply of CO_2 for photosynthesis and expose chloroplasts to excess energy excitation, especially under high light intensity [25].

The low potentials in the soil and in the plant inhibit their growth, reduce the development activities of cells and tissues, decrease the uptake of nutrients, and cause morphological and biochemical changes [28]. To maintain water uptake, the roots have to grow deeper or increase their density. A characteristic of drought-resistant species is that they have a large proportion of their total mass consisting of roots and a deep-rooted habit. A high root/shoot ratio does not indicate in itself great ability to absorb water: water deficiency invariably increases the root/shoot ratio, but this is due to the loss of plant shoot weight without loss of root mass [1].

Photosynthesis is the driving force of plant productivity. The ability to maintain the rate of photosynthetic carbon dioxide and the assimilation of nitrate under environmental stresses is fundamental for the maintenance of plant growth and production. It is known that when water stress becomes extreme, non-stomatal factors can become even more limiting for photosynthesis [17].

The water deficit often decreases the number of photons captured by the leaves because withered leaves are at a more acute angle to the sun's rays. Changes in the absorption characteristics of the leaves occur due to the shrinkage of the cells. However, changes in chloroplasts and thylakoid during light capture and energy transfer centers are relatively small under water deficit conditions [29].

3.1 Reactive oxygen species and water stress

The diatomic oxygen (O_2) molecules in the Earth's atmosphere are the major promoters of reactions in cells. Except for those organisms that are specially adapted to live under anaerobic conditions, all animals and plants require oxygen for efficient energy production [30].

Aerobic organisms use diatomic oxygen as a terminal electron receptor, providing a high-energy field compared to fermentation and anaerobic respiration. In this base stage, molecular oxygen is relatively nonreactive, but it is capable of giving rise to excited reactive and lethal states, such as free radicals and their derivatives [31].

Superoxide, produced by electron transport to oxygen, is not compatible with cellular metabolism; hence, all organisms that are involved in aerobic environments must have an efficient mechanism capable of removing or neutralizing free radicals from cellular components. The balance between oxidative and antioxidant capabilities determines the fate of the plant [32]. Without this defense mechanism, plants may not efficiently convert solar energy into chemical energy [33].

The formation of reactive oxygen species occurs primarily through the superoxide radical ($O_2^{\bullet -}$), which can be dismutated into hydrogen peroxide (H_2O_2), or even through catalytic action, by the action of the superoxide dismutase (SOD) enzyme.

Antioxidant systems in plants act as mechanisms of resistance to stress by protecting the membranes against damage caused by these oxygen species produced under conditions of environmental and xenobiotic stress [34].

The fate of cells under stressful environments is determined by the duration of the stress, as well as the plant's protective capacity. Reactive oxygen species (ROS) play a crucial role in causing cellular damage to plants under stress. The sequence of events in plant tissues subjected to stress is increased production of ROS; increased levels of antioxidants; and increase in the capacity to "sweep" ROS, resulting in the plant's tolerance against water stress [35].

The detoxification mechanisms of ROS exist in all plants and can be categorized into enzymatic (superoxide dismutase, SOD; catalase, CAT; ascorbate peroxidase, APX, among others) and nonenzymatic (carotenoids, ascorbic acid, among others). The degree to which the amount and activities of antioxidant enzymes increases under water stress is extremely variable between many plant species and even between two cultivars of the same species. The level of response depends on the species, the development of the plant, as well as the duration and intensity of the stress [35].

The superoxide produced by the thylakoid can spontaneously be dismutated into molecular oxygen and hydrogen peroxide. In chloroplasts, this reaction is catalyzed enzymatically via superoxide dismutase (SOD). Chloroplasts also contain large amounts of ascorbic acid, which can efficiently reduce superoxide to hydrogen peroxide via ascorbate peroxidase [4].

Plants have the superoxide dismutase enzyme containing Cu and Zn, Fe, or Mn as prosthetic metals. Zn is found in superoxide dismutase present in chloroplasts and cytosol, while Mn is found in superoxide dismutase in mitochondria and Fe in superoxide dismutase is present in chloroplasts and mitochondria [36].

Reactive oxygen species can react with unsaturated fatty acids, causing the peroxidation of essential lipid membranes in plasmalemma or intracellular organelles [33]. The damage caused by the peroxidation of plasmalemma leads to extravasation of cellular content and rapid dissection and cell death. The damaged intracellular membrane affects the respiratory activity in the mitochondria, in addition to depigmentation and loss of the ability to fix carbon in chloroplasts [34].

Under normal conditions, antioxidant systems eliminate or slow the reaction of reactive oxygen, preventing its transformation into products more toxic to cells. Photosynthetic cells can tolerate high levels of oxygen because endogenous mechanisms sweep and remove toxic products before cell damage occurs [32]. However, oxidative damage is evident under conditions where the rate of production of ROS is high and the removal ability is low [37].

Water stress conditions can trigger an increase in the production of various forms of reactive oxygen, which can explain the damage to chloroplasts, lipids, and proteins and the alteration of the structural integrity of cell membranes. During the reduction of water inside the plant, the superoxide radical ($O_2^{\bullet-}$) can also react nonenzymatically with hydrogen peroxide (H_2O_2), giving rise to products such as hydroxyl radicals (OH^-) and singlet oxygen ($_1O_2$), which are more reactive than the superoxide radical ($O_2^{\bullet-}$) [32].

Although a number of regulatory mechanisms have been evolved within the plant cell to limit the production of these toxic molecules, oxidative damage remains a potential problem, as it causes disturbances in metabolism, such as loss of coordination between production processes (source) and energy use (drain) during photosynthesis on green leaves under stressful environments [38].

When plants are under stress, free radicals or ROS damage plant cells, and antioxidants decrease the toxicity of these radicals. Plants with high levels of antioxidants produce better root and shoot growth, maintaining a high water content in the leaves and low incidence of disease, both occurring when they are under ideal growing conditions and under environmental stress [18].

3.2 Biostimulants and reactive oxygen species

The use of biostimulants in plant breeding could change the activity of enzymes and antioxidant properties. Lycopene, ascorbic acid, phenolic compounds, and others have antioxidant properties. Antioxidant compounds (e.g., phenols, ascorbic acid) and enzymes (e.g., catalase, peroxidase, superoxide dismutase) detoxify reactive oxygen molecules [20].

Biostimulants stimulate root production and growth when applied to seeds or early plant development, especially in soils with low fertility and low water availability. Biostimulants act in accelerating the recovery of the seedlings in unfavorable conditions, such as water deficit. In addition, biostimulants reduce the need of fertilizers to the plants and increase their productivity and resistance to water stress, since they act as a hormonal and nutritional increment [15].

The application of humic acid extracts seems to be beneficial for field crop monocots. In a study conducted by [39], extracts from vermicompost applied to rice (*Oryza sativa* L.) played a role in activating antioxidative enzymatic function and increased ROS-scavenging enzymes. These enzymes are required to inactivate toxic-free oxygen radicals produced in plants under drought stress. Humic acid extracts may stimulate plant growth by improving nutrient uptake by exerting hormone-like effects as auxins, stimulating shoot elongation and increasing leaf nutrient accumulation and chlorophyll biosynthesis [40].

According to [41], humic acids improve root and shoot growth by increasing the concentrations of antioxidants in tall fescue (*Festuca arundinacea*) and creeping bent grass (*Agrostis palustris*) grown under conditions of low water availability. The authors also claim that exogenous applications of seaweed extracts together with humic acids promote root and shoot growth through the action of antioxidants in plants under water stress conditions.

A study carried out using a biostimulant based on salicylic acid and chitosan nanoparticles had an effect on the enzyme and antioxidant activity in maize leaves under water shortage [42]. The enzyme activity in leaves treated with chitosan, salicylic acid, and a control was comparable, and the activity of superoxide dismutase and peroxidase activity in plants treated with a biostimulant was 7.7 (after 2 days) and 5.2 (after 3 days) times higher than for plants treated with only salicylic acid.

The activities of antioxidant enzymes in plants are normally favored when plants are subjected to some kind of improvement in the conditions in which they are grown. The superoxide dismutase (SOD) antioxidant enzyme is the first line of defense against ROS caused by environmental stresses. Increases in SOD values provide an increase in plant resistance when subjected to environmental stresses [43].

In an experiment with Kentucky bluegrass (*Poa pratensis*) subjected to water stress and humic acid applications, [44] observed an increase in superoxide dismutase activities related to the applied doses of humic acids. However, a decrease in the activity of superoxide dismutase related to soil moisture content was observed. The authors justify this decrease by the increase in nonenzymatic antioxidants favored by the action of humic acids, which caused a decrease in the reactive oxygen species present in the cells.

The activity of superoxide dismutase responds differently to water deficit in different experiments and species: it can be increased [45] or decreased [46], or it cannot be altered [45]. Due to the presence of multiple enzymatic forms of the superoxide dismutase enzyme [33], only the investigation of the responses of each of its enzymatic forms can provide more information about the behavior of this enzyme in plants subjected to water stress.

Some authors mention that catalase activity has little affinity for hydrogen peroxide, a reason why it is common not to have a significant increase in its activity when evaluated in plants under stress [7]. [47] examined the activity of catalase in rice seedlings (*Oryza sativa*) under water stress and found that the increase of this enzyme in plants was not significant. Likewise, [48] did not find a significant increase for catalase in tomato plants (*Lycopersicon esculentum* Mill. cv. Nikita) submitted to three different levels of water stress. However, the extract of *Moringa oleifera* used as a biostimulant in rocket plants (*Eruca vesicaria subsp. sativa*) under water stress presented a decrease in the activity of the antioxidant enzymes (catalase, peroxidase, and superoxide dismutase) [49].

Several seaweed species influence ROS-scavenging systems in the plant tissue. Seaweed extracts controlled oxidative stress under drought conditions, by reducing lipid peroxidation, increasing total phenolic content, and enhancing superoxide dismutase, catalase, and ascorbate peroxidase activity in green bean (*Phaseolus vulgaris*) [50]. Extracts from *Sargassum* and *Ulva*, applied as seed presoaking, activated antioxidant systems by enhancing catalase and peroxidase activities, increasing ascorbic acid content, and therefore alleviating stress symptoms in wheat grown under drought conditions [51]. *Ascophyllum nodosum* extract applied to roots increased the total phenolic and flavonoid content and total antioxidant activity in spinach (*Spinacia oleracea*) [52]. In tall fescue (*Festuca arundinacea*), *A. nodosum* extract increased the activity of superoxide dismutase and in another study additionally enhanced glutathione reductase and ascorbate peroxidase activities [36]. Similarly applied seaweed extract increased the antioxidant capacity and enhanced flavonoid and tannin content in plant leaves of the ornamental hybrid *Calibrachoa* x *hybrida* under normal conditions [53].

Seaweed extracts have also been applied in combination with other compounds to enhance antioxidant activity in plants under water stress, such as a mixture of seaweed extracts from *A. nodosum, Fucus* spp., and *Laminaria* spp. with zinc and manganese and *A. nodosum* extract with free amino acids. These mixtures increased superoxide dismutase activity in shoots and roots of maize (*Zea mays*) and soybean (*Glycine max*). Collectively, these studies demonstrate that seaweed extracts enhance antioxidant activity, indicating their potential to scavenge damaging ROS molecules and improve plant stress tolerance [54].

Humic acids have also been shown to alleviate water deficit stress. Faba bean (*Vicia faba*) plants were protected from lead-induced oxidative damage by fulvic acids, which reduced lipid peroxidation, hydrogen peroxide, and pigment content [55]. The foliar application of fulvic acid ameliorated drought stress symptoms of reduced chlorophyll content, gas exchange, and yield while enhancing activities of superoxide dismutase, peroxidase, and catalase and increasing proline content in a study with maize [56]. Humic and fulvic acid based biostimulants, applied to the soil, enhanced superoxide dismutase, ascorbate peroxidase, and catalase activities in leaves of maize grown under well-watered and drought conditions. However, the effect of these biostimulants was less pronounced in soybeans [7].

Humic substances can also increase activity of antioxidant enzymes. Activity of superoxide dismutase, peroxidase, and catalase was higher after foliar application of fulvic acid in maize grown under drought conditions. Biostimulant containing humic and fulvic acids and amino acids increased activity of antioxidant enzymes, specifically superoxide dismutase and ascorbate peroxidase in maize subjected to drought stress, but did not affect catalase activity [7].

4. Conclusions

The composition of biostimulants should present a variety of organic materials such as humic substances, seaweed extracts, organic matter, and amino acids in order to improve stress tolerance. The literature on biostimulants have been reporting an increase in enzyme activities involved in antioxidant functions, especially under stress conditions.

Investigations on the role of biostimulants in the physiological mode of action in plants subjected to drought stress should be continued, since considerable researches remain to be completed to gain a clearer understanding of how these products increase the physiological health of plants under water stress.

Author details

Ana Carolina Feitosa de Vasconcelos
Federal University of Campina Grande, Campina Grande, Brazil

*Address all correspondence to: ana3carol@yahoo.com.br

References

[1] Taiz L, Zeiger E. Plant Physiology. 3rd ed. Sunderland, Mass: Sinauer Associates; 2002. p. 690

[2] Bodner G, Nakhforoosh A, Kaul HP. Management of crop water under drought: A review. Agronomy for Sustainable Development. 2015;**35**:401-442. DOI: 10.1007/s13593-015-0283-4

[3] Artlip TS, Wisniewski ME. Induction of proteins in response to biotic and abiotic stresses. In: Pessarakli M, editor. Handbook of Plant and Crop Physiology. 2nd ed. New York: M. Dekker; 2002. pp. 657-679

[4] Foyer CH. The contribution of photosynthetic oxygen metabolism to oxidative stress in plants. In: Inze D, Van Montagu M, editors. Oxidative Stress in Plants. New York: Taylor & Francis; 2002. pp. 33-68

[5] Xie X, He Z, Chen N, Tang Z, Wang Q, Cai Y. The roles of environmental factors in regulation of oxidative stress in plant. BioMed Research International. 2019;**2019**:9732325. DOI: 10.1155/2019/9732325

[6] Bulgari R, Cocetta G, Trivellini A, Vernieri P, Ferrante A. Biostimulants and crop responses: A review. Biological Agriculture and Horticulture. 2014;**31**(1):1-17. DOI: 10.1080/01448765.2014.964649

[7] Vasconcelos ACF, Zhang X, Ervin EH, Kiehl JC. Enzymatic antioxidant responses to biostimulants in maize and soybean subjected to drought. Scientia Agricola. 2009;**66**(3):395-402. DOI: 10.1590/S0103-90162009000300015

[8] Das K, Roychoudhury A. Reactive oxygen species (ROS) and response of antioxidants as ROS-scavengers during environmental stress in plants. Frontiers in Environmental Science. 2014;**2**:53. DOI: 10.3389/fenvs.2014.00053

[9] Caverzan A, Passaia G, Rosa SB, Ribeiro CW, Lazzarotto F, Margis-Pinheiro M. Plant responses to stresses: Role of ascorbate peroxidase in the antioxidant protection. Genetics and Molecular Biology. 2012;**35**(4):1011-1019. DOI: 10.1590/S1415-47572012000600016

[10] Laxa M, Liebthal M, Telman W, Chibani K, Dietz KJ. The role of the plant antioxidant system in drought tolerance. Antioxidants. 2019;**8**(4):94. DOI: 10.3390/antiox8040094

[11] Gajc-Wolska J, Spizewski T, Grabowska A. The effect of seaweed extracts on the yield and quality parameters of broccoli (*Brassica oleracea* var. cymosa L.) in open field production. Acta Horticulturae. 2013;**1009**:83-89

[12] Sharma HS, Fleming C, Selby C, Rao JR, Martin T. Plant biostimulants: A review on the processing of macroalgae and use of extracts for crop management to reduce abiotic and biotic stresses. Journal of Applied Phycology. 2014;**26**:465-490

[13] Hamza B, Suggars A. Biostimulants: Myths and realities. Turf grass trends. 2001;**10**:6-10

[14] O'Donnell R. W. the auxin-like effects of humic preparations from leonardite. Soil Science. 1973;**116**(2):106-112

[15] Vasconcelos ACF, Chaves LHG. In: Mirmajlessi SM, editor. Biostimulants and Their Role in Improving Plant Growth under Abiotic Stresses. IntechOpen; 2019. DOI: 10.5772/intechopen.88829. Available from: https://www.intechopen.com/online-first/biostimulants-and-their-role-in-improving-plant-growth-under-abiotic-stresses

[16] Van Oosten MJ, Pepe O, De Pascale S, Siletti S, Maggio A. The

role of biostimulants and bioeffectors as alleviators of abiotic stress in crop plants. Chemical and Biological Technologies in Agriculture. 2017;**4**:5. DOI: 10.1186/s40538-017-0089-5

[17] Richardson AD, Aikens M, Berlyn GP, Marshall P. Drought stress and paper birch (*Betula papyrifera*) seedlings: Effects of an organic biostimulant on plant health and stress tolerance, and detection of stress effects with instrument-based, noninvasive methods. Journal of Arboriculture. 2004;**30**:52-61

[18] Su Y, Xia S, Zhong R, Wang L. Phytohormonal quantification based on biological principles. Hormone Metabolism and Signaling in Plants. 2017;**13**:431-470

[19] Nardi S, Pizzeghello D, Schiavon M, Ertani A. Plant biostimulants: Physiological responses induced by protein hydrolyzed-based products and humic substances in plant metabolism. Scientia Agricola. 2016;**73**(1):18-23. DOI: 10.1590/0103-9016-2015-0006

[20] Drobek M, Frac M, Cybulska J. Plant biostimulants: Importance of the quality and yield of horticultural crops and the improvement of plant tolerance to abiotic stress—A review. Agronomy. 2019;**9**(6):335. DOI: 10.3390/agronomy9060335

[21] Abbas SM. The influence of biostimulants on the growth and on the biochemical composition of *Vicia faba* CV. Giza 3 beans. Romanian Biotechnological Letters. 2013;**18**(2):8061-8068

[22] Du Jardin P. The science of plant biostimulants - A bibliographic analysis, Ad hoc study report. 2012. Brussels: European Commission. Available from: http://hdl.handle.net/2268/169257 [Accessed: 20 February 2020]

[23] Levit J. Plant responses to environmental stress. New York: Academic Press; 1980. p. 486

[24] Salehi-Lisar SY, Bakhshayeshan-Agdam H. Drought stress in plants: Causes, consequences, and tolerance. In: Hossain MA, Wani SH, Bhattacharjee S, Burritt DJ, Tran LP, editors. Drought Stress Tolerance in Plants: Physiology and Biochemistry. Switzerland: Springer International Publishing; 2016. pp. 1-16. DOI: 10.1007/978-3-319-28899-4

[25] Lawlor DW. The effects of water deficit on photosynthesis. In: Smirnoff N, editor. Environment and Plant Metabolism: Flexibility and Acclimation. Oxford; Herndon: BIOS Scientific Publishers; Books International; 1995. pp. 129-160

[26] Mckersie BD, Leshem YY. Stress and Stress Coping in Cultivated Plants. Dordrecht; Boston: Kluwer Academic Publishers; 1994. p. 256

[27] Jones MM, Turner NC, Osmond CB. Mechanisms of drought resistance. In: Paleg LG, Aspinall D, editors. The Physiology and Biochemistry of Drought Resistance in Plants. Sydney; New York: Academic Press; 1981. pp. 15-38

[28] Dubey RS, Pessarakli M. Physiological mechanisms of nitrogen absorption and assimilation in plants under stressful conditions. In: Pessarakli M, editor. Handbook of Plant and Crop Physiology. New York: M. Dekker; 2002. pp. 637-655

[29] Urban L, Aarrouf J, Bidel L. Assessing the effects of water deficit on photosynthesis using parameters derived from measurements of Leaf gas exchange and of chlorophyll a fluorescence. Frontiers in Plant Science. 2017;**8**:2068. DOI: 10.3389/fpls.2017.02068

[30] Halliwell B, Gutteridge JMC. Free Radicals in Biology and Medicine. Oxford: Clarendon Press; 1989. p. 543

[31] Perl-Treves R, Perl A. Oxidative stress. In: Inze D, Van Montagu M,

editors. Oxidative Stress in Plants. New York: Taylor & Francis; 2002. pp. 1-32

[32] Navari-Izzo F, Rascio N. Plant response to water-deficit conditions. In: Pessarakli M, editor. Handbook of Plant and Crop Stress. New York: M. Dekker; 1999. pp. 231-270

[33] Scandalios JG. Oxygen stress and superoxide dismutase. Plant Physiology. 1993;**101**:7-12

[34] Malan C, Greyling MM, Gressel J. Correlation between Cu/Zn superoxide dismutase and glutathione reductase, and environmental and xenobiotic stress tolerance in maize inbreeds. Plant Science. 1990;**69**:157-166

[35] Mano J. Early events in environmental stresses in plants – Induction mechanisms of oxidative stress. In: Inze D, Van Montagu M, editors. Oxidative Stress in Plants. New York: Taylor & Francis; 2002. pp. 216-245

[36] Zhang XZ. Influence of Plant Growth Regulators on Turfgrass Growth, Antioxidant Status, and Drought Tolerance. Blacksburg, VA: Virginia Polytechnic Institute and State University; 1997

[37] Asada K, Takahashi M. Production and scavenging of active oxygen in photosynthesis. In: Kyle DJ, Osmond CB, Arntzen CJ, editors. Photoinhibition. New York: Elsevier; 1987. pp. 228-287

[38] Reddy AR, Chaitanya KV, Vivekanandan M. Drought-induced responses of photosynthesis and antioxidant metabolism in higher plants. Journal of Plant Physiology. 2004;**161**:1189-1202

[39] García AC, Santos LA, Izquierdo FG, Sperandio MVL, Castro RN, Berbara RLL. Vermicompost humic acids as an ecological pathway to protect rice plant against oxidative stress. Ecological Engineering. 2012;**47**:203-208

[40] Baldotto MA, Baldotto LEB. Gladiolus development in response to bulb treatment with different concentrations of humic acids. Revista Ceres. 2013;**60**:138-142

[41] Zhang X, Schmidt RE. Hormone-containing products' impact on antioxidant status of tall fescue and creeping bentgrass subjected to drought. Crop Science. 2000;**40**:1344-1349

[42] Kumaraswamy RV, Kumari S, Choudhary RC, Sharma SS, Pal A, Raliya R, et al. Salicylic acid functionalized chitosan nanoparticle: A sustainable biostimulant for plant. International Journal of Biological Macromolecules. 2019;**123**:59-69

[43] Zhang X, Ervin E, Evanylo G, Sherony C, Peot C. Biosolids impact on tall fescue drought resistance. Journal of Residuals Science and Technology. 2005;**2**:173-180

[44] Zhang X, Schmidt RE. Antioxidant response to hormone-containing product in Kentucky bluegrass subjected to drought. Crop Science. 1999;**39**:545-551

[45] Luna M, Badiani M, Felici M, Artemi F, Sermanni GG. Selective enzyme inactivation over water stress in maize (*Zea mays* L.) and wheat (*Triticum aestivum* L.) seedlings. Environmental and Experimental Botany. 1985;**25**:153-156

[46] Quartacci MF, Navari-Izzo F. Water stress and free radical mediated changes in sunflower seedlings. Journal of Plant Physiology. 1992;**139**:621-625

[47] Sharma P, Dubey RS. Drought induces oxidative stress and enhances the activities of antioxidant enzymes in

growing rice seedlings. Plant Growth Regulation. 2005;**46**:209-211

[48] Zgallaï H, Steppe K, Lemeur R. Effects of different levels of water stress on leaf water potential, stomatal resistance, protein and chlorophyll content and certain antioxidative enzymes in tomato plants. Journal of Integrative Plant Biology. 2006;**48**:679-685

[49] Abdalla MM. Boosting the growth of rocket plants in response to the application of *Moringa oleifera* extracts as a biostimulant. Life Sciences. 2014;**11**:1097-8135

[50] Mansori M, Chernane H, Latique S. Seaweed extract effect on water deficit and antioxidative mechanisms in bean plants (*Phaseolus vulgaris* L.). Journal of Applied Phycology. 2015;**27**(4):1689-1698

[51] Kasim WA, Hamada EAM, El-Din NGS, Eskander S. Influence of seaweed extracts on the growth, some metabolic activities and yield of wheat grown under drought stress. International Journal of Agronomy and Agricultural Research. 2015;**7**(2):173-189

[52] Fan D, Hodges DM, Zhang JZ. Commercial extract of the brown seaweed *Ascophyllum nodosum* enhances phenolic antioxidant content of spinach (*Spinacia oleracea* L.) which protects *Caenorhabditis elegans* against oxidative and thermal stress. Food Chemistry. 2015;**124**(1):195-202

[53] Elansary HO, Norrie J, Ali HM. Enhancement of Calibrachoa growth, secondary metabolites and bioactivity using seaweed extracts. BMC Complementary and Alternative Medicine. 2016;**16**:341

[54] Wozniak E, Blaszczak A, Wiatrak P, Canady M. Biostimulant mode of action. In: The Chemical Biology of Plant Biostimulants. 1st ed. John Wiley & Sons Ltd; 2020. pp. 229-243. DOI: 10.1002/9781119357254.ch9

[55] Shahid M, Dumat C, Silvestre J, Pinelli E. Effect of fulvic acids on lead-induced oxidative stress to metal sensitive *Vicia faba* L. plant. Biol. Fertil. The Soil. 2012;**48**(6):689-697

[56] Anjum SA, Wang L, Farooq M. Fulvic acid application improves the maize performance under well-watered and drought conditions. Journal of Agronomy and Crop Science. 2011;**197**(6):409-417

Chapter 19

Potential Role of Plants *Hordeum vulgare* L. and *Panax ginseng* L. in Resolving the Fertility Disorders and Stress-Induced Oxidative Stress Arises from Hypothyroidism in Adult Female Rats

Lobna F. Wahman, Marwa M. Abd Rabo, Amany Hanafy M. Elgoly and Magda H.M. Yousef

Abstract

Hordeum vulgare (Barley) and *Panax ginseng* have antioxidant activity referring to their diverse phytonutrient. Hypothyroidism in adult female rats was induced by pituitary-gonadal-adrenal disturbance, depleting the serum FSH levels with the elevation of corticosterone, prolactin, progesterone and testosterone hormones as well as (ERK1/2). Hypothyroidism evoked an oxidative stress status by increasing 8-hydroxy guanosine, which initiated apoptosis by uplifting apoptotic marker Caspase-3 both in serum and brain tissues. This is confirmed by the increase in the percentage of DNA-damage in the brain tissues. Significant decrease in all monoamines' levels in different brain areas, downregulation of dopamine and 5-hydroxytryptamine receptors transcription, with a significant increase in excitatory amino acids was noted. Barley and ginseng renormalized cortisol and oxidative stress markers by increasing cellular resistance to stress and potentiated the role of the immune system through phytosterol and ginsenosides, so they considered potent free radical scavengers. Barley and *Panax ginseng* ameliorate the hormonal and neural dysfunction resulting from hypothyroidism, so they are recommended for relieving stress and improving mood and depression.

Keywords: barley, *Panax ginseng*, oxidative stress, antioxidant, hypothyroidism: gonadal-neural dysfunction

1. Introduction

There are many evidences revealed that food intake enriched with whole grain reduces the susceptibility of the incidence of many chronic diseases. Barley (*Hordeum vulgare*) is a food source deemed to be available for all disparate social classes of humankind.

IntechOpen

In the Arab culture, *Hordeum vulgare* or barley syrup is used to relieve depression. It is categorized into the spring and winter types, which are considered two-rowed or six-rowed depending on the number of seed rows on each spike. Based on its grain composition, barley is further classified into normal, waxy or high amylose starch type [1].

Barley has found to be enriched with valuable minerals (iron, selenium, potassium, calcium, phosphorous; zinc), phytoestrol (β-sitosterol, campesterol, stigmasterol), polyphenol (ferulic, p-coumaric, sinapic, vanillic and p-hydroxybenzoic acids, cinnamic acid derivatives, proanthocyanidins, quinines, flavonols, chalcones, flavones, flavanones, and amino phenolic compound), water-soluble vitamins (C; B1; B2; folic acid and B12), β-glucan, dietary soluble fiber, vitamin E; nicotinic acid; pyridoxine; folic acid; essential amino acids, such as tryptophan and phenylalanine; neutral amino acids (LNAA), such as the three branched-chain aromatic amino acids leucine, isoleucine, and valine [2, 3]. So, barley grain exhibits potential antioxidant and antiproliferative actions because its powerful phytochemical compounds that have been shown to lower the risk of many diseases [4, 5]. The diverse phytonutrient of barley implicates its protection activity against certain types of cancers, cardiovascular disease, arthritis, diabetic, and hypercholesterolemia. It also increases cellular energy to sustain the body homeostasis [6, 7] and modulating endocrine and neurotransmitters functions [8].

Red ginseng represents an important position as a health functional food. It belongs to the Panax genus of the *Araliaceae* family, ginseng characterized by a complex activity profile that includes antioxidant, anti-inflammatory, anti-apoptotic, and immune-stimulatory properties and has the effects of stabilizing and balancing the entire physiology [9]. So, in Asian countries; Korea, China, and Japan, ginseng used as a therapeutic agent for a variety of diseases [10].

The major active ingredients of *Panax ginseng* are saponins, which are triterpene glycosides called "ginsenosides". Other active components include proteins, peptides, and alkaloids, which are nitrogenous compounds; polyacetylene, which is a fat-soluble component; polysaccharides and other flavonoids; fatty acids, organic acids, vitamins, sugars, inorganic salts, sterols, oligopeptides [11, 12]. Ginsenosides can be classified into three categories: the panaxadiol group (e.g. Rb1, Rb2, Rb3, Rc, Rd., Rg3, Rh2, Rs1), the panaxatriol group (e.g. Re, Rf, Rg1, Rg2, Rh1), and the oleanolic acid group (Ro) [13, 14].

Ginsenosides are lipophilic compounds so they can pass easily through the cell membrane by simple diffusion and bind to its intracellular target proteins in the cytoplasm and nucleus. Ginseng also contains more than 10 phenolic compounds that possess antioxidant biological properties that have ability to lower the effect of oxidative stress [15]. Phytoestrogens, such as genistein is an important component of ginseng, have shown protective effects on conditions related to decreased estrogen, including menopause, osteoporosis, and cognitive disorders [13, 16].

Thyroid-stimulating hormone (TSH) is synthesized and secreted by the adenohypophysis lobe and exerts its effect by binding to the cognate thyrotropin receptor (TSHR) to stimulates the production of thyroglobulin and thyroid peroxidase proteins, which are essential for the synthesis and secretion of thyroid hormones (THs) [17]. THs bind their nuclear receptors (TRs), which are present in many tissues and organs in the human body and hence regulate their functions [18].

Thyroid hormones influence a wide range of brain developmental processes, such as myelination, neuronal and glial cell differentiation by regulating the gene involved in these processes thus hypothyroidism may reduce axonal growth and dendritic arborization in the cerebral cortex, visual cortex, auditory cortex, hippocampus, and cerebellum, as well as impaired memory, cognitive function and attentiveness [3, 19].

THs plays a role in neurotransmitter release from their storage vesicles such as norepinephrine (NE), epinephrine (E), serotonin (5-HT) and dopamine (DA) and hence maintaining good mental state, mood regulation, modulating post-receptor signal transduction, gene expression and preventing depression [19]. So, hypothyroidism is a highly prevalent condition that impairs learning, memory, induce delayed skeletal development, cardiovascular diseases, secondary hypertension, the deterioration of human reproductive health and brain dysfunction [20].

2. Materials and methods used in the research

2.1 Animals

The study was carried out by using adult female Wistar albino rats weighing 180–200 g. Animals were housed at 23 ± 2°C and 55 ± 5% humidity with a 12 h light/dark cycle rats were provided a standard diet and water *ad libitum*.

2.2 Preparation of *Hordeum vulgare* (barley)

Barley was prepared as an emulsion in water (1 g ground barley soaked in 10 ml of distilled water) and administered daily *per* [21]. The nutritional facts of barley per 100 g are presented in **Table 1**.

2.3 Preparation of *Panax ginseng*

Dried roots of the Korean *Panax ginseng* were obtained as a brown powder and dissolved in distilled water. Animals received a daily oral dose of 1.8 mg/200 g body weight (equivalent to the therapeutic dose [22]) for 30 days.

2.4 Induction of hypothyroidism

Neo-Mercazole is the least toxic anti-thyroid agent within therapeutic dose ranges [23] therefore it was selected for hypothyroidism induction. The animals were orally administered a daily dose of 5.0 mg.kg—of Neo-Mercazole for 1 month [24]. Hypothyroidism was manifested by the increased level of serum TSH associated with low level of f T4.

Carbohydrates	(78.2 g)	Vitamin B6	(0.29 μg)	Choline	(38 mg)
Fibers	(15.5 g)	Vitamin K	(2.5 μg)	Riboflavin (B2)	(0.124 μg)
Energy	(350 kcal)	Niacin (B3)	(4.8 μg)	Calcium	(30 mg)
Fat	(1.2 g)	Pantothenic acid (B5)	(0.29 μg)	Iron	(3.5 mg)
Protein	(10 g)	Thiamine (B1)	(0.2 μg)	Magnesium	(80 mg)
Vitamin A	(15 μg)	Folic acid	(25 μg)	Phosphorus	(200 mg)
Zinc			(2.5 mg)		
Potassium			(250 mg)		

Table 1.
Nutrition facts in barley/100 g.

2.5 Experimental design of barley work

Animals randomly divided into equally four-treatment groups. Except for euthyroid animals (EU) (groups 1&2), hypothyroid animals (H) (group 3&4) were orally administered 5.0 mg kg^{-1} bwt Neo-Mercazole until the end of the study. Following 30 days of Neo-Mercazole administration, groups 2 and 4 orally administered 100 mg kg^{-1} bwt barley (B) [21] water suspension for 4 weeks. The four groups named: EU; EU + B; H; H + B.

2.6 Experimental design of *Panax ginseng* work

Rats were divided equally into four groups. First group was an intact control group that received distilled water. The second group was the hypothyroid group (H group) which orally treated with 5 mg kg^{-1} body- weight Neo-Mercazole for 30 days for induction of hypothyroidism. The third group was orally administered *Panax ginseng* (G group) in a daily oral dose of 1.8 mg/200 g body weight for 30 days. The fourth group receiving both Neo-Mercazole for 30 days and followed by *Panax ginseng* (H + G group) for another 30 days.

2.7 Blood and tissue collection

At the end of treatment, the animals were anesthetized with 1% isoflurane followed by decapitation [25], blood was collected into serum preparation tube and the separated serum was collected and divided into aliquots, stored at −20°C for further hormones assay. The whole brains were removed from 10 rats from each group and the hypothalamus, hippocampus, cerebral cortex, midbrain, and cerebellum were dissected using a sharp blade. From another 10 rats' whole brain and thyroid gland were immediately removed and stored in ice-cold saline at −20°C for further biochemical and comet assay.

2.8 Methods

Levels of f T3, f T4, ERK1/2, 8-hydroxy-2′-de-oxyguanosine (8-OhdG) and apoptotic marker Caspase-3 were determined using ELISA kit specific for rats according to manufacturer's instruction (Glory Science Co., Ltd., USA). Serum corticosteroid and gonadal hormones were determined using the ELISA kit according to the instruction of BioCheck (BioCheck Co., Ltd., USA). DNA degradation in brain and thyroid homogenates was determined by using the Comet technique according to the method [26]. Determination of monoamines in brain areas were carried according to methods of [19, 27] while free amino acids were done according to the method of [28] using the precolumn phenylisothiocyanate (PTC) derivatization technique.

3. The antioxidant effect of barley on fertility disorders and oxidative stress induced by hypothyroidism

The study was conducted to address the potential ameliorative effect of barley on the disturbance in adrenal pituitary-gonadal hormones, as well as oxidative stress following hypothyroid induction. Hypothyroidism induction caused disturbances in adrenal, pituitary and gonadal hormones (**Figures 1–3**). Barley reversed the effect of the antithyroid drug on the levels of thyroid hormones (TSH, f T4) and their transporting proteins (TBG, TTR) as shown in **Table 2** due to its higher iron (Fe) content which plays a crucial role in modulating thyroid peroxidase

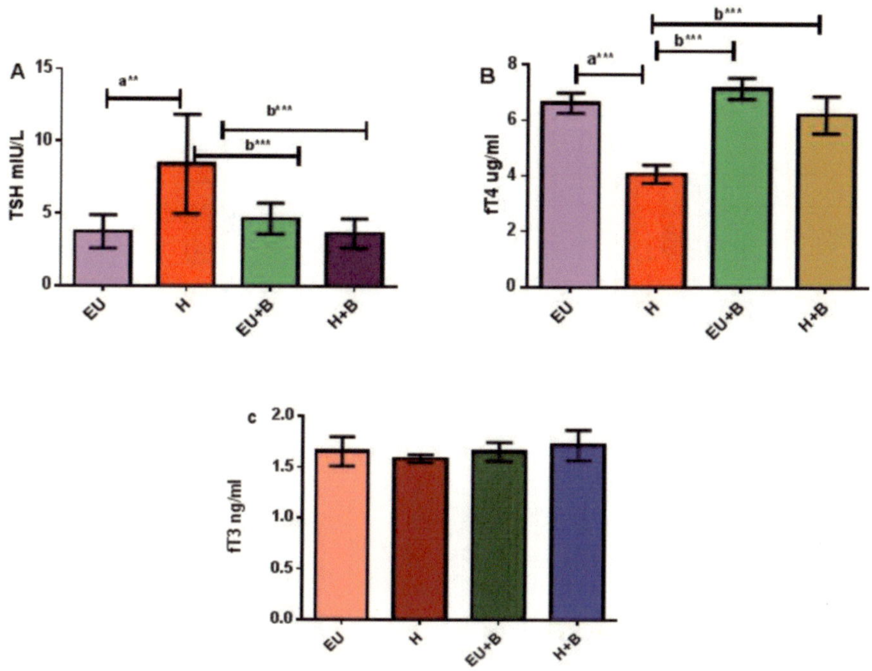

Figure 1.
(A) Serum TSH mIU/L, (B) serum fT4 µg/ml, and (C) serum f T3 in control (CO) group, hypothyroid (H) group, barley treated (T) group and hypothyroid-barley-treated group (HT).

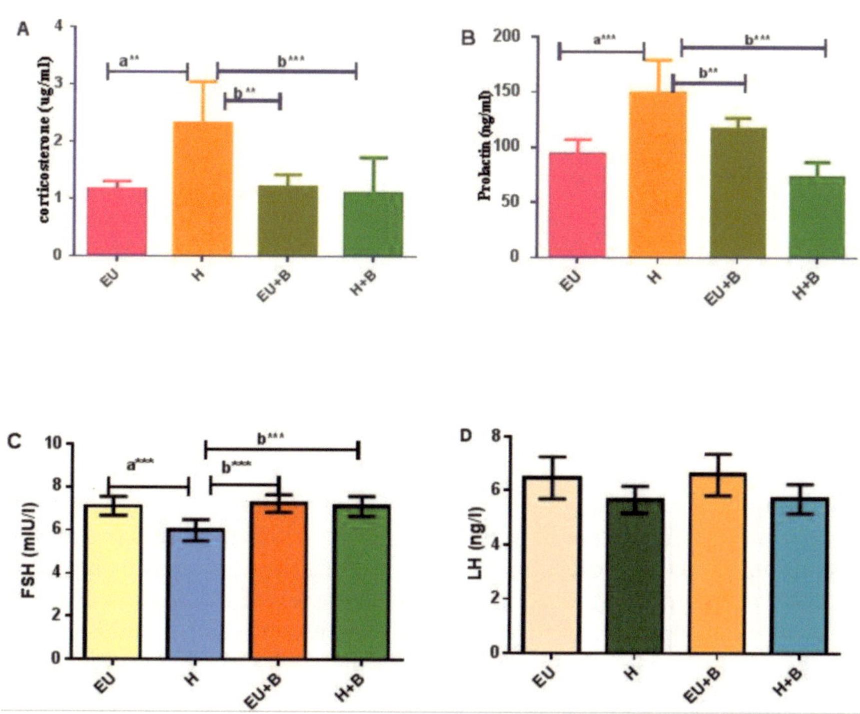

Figure 2.
Serum corticosterone (ug/dl) (A), serum prolactin (ng/ml) (B), serum FSH (mIU/ml) (C) and serum LH (ng/l) (D) in control (CO) group, hypothyroid (H) group, barley-treated (T) group and hypothyroid-barley-treated group (HT).

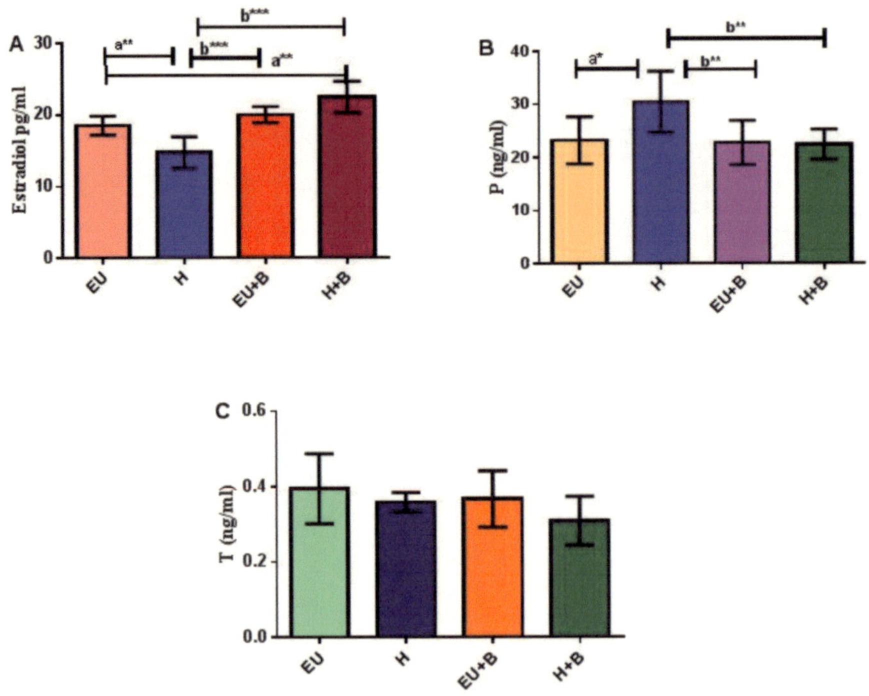

Figure 3.
(A) Serum estradiol (pmol/L), (B) serum P (ng/ml) and (C) serum T (ng/ml), in control group (CO), hypothyroid group (H), barley-treated group (T) and hypothyroid-barley-treated group (HT).

	EU	EU + B	H	H + B
f T4 (μg/ml)	4.43 ± 0.11	4.21 ± 0.14	3.09 ± 0.25 (a*)	4.46 ± 0.19(b**)
TSH (mIU/ml)	12.47 ± 0.42	12.73 ± 0.62	15.97 ± 0.26 (a*)	12.53 ± 0.29 (b*)
TTR (ng/ml)	51.23 ± 2.39	45.64 ± 6.53 (a*)	16.85 ± 1.88 (a***)	34.25 ± 1.65 (a** b***)
TBG (pg/ml)	1.58 ± 0.04	2.35 ± 0.16 (a**)	2.09 ± 0.04 (a*)	2.57 ± 0.14 (a***b*)
ERK1/2 (pg/ml)	43.67 ± 1.53	46.33 ± 1.53(a*)	33.19 ± 2.16 (a**)	47.94 ± 0.60(b**)

All data in tables represented by mean ± SD, n = 10 animals.
$^{*}p < 0.05$, $^{**}p < 0.01$ and $^{***}p < 0.001$.
a: mean significance difference from control group. B: mean significance difference from hypothyroid group.

Table 2.
Effect of barley on THs, TTR and TBG in brain tissue of EU- and H-groups.

(TPO) enzyme activity [2, 29]. Oxidative stress is related to hormonal disorders in a reciprocal way so in our study, the hyper TSH level stimulated the synthesis of corticosterone, and generated a state of oxidative stress which inhibited the pituitary gonadotropin [30, 31], and cause FSH depletion with the non-significant decrease of LH levels in the hypothyroid group. Also, lower levels of estradiol in the hypothyroid group associated with high progesterone and prolactin levels could be attributed to high ERK1/2 level (**Figure 4**). Barley, with its high content of phytosterol, could modulate ER-α, and β expression, augmented estradiol levels, in turn, led to activate negative feedback mechanism of pituitary-gonadal adrenal axis function and renormalize the disturbances of endocrine gland elicited by hypothyroidism.

Oxidative stress (ROS) is an imbalance between the production of pro-oxidant substances and antioxidant defense. Hypothyroidism augments the oxidative

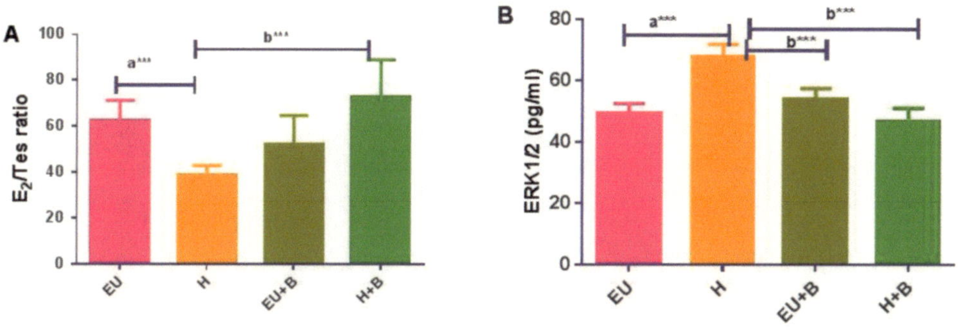

Figure 4.
(A) Serum E2/T ratio (B) serum ERK1/2 (pg/ml) in control group (CO), group hypothyroid (H), barley-treated group (T) and hypothyroid-barley-treated group (HT).

insult, impairing the brain by increasing nitric oxide (NO) and NO synthase (NOS) levels in the hippocampus, which affects the lipid composition of rat brain tissues and induces DNA damage [32, 33]. In present work, hypothyroidism was associated with the high significant increase in 8-hydroxyguanosine (an oxidative stress marker), together with marked elevation in Caspase-3 (an apoptotic marker) in serum and brain tissue (**Figures 5** and **6**). These findings were confirmed by alkaline comet assays of thyroid and brain tissue homogenates. As DNA was degraded, it converted from a supercoiled form to a comet-like shape with a measurable tail length so our study revealed that the hypothyroid status induced a significant increase in the tail length, in the thyroid and brain tissues. Treatment with barley attenuated the oxidative stress status induced by hypothyroid status; it significantly decreased 8-OH guanosine levels and Caspase-3 activity. This antioxidant activity of barley could be attributed to flavonoids, ferulic, sinapic and ß- hydroxy acids (BHA) content, the major predominant polyphenol, in barley with their potent free radical scavengers by absorbing and neutralizing oxygen radicals [21, 34, 35]. Also, DNA damage was repaired by the antioxidant activity of barley as illustrated in **Figures 7** and **8**, which refer to vitamins A and E that essential for nucleotide, DNA biosynthesis, DNA repair, and methylation [1, 2, 19, 36]. Additionally, zinc protects against oxidative stress by stabilizing membranes through the inhibition of the enzyme nicotinamide adenine dinucleotide phosphate oxidase (NADPH oxidase) and the stimulating of the synthesis of metallothioneins, which reduce the levels of hydroxyl radicals and sequester ROS produced in response to hypothyroidism.

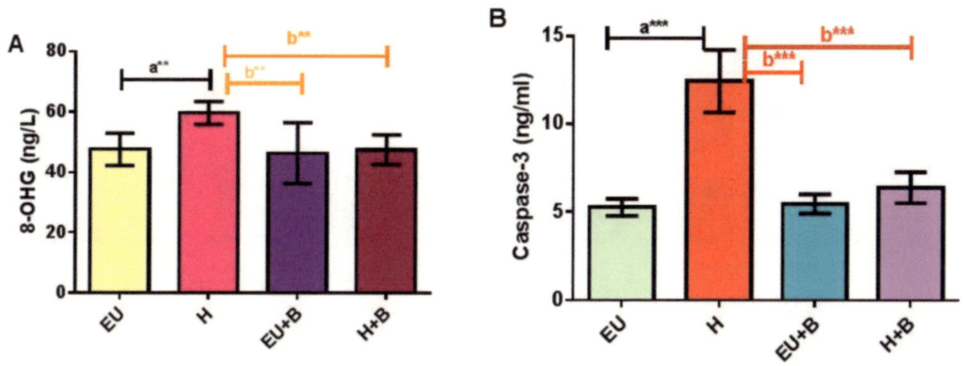

Figure 5.
Serum 8-hydroxy guanosine (8-OHG) (ng/L) (A), Caspase-3 (ng/ml) (B) in control group (CO), hypothyroid group (H), barley-treated group (T) and hypothyroid-barley-treated group (HT).

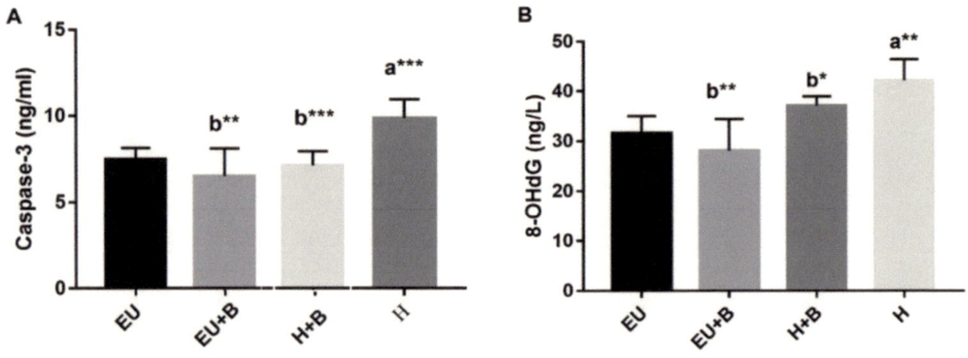

Figure 6.
Effect of barley on caspase-3 (ng/ml) and 8-OhdG (ng/ml)in brain tissue.

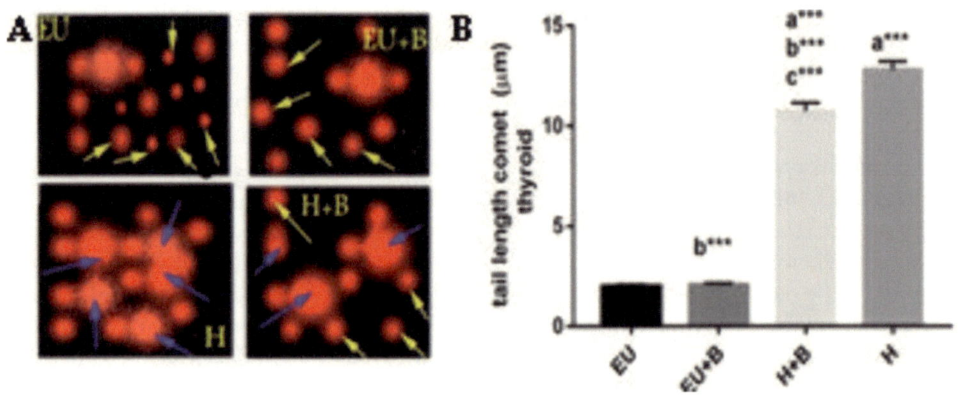

Figure 7.
Effect of barley on DNA damage in thyroid tissue (A) fluorescence photomicrograph showing comets in EU-, H-, EU + B and H + B-groups. The → indicated the intact DNA and → indicated the degree of damaged DNA (B) tail length expressed in µm in thyroid tissue of all treated groups.

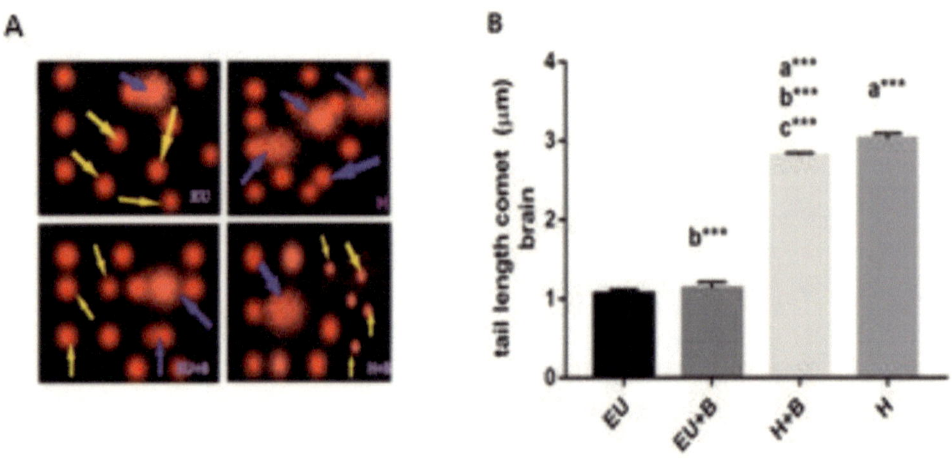

Figure 8.
Effect of barley on DNA damage in whole brain tissue (A) fluorescence photomicrograph showing comets in EU-, H-, EU + B and H + B-groups. The → indicated the intact DNA and → indicated the degree of damaged DNA (B) tail length expressed in µm in whole brain tissue of all treated groups.

Neurotransmitters NE, DA, and 5-HT levels were significantly reduced in all brain areas (cerebellum, midbrain, cerebral cortex, hypothalamus, and hippocampus) with hypothyroidism induction as shown in **Table 3**, these could be attributed to the reduced oestradiol level as mentioned above [37, 38]. The administration of *Hordeum vulgare* (barley) improved the disturbances in the dopaminergic, serotonergic and noradrenergic pathways via two different mechanisms; first, the high phytosterol content modulates estrogen receptors (Erα and β) expression and elevate oestradiol levels. Second, barley is enriched with tryptophan and phenylalanine and could regulate the synthesis of 5-HT, DA and NE through the conversion of tryptophan to 5-hydroxytryptophan (5-HTP) to 5-HT and hydrolysis of phenylalanine to generate tyrosine that ultimately produces DA and NE [2, 36, 39, 40]. In the present study, hypothyroidism induced a significant increase in inhibitory amino acid, including GABA and histidine, which is an excitatory amino acid (**Figures 9** and **10**).

These results could explain the increase in Caspase-3, which may be attributed to reduced blood oxygen–glucose levels in several brain regions as a result of increased GABA levels. The study also revealed an increase in dopamine receptors, whereas serotonin receptors were significantly decreased. The *Hordeum vulgare* (barley) treatment in the present study caused a renormalization the observed disturbances in the amino acid and neurotransmitter levels because it is enriched with folic acid, which is involved in the synthesis of monoamine neurotransmitters and modulate serotonergic, dopaminergic and noradrenergic systems by acting as a cofactor for enzymes that convert tryptophan to 5-HT and enzymes that convert tyrosine to noradrenaline [19]. The alteration in serotonin receptor densities was restored by the barley administration, due to its enriched levels of tryptophan, which is metabolized to serotonin [36, 40] and activates these receptors (**Figure 11**).

	Frontal cortex	Hippocampus	Hypothalamus	Mid brain	Cerebellum
NE (μg g−1 tissue)					
EU	0.52 ± 0.03	0.69 ± 0.02	0.40 ± 0.01	0.70 ± 0.01	0.58 ± 0.01
EU + B	0.45 ± 0.02	0.65 ± 0.01	0.40 ± 0.01	0.69 ± 0.01	0.60 ± 0.01
H	0.21 ± 0.09a***	0.31 ± 0.01a***	0.13 ± 0.01a***	0.35 ± 0.01a***	0.29 ± 0.01a**
H + B	0.35 ± 0.01a**b*	0.46 ± 0.01a**b**	0.22 ± 0.01a**b**	0.48 ± 0.01a**b**	0.40 ± 0.01a**b**
DA (μg g−1 tissue)					
EU	0.59 ± 0.02	2.40 ± 0.07	1.473 ± 0.10	1.31 ± 0.01	0.60 ± 0.01
EU + B	0.55 ± 0.01	2.39 ± 0.09	1.32 ± 0.06	1.32 ± 0.01	0.60 ± 0.01
H	0.26 ± 0.01a***	0.93 ± 0.03a***	0.90 ± 0.03a*	0.85 ± 0.01a**	0.27 ± 0.05a***
H + B	0.35 ± 0.01a**b**	1.17 ± 0.01a**b**	1.07 ± 0.02a*b*	1.0 ± 0.01a*b*	0.40 ± 0.01a**b**
5-HT (μg g−1 tissue)					
EU	0.57 ± 0.01	0.38 ± 0.01	0.78 ± 0.01	0.72 ± 0.01	0.47 ± 0.01
EU + B	0.56 ± 0.02	0.38 ± 0.01	0.76 ± 0.02	0.65 ± 0.05	0.50 ± 0.01
H	0.23 ± 0.01a***	0.11 ± 0.01a***	0.45 ± 0.01a**	0.30 ± 0.01a ***	0.18 ± 0.01a***
H + B	0.35 ± 0.01a**b**	0.18 ± 0.01a**b*	0.54 ± 0.01a*b*	0.41 ± 0.01a**b**	0.28 ± 0.01a**b**

All data in tables represented by mean ± SD, n = 10 animals.
$^{*}p < 0.05$, $^{**}p < 0.01$ and $^{***}p < 0.001$.
a: Mean significance difference from control group. b: Mean significance difference from hypothyroid group.

Table 3.
Effect of barley on neurotransmitters level in discrete brain regions in control and treated groups.

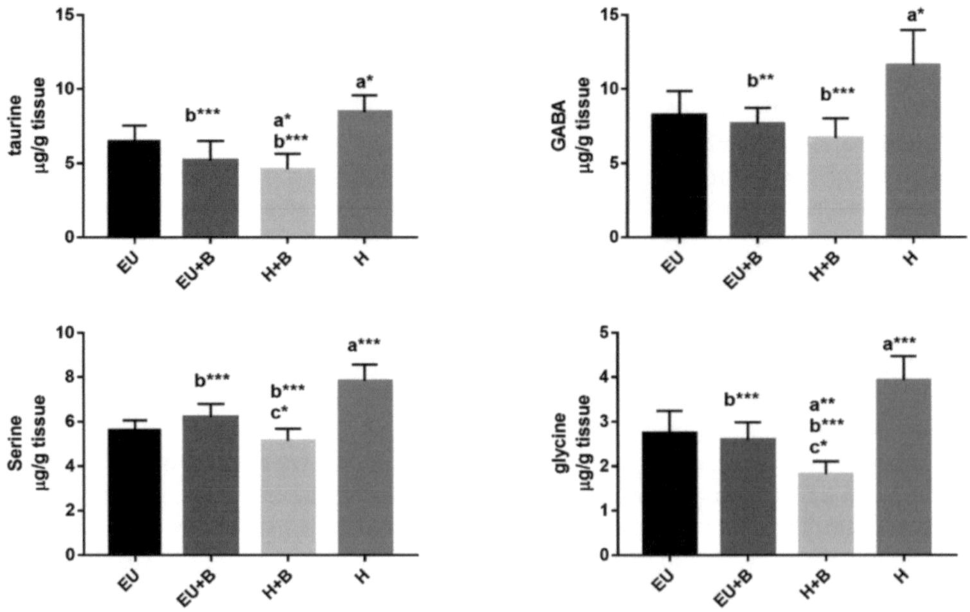

Figure 9.
*Effect of barley on inhibitory amino acids in EU and H groups. All data represented by mean ± SD, n= 10 animals, *p<0.05, **p<0.01 and ***p<0.001, (a) mean significance difference from control group, (b) mean significance difference from hypothyroid group. (C) mean significance difference from EU+B.*

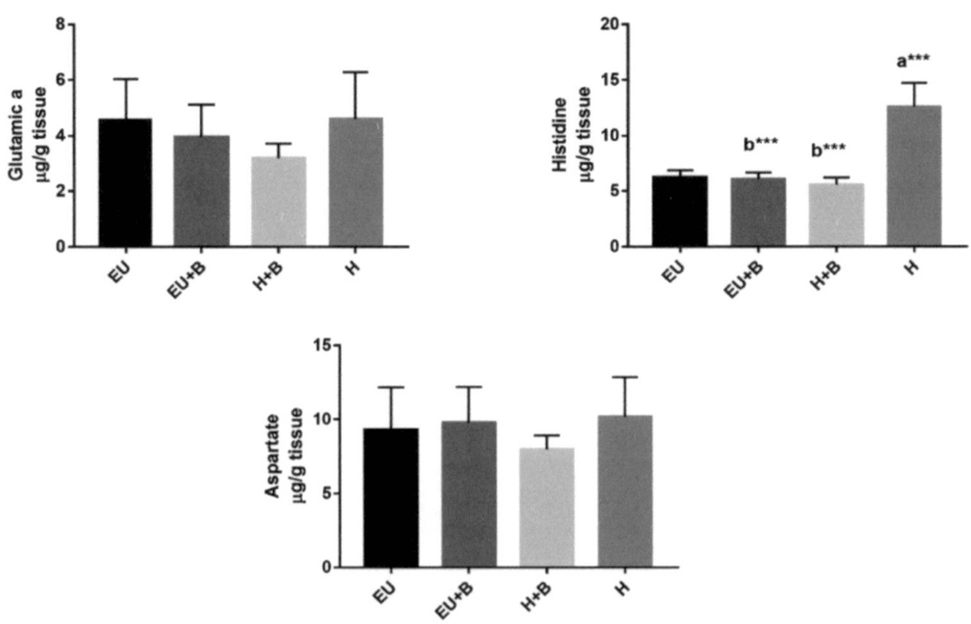

Figure 10.
*Effect of barley on excitatory amino acids in EU and H groups. All data represented by mean ± SD, n= 10 animals, *p<0.05, **p<0.01 and ***p<0.001, (a) mean significance difference from control group. (B) mean significance difference from hypothyroid group. (C) mean significance difference from EU+B.*

The elevation in serotonin levels after barley administration in the present study also resulted in ERK1/2 improvement in brain tissue, which was reduced by hypothyroidism induction. The binding of serotonin to 5-TH2 receptors stimulates ERK1/2 phosphorylation via the release of epidermal growth factor (EGF) agonist

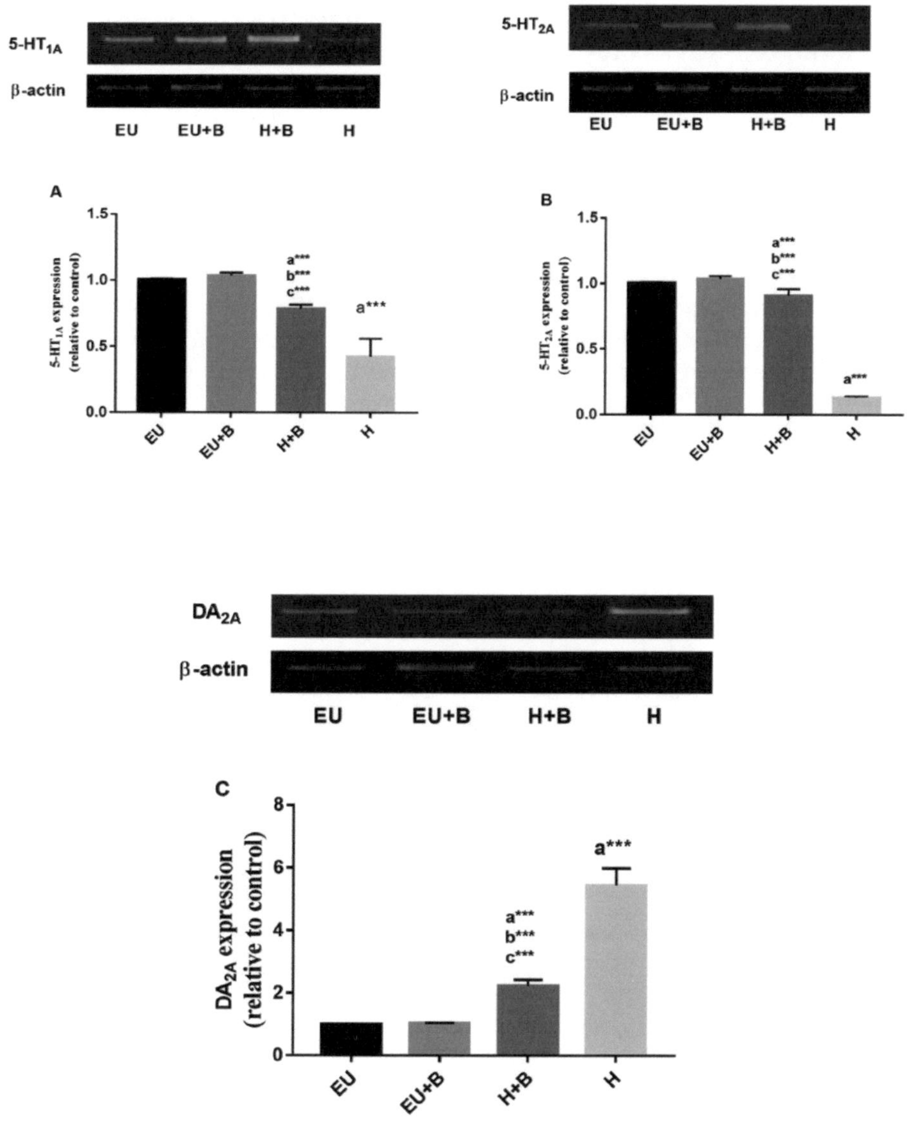

Figure 11.
Effect of barley on relative expression of (A) mRNA 5-HT1A, (B) mRNA 5-HT2A and (C) mRNA DA2A.

and transactivation of (EFG) receptors [41]. The improvement of serum oestradiol in hypothyroid-barley-treated groups as mentioned before could explain the positive effect of barley on restoring dopamine levels in brain tissues because the reproductive hormones, estrogen and progesterone, modulate the dysregulated serotonergic, dopaminergic, and glutamatergic neurotransmission by regulating the expression of receptors, the synthesis, reuptake, and release of the neurotransmitter serotonin and dopamine, which interact with dopaminergic neurons directly to downregulate D2 autoreceptors and indirectly by inhibiting GABAergic transmission [42].

Based on the above findings, we conclude that barley (*Hordeum vulgare*) is a nutritious food with high carbohydrate, zinc, magnesium content, and a high amino acids Trp:BCAA ratio has a positive effect on ameliorating the neural dysfunction induced by hypothyroidism and recommended for relieving stress, improving mood and depression.

4. The antioxidant effect of *Panax ginseng* on fertility disorders and oxidative stress induced by hypothyroidism

The present study revealed a reduction in fT3, fT4, and elevation in TSH levels as well as deterioration in THs transporting protein (TBG and TTR) in brain tissues of hypothyroid rats as shown in **Table 8**. This disturbance attributed to Neo-Mercazole which is an antithyroid agent that blocks thyroid hormonogenesis by inhibiting thyroid peroxidase (TPO) activity and preventing the formation of thyroglobulin from tyrosine [43, 44]. Hypothyroidism also causes elevation of cortisol that leads to inhibition of the deiodinase enzyme type 2 (D2) enzyme, responsible for the conversion of T4 into T3 [45]. Ginseng treatment improves the levels of thyroid hormones in serum and brain tissues through restoration of the impairment transporting protein (TTR & TBG) as shown in **Tables 4** and **8**. Moreover, ginseng boosts the activity of the enzyme responsible for converting T4 to active T3 and reduces thyroid hormone-blocking reverse T3 (rT3) which inhibits active T3 from binding to its functioning T3 receptors [46, 47].

The sexual dysfunction may be developed from psychological stress state exerted by hypothyroidism induction and this confirmed the role of the pituitary–adrenal gonadal axis (HPA) as a defense mechanism carried out by the organism against stress event [48]. So the reduction in trophic hormone (FSH & LH) associated with hypothyroidism led to decreasing the E2 hormone level and elevation of progesterone and testosterone. The inhibition in gonadal activity in hypothyroid rats in the present model as documented in **Table 5** was confirmed by the lowering of E2/T (**Table 6**) ratio which is a marker of the aromatase enzyme activity

Parameter	Control	G	H	H + G
f T3 (ng/ml)	1.99 ± 0.06	1.54 ± 0.05	1.59 ± 0.02	1.80 ± 0.04 (b*)
f T4 (μg/ml)	6.61 ± 0.14	7.32 ± 0.36	4.07 ± 0.12 (a*)	6.78 ± 0.34 (b**)
TSH (mIU/ml)	12.31 ± 0.62	12.81 ± 0.69	14.09 ± 0.42 (a*)	10.75 ± 0.26 (b**)
TTR (ng/ml)	32.56 ± 2.04	36.93 ± 2.95	44.19 ± 3.34 (a*)	27.41 ± 1.96 (b*)
TBG(pg/ml)	2.54 ± 0.19	2.49 ± 0.06	2.26 ± 0.07	2.37 ± 0.13

All data in tables represented by mean ± SD, n= 10 animals.
**p<0.05, **p<0.01 and ***p<0.001.*
a: mean significance difference from control group. B: mean significance difference from hypothyroid group.

Table 4.
Thyroid hormones and their carrying proteins levels of in the studied groups.

Parameter	Control	G	H	H + G
FSH (mIU/L)	7.11 ± 0.16	7.98 ± 0.14	5.99 ± 0.18 (a*)	7.03 ± 0.12
LH (ng/L)	6.47 ± 0.29	7.38 ± 0.38	4.35 ± 0.19 (a**)	6.73 ± 0.28 (b**)
E2 (Pg/ml)	20.76 ± 0.64	20.3 ± 0.39	14.7 ± 0.8 (a*)	18.00 ± 0.77 (a*,b**)
P (ng/ml)	23.24 ± 1.66	20.66 ± 1.51	30.41 ± 1.18 (a*)	20.56 ± 1.67 (b*)
T (ng/ml)	0.32 ± 0.06	0.35 ± 0.03	0.39 ± 0.07 (a*)	0.35 ± 0.04
PRL (ng/ml)	94.06 ± 4.89	104.12 ± 4.41	149.11 ± 11.33 (a**)	101.02 ± 6.58 (b*)

All data in tables represented by mean ± SD, n= 10 animals.
**p<0.05, **p<0.01 and ***p<0.001.*
a: mean significance difference from control group. B: mean significance difference from hypothyroid group.

Table 5.
Serum fertility hormones levels in hypothyroid and treated adult female albino rats.

(estrogen synthase, CYP19A1), this is a key enzyme converted testosterone into estrogen in granulosa cell [49]. Hyperprolactinemia, as recorded in the present study, negatively affected the activity of aromatase enzyme and this lead to hypoestrogenism, inhibited the release of gonadotrophic hormones (LH & FSH) from the pituitary gland and potentiated the inhibitory action of inhibin hormone that stimulated negative feedback and lowered estradiol level [50]. The administration of *Panax ginseng* was ameliorated these changes in trophic and steroidal gonadal hormones due to the triterpenoid saponins which steroidal in nature and considered the precursors of steroidal hormones in both plants and animals [51].

The current study exhibited a significant increase in serum corticosterone hormone. Also, the elevation of 8-hydroxyguanosine, an oxidative stress marker, and Caspase-3, an apoptotic marker in serum and brain tissues (**Table 7** and **Figure 12**). The administration of *Panax ginseng* was renormalized the cortisol and oxidative stress markers by increasing the cellular resistance to stress and potentiated the role

Parameter	Control	G	H	H + G
ERK1/2 (pg/ml)	47.38 ± 1.93	50.17 ± 2.57	63.17 ± 3.13 (a*)	42.26 ± 1.7 (a*b**)
E2/T ratio	64.87 ± 10.3	58.0 ± 13.0	37.6 ± 11.4(a***)	51.4 ± 19.25 (b*)
Corticosterone (μg/dl)	1.8 ± 0.06	1.7 ± 0.11	2.82 ± 0.21 (a**)	1.73 ± 0.10 (b*)

All data in tables represented by mean ± SD, n= 10 animals.
**p<0.05, **p<0.01 and ***p<0.001.*
a: mean significance difference from control group. B: mean significance difference from hypothyroid group.

Table 6.
ERK1/2, E2/T ratio and cortisol level in serum of hypothyroid and treated female albino rats.

Parameter	Control	G	H	H + G
8-OH Guanosine (ng/L)	48.46 ± 2.21	32.79 ± 0.05	57.99 ± 0.02 (a*)	48.63 ± 0.08 (b*)
Caspases (ng/ml)	5.26 ± 0.20	5.94 ± 0.15	12.44 ± 0.67 (a***)	5.48 ± 0.14 (b***)

All data in tables represented by mean ± SD, n= 10 animals.
**p<0.05, **p<0.01 and ***p<0.001.*
a: mean significance difference from control group. B: mean significance difference from hypothyroid group.

Table 7.
Oxidative stress markers in serum of hypothyroid and ginseng-treated female albino rats.

	Control	H	G	H + G
f T4 (μg/ml)	4.44 ± 0.11	3.97 ± 0.10 (a*)	3.65 ± 0.34 (a*b*)	4.08 ± 0.58 (c*)
f T3 (μg/ml)	1.62 ± 0.08	1.40 ± 0.08 (a*)	1.49 ± 0.12 (a*b*)	1.51 ± 0.11 (a*b*)
TSH (mIU/ml)	12.47 ± 0.42	15.81 ± 0.53 (a**)	10.87 ± 0.48 (a***b***)	13.57 ± 0.87 (b***c***)
TTR (ng/ml)	51.23 ± 2.39	15.07 ± 0.14 (a***)	42.1 ± 2.88 (a**b***)	33.28 ± 1.88 (a***b***c**)
TBG (pg/ml)	1.58 ± 0.04	2.09 ± 0.08 (a*)	1.75 ± 0.39 (a*b*)	1.73 ± 0.35 (a*b*)

All data in tables represented by mean ± SD, n= 10 animals.
**p<0.05, **p<0.01 and ***p<0.001.*
a: mean significance difference from control group. B: mean significance difference from hypothyroid group.

Table 8.
Effect of ginseng on f T3, f T4, TSH, TBG and TTR in brain tissue of control and hypothyroid-treated rats.

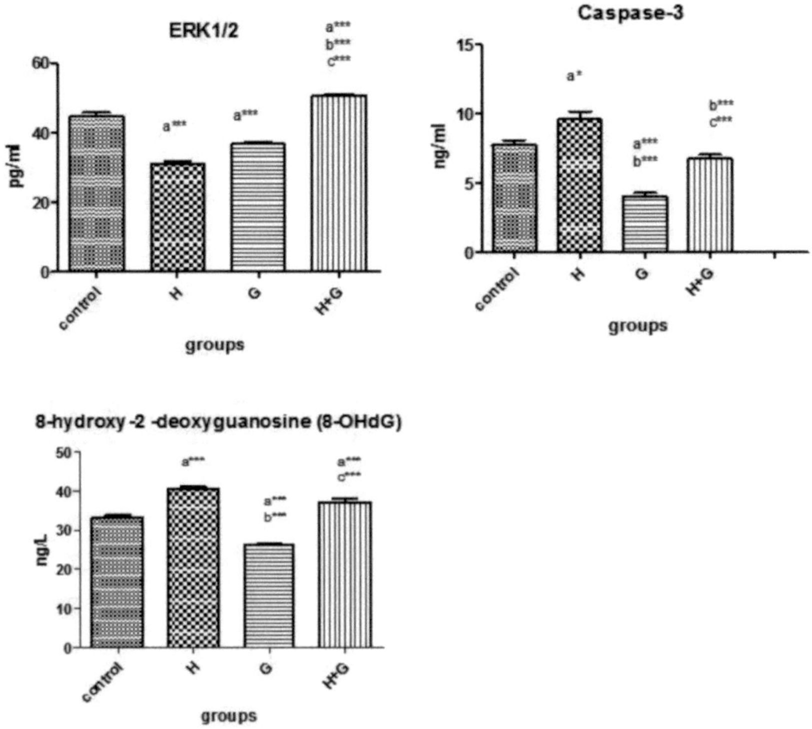

Figure 12.
Effect of ginseng on ERK1/2, Caspase-3 and 8-OHdG in all studied groups.

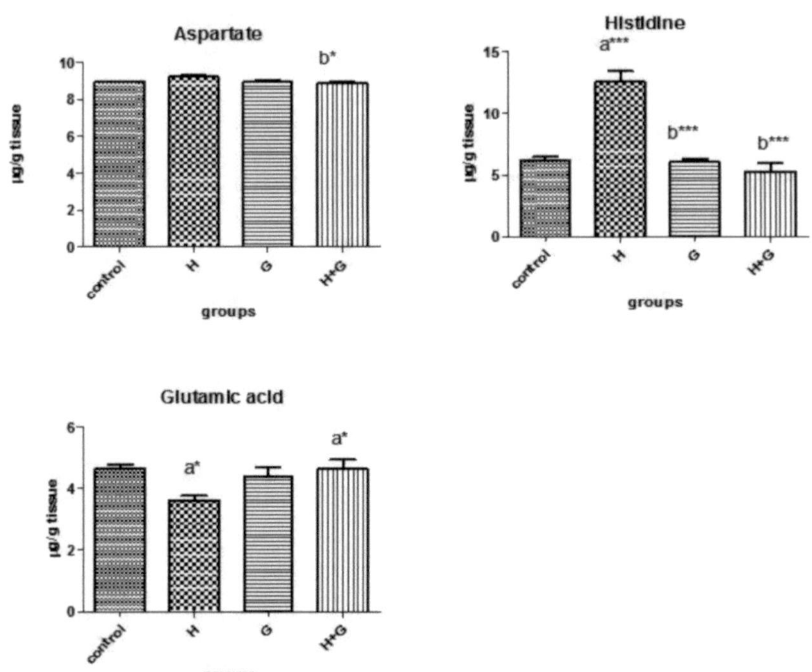

Figure 13.
Effect of ginseng on excitatory amino acids in all studied groups.

of the immune system through triterpenes of ginsenosides [48, 52]. Serum ERK1/2 was activated in response to the elevation of oxidative stress and cell death apoptotic markers as underlined in the present study. Treatment with ginseng ameliorates 8-OHdG, Caspase-3 and ERK1/2 levels referring to its neuroprotective effect and retrieval homeostasis. Also, ginsenosides which is the pharmacologically active constituents with its adaptogenic, powerful antioxidant and radical scavenging activities, regulate the function of HPA, support neurogenesis, synaptogenesis, neuronal growth, and neurotransmission, in turn, protect the central nervous system [53–56]. The study of [57] showed that the panaxatriol group of ginsenosides blocked Caspase-3 expression and increased anti-apoptotic Bcl-2 and p53, indicating that RG repressed cellular apoptosis otherwise, ginsenosides Rd and Re have neuroprotective properties by modulation of ERK1/2 signaling pathway [58]. The elevation of Caspase-3 in the present hypothyroid modal was confirmed by studying the Comet tailed DNA damage of the brain and thyroid tissues as illustrated in **Figures 15** and **16**. Ginseng treatment repair DNA damage in brain and thyroid tissues, this denoted to its highest content of phenolic compounds which act as antioxidants.

Thyroid hormones control the levels of these neurotransmitters which are responsible for maintaining a good mental state and preventing depression [19]. THs regulate both the release of the neurotransmitters and their post-receptor signaling to promote mood stabilization so, their deficiency may weaken the neurogenesis, maturation and synaptic transmission (**Figures 13** and **14**).

The present data, exhibited that the induction of hypothyroidism resulted in a significant decrease of NE, DA and 5-HT concentrations in all studied brain areas

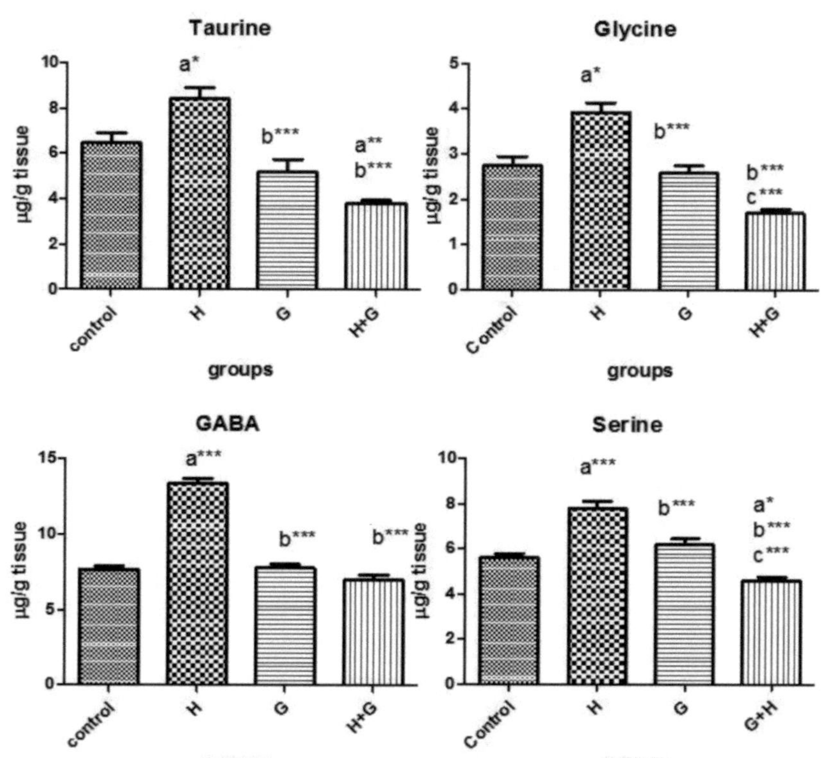

Figure 14.
Effect of ginseng on inhibitory amino acids in all studied groups.

	Frontal cortex	Hippocampus	Hypothalamus	Mid brain	Cerebellum
DA (μg g−1 tissue)					
Control	0.59 ± 0.04	2.39 ± 0.18	1.47 ± 0.27	1.31 ± 0.01	0.60 ± 0.02
H	$0.25 \pm 0.02\ a^{***}$	$0.93 \pm 0.07\ a^{***}$	$0.89 \pm 0.08\ a^{***}$	$0.89 \pm 0.03\ a^{***}$	$0.27 \pm 0.01\ a^{***}$
G	$0.58 \pm 0.01b^{***}$	$2.74 \pm 0.08b^{***}$	$1.69 \pm 0.17b^{***}$	$1.31 \pm 0.01b^{**}$	$0.59 \pm 0.01b^{***}$
H + G	$0.44 \pm 0.02a^{***}b^{**}c^{***}$	$1.67 \pm 0.07a^{***}b^{***}c^{***}$	$1.12 \pm 0.04a^{**}c^{***}$	$1.18 \pm 0.01a^{***}b^{***}c^{***}$	$0.50 \pm 0.01a^{**}b^{***}c^{**}$
NE (μg g−1 tissue)					
Control	0.51 ± 0.07	0.68 ± 0.04	0.39 ± 0.01	0.70 ± 0.01	0.58 ± 0.03
H	$0.21 \pm 0.02\ a^{***}$	$0.31 \pm 0.02\ a^{***}$	$0.13 \pm 0.02\ a^{***}$	$0.35 \pm 0.02\ a^{***}$	$0.29 \pm 0.01\ a^{***}$
G	$0.56 \pm 0.08b^{***}$	$0.67 \pm 0.06\ b^{***}$	$0.38 \pm 0.02b^{***}$	$0.69 \pm 0.01b^{***}$	$0.6 \pm 0.01b^{***}$
H + G	$0.35 \pm 0.03a^{***}b^{**}c^{***}$	$0.54 \pm 0.03a^{***}b^{***}c^{***}$	$0\ .29 \pm 0.01a^{***}b^{***}c^{***}$	$0.56 \pm 0.03a^{***}b^{***}c^{***}$	$0.49 \pm 0.01a^{***}b^{***}c^{***}$
5-HT (μg g−1 tissue)					
Control	0.57 ± 0.03	0.38 ± 0.02	0.78 ± 0.03	0.72 ± 0.03	0.47 ± 0.02
H	$0.23 \pm 0.01\ a^{***}$	$0.11 \pm 0.01\ a^{***}$	$0.45 \pm 0.02\ a^{***}$	$0.29 \pm 0.01\ a^{***}$	$0.18 \pm 0.01\ a^{***}$
G	$0.57 \pm 0.03b^{***}$	$0.39 \pm 0.01b^{***}$	$0.77 \pm 0.02b^{***}$	$0.70 \pm 0.03b^{***}$	$0.47 \pm 0.01b^{***}$
H + G	$0.41 \pm 0.02a^{***}b^{***}c^{***}$	$0.29 \pm 0.01a^{***}b^{***}c^{***}$	$0.56 \pm 0.02a^{***}b^{***}c^{***}$	$0.5 \pm 0.01a^{***}b^{***}c^{***}$	$0.34 \pm 0.01a^{***}b^{***}c^{***}$

Data in tables given are mean ± S.D. the number of animals was 10 in each group.
***Significant at p 0.01 and ***significant at p 0.001.*
(a) Significant versus control group; (b) significant versus hypothyroid (H) group and (c) significant versus ginseng (G) treated group.

Table 9.
Effect of ginseng on monoamines levels of discrete brain regions in control and hypothyroid-treated rats.

(frontal cortex, hippocampus, hypothalamus, midbrain and cerebellum) as shown in **Table 9**. Monoamines reduction after hypothyroidism refereed to the disturbance in the synthesis and release of these amines from impairment neurons or may be due to an alteration pattern of their synthesizing and/or degradative enzymes [59]. Ginseng treatment ameliorate the reduced monoamine levels of hypothyroid rats and this refers to its powerful ability to maintain homeostasis and modulating neurotransmitter levels hence can amend the neurodegenerative diseases [54, 60]. Also, ginseng saponins modulate dopaminergic activity at both presynaptic and postsynaptic receptors [54, 55].

Improvement of monoamines after ginseng treatment refer also to gintonin which is one of the important ginseng constituents that increased the expression of learning and memory and modulate cholinergic, glutaminergic and other molecular signaling pathways that are vital for cognitive activity as stated in [16]. Ginseng which considered a potential phytoestrogen exhibits antidepressant so, ginsenosides Rb1 enhances the serotonergic system by increasing 5-HT synthesis,

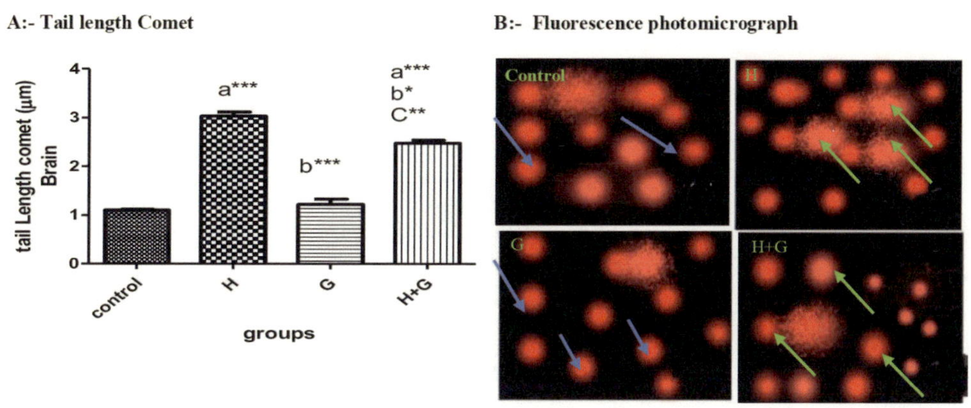

Figure 15.
*Effect of ginseng on DNA damage in whole brain tissue, (A) Tail length expressed in μm in brain tissue of all treated groups. *Significant at p. 0.05, **significant at p. 0.01 and ***significant at p. 0.001. (a), significant versus control group, (b) significant versus hypothyroid, (H) group and (c) significant versus ginseng, (G) treated group, (B) Fluorescence photomicrograph showing comets in control, H, G and H+G -groups. (blue arrows) indicated the intact DNA and (green arrows) indicated the degree of damaged DNA (tail).*

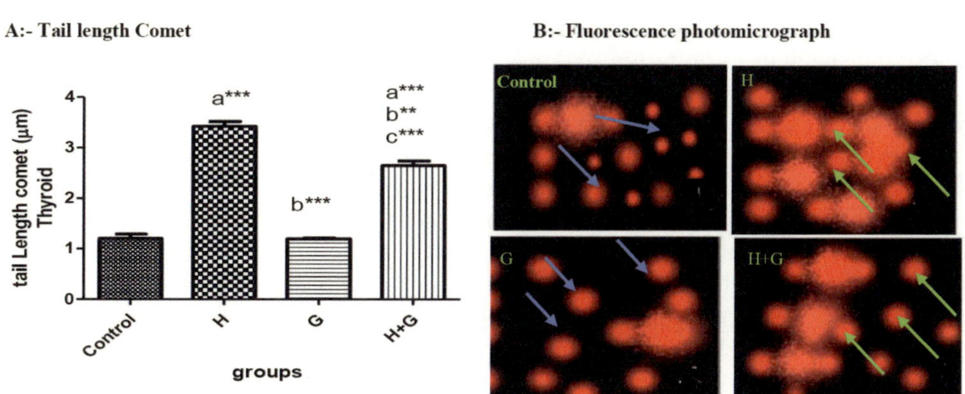

Figure 16.
*Effect of ginseng on DNA damage in thyroid tissue. (A) Tail length expressed in μm in thyroid tissue of all treated groups. ** significant at p 0.01 and *** significant at p 0.001. (a) significant versus control group, (b) significant versus hypothyroid, (H), group and (c), significant versus ginseng (G), treated group (B), Fluorescence photomicrograph showing comets in control, H, G and H+G -groups. (blue arrows) indicated the intact DNA and (green arrows) indicated the degree of damaged DNA (tail).*

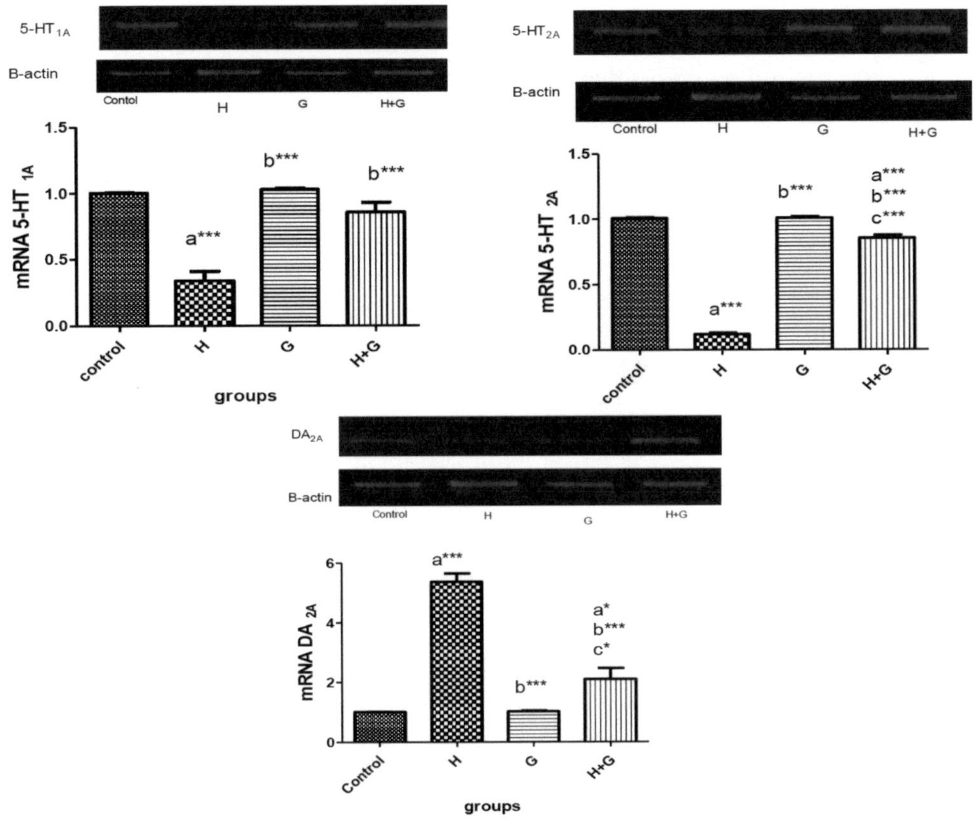

Figure 17.
*Effect of ginseng on expression of mRNA 5-HT1A, mRNA 5-HT2A and mRNA DA2A. *Significant at p 0.05, and ***significant at p 0.001. (a) Significant versus control group (b) significant versus hypothyroid (H) group and (c) significant versus ginseng (G) treated group.*

decreasing 5-HT degradation, stimulating 5-HT2A receptor and suppress the activity of the inhibitory 5-HT3A receptor in the brain. Also, effect via increasing 5-HT activity. This effect is mediated by the activation of estrogen receptor [61, 62].
In the present study, hypothyroidism induction by Neo-Mericazole lead to a significant increase of excitatory amino acid, histidine, and all inhibitory amino acids while excitatory glutamic acid was significantly decreased in brain tissues (**Figures 15** and **16**). *Panax ginseng* with its powerful components ameliorates the disturbance of amino acids and in turn monoamines. The induction of hypothyroidism revealed an elevation in the concentration of dopamine receptors and the reduction of serotonin receptors density (**Figure 17**). Treatment with ginseng restores the level of dopamine and serotonin receptors density towards the control value. This refers firstly to the genomic pathway of ginsenosides which bind to intracellular nuclear hormone receptors like androgen receptor (AR), estrogenic receptor (ER) and progesterone receptor [13].

5. Conclusion

In conclusion, the present study is pointed out to the pituitary-gonad-adrenal disturbances aroused from the hypothyroidism induction by Neo-Mericazole and how ginseng, one of the most Asian medicinal traditional plants, significantly normalized the fertility disorders and stress by acting as free radicals' scavenger.

Conflict of interest

The authors declare no conflict of interest.

Author details

Lobna F. Wahman[1*], Marwa M. Abd Rabo[1], Amany Hanafy M. Elgoly[1]
and Magda H.M. Yousef[2,3]

1 National Organization for Drug Control and Research (NODCAR), Egypt

2 Faculty of Medicine, Taibah University, Saudi Arabia

3 Faculty of Medicine, Ain-Shams University, Egypt

*Address all correspondence to: lwahman62@gmail.com

References

[1] Gangopadhyay N, Hossain MB, Rai DK, Brunton NP. A review of extraction and analysis of bioactives in oat and barley and scope for use of novel food processing technologies. Molecules. 2015;**20**:10884-10909

[2] Youssef MKE, El-fishawy FAE, Ramadan EAE. Nutritional assessment of barley, talbina and their germinated products. Scientific Journal of Crop Science. 2013;**3**:8-19

[3] Rahman MH, Ali MY. The relationships between thyroid hormones and the brain serotonin (5-Ht) system and mood: Of synergy and significance in the adult brain—A review. Faridpur Medical College Journal. 2015;**9**:98-101

[4] Zhu Y, Li T, Fu X, Abbasi AM, Zheng B, Liu RH. Phenolics content, antioxidant and antiproliferative activities of dehulled highland barley (*Hordeum vulgare* L.). Journal of Functional Foods. 2015;**19**:439-450

[5] Idehen E, Tang Y, Sang S. Bioactive phytochemicals in barley. Journal of Food and Drug Analysis. 2017;**25**: 148-161

[6] Choi JS et al. Consumption of barley β-glucan ameliorates fatty liver and insulin resistance in mice fed a high-fat diet. Molecular Nutrition & Food Research. 2010;**54**:1004-1013

[7] Jonnalagadda SS et al. Putting the whole grain puzzle together: Health benefits associated with whole grains—Summary of American Society for Nutrition 2010 Satellite Symposium. The Journal of Nutrition. 2011;**141**:1011S–1022S

[8] Schlemmer U, Frølich W, Prieto RM, Grases F. Phytate in foods and significance for humans: Food sources, intake, processing, bioavailability, protective role and analysis. Molecular Nutrition & Food Research. 2009;**53**: S330–S375

[9] Yeo HB, Yoon HK, Lee HJ, Kang SG, Jung KY, Kim L. Effects of Korean Red Ginseng on cognitive and motor function: A double-blind, randomized, placebo-controlled trial. Journal of Ginseng Research. 2012;**36**(2):190-197

[10] So S, Lee JW, Kim YS, Hyun SH, Han CK. Red ginseng monograph. Journal of Ginseng Research. 2018;**42**(4):549-561

[11] Kim J, Cho SY, Kim SH, Kim S, Park CW, Cho D, et al. Ginseng berry and its biological effects as a natural phytochemical. Natural Products Chemistry & Research. 2016;**4**(2):2-5

[12] Rokot NT, Kairupan TS, Cheng KC, Runtuwene J, Kapantow NH, Amitani M, et al. A role of ginseng and its constituents in the treatment of central nervous system disorders. Evidence-based Complementary and Alternative Medicine. 2016. ID 2614742

[13] Park J, Song H, Kim SK, Lee MS, Rhee DK, Lee Y. Effects of ginseng on two main sex steroid hormone receptors: Estrogen and androgen receptors. Journal of Ginseng Research. 2017;**41**(2):215-221

[14] Kim KH, Lee D, Lee HL, Kim CE, Jung K, Kang KS. Beneficial effects of Panax ginseng for the treatment and prevention of neurodegenerative diseases: Past findings and future directions. Journal of Ginseng Research. 2018;**42**(3):239-247

[15] Chung IM, Lim JJ, Ahn MS, Jeong HN, An TJ, Kim SH. Comparative phenolic compound profiles and antioxidative activity of the fruit, leaves, and roots of Korean ginseng (*Panax ginseng* Meyer) according to cultivation years. Journal of Ginseng Research. 2016;**40**(1):68-75

[16] Jakaria M, Haque MDE, Kim J, Cho D-Y, Kim I-S, Choi DK. Active ginseng components in cognitive impairment: Therapeutic potential and prospects for delivery and clinical study. Oncotarget. 2018;**9**(71):33601-33620

[17] Goel R, Rajesh R. A signaling network of thyroid-stimulating hormone. Journal of Proteomics & Bioinformatics. 2011;**04**:238-241

[18] Bernal J. Thyroid hormone receptors in brain development. Nature Clinical Practice. Endocrinology & Metabolism. 2007;**3**:249-259

[19] Tousson E, Ibrahim W, Arafa N, Akela MA. Monoamine concentrations changes in the PTU-induced hypothyroid rat brain and the ameliorating role of folic acid. Human & Experimental Toxicology. 2012;**31**:282-289

[20] Wu S, Tan G, Dong X, Zhu Z, Li W, Lou Z, et al. Profiling provides a system understanding of hypothyroidism in rats and its application. PLoS One. 2013;**8**:e55599

[21] Bawazir AE. Investigations on the chronic effect of Talbina (barley water) on hormone (cortisol and testosterone), reproductive system and some neurotransmitter. American Eurasian Journal of Scientific Research. 2010;**5**:134-142

[22] Paget GE, Barnes JM. Toxicity tests: Evaluation of drug activities. In: Laurence DR, Bandrakoach AL, editors. Pharmacometrics. Vol. 1. 1964. p. 135. Chapter 6

[23] Fraser TR et al. Antithyroid activity and toxicity of mercazole and neo-mercazole. The Journal of Clinical Endocrinology and Metabolism. 1954;**14**:1230-1244

[24] El-Bakry AM, El-Gareib AW, Ahmed RG. Comparative study of the effects of experimentally induced hypothyroidism and hyperthyroidism in some brain regions in albino rats. International Journal of Developmental Neuroscience. 2010;**28**:371-389

[25] Karmarkar SW, Bottum KM, Tischkau SA. Considerations for the use of anaesthetics in neurotoxicity studies. Comparative Medicine. 2010;**60**: 256-262

[26] Singh NP, McCoy MT, Tice RR, Schneider EL. A simple technique for quantitation of low levels of DNA damage in individual cells. Experimental Cell Research. 1988; **175**(1):184-191

[27] Pagel P, Blome J, Wolf HU. High-performance liquid chromatographic separation and measurement of various biogenic compounds possibly involved in the pathomechanism of Parkinson's disease. Journal of Chromatography. B, Biomedical Sciences and Applications. 2000;**46**(2):297-304

[28] Heinrikson RL, Meredith SC. Amino acid analysis by reverse-phase high-performance liquid chromatography: Precolumn derivatization with phenylisothiocyanate. Analytical Biochemistry. 1984;**136**(1):65-74

[29] Borg S et al. Iron transport, deposition and bioavailability in the wheat and barley grain. Plant and Soil. 2009;**325**:15-24

[30] Sun Q et al. Effects of forced swimming stress on thyroid function, pituitary thyroid-stimulating hormone and hypothalamus thyrotropin releasing hormone expression in adrenalectomy Wistar rats. Experimental and Therapeutic Medicine. 2016

[31] Faccio L et al. Serum levels of LH, FSH, estradiol and progesterone in female rats experimentally infected by *Trypanosoma evansi*. Experimental Parasitology. 2013;**135**:110-115

[32] Hosseini M, Dastghaib SS, Rafatpanah H, Hadjzadeh MA-R, Nahrevanian H, Farrokhi I. Nitric oxide contributes to learning and memory deficits observed in hypothyroid rats during neonatal and juvenile growth. Clinics. 2010;**65**:1175-1181

[33] Mancini A, Di Segni C, Raimondo S, Olivieri G, Silvestrini A, Meucci E, et al. Thyroid hormones, oxidative stress, and inflammation. Mediators of Inflammation. 2016:1-12

[34] Gamel TH, Abdel-Aal ESM. Phenolic acids and antioxidant properties of barley wholegrain and pearling fractions. Agricultural and Food Science. 2012;**21**:118-131

[35] Leitao C et al. Fate of polyphenols and antioxidant activity of barley throughout malting and brewing. Journal of Cereal Science. 2012;**55**:318-322

[36] Badrasawi MM, Shahar S, Manaf ZA, Haron H. Effect of Talbinah food consumption on depressive symptoms among elderly individuals in long term care facilities, randomized clinical trial. Clinical Interventions in Aging. 2013;**8**:279-285

[37] Gillies GE, McArthur S. Estrogen actions in the brain and the basis for differential action in men and women: A case for sex-specific medicines. Pharmacological Reviews. 2010;**62**:155-198

[38] Benmansour S, Arroyo LD, Frazer A. Comparison of the antidepressant-like effects of estradiol and that of selective serotonin reuptake inhibitors in middle-aged ovariectomized rats. Frontiers in Aging Neuroscience. 2016;**8**:1-13

[39] Cansev M, Wurtman RJ. Aromatic amino acids in the brain. In: Handbook of Neurochemistry and Molecular Neurobiology. 2007. pp. 59-97

[40] Delisi LE. Handbook of Neurochemistry and Molecular Neurobiology. Handbook of Neurochemistry and Molecular Neurobiology. Schizoprenia. 2009;**27**:107-241

[41] Li B, Zhang S, Li M, Hertz L, Peng L. Serotonin increases ERK1/2 phosphorylation in astrocytes by stimulation of 5-HT2 Band 5-HT2C receptors. Neurochemistry International. 2010;**57**:432-439

[42] Karpinski M, Mattina GF, Steiner M. Effect of gonadal hormones on neurotransmitters implicated in the pathophysiology of obsessive-compulsive disorder: A critical review. Neuroendocrinology. 2017;**105**:1-16

[43] Manna D, Roy G, Mugesh G. Antithyroid drugs and their analogues: Synthesis, structure, and mechanism of action. Accounts of Chemical Research. 2013;**46**(11):2706-2715

[44] Singh RP, Singh A, Sirohi HV, Singh AK, Kaur P, Sharma S, et al. Dual binding mode of antithyroid drug methimazole to mammalian heme peroxidases—Structural determination of the lactoperoxidase-methimazole complex at 1.97 Å resolution. FEBS Open Bio. 2016;**6**:640-650

[45] Hage MP, Azar ST. The link between thyroid function and depression. Journal of Thyroid Research. 2012:1-8 Article ID 590648

[46] McCormack PD, Thomas J, Malik M, Staschen CM. Cold stress, reverse T3 and lymphocyte function. Alaska Medicine. 1998;**40**:55-62

[47] Dai X, Zhou Y, Yu X. Effect of ginseng injection in treating congestive heart failure and its influence on thyroid hormones. Zhongguo Zhong Xi Yi Jie He Za Zhi. 1999;**19**(4):209-211

[48] Kim Y, Choi EH, Doo M, Kim JY, Kim CJ, Kim CT, et al. Anti-stress

effects of ginseng via down-regulation of tyrosine hydroxylase (TH) and dopamine β-hydroxylase (DBH) gene expression in immobilization-stressed rats and PC12 cells. Nutrition Research and Practice. 2010;**4**(4):270-275

[49] Wen L, Jiang X, Sun J, Li X, Li X, Tian L, et al. Cyanidin-3-O-glucoside promotes the biosynthesis of progesterone through the protection of mitochondrial function in Pb-exposed rat Leydig cells. Food and Chemical Toxicology. 2018;**112**:427-434

[50] Jeyaraj DA, Mani Maran RR, Aruldhas MM, Govindarajulu P. Progesterone induced modulations of serum hormonal profiles in adult male and female rats. Endocrine Research. 2001;**27**(1-2):223-232

[51] Kuete V. Medicinal Plant Research in Africa: Pharmacology and Chemistry. 1st ed. Elseiver; 2013. p. 890

[52] Augustin JM, Kuzina V, Andersen SB, Bak S. Molecular activities, biosynthesis and evolution of triterpenoid saponins. Phytochemistry. 2011;**72**(6):435-457

[53] Van Kampen JM, Baranowski DB, Shaw CA, Kay DG. Panax ginseng is neuroprotective in a novel progressive model of Parkinson's disease. Experimental Gerontology. 2014;**50**:95-105

[54] Al-Hazmi MA, Rawi SM, Arafa NM, Wagas A, Montasser AO. The potent effects of ginseng root extract and memantine on cognitive dysfunction in male albino rats. Toxicology and Industrial Health. 2015;**31**(6):494-509

[55] Hussein J, El-Khayat Z, El-Toukhy S, El-Bana M, Medhat D, Morsy S. Panax ginseng regulates brain monoamines in lipopolysaccharide-induced experimental brain injury. Der Pharma Chemica. 2016;**8**(6):116-121

[56] Lee S, Rhee DK. Effects of ginseng on stress-related depression, anxiety, and the hypothalamic pituitary adrenal axis. Journal of Ginseng Research. 2017;**41**(4):589-594

[57] Kim EH, Kim IH, Ha JA, Choi KT, Pyo S, Rhee DK. Antistress effect of red ginseng in brain cells is mediated by TACE repression via PADI4. Journal of Ginseng Research. 2013;**37**(3):315-323

[58] Kang H, Lim JW, Kim H. Inhibitory effect of Korean red ginseng extract on DNA damage response and apoptosis in *Helicobacter pylori*-infected gastric epithelial cells. Journal of Ginseng Research. 2018:1-7

[59] Hassan WA, Aly MS, Rahman TA, Shahat AS. Impact of experimental hypothyroidism on monoamines level in discrete brain regions and other peripheral tissues of young and adult male rats. International Journal of Developmental Neuroscience. 2013;**31**(4):225-233

[60] Lee DCW, Lau ASY. Effects of Panax ginseng on tumour necrosis factor-α-mediated inflammation: A mini-review. Molecules. 2011;**16**(4):2802-2816

[61] Hao K, Gong P, Sun SQ, et al. Beneficial estrogen-like effects of ginsenoside Rb1, an active component of Panax ginseng, on neural 5-HT disposition and behavioural tasks in ovariectomized mice. European Journal of Pharmacology. 2011;**659**(1):15-25

[62] Jang D, Lee HLK, Kim KR, Won R, Lee SE, Shim I. White ginseng ameliorates depressive behavior and increases hippocampal 5-ht level in the stressed ovariectomized rats. BioMed Research International. 2019. Article ID 5705232

Section 5

Micoorganisms Mediated Adaptive Mechanisms to Abiotic Stresses

Chapter 20

Plant Growth-Promoting Bacteria as a Natural Resource for Sustainable Rice Production under the Soil Salinity, Wastewater, and Heavy Metal Stress

Raghad S. Mouhamad and Michael Alabboud

Abstract

Rice is a cereal plant that is consumed in a grain form; however, its prolonged contact with irrigation wastewater might pose a threat to the consumers despite the following milling processes to eliminate the grain surface contamination which means that it needs further cooking to be suitable for human use. Additionally, excessive salt levels in wastewater can cause plant toxicity. Therefore, wastewater disposal can be handled by farm remediation. *Rhizobacteria* can also be used in this stressful environment to alleviate the problem by triggering a plant growth-promoting response (PGPR). The importance of promoting and biocontrol plant growth is based upon its long-term stability, as well as the numerous generated secondary metabolites, besides its ability to remove heavy metal. The current study revealed that PGPR allowed such toxic effects on sewage to encourage and define the characteristics of plant growth through urban environments.

Keywords: heavy metals, wastewater, PGPR

1. Introduction

1.1 Relationship between PGPB and rice production under nutrient and salinity

As a consequence of the continuous population growth worldwide along with the shortage of food sustainability [1], it is necessary to create an alternative agricultural productivity systems [2, 3]. One of the sustainable alternative strategies is the utilization of plant growth-promoting bacteria (PGPB) in agricultural practices [4]. Promoting plant growth (PGP) has numerous correlation capabilities either by endophyte in plant tissue [5], rhizosphere in seed surface as well as plant root [6], symbiosis in root nodules, and phyllosphere in stem and/or leaf surface (Turner). PGPB involve 1-aminocyclopropane-l-carboxylic acid (ACC) deaminase that is applied to seedling which could effectively stimulate plant growth by reducing plant ethylene rates [7] under drought, salinity [8, 9], flooding, and contaminant condition [10] and increasing phosphate solubility and availability in soil, along with the increase in plant biomass, root area, and total N and P contents in rice [11].

Rice production is reduced under saline agriculture system (**Figure 1**); therefore, it is becoming increasingly important to imply plant growth-promoting traits for mitigation of salt stress [12–14]. Promoting plant growth was shown to enhance growth effectively, and the growth-stimulating effect was also suggested to be beneficial in crop production under stressful conditions. Mechanisms for inducing plant growth-promoting response (PGPR) toward abiotic stress are usually interpreted as the result of certain phytohormone production, including ABA, GA, or IAA, or lower ethylene levels in roots of the ACC, which generates systemic bacterial resistance and enhances exopolysaccharides.

A wide spectrum of endophyte bacteria is well adjusted to the rice niche under abiotic stress condition. The emergence of rice seedlings and growth and development parameters were previously reported to be significantly affected by many PGPR strains [15]. Beneduzi et al. [16] evaluated efficient bioinoculant for rice growth improvement by bacillus strain (SVPR30). Bisht and Mishra [17] reported that rice root length and shoot length increased by 9.7 and 13.9%, respectively, when inoculated with *B. thuringiensis* (VL4C); Nautiyal et al. [18] reported that rice inoculation with *B. amyloliquefaciens* (SN-13) under saline conditions in hydroponic/saline soils has improved stress sensitivity due to an altered transcription of 14 genes, including SERK1, ethylene-responding factor EREBP, NADP-malic enzyme (NADP-Me2), and SOS1. Additionally, downregulated expression of glucose-insensitive growth (IGG) and serine–threonine (Sapk4) protein kinase in

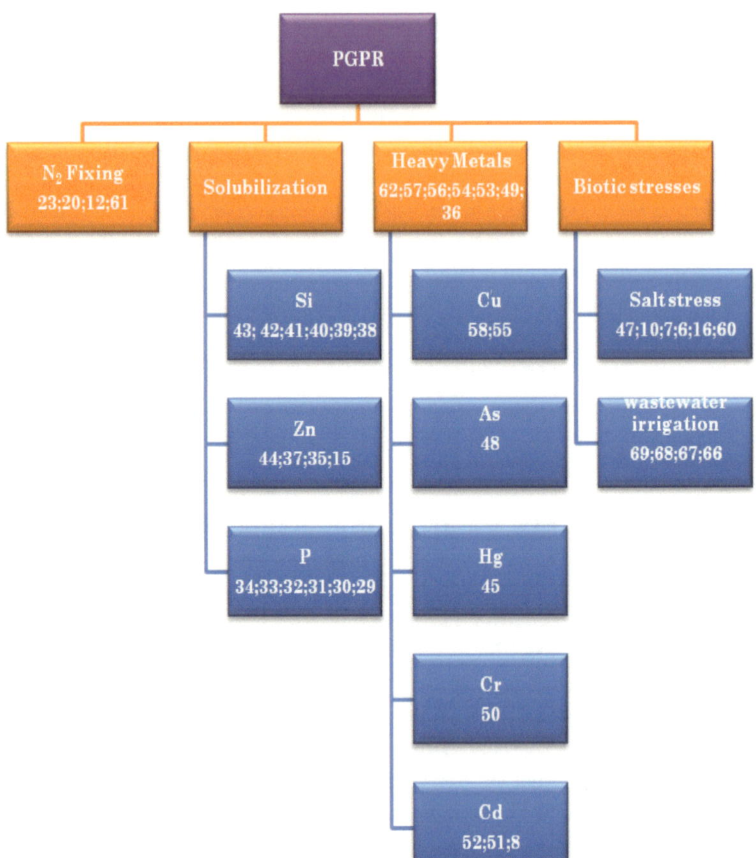

Figure 1.
Schematic description of the different plant promotion processes by PGPR.

the hydroponic setup and upregulated MAPK5 were observed in the greenhouse experiments [19]. The inoculation of SN13 improved the gene transcription involved in the sensitivity of ionic and salt stresses [20]. Endophytic bacteria can give N to rice without loss compared with other bacteria, because of their strong relationship with the plant [21]. Endophytic bacteria are a better N supplier to rice than other bacteria. Endophytic bacteria are the bacteria derived from the plants' inner tissues or extracted from plants with a sterilized layer, which have no infection symptoms [22]. The rice yield achieved by N_2-fixing *Pseudomonas* sp. was improved by 23% by Mäder et al. [23]. Several studies showed significantly greater K, N, and P levels with an increased rice output of 9.2% in co-inoculation with N_2-fixing microbes relative to the use of prescribed amounts of fertilizers as N, P, and K [24, 25]. There have been detailed documentations that rice is generally infected with a large variety of endophytic bacteria (*Azospirillum*, *Herbaspirillum*, *Rhizobium*, *Pantoea*, *Methylobacterium*, and *Burkholderia*, among others) [22]. Diazotrophs colonized effectively in the roots of rice may have a higher N fixation potential [26]. Endorhizosphere bacteria contribute far more than rhizospheric bacteria to N fixation since there is no competition with other rhizospheric microorganisms in the endorhizosphere and under low oxygen; carbon sources are provided [27, 28].

The bacterial IAA was shown in Etesami and Alikhani [29] to have significant roles in improving efficiency in the use of N and in increasing nitrogen-based substances in rice. Estrada et al. [30] also found that diazotrophic P-solubilizing bacteria improved the absorption of nutrients in rice, while Rangjaroen et al. [31] suggested that *Novosphingobium* diazotrophic is an important microbial tool of nitrogen providing for further production which renders it as a healthy biomonitor to improve organic rice cultivation.

De Souza et al. [32] demonstrated the decrease of in vitro phosphate solubility and minimization of acetylene (low reduction in acetylene) in rice shoots by bacteria, including *Herbaspirillum* sp., *Burkholderia* sp., *Pseudacidovorax*, and *Rhizobium* sp. Therefore, non-N_2 fixation growth promotion mechanisms include an IAA development and improved nutrient balanced absorption. Glick [7] shows that if a bacterium is used to produce nitrogen-solubilizing for plants, which have PGP traits (IAA, ACC deaminase, siderophore, and phosphate solubility), it should be used, and the genetic characteristics in plants should be transferred. The application of P fertilizers in rice production has continuously increased [33]. Sahrawat et al. [34] show that the use of rice P fertilizers has been continuously increased since it is one of the key restrictive factors in many regions of the world for the production of upland rice. Othman and Panhwar [35] detected that the sum of nutrition provided by aerobic rice is the same as the flooded rice, but the abundance of P is a challenge due to its immediate immobilizing and fixing with calcium (Ca^{2+}), iron (Fe^{3+}), and aluminum (Al^{3+}) elements. P deficiency in aerobic crops is also widely seen as a phenomenon [36]. The secretion of organic acids and the interaction of mycorrhizal fungi are among these methods that are very weak in rice under flooding conditions. Islam and Hossain [37] have stated that P deficiency is quite normal which increases the demand for mycorrhizal fungal interactions under flood conditions. Panhwar et al. [38] detected that the rice plants need an ancillary structure that quickly goes beyond such degraded regions and receives P for exorbitant neighboring soil composition through the development of a vast network of phosphate-solubilizing bacteria (PSB) which might satisfy some of the nutrient needs.

The growth of many plants including staple rice is hindered by micronutrient-deficient soils [39]. The toxicity of Fe is also important as Fe is one of the major constraints on the production of lowland rice. Furthermore, the scarcity of Mn in upland rice is also commonly seen [40].

A significant increase in the number of tilers provided by plan (15.1%), crop panicles (13.3%), overall grain intake Zn (52.5%), and a modest yield of the dry product by pot (12.9%) has been shown by Vaid et al. [41]. This rise was detected through soil solubilization of insoluble Zn, all of which as a result of the production of bacterial gluconic acid.

Fe, Zn, Cu, and Mn concentrations were increased by 13–16% (*Brevundimonas diminuta* PR7) and in rice co-inoculation (*Providencia* sp. PR3) (*Ochrobactrum anthropi* PR10); Adak et al. [42] detected that Fe absorbance was enhanced by 13–46% using cyanobacterial inoculants and 15–41% in Zn with the use of cyanobacterial inoculums, in rice cultivation for various modes.

1.2 Relationship between PGPB and rice production under wastewater and heavy metals

Metals as zinc (Zn), molybdenum (Mo), cobalt (Co), chromium (Cr), selenium (Se), copper (Cu), iron (Fe), manganese (Mn), magnesium (Mg), and nickel (Ni) have essential nutrients necessary for a diversity of biological and physiological functions [43]. Biological functions that are not identified are identified as nonessentials: bismuth (Bi), antimony (Sb), platinum (Pt), indium (In), arsenic (As), beryllium (Be), mercury (Hg), barium (Ba), gallium (Ge), gallium (G), gold (Au), lead (Pb), barium (Be), nickel (Ni), silver (Ag), aluminum (Al), as well as uranium (U) [44].

Ma and Takahashi [45] demonstrate that the rice PGPB ability can be used to resolve deficits in micronutrients and to biofertilize (**Table 1** and **Figure 1**). Rice is a plant that accumulates Si and considered an Si accumulator as silicon content in dry weight of the shoots may reach up to 10%, and therefore, the plants require high Si content. Rice is associated with Si depletion in its unit area; due to the removal from the earth of 100 kg of Si for brown rice (about 20 kg/hm^2 SiO_2) and exports to the farm by the extraction of straw residues during harvest and the conniving for exogenous use of Si in rice growing, Si in paddy field is available [66].

Bocharnikova et al. [67] and Ning et al. [68] previously reported that Si-deficient paddy soils may be needed to generate an economically sustainable rice crop capable of producing high yield and disease resistance. Si fertilizers are being used for growing rice production in many countries and have positive effects. Vasanthi et al. [69] detected that the *Bacillus globisporus*, *B. crustacea*, *B. flexus*, *B. megaterium*, *Pseudomonas fluorescens*, and *Burkholderia eburnean* can activate K and Si in feldspar, muscovite, and biotite silicate mineral resources. Specific pathways are used to generate disproportionate protons, organic ligand, organic acid, anion, hydroxyl, EPS, and enzymes. However, the solubilizing Si, K, and P in soil might be accompanied by an increased supply of Fe and Mn metals in plants by interacting with P-fixing sites.

Gandhi and Muralidharan [19] show that the rice growth, development, yield, and Zn solubility from ZnO and $ZnCO_3$ to *Acinetobacter* sp. have been greatly increased.

This gene recombination processing was also extended to rice, which produces rice transgenics generated via a partial weapon bombardment containing a 250 lM $HgCl_2$-resistant merA gene [65]. Recently, mercury toxicity has been identified as a triggering factor in aromatic amino acid biosynthesis (tryptophan and phenylalanine), aggregation of calcium, and activation of MAPK in rice [70]. The synthesis and accumulation of the Glybet were stimulated by *Pseudomonas* alkaline inoculation in rice plants [64]. Chakrabarty et al. [63] detected that the As (III)-treated rice seedlings proposed signal transduction regulation and hormonal and crop defense signaling mechanisms (ABA metabolism). Comparative rice-treated transcriptomic study showed explicitly the shifts in plant reaction to metal pressure

Results of bacteria added to plants		References
Mutation	Physicochemical	[3]
PGPR; *Novosphingobium*	Optimize rice cultivation	[31]
Bioindicator	Wastewater irrigation	[43, 44, 46, 47]
Indicators	Sustainable rice cultivation	[2]
Plant microbiome and *Herbaspirillum seropedicae* and *Bacillus amyloliquefaciens*	Plant growth	[1, 4, 5, 11, 18, 28, 48]
Seed endosphere; PGPR and ACC Deaminase and *Corynebacterium* and diazotrophic spp.	Plant growth	[7, 15, 21, 22, 25, 26, 49]
Soil *Rhizobacteria*	Heavy metals	[50–54]
Azospirillum	N_2 fixing	[55]
Arbuscular mycorrhizal symbiosis and *Pseudomonas putida*	Salinity stress; biological control; drought stress	[20, 29, 56, 57]
PGPR	Cu-contaminated	[43, 58]
Exogenous application	Cd-contaminated	[10, 59, 60]
Genomic rice	Cr-contaminated	[61]
Ochrobactrum sp. and *Bacillus* spp. and biofortification	Heavy metals	[40, 62]
	Ar-contaminated	[63]
Endophytic and PGPR and *Bacillus safensis*	Salt stress	[8, 9, 12, 64]
Genetically engineered	Hg	[65]
Acinetobacter sp. and PGPR	Zinc solubilizing	[19, 39, 41]
Bacterial species	Si solubilization	[42, 45, 66–69]
Phosphate-solubilizing bacteria	Phosphate solubilization	[33–38]

Table 1.
Plant growth-promoting Rhizobacteria *used in rice production.*

in the rates of phytohormones: As and Pb resistant by *Bacillus* spp. There are various PGPR features that contribute to the bioremediation and rice cultivar growth promotion; Cd-resistant *Ochrobactrum* sp. was first reported by Pandey et al. [62]. The presence of CDPKs was demonstrated by Cr pressure as their activity increased with increasing Cr (VI) concentration. Huang et al. [61] showed that rice roots have long- and short-term stress transcription profiling. Yeh et al. [59] have demonstrated Cd-induced gene transcription of OsMAPK2 and MBP kinase in rice plant. The activation of heavy metal mediated MAPK by ROS production, build-up, and alteration of the antioxidant system in the rice; ROS is well-rated for its disruption specific pathways such as auxin, ethylene, and jasmonate (JA) phytohormone. However, exposure to JAs has shown that antioxidant reaction has been enhanced due to rice stress sensitivity of Cd [60]. However, an extensive study on heavy metal in plants has shown great interest in the extensive study of the plant microbial-metal relationship due to its direct impact on enhanced production of biomass and improved metal tolerances [50].

Plants have developed a number of defense mechanisms to resist heavy metal stresses and toxicities such as reducing heavy metal consumption, sequestering metal into vacuoles, binding phytochelatins or metallothionein, and antioxidant activation [51]. The toxic substances As, Pb, Cd, and Hg are considered by Disease Registry Agency as the most toxic metals (**Figure 1**) for their toxicity frequency and above all their flora and fauna exposure potential. Pb toxicity leads to ATP inhibition, lipid peroxidation, and damage to DNA through the production of ROS [43].

In recent decades there has been rapid progress in the area of plant reactions and the tolerance of stress of metal when related bacteria are present with plants. The activation of these genes, which are crucial to heavy metal stress signaling, also suggests dynamic crosspieces of stress and resistance between plant, microbes, and heavy metals [52]. Heavy metal remediation is necessary to protect and preserve the environment. There are only a small number of evidence that heavy metals are remediated by extracellular capsules, heavy metal precipitation, and oxidation reduction [53].

It will be used in the immediate future for remediation of contaminated soils, as shown by the beneficial effects of microbe causes and the planned interconnection between heavy metal resistance and plant growth abilities [58]. Additionally, arbuscular mycorrhizal fungi (AMF) ecological species and ecotypes, metal and edaphic conditions of its availability, and soil and water, including soil fertilizer and requirements of plants for growing under light or root conditions, depend on various factors of exposure to heavy metals in the environment [56].

AMF changes salt stress toxicity. AMF exists due to enhanced mineral nutrition and as a result of various physiological processes such as photosynthesis, water usage efficiency, osmoregulator production, higher $K+/Na +$ ratio, and molecular changes caused by the expression of genes [57].

The synergistic effects on plant growth, particularly in growth restrictions, of the co-inoculation with PGPR and AMF, have shown that the growth responses are significant when rice plants are inoculated with AMF and *Azospirillum*. All of these findings thus show that rice mycorrhization is important [55].

The methods employed by PGPB to promote plant remediation cycle include enhancing plant metal resistance and increasing plant growth as well as altering plant metal accumulation; however, the recent PGPB studies in metal phytoremediation showed that plant inoculation with plant-building bacteria-induced metal phytotoxicity can be alleviated and the production of plant biomass produced in metal-contaminated soils can be strengthened [48, 49, 54]. The reuse of wastewater as a strategy to adjust to climate change is shown in Vietnam. Chung et al. [46] illustrated that rice wastewater effluents can be irrigated for at least 22,719 ha (16% of the urban rice area) in plants annually. Additionally, Jang et al. [47] found that there is no significant environmental risk to rice paddy agroecosystems that were associated with wastewater irrigation (**Table 1** and **Figure 1**).

2. Conclusion

The main limiting factors for cultivation worldwide are water stress conditions [71]. Wastewater water has a negative effect on the production and yield of rice. Selected PGPR might be the perfect candidate for heavy metal pollution and related surface constraints for growth and yields of rice plants irrigated with wastewater as PGPR extracted wastewater strains of bioremediation products show positive results in the literature.

Author details

Raghad S. Mouhamad[1*] and Michael Alabboud[2]

1 Department of Soil and Water Resources, Ministry of Science and Technology, Baghdad, Iraq

2 Department of Horticultural Sciences, UTCAN, University of Tehran, Iran

*Address all correspondence to: raghad1974@yahoo.com

References

[1] Kakar K, Xuan TD, Haqani MI, Rayee R, Wafa IK, Abdiani S, et al. Current situation and sustainable development for rice cultivation and production in Afghanistan. Agriculture. 2019;**9**:49

[2] SRP. Performance Indicators for Sustainable Rice Cultivation, Sustainable Rice Platform. Bangkok: SRP; 2015. Available from: http://www.sustainablerice.org

[3] Kakar K, Xuan TD, Quan NV, Wafa IK, Tran H-D, Khanh TD, et al. Efficacy of *N*-methyl-*N*-nitrosourea mutation on physicochemical properties, phytochemicals, and momilactones A and B in Rice. Sustainability. 2019;**11**:6862

[4] Prasad M, Srinivasan R, Chaudhary M, Choudhary M, Jat LK. Plant growth promoting rhizobacteria (PGPR) for sustainable agriculture: Perspectives and challenges. In: PGPR Amelioration in Sustainable Agriculture. Woodhead Publishing (Elsevier); 2019. pp. 129-157

[5] Khare E, Mishra J, Arora NK. Multifaceted interactions between endophytes and plant: Developments and prospects. Frontiers in Microbiology. 2018;**9**:2732. DOI: 10.3389/fmicb.2018.02732

[6] Souza R, de Ambrosini A, Passaglia LMP. Plant growth-promoting bacteria as inoculants in agricultural soils. Genetics and Molecular Biology. 2015;**38**:401-419

[7] Glick BR. Bacteria with ACC deaminase can promote plant growth and help to feed the world. Microbiological Research. 2014;**169**:30-39

[8] Shah G, Jan M, Afreen M, et al. Halophilic bacteria mediated phytoremediation of salt-affected soils cultivated with rice. Journal of Geochemical Exploration. 2017;**174**:59-65

[9] Khan MHU, Khattak JZK, Jamil M, et al. *Bacillus safensis* with plant-derived smoke stimulates rice growth under saline conditions. Environmental Science and Pollution Research. 2017;**24**(30):23850-23863

[10] Jan M, Shah G, Masood S, et al. *Bacillus cereus* enhanced phytoremediation ability of rice seedlings under cadmium toxicity. BioMed Research International. 2019;**2019**:12. Article ID 8134651. Available from: https://doi.org/10.1155/2019/8134651

[11] Yasmin S, Rahman M, Hafeez FY. Isolation, characterization and beneficial effects of rice associated plant growth promoting bacteria from Zanzibar soils. Journal of Basic Microbiology. 2004;**44**:241-252

[12] Sulastri, Wiyono S, Soepandie D, Santosa DA. IOP Conference Series: Earth and Environmental Science. The 4th International Seminar on Sciences, Vol. 187. 19-20 October 2017, Bogor, Indonesia: IOP Publishing Ltd; 2018

[13] Mouhamad RS, Mutlag LA, Al-Khateeb MT, Iqbal M, Nazir A, Ibrahim KM, et al. Salinity tolerance at seedling stage for rice genotypes: In vitro analysis. Netherlands Journal of Agricultural Science. 2017;**5**:114-120

[14] Mouhamad RS, Jaafar ZM, EAJ E-K, Iqbal M, Arif N. Evaluation of agronomic traits and inorganic nutritional composition of rice seed from IRSSTN genotypes in Iraq. Journal of Rice Research. 2018;**6**:189

[15] Ashrafuzzaman M, Hossen FA, Ismail MR, Hoque A, Islam MZ, Shahidullah SM, et al. Efficiency of plant growth-promoting rhizobacteria

(PGPR) for the enhancement of rice growth. African Journal of Biotechnology. 2009;**8**(7):1247-1252

[16] Beneduzi A, Peres D, Vargas LK, Bodanese-Zanettini MH, Passaglia LM. Evaluation of genetic diversity and plant growth promoting activities of nitrogen-fixing bacilli isolated from rice fields in South Brazil. Applied Soil Ecology. 2008;**39**(3):311-320

[17] Bisht SC, Mishra PK. Ascending migration of endophytic *Bacillus thuringiensis* and assessment of benefits to different legumes of NW Himalayas. European Journal of Soil Biology. 2013;**56**:56-64

[18] Nautiyal CS, Srivastava S, Chauhan PS, Seem K, Mishra A, Sopory SK. Plant growth-promoting bacteria *Bacillus amyloliquefaciens* NBRISN13 modulates gene expression profile of leaf and rhizosphere community in rice during salt stress. Plant Physiology and Biochemistry. 2013;**66**:1-9

[19] Gandhi A, Muralidharan G. Assessment of zinc solubilizing potentiality of Acinetobacter sp. isolated from rice rhizosphere. European Journal of Soil Biology. 2016;**76**:1-8

[20] Tiwari S, Lata C, Chauhan PS, Nautiyal CS. *Pseudomonas putida* attunes morphophysiological, biochemical and molecular responses in *Cicer arietinum* L. during drought stress and recovery. Plant Physiology and Biochemistry. 2016;**99**:108-117

[21] Kandel SL, Herschberger N, Kim SH, Doty SL. Diazotrophic endophytes of poplar and willow for growth promotion of rice plants in nitrogen-limited conditions. Crop Science. 2015;**55**:1765-1772

[22] Mano H, Morisaki H. Endophytic bacteria in the rice plant. Microbes and Environments. 2008;**23**:109-117

[23] Mäder P, Kaiser F, Adholeya A, Singh R, Uppal HS, Anil K, et al. Inoculation of root microorganisms for sustainable wheat-rice and wheat-black gram rotations in India. Soil Biology and Biochemistry. 2011;**43**:609-619

[24] Ladha JK, Reddy PM. Nitrogen fixation in rice systems: State of knowledge and future prospects. Plant and Soil. 2003;**252**:151-167

[25] Yasmin S, Bakar MAR, Malik KA, Hafeez FY. Isolation, characterization and beneficial effects of rice associated plant growth promoting bacteria from Zanzibar soils. Journal of Basic Microbiology. 2004;**3**:241-252

[26] Naher UA, Othman R, Shamsuddin ZHJ, Saud HM, Ismail MR. Growth enhancement and root colonization of rice seedlings by rhizobium and *Corynebacterium* spp. International Journal of Agriculture and Biology. 2009;**11**:586-590

[27] Rosenblueth M, Ormeño-Orrillo E, López-López A, Rogel MA, Jazmín Reyes-Hernández B, Martínez-Romero JC, et al. Nitrogen fixation in cereals. Frontiers in Microbiology. 2018;**9**:1794

[28] James EK, Gyaneshwar P, Mathan N, Barraquio WL, Reddy PM, Iannetta PP, et al. Infection and colonization of rice seedlings by the plant growth-promoting bacterium *Herbaspirillum seropedicae* Z67. Molecular Plant-Microbe Interactions. 2002;**15**:894-906

[29] Etesami H, Alikhani HA. Evaluation of gram-positive rhizosphere and endophytic bacteria for biological control of fungal rice (*Oryzia sativa* L.) pathogens. European Journal of Plant Pathology. 2017;**147**(1):7-14

[30] Estrada-de los Santos P, Vinuesa P, Martínez-Aguilar L, Hirch AM, Caballero-Mellado J. Phylogenetics analysis of *Burkholderia* species by

multilocus sequence analysis. Current Microbiology. 2013;**67**:51-60

[31] Rangjaroen C, Sungthong R, Rerkasem B, Teaumroong N, Noisangiam R, Lumyong S. Untapped endophytic colonization and plant growth-promoting potential of the genus *Novosphingobium* to optimize rice cultivation. Microbes and Environments. 2017;**32**(1):84-87

[32] De Souza V, Piro VC, Faoro H, Tadra-Sfeir MZ, Chicora VK, Guizelini D, et al. Draft genome sequence of *Herbaspirillum huttiense* subsp. *putei* IAM 15032, a strain isolated from well water. Genome Announcements. 2013;**1**:e00252-e00212

[33] Syers JK, Johnston AE, Curtin D. Efficiency of soil and fertilizer phosphorus use. Rome, Italy: FAO Fertilizer and Plant Nutrition Bulletin 18; 2008

[34] Sahrawat KL, Abekoe MK, Diatta S. Application of inorganic phosphorus fertilizer. In: Tian G, Ishida F, Keatinge D, editors. Sustaining Soil Fertility in West Africa. Madison, Wisconsin, USA: Soil Science Society of America Special Publication Number 58. Soil Science Society of America and American Society of Agronomy; 2001. pp. 225-246

[35] Othman R, Panhwar QA. Phosphate-solubilizing bacteria improves nutrient uptake in aerobic rice. In: Khan MS, editor. Phosphate Solubilizing Microorganisms. Cham: Springer; 2014. pp. 207-224

[36] Fageria NK. Nutrient management for improving upland rice productivity and sustainability. Communications in Soil Science and Plant Analysis. 2001;**32**:2603-2629

[37] Islam MT, Hossain MM. Plant probiotics in phosphorus nutrition in crops, with special reference to rice. In: Maheshwari DK, editor. Bacteria in Agrobiology: Plant Probiotics. Berlin: Springer; 2012. pp. 325-363

[38] Panhwar QA, Jusop S, Naher UA, Othman R, Razi MI. Application of potential phosphate-solubilizing bacteria and organic acids on phosphate solubilization from phosphate rock in aerobic rice. Scientific World Journal. 2013;**2013**:10. Article ID: 272409. Available from: https://doi.org/10.1155/2013/272409

[39] Kamran S, Shahid I, Baig DN, Rizwan M, Malik KA, Mehnaz S. Contribution of zinc solubilizing bacteria in growth promotion and zinc content of wheat. Frontiers in Microbiology. 2017;**8**:2593

[40] Bouis HE, Welch RM. Biofortification—A sustainable agricultural strategy for reducing micronutrient malnutrition in the global south. Crop Science. 2010;**50**:S-20

[41] Vaid SK, Kumar B, Sharma A, Shukla AK, Srivastava PC. Effect of Zn solubilizing bacteria on growth promotion and Zn nutrition of rice. Journal of Soil Science and Plant Nutrition. 2014;**14**:889-910

[42] Adak A et al. Micronutrient enrichment mediated by plant-microbe interactions and rice cultivation practices. Journal of Plant Nutrition. 2016;**39**:1216-1232

[43] Tchounwou P, Newsome C, Williams J, Glass K. Copper-induced cytotoxicity and transcriptional activation of stress genes in human liver carcinoma cells. Metal Ions in Biology and Medicine. 2008;**10**:285-290

[44] Chang LW, Magos L, Suzuki T, editors. Toxicology of Metals. Boca Raton, FL, USA: CRC Press; 1996

[45] Ma JF, Takahashi E. Soil, Fertilizer, and Plant Silicon Research in Japan.

Amsterdam, the Netherlands: Elsevier; 2002. pp. 1-294

[46] Chung BY, Song CH, Park BJ, Cho JY. Heavy metals in brown rice (*Oryza sativa* L.) and soil after long-term irrigation of wastewater discharged from domestic sewage treatment plants. Pedosphere. 2011;**21**(5):621-627

[47] Jang T, Jung M, Lee E, Park S, Lee J, Jeong H. Assessing environmental impacts of reclaimed wastewater irrigation in paddy fields using bioindicator. Irrigation Science. 2013;**31**(5):1225-1236

[48] Turner TR, James EK, Poole PS. The plant microbiome. Genome Biology. 2013;**14**(6):209

[49] Walitang DI, Kim K, Madhaiyan M, Kim YK, Kang Y, Sa T. Characterizing endophytic competence and plant growth promotion of bacterial endophytes inhabiting the seed endosphere of rice. BMC Microbiology. 2017;**17**(1):209

[50] Suman J, Uhlik O, Viktorova J, Macek T. Phytoextraction of heavy metals: A promising tool for clean-up of polluted environment? Frontiers in Plant Science. 2018;**9**:1476

[51] Morkunas I, Woźniak A, Mai V, Rucińska-Sobkowiak R, Jeandet P. The role of heavy metals in plant response to biotic stress. Molecules. 2018;**3**:2320

[52] Tiwari S, Lata C. Heavy metal stress, signaling, and tolerance due to plant-associated microbes: An overview. Frontiers in Plant Science. 2018;**9**:452. Published 6 April 2018. DOI: 10.3389/fpls.2018.00452

[53] Babalola A. A new strategy for heavy metal polluted environments: A review of microbial biosorbents. International Journal of Environmental Research and Public Health. 2017;**14**:94

[54] Jing YD, He ZL, Yang XE. Role of soil rhizobacteria in phytoremediation of heavy metal contaminated soils. Journal of Zhejiang University. Science. 2007;**8**(3):192-207

[55] Fukami J, Cerezini P, Hungria M. Azospirillum: Benefits that go far beyond biological nitrogen fixation. AMB Express. 2018;**8**:1-1

[56] Gai JP, Cai XB, Feng G, Christie P, Li XL. Arbuscular mycorrhizal fungi associated with sedges on the Tibetan plateau. Mycorrhiza. 2006;**16**(3):151-157

[57] Evelin H, Devi TS, Gupta S, Kapoor R. Mitigation of salinity stress in plants by arbuscular mycorrhizal symbiosis: Current understanding and new challenges. Frontiers in Plant Science. 2019;**10**:470

[58] Ren XM, Guo SJ, Tian W, et al. Effects of plant growth-promoting bacteria (PGPB) inoculation on the growth, antioxidant activity, Cu uptake, and bacterial community structure of rape (*Brassica napus* L.) grown in Cu-contaminated agricultural soil. Frontiers in Microbiology. 2019;**10**:1455

[59] Yeh CM, Hsiao LJ, Huang HJ. Cadmium activates a mitogen-activated protein kinase gene and MBP kinases in rice. Plant & Cell Physiology. 2004;**45**:1306-1312

[60] Singh I, Shah K. Exogenous application of methyl jasmonate lowers the effect of cadmium-induced oxidative injury in rice seedlings. Phytochemistry. 2014;**108**:57-66

[61] Huang T-L, Huang L-Y, Fu S-F, Trinh NN, Huang H-J. Genomic profiling of rice roots with short- and long-term chromium stress. Plant Molecular Biology. 2014;**86**:157-170

[62] Pandey S, Ghosh PK, Ghosh S, De TK, Maiti TK. Role of heavy metal resistant *Ochrobactrum* sp. and *Bacillus*

spp. strains in bioremediation of a rice cultivar and their PGPR like activities. Journal of Microbiology. 2013;**51**:11-17

[63] Chakrabarty D, Trivedi PK, Misra P, Tiwari M, Shri M, Shukla D, et al. Comparative transcriptome analysis of arsenate and arsenite stresses in rice seedlings. Chemosphere. 2009;**74**:688-702

[64] Jha Y, Subramanian RB, Patel S. Combination of endophytic and rhizospheric plant growth promoting rhizobacteria in *Oryza sativa* shows higher accumulation of osmoprotectant against saline stress. Acta Physiologiae Plantarum. 2011;**33**:797-802

[65] Heaton ACP, Rugh CL, Kim T, Wang NJ, Meagher RB. Toward detoxifying mercury-polluted aquatic sediments with rice genetically engineered for mercury resistance. Environmental Toxicology and Chemistry. 2003;**22**:2940-2947

[66] Savant NK, Datnoff LE, Snyder GH. Depletion of plant-available silicon in soils: A possible cause of declining rice yields. Communications in Soil Science and Plant Analysis. 1997;**28**:1245-1252

[67] Bocharnikova EA, Loginov SV, Matychenkov VV, Storozhenko PA. Silicon fertilizer efficiency. Russian Agricultural Sciences. 2010;**36**:446-448

[68] Ning D, Song A, Fan F, Li Z, Liang Y. Effects of slag-based silicon fertilizer on rice growth and brown-spot resistance. PLOS One. 2014;**9**:e102681

[69] Vasanthi N, Saleena LM, Raj SA. Silica solubilization potential of certain bacterial species in the presence of different silicate minerals. Silicon. 2018;**10**:267-275

[70] Chen X, Zuo S, Schwessinger B, Chern M, Canlas PE, Ruan D, et al. An XA21-associated kinase (OsSERK2) regulates immunity mediated by the XA21 and XA3 immune receptors Mol. Planta. 2014;7(2014):874-892

[71] Mouhamad RS. Morphological study of different varieties of rice traits influencing nitrogen and water uptake efficiency. Revista Bionatura. 2020;**5**(1):1039-1043. DOI: 10.21931/RB/2020.05.01.5

Chapter 21

Actinobacteria: Potential Candidate as Plant Growth Promoters

Sumreen Hayat, Asma Ashraf, Bilal Aslam, Rizwan Asif, Saima Muzammil, Muhammad Asif Zahoor, Muhammad Waseem, Imran Riaz Malik, Mohsin Khurshid, Muhammad Afzal, Muhammad Saqalein, Muhammad Hussnain Siddique, Aqsa Muzammil and Sumera Sabir

Abstract

Plant growth enhancement using plant beneficial bacteria has been viewed in the sustainable agriculture as an alternative to chemical fertilizers. Actinobacteria, among the group of important plant-associated bacteria, have been widely studied for its plant growth promotion activities. Actinobacteria are considered as a limelight among agriculturists for their beneficial aspects toward plants. They are naturally occurring spore-forming bacteria inhabiting the soil and known for their plant growth-promoting and biocontrol properties. The mechanisms behind these activities include nitrogen fixation, phosphate solubilization, siderophore production, and other attributes such as antifungal production of metabolites, phytohormones, and volatile organic compound. All these activities not only enhance the plant growth but also provide resistance in plants to withstand unfavorable conditions of the environment. Hence, this chapter emphasizes on the plant growth traits of actinobacteria and how far it was studied for enhanced growth and bio-fortification.

Keywords: actinobacteria, rhizosphere, PGPR, growth promotion

1. Introduction

Plant growth-promoting (PGP) microbes (epiphytic, endophytic, and rhizospheric) are likely to enhance the growth and productivity of crop by increasing the nutrient content. These plant microbiomes have been sorted out from diverse sources belonging to all three domains: archaea, bacteria, and fungi. The microbes associated with the plant rhizosphere are termed as rhizospheric microbes, and among them, actinobacteria are most dominant in nature [1]. As many researches stated actinobacteria as major microbial population present in the soil. The actinobacteria are known to have high G-C (57–75%) contents and comprise a broad group of filamentous, spore-forming, gram-positive, and aerobic bacteria that form branching filaments or hyphae and play a fundamental function in ecology along with soil nutrient cycle. Actinobacteria resemble to unicellular bacteria, they are different by not having distinct cell wall; instead they produce mycelium, a nonseptate and more slender [2]. Actinobacteria are widely dispersed in both terrestrial ecosystem, as present in soil, and aquatic ecosystems as in fresh and marine water. The terrestrial actinobacteria contribute in recycling process and are essential to the decomposition of many complex mixtures of polymers and organic material, located in dead plants, animals, and fungal materials. The phylum actinobacteria is currently recognized as the largest taxonomic units within the bacterial domain and recognized for its economic importance because it produces various biological active substances like vitamins, antibiotics, and enzymes. It is estimated that almost 23,000 bioactive secondary metabolites are produced by many microorganisms and almost 45% (10,000 out of 23,000) of these bioactive microbial metabolites are produced by actinomycetes. Among these actinomycetes, *Streptomyces* are classified as most abundantly occurring *Actinomycete* in the soil, while *Nocardia*, *Micromonospora*, and *Streptosporangium* are the less abundant [3]. Streptomyces has established its importance in numerous sectors like health and agriculture, and it is also considered as most dominant actinobacteria, for root colonization and close association with plant roots. Actinomycetes are diversified organisms having various applications on many fields. Plant growth-promoting rhizobacteria (PGPR) actinobacteria promote the plant growth through a variety of mechanisms including production of phytohormones, antibiotics, siderophore, volatile organic compound, and different hydrolytic enzymes. These also promote nutrient fixation for easy

Actinobacteria	Host plant	Target pathogen	References
Streptomyces sp. S30, *Streptomyces* sp. R18	*Lycopersicon esculentum*	*Rhizoctonia solani*	[5]
Actinoplanes	*Cucumis* sp.	*Pythium aphanidermatum*	[6]
Streptomyces diastaticus, *Streptomyces fradiae*, *S. collinus*	Medicinal plants	*Fusarium oxysporum*, *Alternaria solani*, *Sclerotium rolfsii*	[7]
Streptomyces sp. DBT204	*Solanum lycopersicum*	*Fusarium proliferatum*	[8]
Leifsonia xyli, BPSAC24, *Streptomyces* sp. BPSAC34	*Curcuma longa*, *Eupatorium odoratum*, *Mirabilis jalapa*	*Fusarium oxysporum* ciceri, *F. oxysporum*, *F. graminearum*, *Rhizoctonia solani*	[9]
Microbispora spp.	*Solanum tuberosum* L.	*Streptomyces scabies*	[10]
Streptomyces spp. R-5	*Rhododendron* sp.	*Phytophthora cinnamomi*	[11]
Streptomyces sp. MBPu-75	*Cucumis* sp. and *Cucurbita* sp.	*Colletotrichum orbiculare*	[12]
Streptomyces sp. RM 365	Leguminous plants	*Xanthomonas campestris*	[13]
Streptomyces sp. PRY-2RB2	*Pseudowintera colorata*	*Nocardia parvum* MM562	[14]
Nocardiopsis sp. ac 9, *Streptomyces* sp. ac19	*Elaeis guineensis* Jacq.	*Ganoderma boninense*	[15]
Streptomyces sp. AzR-051, *Streptomyces* sp. AzR-010	*Azadirachta indica*	*Alternaria alternate*	[16]
Mutabilis CA-2	*Cleome arabica*, *Aristida pungens*	*Rhizoctonia solani*	[17]

Table 1.
Biological activities of some isolated actinobacteria.

uptake by the plant and develop abiotic stress tolerance in plants. Actinobacteria are also considered to have the potential to be used as promising biocontrol agents because they produce spore which can resist environmental stress. The actinobacteria help plants by suppressing disease-causing microbes and enhancing nutrient availability and assimilation which subsequently have beneficial impact on the agricultural sector by accelerating plant growth [4]. Although actinomycetes have many applications in different sectors including health and agriculture, this chapter will only focus on actinomycetes' role as PGPR. Some commonly isolated actinobacteria, its host plant, and targets are mentioned in **Table 1**.

2. Actinobacteria's role in PGPR activity

Plant growth and development of important organs in plant are facilitated by plant hormones called plant growth regulators (PGR). These PGR can influence plant growth even at very low concentration. Actinobacteria act as PGPR, and its impact is determined by considering the effectiveness and ability to influence PGR in root system. Different mechanisms used for promoting PGR by actinobacteria are classified into direct and indirect method (**Figure 1**).

The direct method exhibits various activities including solubilization of phosphorus, nitrogen fixation, iron acquisition, and production of different phytohormones, for instance, indole acetic acid (IAA), cytokinins, and gibberellins. In indirect method, actinobacteria promote plant growth in many ways such as synthesizing extracellular enzymes for fungal cell wall degradation producing antibiotics, volatile compounds (VOCs), inducting systemic resistance, as well as competition for nutrients [18].

2.1 Direct method for plant growth promoter by actinobacteria

Actinobacteria promote the growth of plants by involving in various direct activities as shown in **Table 2**.

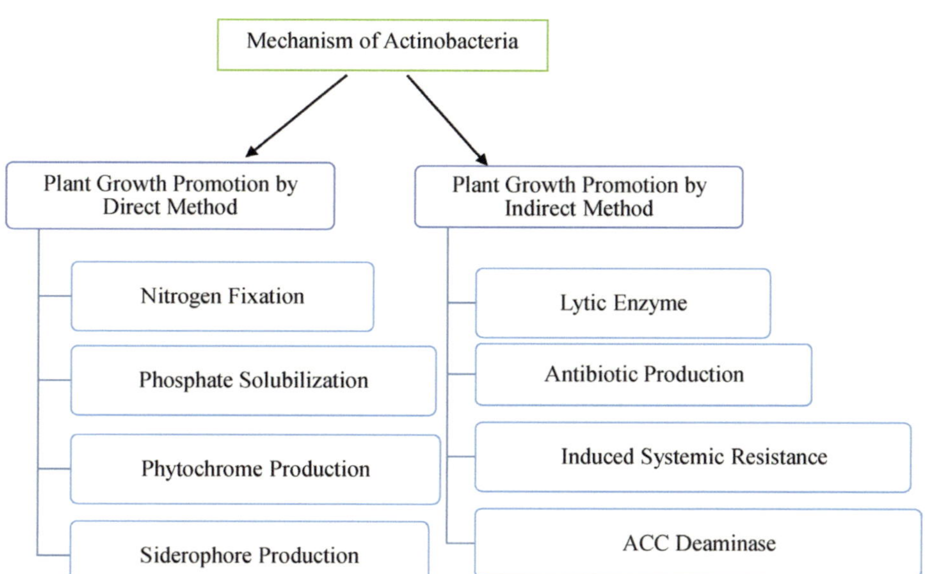

Figure 1.
Mode of action of actinobacteria.

Actinobacteria	PGP traits	Host plant	Reference
Streptomyces sp. GMKU 3100	Siderophore production	Rice (*Oryza sativa L.)*	[19]
Streptomyces sp.	IAA production Phosphate solubilization	Wheat	[20]
Streptomyces griseoflavus P4	Nitrogen fixation	Soybean (*Glycine max*)	[21]
Microbispora spp. *Micromonospora* spp. *Nocardia* spp.	IAA production	Mandarin (*Citrus reticulata* L.)	[22]
Streptomyces spp.	IAA production	Sorghum Rice	[23]
Arthrobacter sp. strain EZB4	ACC deaminase activity	Pepper (*Capsicum annuum* L.)	[24]
Micromonospora endolithica	Phosphate solubilization	Carrot (*Daucus carota*)	[25]
Streptomyces sp.	Nutrient uptake	Clover	[26]
Streptomyces sp.	Gibberellic acid, IAA,	Marine environments	[27]
Streptomyces olivaceoviridis, *S. rochei*	Auxin, gibberellin, and cytokinin synthesis	Wheat	[28]
Brevibacterium epidermidis RS15, *Micrococcus yunnanensis* RS222	Nitrogen fixation IAA production ACC deaminase activity	Canola	[29]
Streptomyces spp.	Phosphate solubilization	—	[30]
Actinobacteria	Phosphate solubilization, Nitrogen fixation	Soya bean	[31]

Table 2.
Direct plant growth-promoting (PGP) properties of actinobacteria.

2.1.1 Nitrogen fixation

Nitrogen is a well-known and key element of nucleic acids and proteins, and it is also an indispensable nutrient for plant growth. Nitrogen gas is abundantly found in the air, constituting 78% of the atmosphere, but it is not directly available to plants for uptake unless it is converted into its soluble form [32]. Biological nitrogen fixer (BNF) used nitrogenase enzyme system which converts the atmospheric nitrogen required by plants into ammonium and nitrates [33]. Additionally, synthetic nitrogen fertilizers are also supplied to balance the limited availability of nitrogen provided by biological nitrogen fixer. But these fertilizers might be harmful to health and agricultural sustainability. Therefore, actinobacteria are good choice to be utilized as BNF to improve the plant growth for sustainable agriculture. *Frankia* is a versatile actinobacterium which enters in the root cell through different ways, such as intracellular using root-hair or intercellular by means of root invasion, and fixes nitrogen under both free-living and symbiotic conditions in nonlegume plants [34]. In addition to *Frankia*, many other endophytes like *Agromyces*, *Arthrobacter*, *Micromonospora*, *Corynebacterium*, *Propionibacterium*, *Mycobacterium*, and *Streptomyces* also demonstrated N-fixing ability [35].

2.1.2 Phosphorus solubilization

After nitrogen, the second major element, for plant growth, is phosphorus [36]. Phosphorus exists in soil as both inorganic and organic forms [37], but 0.1% phosphorus is available as soluble form to be absorbed by plants. An immediate need of phosphorus is fulfilled by chemical fertilizers, like nitrogen, but the majority of these applied chemical fertilizers are not only expensive but also wasted because it retains in soil as an insoluble form just after the application [37]. In the past decades, many microbes have been described which can solubilize the insoluble phosphorus, and since then, numerous studies by many researchers have been carried out to investigate the phosphate-solubilizing potential of different microbes such as bacteria, fungi, and actinobacteria [38]. Different in vitro and in vivo studies have been executed, which highlight capability to solubilize soil phosphorus by PGP actinobacteria, for instance, *Streptomyces*, *Rhodococcus*, *Arthrobacter*, *Gordonia*, and *Micromonospora* [39]. *Micromonospora endolithica* a non-streptomycete enhances the phosphate-solubilizing ability in bean plants which subsequently increase the growth of bean plants [40]. *Micromonospora aurantiaca*, *Streptomyces* sp., and *Streptomyces griseus* also showed similar results on wheat plants when grown under phosphorus-deficient soil. Many actinobacterial strains also aid in phosphorus solubilization by producing several organic acids such as citric acid, gluconic acid, oxalic acid, lactic acid, malic acid, succinic acid, and propionic acid. Therefore, it is more viable to utilize microorganisms like actinobacteria as biofertilizers economically. Some plant growth-promoting characteristic of actinobacteria is shown in **Figure 2**.

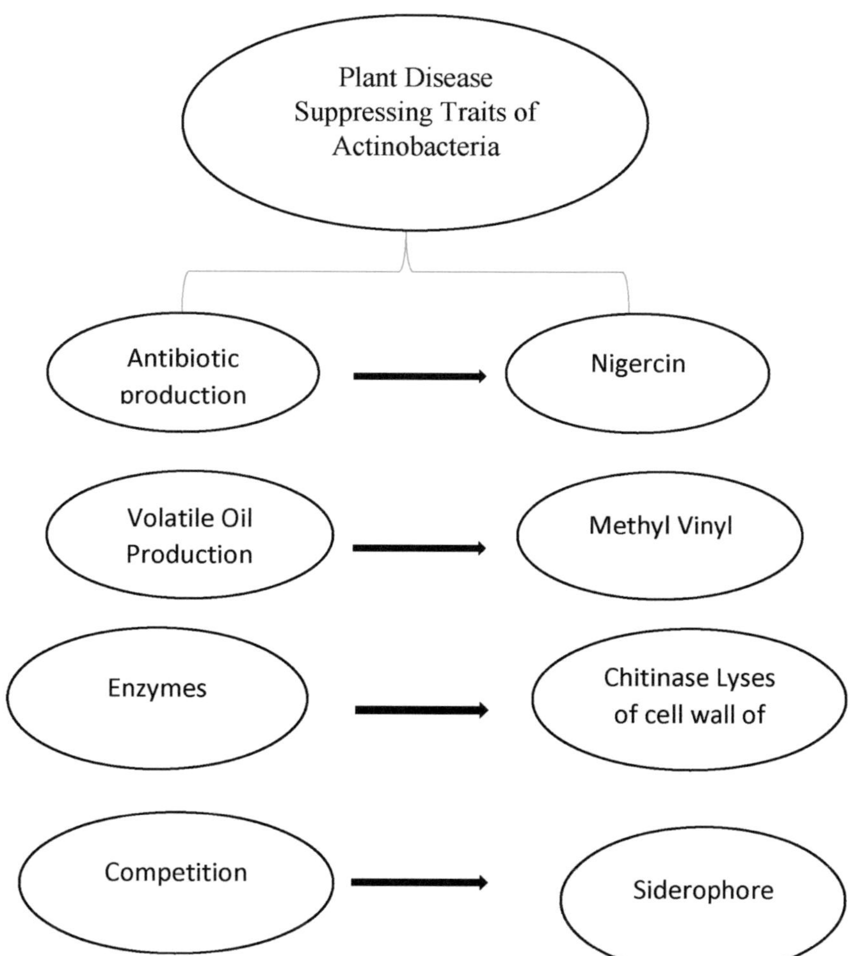

Figure 2.
Plant disease suppressing trait adopted by actinobacteria.

2.1.3 Siderophore production by actinobacteria

Iron is an essential nutrient element for all organisms, which acts as a necessary co-factor for several enzymatic reactions. Many beneficial actinobacteria including *Streptomyces*, *Micrococcus*, *Microbacterium*, *Kocuria*, *Corynebacterium*, and *Arthrobacter* improve the plant growth through production of Fe-chelating compounds. Normally, soil iron resides in insoluble hydroxides and oxy-hydroxides forms that are not available for microbes and plants. For iron availability siderophore, a compound of high iron affinity and low molecular weight is needed to be synthesized [41]. Two major classes of siderophore are catechols and hydroxamate, produced by microbes and various actinobacterial strains that have been nominated as producers of siderophore [42]. Some well-known siderophores that are produced by the genus *Streptomyces* are hydroxamate, desferrioxamines, and coelichelin [43], and other members of actinobacteria like *Rhodococcus* and *Nocardia* produced heterobactin siderophore [44]. Siderophore also plays a role in plant protection from phytopathogens besides its role in plant nutrition. Both siderophore and pathogenic microbe require iron so a competitive environment creates between them in the root vicinity [45]. As actinobacteria produce high-affinity siderophore and fungal pathogen produces low-affinity siderophores, therefore actinobacteria can eliminate the fungal pathogen. *Streptomyces* produce siderophore that is also found to be effective against wilt disease on chickpea caused by *F. oxysporum* f. sp. *ciceri* [46].

2.1.4 Production of hormone

Several rhizospheric and endophytic actinobacteria have been noticed to yield several phytohormones, namely, indole acetic acid (IAA), cytokinins, and gibberellins. These phytohormones show a significant role in the plant growth [47]. The most important phytohormone is indole-3-acetic acid, a principal form of auxin that shows the useful impact on plants by various cellular processes like cell division, elongation, and differentiation. Recently, endophytic actinobacteria are getting more attention because of their role in the production of phytohormones. It has been reported that *Nocardiopsis*, an endophytic actinobacterium, produces the highest percentage of IAA [22]. Many researchers studied that *Streptomyces* endophytes like *S. olivaceoviridis*, *S. rimosus*, *S. atrovirens*, *S. rochei*, and *S. viridis* also produce IAA that is responsible for improved seed germination, root elongation, and growth in different plants [48]. Hence, actinobacteria have the ability to boost the production and growth of plants by producing the phytohormone as shown in **Table 2**.

2.2 Indirect plant growth mechanism of actinobacteria

In indirect plant growth mechanism, actinobacteria also enhance the growth of plants like direct mechanism which is mentioned in **Figure 3**.

2.2.1 Cell wall-degrading enzymes

Actinobacteria synthesize many different extracellular enzymes that help to decompose material in soil. Some of these enzymes include xylanases, chitinases, hemicellulose, nucleases, amylases, lipases, glucanases, pectinase, proteinases, cellulases, ligninases, and keratinase. Mainly soil-living actinobacteria are saprophytic and play a central role in decomposition. Actinobacteria use this mixture of enzymes for decomposition against a variety of phytopathogens and majorly contribute to biocontrol potential by damaging cell wall of these pathogens. Cell wall of many bacteria and fungi is made up of polymers like glycan, cellulose, chitin, protein, and lipids [49]. Actinobacteria are regarded as the dominant organisms that decompose chitin in soil and also considered as promising antagonistic agents for biocontrol because of the hydrolytic reaction on mycelium of the fungi. Acctinobacteria are also observed to produced chitinase enzyme that inhibit fungal growth by cell wall chitin hydrolysis. Many species of *Streptomyces* genus have the potential to degrade the chitin polymer and are, therefore, known as a principal chitinolytic microbial group in soil [50, 51]. A list of some important enzyme secreted by actinobacteria is shown in **Table 3**.

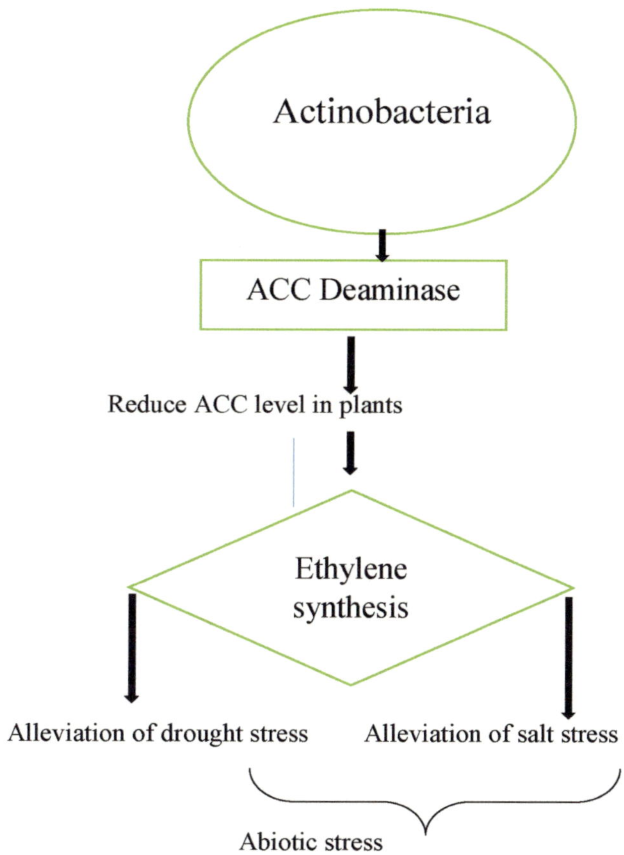

Figure 3.
Alleviation of abiotic stress in actinobacteria.

Enzymes	Actinobacteria	References
Chitinase	*Streptomyces viridificans*, *S. coelicolor*, *S. griseus*, *S. albovinaceus*, *S. caviscabies*, *S. setonii*, *S. virginiae*	[52]
Chitinase, glucanase	*S. cavourensis* SY224	[53]
Cellulose	*Thermomonospora* spp. *Actinoplanes philippinensis*, *A. missouriensis*, *Streptomyces clavuligerus*	[54]
Ligase	*Nocardia autotrophica*	[25]
Amylases, lipases, β-1-3-glucanase	*Thermomonospora curvata*, *Streptomyces* spp.	[55]
Chitinase, glucanase and protease	*Streptomyces* spp. 80	[56]

Table 3.
Production of hydrolytic enzymes by actinobacteria.

2.2.2 Actinobacteria's role as nutrient promoter

As PGP, actinobacteria also act to raise the soil fertility by exhibiting various activities; hence, it is acknowledged as a main natural nutrient enhancer. Besides siderophore producer and phosphate solubilizer, actinobacteria also produce many kinds of enzymes like lipase, amylase, peroxidase, xylanase, chitinase, keratinase, pectinase, cellulase, and protease. This cocktail of enzymes helps to convert nutrients into simple mineral forms, and due to this nutrient cycling ability of actinobacteria, it is considered as an optimal candidate for natural fertilizers [38]. These actinobacteria also promote the soil metal-mobilizing ability like Fe, Zn, and Se, which ultimately increase the germination of seeds and plant growth. Current research has exposed that the root colonization of arbuscular mycorrhizal fungi increases growth of crop and zinc and iron content of chickpea grains [57]. Under greenhouse and field conditions, two PGPR, namely, *Mesorhizobium* sp. and *Pseudomonas* sp., also enhance the production and acquisition of Fe in chickpea [58]. Some previous studies elaborated that actinobacteria enhance plant growth in various crops like cereals, oilseeds, and leguminous by mobilizing the minerals. PGP *Streptomyces* were also observed to increase Fe and Zn quantity by 38% and 30%, respectively, in grains of chickpea [59].

2.2.3 Actinobacteria in bioremediation of metals

Anthropogenic activities are the main cause of metal pollution of agricultural lands which led to a decrease in the fruitful agricultural cropland. As reported by the Environmental Protection Agency (EPA), nearly more than 40,000 contaminated sites are present in the United States. Furthermore, due to heavy metal contamination, 50,000 hectare of forest, 55,000 hectare of pasture, and 100,000 hectare of cropland have vanished, and these need retrieval process [60]. PGP like actinobacteria stay in metal-contaminated soil and increase the bioremediation process by extracting and solubilizing mineral. Different reactions like oxidation, metal reduction, and biosorption as well as several substances like organic acids, siderophores, polymeric substances, glycoprotein, and bio-surfactants are released by the microbes which aid in the metal-mobilizing property of these microbes. Many studies have been performed by researchers which demonstrated the metal-mobilizing mechanism [61].

2.2.4 Reduction of plant-pathogen stress by actinobacteria

Primarily, plants use beneficial microorganisms and plant integrated defense mechanism to protect themselves from phytopathogens [62]. Beneficial microorganisms (pathogen antagonistic) alleviate the pathogen stress in plants through different mechanisms like secretion of anti-pathogenic metabolites, competition for space, and nutrients [8]. Actinobacteria also play vital role in plant protection against plant pathogens utilizing nutrients, required by pathogens for growth. Meanwhile, actinobacteria produce different volatile compounds, antibiotics and cell wall degrading enzymes against phytopathogens [63]. Actinobacteria have been reported to produce various antifungal volatile organic compounds against fungal disease [64]. *Streptomyces* actinobacteria also produce many kinds of volatile compounds which have antifungal activities against *Rhizoctonia solani* and *Botrytis cinerea* [65]. Actinobacteria have the ability to produce different hydrolytic enzymes that degrade fungal and bacteria pathogens cell wall, so protect the plants against phytopathogens [66]. A nonspecific (indirect) mechanism has also been developed by plants which provide the long-term protection against a wide range of phytopathogens. PGP actinobacteria have played an important role in developing disease resistance in plants by inducing gene expression related to defense pathway [67]. Plants display two types of indirect or nonspecific defensive mechanism: the one involves salicylic acid (SA) signaling pathway and pathogenesis-related (PR) protein genes, called systemic acquired resistance (SAR), and the other is induced systemic resistance (ISR) that involves two pathways, ethylene (ET) and jasmonic acid (JA) signaling pathways [68]. A study described that *Streptomyces bikiniensis* HD-087 produces metabolites which cause systemic resistance and suppress the *Fusarium* wilt in cucumber raised by *F. oxysporum* f. sp. *cucumerinum* [69]. Some important metabolites which are synthesized by actinobacteria against phytopathogens are shown in **Table 4**.

Endophytic actinobacteria	Host plant	Metabolite	Target pathogen(s)	Reference
Streptomyces sp. NRRL 3052	*Kennedia nigriscans*	Munumbicins A, B, C and D	*Pythium ultimum*, *Rhizoctonia solani*, *Phytophthora cinnamomi*	[70]
S. melanosporofaciens EF-76 and FP-54	Potato	Geldanamycin	*Streptomyces scabiei*	[71]
Micromonospora sp. M39	Rice	2,3-Dihydroxybenzoic acid, phenylacetic acid, cervinomycin A1 and A2	*P. oryzae*	[72]
S. malaysiensis	Wheat	Malayamycin	*Stagonospora nodorum*	[73]
S. cavourensis subsp. *cavourensis* SY224	Pepper	2-Furancarboxaldehyde	*Colletotrichum gloeosporioides*	[74]
Streptomyces chryseus	*Potentilla discolor*	Saadamycin/5,7-Dimethoxy-4-pmethoxylphenyl coumarin	*Botrytis cinerea*	[75]
Streptomyces sp. MSU-2110	*Monstera* sp.	Coronamycin	*Pythium ultimum*, *Fusarium solani*, *Rhizoctonia solani*	[76]
Microbacterium sp. S4S17	*Ferula sinkiangensis*	Coumarin	*Alternaria alternate*	[77]
Streptomyces olivaceus, *Streptomyces* sp. BPSA 121	*Rhynchotechum ellipticum*	Ketoconazole, fluconazole, miconazole	*Fusarium oxysporum*, *Fusarium proliferatum*	[78]
S. miharaensis 100%	Tomato	Filipin III (purified antibiotic)	*F. oxysporum* f. sp. *lycopersici*	[79]
Streptomyces sp. G10	Banana		*F. oxysporum* f. sp. *cubense*	[80]
Streptomyces sp. AMA49	Rice	Bonactin	*Pyricularia oryzae*	[81]
Streptomyces angustmyceticus NR8–2	*Brassica rapa*	Benzaldehyde, butanoic acid	*Colletotrichum* sp. *Curvularia lunata*	[53]

Table 4.
Metabolites produced by actinobacteria used to suppress disease.

2.2.5 Actinobacteria's role against stress

Several abiotic stress factors including flooding, extreme temperatures, salinity, nutrient stress, drought, and metal stress impose a harmful impact on yields of the crop, as well as it also severally damaged the soil. As described by the Food and Agriculture Organization (FAO), if precautionary steps are not implemented, in the next 25 years 30% land degradation will happen due to abiotic stress factors, and this will rise to 50% in 2050 [16]. Strains of actinobacteria have better tolerance against abiotic stress factors like temperature, salinity, and metal stress, and inoculation of tolerant actinobacteria strain was noticed to encourage the plant growth. Useful effects of PGP *Streptomyces* sp. were observed on maize and wheat under saline conditions [82]. In another in vitro study, *Streptomyces* sp. PGPA39 inoculation showed similar results under saline conditions and ultimately increase the biomass and secondary growth of *Arabidopsis* seedlings (Palaniyandi et al. 2014). Actinobacteria stress tolerance potential was also studied in chickpea [49]. Treatment with *Streptomyces rochei* SM3 in chickpea under stress salt condition decreases mortality (48%) toward *Sclerotinia sclerotiorum* infection and increases biomass (20%). Physiological studies of SM3-treated plants showed increased accumulation of phenolics and proline along with increased catalase and phenylalanine ammonia lyase activities. Further genetic level investigation showed that ET-responsive ERF transcription factor (CaTF2) is triggered by strain SM3 under challenging conditions. Moreover, *Streptomyces padanus* tolerate drought situations by induction of increased osmotic pressure of plant cells and cell wall lignification. Co-inoculation of drought-tolerant *Streptomyces olivaceus* DE10 and *Streptomyces geysiriensis* DE27 endophytic actinobacteria verified the highest yield in wheat [83]. In response to stress, plants produce stress ethylene also known as ET which leads to premature plant death [84]. Microbes synthesize an enzyme known as 1-aminocyclopropane-1-carboxylate (ACC) deaminase that prevents the effect of ethylene due to ethylene precursor ACC conversion to ammonia and a-ketobutyrate which is shown in **Figure 3**. Currently effects of this enzyme on stress management are considered as a central phenomenon of PGP traits and are studied for the past two decades [85]. Some famous actinobacteria that are known to produce ACC deaminase include *Amycolatopsis*, *Streptomyces*, *Nocardia*, *Rhodococcus*, and *Mycobacterium* [86]. Many halo-tolerant actinobacteria having ACC deaminase are isolated from rhizosphere of naturally growing halophytic plants and soil of barren land [29].

3. Conclusion

Production of food to fulfill the need of an increasing population and mimic the reliance on nonrenewable resources and also environmental effect is the greatest challenge of this century. To complete this challenge, the use of plant growth microbes such as actinobacteria is a good choice as an alternative tool for sustainable agriculture. Various studies highlight the abilities of actinobacteria as a plant growth promoter and their additive impact on plant growth and protection. Actinobacteria isolates have shown the multidimensional way to be effective on plant growth. They promote plant growth by involving various activities like production of phytohormones, siderophore production, solubilization of phosphate, fixation of nitrogen, complementing mycorrhizal fungi, and also balancing the ecology of the soil system. Additionally, many studies also have proven the potential of actinobacteria as a biocontrol agent. These characteristics of the actinobacteria group have proved them as inevitable tools for increasing productivity and quality in agriculture. Keeping in mind all these aspects, it is a need of time that we focus on the use of actinobacteria as an alternative tool and reduce the use of harmful chemicals. The studies referred in this chapter also support the belief that the use of eco-friendly microorganisms and designing new formulations with cooperative microbe might contribute to plant growth improvement.

Author details

Sumreen Hayat[1,3], Asma Ashraf[2], Bilal Aslam[1], Rizwan Asif[1], Saima Muzammil[1*], Muhammad Asif Zahoor[1], Muhammad Waseem[1], Imran Riaz Malik[3], Mohsin Khurshid[1], Muhammad Afzal[4], Muhammad Saqalein[1], Muhammad Hussnain Siddique[4], Aqsa Muzammil[4] and Sumera Sabir[1]

1 Department of Microbiology, Government College University, Faisalabad, Pakistan

2 Department of Zoology, Government College University, Faisalabad, Pakistan

3 Department of Biotechnology, University of Sargodha, Sargodha, Pakistan

4 Department of Bioinformatics and Biotechnology, Government College University, Faisalabad, Pakistan

*Address all correspondence to: saimamuzammil83@gmail.com

References

[1] Lata RK, Divjot K, Nath YA. Endophytic microbiomes: Biodiversity, ecological significance and biotechnological applications. Research Journal of Biotechnology. 2019;**14**:10

[2] Chander J. Textbook of Medical Mycology. Chandigarh, India: JP Medical Ltd; 2017

[3] Binda E et al. Specificity of induction of glycopeptide antibiotic resistance in the producing actinomycetes. Antibiotics. 2018;**7**(2):36

[4] Stamenov D et al. The use of *Streptomyces* isolate with plant growth promoting traits in the production of English ryegrass. Romanian Agricultural Research. 2016;**33**:299-306

[5] Cao L et al. Isolation and characterization of endophytic *Streptomyces* strains from surface-sterilized tomato *(Lycopersicon esculentum)* roots. Letters in Applied Microbiology. 2004;**39**(5):425-430

[6] El-Tarabily KA, Hardy GESJ, Sivasithamparam K. Performance of three endophytic actinomycetes in relation to plant growth promotion and biological control of *Pythium aphanidermatum*, a pathogen of cucumber under commercial field production conditions in the United Arab Emirates. European Journal of Plant Pathology. 2010;**128**(4):527-539

[7] Singh S, Gaur R. Evaluation of antagonistic and plant growth promoting activities of chitinolytic endophytic actinomycetes associated with medicinal plants against *Sclerotium rolfsii* in chickpea. Journal of Applied Microbiology. 2016;**121**(2):506-518

[8] Passari AK et al. Detection of biosynthetic gene and phytohormone production by endophytic actinobacteria associated with *Solanum lycopersicum* and their plant-growth-promoting effect. Research in Microbiology. 2016;**167**(8):692-705

[9] Passari AK et al. In vitro and in vivo plant growth promoting activities and DNA fingerprinting of antagonistic endophytic actinomycetes associates with medicinal plants. PLoS One. 2015;**10**(9):e0139468

[10] Goodman AA. Endophytic Actinomycetes as Potential Agents to Control Common Scab of Potatoes. Nothern Michigan University: NMU Master's Theses; 2014

[11] Shimizu M et al. Identification of endophytic *Streptomyces* sp. R-5 and analysis of its antimicrobial metabolites. Journal of General Plant Pathology. 2004;**70**(1):66-68

[12] Shimizu M, Yazawa S, Ushijima Y. A promising strain of endophytic *Streptomyces* sp. for biological control of cucumber anthracnose. Journal of General Plant Pathology. 2009;**75**(1):27-36

[13] Shivlata L, Satyanarayana T. Actinobacteria in agricultural and environmental sustainability. In: Agro-Environmental Sustainability. New Delhi, India: Springer; 2017. pp. 173-218

[14] Purushotham N et al. Community structure of endophytic actinobacteria in a New Zealand native medicinal plant *Pseudowintera colorata* (Horopito) and their influence on plant growth. Microbial Ecology. 2018;**76**(3):729-740

[15] Ting ASY, Hermanto A, Peh KL. Indigenous actinomycetes from empty fruit bunch compost of oil palm: Evaluation on enzymatic and antagonistic properties. Biocatalysis and Agricultural Biotechnology. 2014;**3**(4):310-315

[16] Verma V, Singh S, Prakash S. Bio-control and plant growth promotion

potential of siderophore producing endophytic *Streptomyces* from *Azadirachta indica A. Juss*. Journal of Basic Microbiology. 2011;**51**(5):550-556

[17] Goudjal Y et al. Biocontrol of *Rhizoctonia solani* damping-off and promotion of tomato plant growth by endophytic actinomycetes isolated from native plants of *Algerian Sahara*. Microbiological Research. 2014;**169**(1):59-65

[18] Majeed A et al. Isolation and characterization of plant growth-promoting rhizobacteria from wheat rhizosphere and their effect on plant growth promotion. Frontiers in Microbiology. 2015;**6**:198

[19] Rungin S et al. Plant growth enhancing effects by a siderophore-producing endophytic streptomycete isolated from a Thai jasmine rice plant (*Oryza sativa* L. cv. KDML105). Antonie Van Leeuwenhoek. 2012;**102**(3):463-472

[20] Aly MM, El Sayed H, Jastaniah SD. Synergistic effect between *Azotobacter vinelandii* and *Streptomyces* sp. isolated from saline soil on seed germination and growth of wheat plant. Journal of American Science. 2012;**8**(5):667-676

[21] Soe KM, Yamakawa T. Low-density co-inoculation of *Myanmar Bradyrhizobium yuanmingense* MAS34 and *Streptomyces griseoflavus* P4 to enhance symbiosis and seed yield in soybean varieties. American Journal of Plant Sciences. 2013;**4**(09):1879

[22] Shutsrirung A et al. Diversity of endophytic actinomycetes in mandarin grown in northern Thailand, their phytohormone production potential and plant growth promoting activity. Soil Science and Plant Nutrition. 2013;**59**(3):322-330

[23] Gopalakrishnan S et al. Plant growth-promoting activities of *Streptomyces* spp. in sorghum and rice. Springerplus. 2013;**2**(1):574

[24] Sziderics A et al. Bacterial endophytes contribute to abiotic stress adaptation in pepper plants (*Capsicum annuum* L.). Canadian Journal of Microbiology. 2007;**53**(11):1195-1202

[25] El-Tarabily KA, Nassar AH, Sivasithamparam K. Promotion of growth of bean (*Phaseolus vulgaris* L.) in a calcareous soil by a phosphate-solubilizing, rhizosphere-competent isolate of *Micromonospora endolithica*. Applied Soil Ecology. 2008;**39**(2):161-171

[26] Franco-Correa M et al. Evaluation of actinomycete strains for key traits related with plant growth promotion and mycorrhiza helping activities. Applied Soil Ecology. 2010;**45**(3):209-217

[27] Rashad FM et al. Isolation and characterization of multifunctional *Streptomyces* species with antimicrobial, nematicidal and phytohormone activities from marine environments in Egypt. Microbiological Research. 2015;**175**:34-47

[28] Aldesuquy H, Mansour F, Abo-Hamed S. Effect of the culture filtrates of *Streptomyces* on growth and productivity of wheat plants. Folia Microbiologica. 1998;**43**(5):465-470

[29] Siddikee MA et al. Isolation, characterization, and use for plant growth promotion under salt stress, of ACC deaminase-producing halotolerant bacteria derived from coastal soil. Journal of Microbiology and Biotechnology. 2010;**20**(11):1577-1584

[30] Nafis A et al. Actinobacteria from extreme niches in Morocco and their plant growth-promoting potentials. Diversity. 2019;**11**(8):139

[31] Amule F et al. Effect of actinobacterial, rhizobium and plant growth promoting rhizobacteria consortium inoculation on rhizosphere soil properties in soybean in

Jabalpur district of Madhya Pradesh. International Journal of Consumer Studies. 2018;**6**(1):583-586

[32] Santi C, Bogusz D, Franche C. Biological nitrogen fixation in non-legume plants. Annals of Botany. 2013;**111**(5):743-767

[33] Kim J, Rees DC. Nitrogenase and biological nitrogen fixation. Biochemistry. 1994;**33**(2):389-397

[34] Benson DR, Silvester W. Biology of *Frankia* strains, actinomycete symbionts of actinorhizal plants. Microbiology and Molecular Biology Reviews. 1993;**57**(2):293-319

[35] Sellstedt A, Richau KH. Aspects of nitrogen-fixing *Actinobacteria*, in particular free-living and symbiotic *Frankia*. FEMS Microbiology Letters. 2013;**342**(2):179-186

[36] Razaq M, Zhang P, Shen H-L. Influence of nitrogen and phosphorous on the growth and root morphology of Acer mono. PLoS One. 2017;**12**(2):e0171321

[37] Bouain N et al. Phosphate and zinc transport and signalling in plants: Toward a better understanding of their homeostasis interaction. Journal of Experimental Botany. 2014;**65**(20):5725-5741

[38] Jog R, Nareshkumar G, Rajkumar S. Enhancing soil health and plant growth promotion by actinomycetes. In: Plant Growth Promoting Actinobacteria. Singapore: Springer; 2016. pp. 33-45

[39] Hamdali H et al. Rock phosphate-solubilizing Actinomycetes: Screening for plant growth-promoting activities. World Journal of Microbiology and Biotechnology. 2008;**24**(11):2565-2575

[40] El-Tarabily KA. Promotion of tomato (*Lycopersicon esculentum Mill.*) plant growth by rhizosphere competent 1-aminocyclopropane-1-carboxylic acid deaminase-producing streptomycete actinomycetes. Plant and Soil. 2008;**308**(1-2):161-174

[41] Crowley DE. Microbial siderophores in the plant rhizosphere. In: Iron Nutrition in Plants and Rhizospheric Microorganisms. Riverside, CA, USA: Springer, University of California; 2006. pp. 169-198

[42] Wang W et al. Siderophore production by actinobacteria. Biometals. 2014;**27**(4):623-631

[43] Challis GL, Ravel J. Coelichelin, a new peptide siderophore encoded by the Streptomyces coelicolor genome: Structure prediction from the sequence of its non-ribosomal peptide synthetase. FEMS Microbiology Letters. 2000;**187**(2):111-114

[44] Lee J et al. Siderophore production by actinomycetes isolates from two soil sites in Western Australia. Biometals. 2012;**25**(2):285-296

[45] Rashid S, Charles TC, Glick BR. Isolation and characterization of new plant growth-promoting bacterial endophytes. Applied Soil Ecology. 2012;**61**:217-224

[46] Gopalakrishnan S et al. Biocontrol of charcoal-rot of sorghum by actinomycetes isolated from herbal vermicompost. African Journal of Biotechnology. 2011;**10**(79):18142-18152

[47] Gopalakrishnan S, Sathya A, Vijayabharathi R. A Book Entitled "Plant Growth-Promoting Actinobacteria: A New Avenue for Enhancing the Productivity & Soil Fertility of Grain Legumes". Singapore: Springer; 2016

[48] Abd-Alla MH, El-Sayed E-SA, Rasmey A-HM. Indole-3-acetic acid (IAA) production by Streptomyces atrovirens isolated from rhizospheric soil in Egypt. Journal of Biology and Earth Sciences. 2013;**3**(2):182-193

[49] Sathya A, Vijayabharathi R, Gopalakrishnan S. Plant growth-promoting actinobacteria: A new strategy for enhancing sustainable production and protection of grain legumes. Biotech. 2017;**7**(2):102

[50] Karthik N, Binod P, Pandey A. Purification and characterisation of an acidic and antifungal chitinase produced by a *Streptomyces* sp. Bioresource Technology. 2015;**188**:195-201

[51] Yandigeri MS et al. Chitinolytic *Streptomyces* vinaceusdrappus S5MW2 isolated from Chilika lake, India enhances plant growth and biocontrol efficacy through chitin supplementation against *Rhizoctonia solani*. World Journal of Microbiology and Biotechnology. 2015;**31**(8):1217-1225

[52] Liotti RG, da Silva Figueiredo MI, Soares MA. *Streptomyces griseocarneus* R132 controls phytopathogens and promotes growth of pepper *(Capsicum annuum)*. Biological Control. 2019;**138**:104065

[53] Wonglom P et al. Streptomyces angustmyceticus NR8-2 as a potential microorganism for the biological control of leaf spots of Brassica rapa subsp. pekinensis caused by *Colletotrichum* sp. and *Curvularia lunata*. Biological Control. 2019;**138**:104046

[54] Saito A, Fujii T, Miyashita K. Distribution and evolution of chitinase genes in *Streptomyces* species: Involvement of gene-duplication and domain-deletion. Antonie Van Leeuwenhoek. 2003;**84**(1):7

[55] Khamna S, Yokota A, Peberdy JF, Lumyong S. Indole-3-acetic acid production by *Streptomyces* sp. isolated from some Thai medicinal plant rhizosphere soils. EurAsian Journal of BioSciences. 2010;**4**(1):23-32

[56] Marsh P, Wellington EMH. Molecular ecology of filamentous actinomycetes in soil. Molecular Ecology of Rhizosphere Microorganisms. Wellington, New Zealand: Wiley-VCH Verlag GmbH; 2007. pp. 133-149

[57] Pellegrino E, Bedini S. Enhancing ecosystem services in sustainable agriculture: Biofertilization and biofortification of chickpea (*Cicer arietinum* L.) by arbuscular mycorrhizal fungi. Soil Biology and Biochemistry. 2014;**68**:429-439

[58] Kaur N, Sharma P. Screening and characterization of native *Pseudomonas* sp. as plant growth promoting rhizobacteria in chickpea (*Cicer arietinum* L.) rhizosphere. African Journal of Microbiology Research. 2013;**7**(16):1465-1474

[59] Sathya A et al. Plant growth-promoting actinobacteria on chickpea seed mineral density: An upcoming complementary tool for sustainable biofortification strategy. Biotech. 2016;**6**(2):138

[60] Mahmood T. Phytoextraction of heavy metals-the process and scope for remediation of contaminated soils. Soil and Environment. 2010;**29**(2):91-109

[61] Sessitsch A et al. The role of plant-associated bacteria in the mobilization and phytoextraction of trace elements in contaminated soils. Soil Biology and Biochemistry. 2013;**60**:182-194

[62] Dangl JL, Jones JD. Plant pathogens and integrated defence responses to infection. Nature. 2001;**411**(6839):826

[63] de Jesus Sousa JA, Olivares FL. Plant growth promotion by streptomycetes: Ecophysiology, mechanisms and applications. Chemical and Biological Technologies in Agriculture. 2016;**3**(1):24

[64] Wang Z et al. Fumigant activity of volatiles from Streptomyces alboflavus TD-1 against *Fusarium moniliforme* Sheldon. Journal of Microbiology. 2013;**51**(4):477-483

[65] Wan M et al. Effect of volatile substances of Streptomyces platensis F-1 on control of plant fungal diseases. Biological Control. 2008;**46**(3):552-559

[66] Pal KK, Gardener BM. Biological Control of Plant Pathogens. Gujarat, India: The Plant Health Instructor; 2006

[67] Conn V, Walker A, Franco C. Endophytic actinobacteria induce defense pathways in *Arabidopsis thaliana*. Molecular Plant-Microbe Interactions. 2008;**21**(2):208-218

[68] Senthilraja G. Induction of systemic resistance in crop plants against plant pathogens by plant growth-promoting actinomycetes. In: Plant Growth Promoting Actinobacteria. Singapore: Springer; 2016. pp. 193-202

[69] Zhao S, Du C-M, Tian C-Y. Suppression of *Fusarium oxysporum* and induced resistance of plants involved in the biocontrol of *Cucumber Fusarium* Wilt by *Streptomyces bikiniensis* HD-087. World Journal of Microbiology and Biotechnology. 2012;**28**(9):2919-2927

[70] Castillo UF et al. Munumbicins E-4 and E-5: Novel broad-spectrum antibiotics from *Streptomyces* NRRL 3052. FEMS Microbiology Letters. 2006;**255**(2):296-300

[71] Clermont N et al. Effect of biopolymers on geldanamycin production and biocontrol ability of *Streptomyces melanosporofaciens* strain EF-76. Canadian Journal of Plant Pathology. 2010;**32**(4):481-489

[72] Ismet A et al. Production and chemical characterization of antifungal metabolites from *Micromonospora* sp. M39 isolated from mangrove Rhizosphere soil. World Journal of Microbiology and Biotechnology. 2004;**20**(5):523-528

[73] Li W et al. Malayamycin, a new streptomycete antifungal compound, specifically inhibits sporulation of *Stagonospora nodorum* (Berk) castell and Germano, the cause of wheat glume blotch disease. Pest Management Science. 2008;**64**(12):1294-1302

[74] Park S et al. Determination of polyphenol levels variation in *Capsicum annuum* L. cv. Chelsea (yellow bell pepper) infected by anthracnose (*Colletotrichum gloeosporioides*) using liquid chromatography-tandem mass spectrometry. Food Chemistry. 2012;**130**(4):981-985

[75] Zhao K et al. The diversity and anti-microbial activity of endophytic actinomycetes isolated from medicinal plants in *Panxi plateau*, China. Current Microbiology. 2011;**62**(1):182-190

[76] Ezra D et al. Coronamycins, peptide antibiotics produced by a verticillate *Streptomyces* sp.(MSU-2110) endophytic on *Monstera* sp. Microbiology. 2004;**150**(4):785-793

[77] Liu Y et al. Endophytic bacteria associated with endangered plant *Ferula sinkiangensis* KM Shen in an arid land: Diversity and plant growth-promoting traits. Journal of Arid Land. 2017;**9**(3):432-445

[78] Passari AK et al. Insights into the functionality of endophytic actinobacteria with a focus on their biosynthetic potential and secondary metabolites production. Scientific Reports. 2017;**7**(1):11809

[79] Kim JD et al. Identification and biocontrol efficacy of *Streptomyces miharaensis* producing filipin III against *Fusarium* wilt. Journal of Basic Microbiology. 2012;**52**(2):150-159

[80] Getha K et al. Evaluation of *Streptomyces* sp. strain g10 for suppression of *Fusarium* wilt and rhizosphere colonization in pot-grown banana plantlets. Journal of Industrial

Microbiology and Biotechnology. 2005;**32**(1):24-32

[81] Buatong J et al. Antifungal metabolites from marine-derived *Streptomyces* sp. AMA49 against *Pyricularia oryzae*. Journal of Pure and Applied Microbiology. 2019;**13**(2):653-665

[82] Sadeghi A et al. Plant growth promoting activity of an auxin and siderophore producing isolate of *Streptomyces* under saline soil conditions. World Journal of Microbiology and Biotechnology. 2012;**28**(4):1503-1509

[83] Yandigeri MS et al. Drought-tolerant endophytic actinobacteria promote growth of wheat (*Triticum aestivum*) under water stress conditions. Plant Growth Regulation. 2012;**68**(3):411-420

[84] Saraf M, Jha CK, Patel D. The role of ACC deaminase producing PGPR in sustainable agriculture. In: Plant Growth and Health Promoting Bacteria. Berlin, Heidelberg: Springer; 2010. pp. 365-385

[85] Etesami H et al. Bacterial biosynthesis of 1-aminocyclopropane-1-carboxylate (ACC) deaminase and indole-3-acetic acid (IAA) as endophytic preferential selection traits by rice plant seedlings. Journal of Plant Growth Regulation. 2014;**33**(3):654-670

[86] Nascimento FX et al. New insights into 1-aminocyclopropane-1-carboxylate (ACC) deaminase phylogeny, evolution and ecological significance. PLoS One. 2014;**9**(6):e99168